PRINCIPLES OF SPREAD-SPECTRUM COMMUNICATION SYSTEMS

PRINCIPLES OF SPREAD-SPECTRUM COMMUNICATION SYSTEMS

By

DON TORRIERI

 Springer

Don Torrieri

Principles of Spread-Spectrum Communication Systems

Library of Congress Cataloging-in-Publication Data

Torrieri, Don J.
    Principles of spread-spectrum communication systems / by Don Torrieri.
       p. cm.
    Includes bibliographical references and index.
    ISBN 0-387-22782-2 (alk. paper)
     1. Spread spectrum communications. I. Title.

  TK5103.45.T67 2004
  621.382--dc22

2004056516

ISBN 0-387-22782-2   e-ISBN 0-387-22783-0   Printed on acid-free paper.

Printed in the United States of America.

9 8 7 6 5 4 3 2 1   SPIN 11055167

springeronline.com

To My Family

Contents

Preface

The goal of this book is to provide a concise but lucid explanation and derivation of the fundamentals of spread-spectrum communication systems. Although spread-spectrum communication is a staple topic in textbooks on digital communication, its treatment is usually cursory, and the subject warrants a more intensive exposition. Originally adopted in military networks as a means of ensuring secure communication when confronted with the threats of jamming and interception, spread-spectrum systems are now the core of commercial applications such as mobile cellular and satellite communication. The level of presentation in this book is suitable for graduate students with a prior graduate-level course in digital communication and for practicing engineers with a solid background in the theory of digital communication. As the title indicates, this book stresses principles rather than specific current or planned systems, which are described in many other books. Although the exposition emphasizes theoretical principles, the choice of specific topics is tempered by my judgment of their practical significance and interest to both researchers and system designers. Throughout the book, learning is facilitated by many new or streamlined derivations of the classical theory. Problems at the end of each chapter are intended to assist readers in consolidating their knowledge and to provide practice in analytical techniques. The book is largely self-contained mathematically because of the four appendices, which give detailed derivations of mathematical results used in the main text.

In writing this book, I have relied heavily on notes and documents prepared and the perspectives gained during my work at the US Army Research Laboratory. Many colleagues contributed indirectly to this effort. I am grateful to my wife, Nancy, who provided me not only with her usual unwavering support but also with extensive editorial assistance.

Chapter 1

Channel Codes

Channel codes are vital in fully exploiting the potential capabilities of spread-spectrum communication systems. Although direct-sequence systems greatly suppress interference, practical systems require channel codes to deal with the residual interference and channel impairments such as fading. Frequency-hopping systems are designed to avoid interference, but the hopping into an unfavorable spectral region usually requires a channel code to maintain adequate performance. In this chapter, some of the fundamental results of coding theory [1], [2], [3], [4] are reviewed and then used to derive the corresponding receiver computations and the error probabilities of the decoded information bits.

1.1 Block Codes

A *channel code* for forward error control or error correction is a set of *codewords* that are used to improve communication reliability. An (n, k) *block code* uses a codeword of n code symbols to represent k information symbols. Each symbol is selected from an alphabet of q symbols, and there are q^k codewords. If $q = 2^m$, then an (n, k) code of q-ary symbols is equivalent to an (mn, mk) binary code. A block encoder can be implemented by using logic elements or memory to map a k-symbol information word into an n-symbol codeword. After the waveform representing a codeword is received and demodulated, the decoder uses the demodulator output to determine the information symbols corresponding to the codeword. If the demodulator produces a sequence of discrete symbols and the decoding is based on these symbols, the demodulator is said to make *hard decisions*. Conversely, if the demodulator produces analog or multilevel quantized samples of the waveform, the demodulator is said to make *soft decisions*. The advantage of soft decisions is that reliability or quality information is provided to the decoder, which can use this information to improve its performance.

The number of symbol positions in which the symbol of one sequence differs from the corresponding symbol of another equal-length sequence is called the *Hamming distance* between the sequences. The minimum Hamming distance

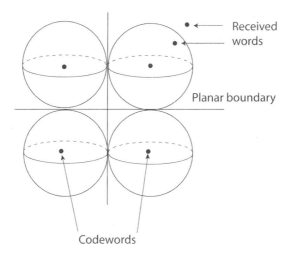

Figure 1.1: Conceptual representation of n-dimensional vector space of sequences.

between any two codewords is called the *minimum distance* of the code. When hard decisions are made, the demodulator output sequence is called the *received sequence* or the *received word*. Hard decisions imply that the overall channel between the output and the decoder input is the classical binary symmetric channel. If the channel symbol error probability is less than one-half, then the maximum-likelihood criterion implies that the correct codeword is the one that is the smallest Hamming distance from the received word. A *complete decoder* is a device that implements the maximum-likelihood criterion. An *incomplete decoder* does not attempt to correct all received words.

The n-dimensional vector space of sequences is conceptually represented as a three-dimensional space in Figure 1.1. Each codeword occupies the center of a *decoding sphere* with radius t in Hamming distance, where t is a positive integer. A complete decoder has decision regions defined by planar boundaries surrounding each codeword. A received word is assumed to be a corrupted version of the codeword enclosed by the boundaries. A *bounded-distance decoder* is an incomplete decoder that attempts to correct symbol errors in a received word if it lies within one of the decoding spheres. Since unambiguous decoding requires that none of the spheres may intersect, the maximum number of random errors that can be corrected by a bounded-distance decoder is

$$t = \lfloor (d_m - 1)/2 \rfloor \tag{1-1}$$

where d_m is the minimum Hamming distance between codewords and $\lfloor x \rfloor$ denotes the largest integer less than or equal to x. When more than t errors occur, the received word may lie within a decoding sphere surrounding an incorrect codeword or it may lie in the interstices (regions) outside the decoding spheres. If the received word lies within a decoding sphere, the decoder selects the incorrect codeword at the center of the sphere and produces an output word of information symbols with undetected errors. If the received word lies in the interstices, the decoder cannot correct the errors, but recognizes their existence. Thus, the decoder fails to decode the received word.

Since there are $\binom{n}{i}(q-1)^i$ words at exactly distance i from the center of the sphere, the number of words in a decoding sphere of radius t is determined from elementary combinatorics to be

$$V = \sum_{i=0}^{t} \binom{n}{i}(q-1)^i \qquad (1\text{-}2)$$

Since a block code has q^k codewords, $q^k V$ words are enclosed in some sphere. The number of possible received words is $q^n \geq q^k V$, which yields

$$q^{n-k} \geq \sum_{i=0}^{t} \binom{n}{i}(q-1)^i \qquad (1\text{-}3)$$

This inequality implies an upper bound on t and, hence, d_m. The upper bound on d_m is called the *Hamming bound*.

A block code is called a *linear block code* if its codewords form a k-dimensional subspace of the vector space of sequences with n symbols. Thus, the vector sum of two codewords or the vector difference between them is a codeword. If a binary block code is linear, the symbols of a codeword are modulo-two sums of information bits. Since a linear block code is a subspace of a vector space, it must contain the additive identity. Thus, the all-zero sequence is always a codeword in any linear block code. Since nearly all practical block codes are linear, henceforth block codes are assumed to be linear.

A *cyclic code* is a linear block code in which a cyclic shift of the symbols of a codeword produces another codeword. This characteristic allows the implementation of encoders and decoders that use linear feedback shift registers. Relatively simple encoding and hard-decision decoding techniques are known for cyclic codes belonging to the class of *Bose-Chaudhuri-Hocquenghem* (BCH) *codes*, which may be binary or nonbinary. A BCH code has a length that is a divisor of $q^m - 1$, where $m \geq 2$, and is designed to have an error-correction capability of $t = \lfloor (\delta - 1)/2 \rfloor$, where δ is the *design distance*. Although the minimum distance may exceed the design distance, the standard BCH decoding algorithms cannot correct more than t errors. The parameters (n, k, t) for binary BCH codes with $7 \leq n \leq 127$ are listed in Table 1.1.

A *perfect code* is a block code such that every n-symbol sequence is at a distance of at most t from some n-symbol codeword, and the sets of all sequences at distance t or less from each codeword are disjoint. Thus, the Hamming bound is satisfied with equality, and a complete decoder is also a bounded-distance decoder. The only perfect codes are the binary repetition codes of odd length, the Hamming codes, the binary Golay (23,12) code, and the ternary Golay (11,6) code. *Repetition codes* represent each information bit by n binary code symbols. When n is odd, the $(n, 1)$ repetition code is a perfect code with

Table 1.1: Binary BCH codes.

n	k	t	D_p	n	k	t	D_p	n	k	t	D_p
7	4	1	1	63	45	3	0.1592	127	92	5	0.0077
7	1	3	1	63	39	4	0.0380	127	85	6	0.0012
15	11	1	1	63	36	5	0.0571	127	78	7	$1.68 \cdot 10^{-4}$
15	7	2	0.4727	63	30	6	0.0088	127	71	9	$2.66 \cdot 10^{-4}$
15	5	3	0.5625	63	24	7	0.0011	127	64	10	$2.48 \cdot 10^{-5}$
15	1	7	1	63	18	10	0.0044	127	57	11	$2.08 \cdot 10^{-6}$
31	26	1	1	63	16	11	0.0055	127	50	13	$1.42 \cdot 10^{-6}$
31	21	2	0.4854	63	10	13	0.0015	127	43	14	$9.11 \cdot 10^{-8}$
31	16	3	0.1523	63	7	15	0.0024	127	36	15	$5.42 \cdot 10^{-9}$
31	11	5	0.1968	63	1	31	1	127	29	21	$2.01 \cdot 10^{-6}$
31	6	7	0.1065	127	120	1	1	127	22	23	$3.56 \cdot 10^{-7}$
31	1	15	1	127	113	2	0.4962	127	15	27	$7.75 \cdot 10^{-7}$
63	57	1	1	127	106	3	0.1628	127	8	31	$8.10 \cdot 10^{-7}$
63	51	2	0.4924	127	99	4	0.0398	127	1	63	1

Table 1.2: Code words of Hamming (7,4) code.

0000000	0001011	0010110	0011101
0100111	0101100	0110001	0111010
1000101	1001110	1010011	1011000
1100010	1101001	1110100	1111111

$d_m = n$ and $t = (n-1)/2$. A hard-decision decoder makes a decision based on the state of the majority of the demodulated symbols. Although repetition codes are not efficient for the additive-white-Gaussian-noise (AWGN) channel, they can improve the system performance for fading channels if the number of repetitions is properly chosen. A *Hamming* (n,k) *code* is a perfect BCH code with $d_m = 3$ and

$$n = \frac{q^{n-k} - 1}{q - 1} \tag{1-4}$$

Since $t = 1$, a Hamming code is capable of correcting all single errors. Binary Hamming codes with $n \le 127$ are found in Table 1.1. The 16 codewords of a Hamming (7,4) code are listed in Table 1.2. The first four bits of each codeword are the information bits. The *Golay* (23,12) *code* is a binary cyclic code that is a perfect code with $d_m = 7$ and $t = 3$.

Any (n, k) linear block code with an odd value of d_m can be converted into an $(n + 1, k)$ *extended code* by adding a parity symbol. The advantage of the extended code stems from the fact that the minimum distance of the block code is increased by one, which improves the performance, but the decoding complexity and code rate are usually changed insignificantly. The *extended Golay* (24,12) *code* is formed by adding an overall parity symbol to the Golay (23,12) code, thereby increasing the minimum distance to $d_m = 8$. As a result, some received sequences with four errors can be corrected with a complete decoder. The (24,12) code is often preferable to the (23,12) code because the *code rate*, which is defined as the ratio k/n, is exactly one-half, which simplifies the system timing.

The *Hamming weight* of a codeword is the number of nonzero symbols in a codeword. For a linear block code, the vector difference between two codewords is another codeword with weight equal to the distance between the two original codewords. By subtracting the codeword **c** to all the codewords, we find that the set of Hamming distances from any codeword **c** is the same as the set of codeword weights. Consequently, in evaluating decoding error probabilities, one can assume without loss of generality that the all-zero codeword was transmitted, and the minimum Hamming distance is equal to the minimum weight of the nonzero codewords. For binary block codes, the Hamming weight is the number of 1's in a codeword.

A *systematic block code* is a code in which the information symbols appear unchanged in the codeword, which also has additional parity symbols. In terms of the word error probability for hard-decision decoding, every linear code is equivalent to a systematic linear code [1]. Therefore, systematic block codes are the standard choice and are assumed henceforth. Some systematic codewords have only one nonzero information symbol. Since there are at most $n-k$ parity symbols, these codewords have Hamming weights that cannot exceed $n-k+1$. Since the minimum distance of the code is equal to the minimum codeword weight,

$$d_m \leq n - k + 1 \qquad (1\text{-}5)$$

This upper bound is called the *Singleton bound*. A linear block code with a minimum distance equal to the Singleton bound is called a *maximum-distance-separable code*

Nonbinary block codes can accommodate high data rates efficiently because decoding operations are performed at the symbol rate rather than the higher information-bit rate. *Reed-Solomon codes* are nonbinary BCH codes with $n = q-1$ and are maximum-distance-separable codes with $d_m = n-k+1$. For convenience in implementation, q is usually chosen so that $q = 2^m$, where m is the number of bits per symbol. Thus, $n = 2^m - 1$ and the code provides correction of 2^m-ary symbols. Most Reed-Solomon decoders are bounded-distance decoders with $t = \lfloor (d_m - 1)/2 \rfloor$.

The most important single determinant of the code performance is its *weight distribution,* which is a list or function that gives the number of codewords with each possible weight. The weight distributions of the Golay codes are listed in Table 1.3. Analytical expressions for the weight distribution are known in a few cases. Let A_l denote the number of codewords with weight l. For a binary Hamming code, each A_l can be determined from the weight-enumerator polynomial

$$A(x) = \sum_{l=0}^{n} A_l x^l = \frac{1}{n+1}[(1+x)^n + n(1+x)^{(n-1)/2}(1-x)^{(n+1)/2}] \qquad (1\text{-}6)$$

For example,the Hamming (7,4) code gives $A(x) = \frac{1}{8}[(1+x)^7 + 7(1+x)^3(1-x)^4] = 1 + 7x^3 + 7x^4 + x^7$, which yields $A_0{=}1$, $A_3{=}7$, $A_4{=}7$, $A_7{=}1$, and $A_l{=}0$,

Table 1.3: Weight distributions of Golay codes.

Weight	Number of Codewords	
	(23,12)	(24,12)
0	1	1
7	253	0
8	506	759
11	1288	0
12	1288	2576
15	506	0
16	253	759
23	1	0
24	0	1

otherwise. For a maximum-distance-separable code, $A_0 = 1$ and [2]

$$A_l = \binom{n}{l}(q-1)\sum_{i=0}^{l-d_m}(-1)^i\binom{l-1}{i}q^{l-i-d_m} \ , \quad d_m \le l \le n \qquad (1\text{-}7)$$

The weight distribution of other codes can be determined by examining all valid codewords if the number of codewords is not too large for a computation.

Error Probabilities for Hard-Decision Decoding

There are two types of bounded-distance decoders: erasing decoders and re-producing decoders. They differ only in their actions following the detection of uncorrectable errors in a received word. An *erasing decoder* discards the received word and may initiate an automatic retransmission request. For a systematic block code, a *reproducing decoder* reproduces the information symbols of the received word as its output.

Let P_s denote the *channel-symbol error probability*, which is the probability of error in a demodulated code symbol. It is assumed that the channel-symbol errors are statistically independent and identically distributed, which is usually an accurate model for systems with appropriate symbol interleaving (Section 1.3). Let P_w denote the *word error probability*, which is the probability that a received word is not decoded correctly due to both undetected errors and decoding failures. There are $\binom{n}{i}$ distinct ways in which i errors may occur among n symbols. Since a received sequence may have more than t errors but no information-symbol errors,

$$P_w \le \sum_{i=t+1}^{n}\binom{n}{i}P_s^i(1-P_s)^{n-i} \qquad (1\text{-}8)$$

for a reproducing decoder that corrects t or few errors. For an erasing decoder, (1-8) becomes an equality. For reproducing decoders, t is given by (1-1) because it is pointless to make the decoding spheres smaller than the maximum allowed by the code. However, if a block code is used for both error correction and error detection, an erasing decoder is often designed with t less than the maximum. If a block code is used exclusively for error detection, then $t = 0$.

Conceptually, a complete decoder correctly decodes when the number of symbol errors exceeds t if the received sequence lies within the planar boundaries associated with the correct codeword, as depicted in Figure 1.1. When a received sequence is equidistant from two or more codewords, a complete decoder selects one of them according to some arbitrary rule. Thus, the word error probability for a complete decoder satisfies (1-8). If $P_s \leq 1/2$, a complete decoder is a maximum-likelihood decoder.

Let P_{ud} denote the probability of an *undetected error*, and let P_{df} denote the probability of a *decoding failure*. For a bounded-distance decoder

$$P_{ud} + P_{df} = \sum_{i=t+1}^{n} \binom{n}{i} P_s^i (1 - P_s)^{n-i} \tag{1-9}$$

Thus, it is easy to calculate P_{df} once P_{ud} is determined. Since the set of Hamming distances from a given codeword to the other codewords is the same for all given codewords of a linear block code, it is legitimate to assume for convenience in evaluating P_{ud} that the all-zero codeword was transmitted. If channel-symbol errors in a received word are statistically independent and occur with the same probability P_s, then the probability of an error in a specific set of i positions that results in a specific set of i erroneous symbols is

$$P_e(i) = \left(\frac{P_s}{q-1} \right)^i (1 - P_s)^{n-i} \tag{1-10}$$

For an undetected error to occur at the output of a bounded-distance decoder, the number of erroneous symbols must exceed t and the received word must lie within an incorrect decoding sphere of radius t. Let $N(l, i)$ is the number of sequences of Hamming weight i that lie within a decoding sphere of radius t associated with a particular codeword of weight l. Then

$$P_{ud} = \sum_{i=t+1}^{n} P_e(i) \sum_{l=\max(i-t,d_m)}^{\min(i+t,n)} A_l N(l,i)$$

$$= \sum_{i=t+1}^{n} \left(\frac{P_s}{q-1} \right)^i (1 - P_s)^{n-i} \sum_{l=\max(i-t,d_m)}^{\min(i+t,n)} A_l N(l,i) \tag{1-11}$$

Consider sequences of weight i that are at distance s from a particular codeword of weight l, where $|l - i| \leq s \leq t$ so that the sequences are within the decoding sphere of the codeword. By counting these sequences and then summing over the allowed values of s, we can determine $N(l, i)$. The counting is done by considering changes in the components of this codeword that can produce one of these sequences. Let j denote the number of nonzero codeword symbols that are changed to zeros, α the number of codeword zeros that are changed to any of the $(q-1)$ nonzero symbols in the alphabet, and β the number of nonzero codeword symbols that are changed to any of the other $(q-2)$ nonzero symbols. For a sequence at distance s to result, it is necessary that $0 \leq j \leq s$. The number of sequences that can be obtained by changing any j of the l nonzero symbols to zeros is $\binom{l}{j}$, where $\binom{b}{a} = 0$ if $a > b$. For a specified value of j, it is necessary that $\alpha = j + i - l$ to ensure a sequence of weight i. The number of sequences that result from changing any α of the $n - l$ zeros to nonzero symbols is $\binom{n-l}{\alpha}$ $(q-1)^\alpha$. For a specified value of j and hence α, it is necessary that $\beta = s - j - \alpha = s + l - i - 2j$ to ensure a sequence at distance s. The number of sequences that result from changing β of the $l - j$ remaining nonzero components is $\binom{l-j}{\beta}$ $(q-2)^\beta$, where $0^x = 0$ if $x \neq 0$ and $0^0 = 1$. Summing over the allowed values of s and j, we obtain

$$N(l,i) = \sum_{s=|l-i|}^{t} \sum_{j=0}^{s} \binom{l}{j}\binom{n-l}{j+i-l}\binom{l-j}{s+l-i-2j}$$
$$\times (q-1)^{j+i-l}(q-2)^{s+l-i-2j} \tag{1-12}$$

Equations (1-11) and (1-12) allow the exact calculation of P_{ud}.

When $q = 2$, the only term in the inner summation of (1-12) that is nonzero has the index $j = (s + l - i)/2$ provided that this index is an integer and $0 \leq (s + l - i)/2 \leq s$. Using this result, we find that for binary codes,

$$N(l,i) = \sum_{s=|l-i|}^{t} \binom{n-l}{\frac{s+i-l}{2}}\binom{l}{\frac{s+l-i}{2}}, \quad q = 2 \tag{1-13}$$

where $\binom{m}{\frac{1}{2}} = 0$ for any nonnegative integer m. Thus, $N(l,l) = 1$ and $N(l,i) = 0$ for $|l - i| \geq t + 1$.

The word error probability is a performance measure that is important primarily in applications for which only a decoded word completely without symbol errors is acceptable. When the utility of a decoded word degrades in proportion to the number of information bits that are in error, the *information-bit error probability* is frequently used as a performance measure. To evaluate it for block codes that may be nonbinary, we first examine the information-symbol error probability.

Let $P_{is}(j)$ denote the probability of an error in information symbol j at the decoder output. In general, it cannot be assumed that $P_{is}(j)$ is independent of j. The *information-symbol error probability*, which is defined as the unconditional error probability without regard to the symbol position, is

$$P_{is} = \frac{1}{k}\sum_{j=1}^{k} P_{is}(j) \tag{1-14}$$

The random variables Z_j, $j = 1, 2, \ldots, k$, are defined so that $Z_j = 1$ if information symbol j is in error and $Z_j = 0$ if it is correct. The expected number of information-symbol errors is

$$E[I] = E\left[\sum_{j=1}^{k} Z_j\right] = \sum_{j=1}^{k} E[Z_j] = \sum_{j=1}^{k} P_{is}(j) \tag{1-15}$$

where $E[\ ]$ denotes the expected value. The *information-symbol error rate* is defined as $E[I]/k$. Equations (1-14) and (1-15) imply that

$$P_{is} = \frac{E[I]}{k} \tag{1-16}$$

which indicates that the information-symbol error probability is equal to the information-symbol error rate.

Let $P_{ds}(j)$ denote the probability of an error in symbol j of the codeword chosen by the decoder or symbol j of the received sequence if a decoding failure occurs. The decoded-symbol error probability is

$$P_{ds} = \frac{1}{n} \sum_{j=1}^{n} P_{ds}(j) \tag{1-17}$$

If $E[D]$ is the expected number of decoded-symbol errors, a derivation similar to the preceding one yields

$$P_{ds} = \frac{E[D]}{n} \tag{1-18}$$

which indicates that the decoded-symbol error probability is equal to the decoded-symbol error rate. It can be shown [5] that for cyclic codes, the error rate among the information symbols in the output of a bounded-distance decoder is equal to the error rate among all the decoded symbols; that is,

$$P_{is} = P_{ds} \tag{1-19}$$

This equation, which is at least approximately valid for linear block codes, significantly simplifies the calculation of P_{is} because P_{ds} can be expressed in terms of the code weight distribution, whereas an exact calculation of P_{is} requires additional information.

An erasing decoder makes an error only if it fails to detect one. Therefore, $P_{ds} = P_{ud}$ and (1-11) implies that the *decoded-symbol error rate for an erasing decoder* is

$$P_{ds} = \sum_{i=t+1}^{n} \left(\frac{P_s}{q-1}\right)^i (1-P_s)^{n-i} \sum_{l=\max(i-t,d_m)}^{\min(i+t,n)} A_l N(l,i)\frac{l}{n} \tag{1-20}$$

The number of sequences of weight i that lie in the interstices outside the decoding spheres is

$$L(i) = (q-1)^i \binom{n}{i} - \sum_{l=\max(i-t,d_m)}^{\min(i+t,n)} A_l N(l,i), \quad i \geq t+1 \tag{1-21}$$

where the first term is the total number of sequences of weight i, and the second term is the number of sequences of weight i that lie within incorrect decoding spheres. When i symbol errors in the received word cause a decoding failure, the decoded symbols in the output of a reproducing decoder contain i errors. Therefore, the *decoded-symbol error rate for a reproducing decoder* is

$$P_{ds} = \sum_{i=t+1}^{n} \left(\frac{P_s}{q-1}\right)^i (1-P_s)^{n-i} \left[\sum_{l=\max(i-t,d_m)}^{\min(i+t,n)} A_l N(l,i)\frac{l}{n} + L(i)\frac{i}{n}\right] \quad (1\text{-}22)$$

Even if $P_{is} = P_{ds}$, two major problems still arise in calculating P_{is} from (1-20) or (1-22). The computational complexity may be prohibitive when n and q are large, and the weight distribution is unknown for many linear or cyclic block codes.

The *packing density* is defined as the ratio of the number of words in the q^k decoding spheres to the total number of sequences of length n. From (2), it follows that the packing density is

$$D_p = \frac{q^k}{q^n} \sum_{i=0}^{t} \binom{n}{i}(q-1)^i \quad (1\text{-}23)$$

For perfect codes, $D_p = 1$. If $D_p > 0.5$, undetected errors tend to occur more often then decoding failures, and the code is considered *tightly packed*. If $D_p < 0.1$, decoding failures predominate, and the code is considered *loosely packed*. The packing densities of binary BCH codes are listed in Table 1.1. The codes are tightly packed if $n = 7$ or 15. For $k > 1$ and $n = 31$, 63, or 127, the codes are tightly packed only if $t = 1$ or 2.

To approximate P_{is} for tightly packed codes, let $A(i)$ denote the event that i errors occur in a received sequence of n symbols at the decoder input. If the symbol errors are independent, the probability of this event is

$$P[A(i)] = \binom{n}{i}P_s^i(1-P_s)^{n-i} \quad (1\text{-}24)$$

Given event $A(i)$ for i such that $d_m \le i \le n$, it is plausible to assume that a reproducing bounded-distance decoder usually chooses a codeword with approximately i symbol errors. For i such that $t+1 \le i \le d_m$, it is plausible to assume that the decoder usually selects a codeword at the minimum distance d_m. These approximations, (1-19), (1-24), and the identity $\binom{n}{i}\frac{i}{n} = \binom{n-1}{i-1}$ indicate that P_{is} for reproducing decoders is approximated by

$$P_{is} \approx \sum_{i=t+1}^{d_m} \frac{d_m}{n}\binom{n}{i}P_s^i(1-P_s)^{n-i} + \sum_{i=d_m+1}^{n}\binom{n-1}{i-1}P_s^i(1-P_s)^{n-i} \quad (1\text{-}25)$$

The virtues of this approximation are its lack of dependence on the code weight distribution and its generality. Computations for specific codes indicate that the accuracy of this approximation tends to increase with P_{ud}/P_{df}. The right-hand side of (1-25) gives an approximate upper bound on P_{is} for erasing bounded-distance decoders, for loosely packed codes with bounded-distance decoders, and for complete decoders because some received sequences with $t + 1$ or more errors can be corrected and, hence, produce no information-symbol errors.

For a loosely packed code, it is plausible that P_{is} for a reproducing bounded-distance decoder might be accurately estimated by ignoring undetected errors. Dropping the terms involving $N(l, i)$ in (1-21) and (1-22) and using (1-19) gives

$$P_{is} \geq \sum_{i=t+1}^{n} \binom{n-1}{i-1} P_s^i (1 - P_s)^{n-i} \tag{1-26}$$

The virtue of this lower bound as an approximation is its independence of the code weight distribution. The bound is tight when decoding failures are the predominant error mechanism. For cyclic Reed-Solomon codes, numerical examples [5] indicate that the exact P_{is} and the approximate bound are quite close for all values of P_s when $t \geq 3$, a result that is not surprising in view of the paucity of sequences in the decoding spheres for a Reed-Solomon code with $t \geq 3$. A comparison of (1-26) with (1-25) indicates that the latter overestimates P_{is} by a factor of less than $d_m/(t + 1)$.

A *q-ary symmetric channel* or *uniform discrete channel* is one in which an incorrectly decoded information symbol is equally likely to be any of the remaining $q - 1$ symbols in the alphabet. Consider a linear (n, k) block code and a q-ary symmetric channel such that q is a power of 2 and the "channel" refers to the transmission channel plus the decoder. Among the $q - 1$ incorrect symbols, a given bit is incorrect in $q/2$ instances. Therefore, the information-bit error probability is

$$P_b = \frac{q}{2(q-1)} P_{is} \tag{1-27}$$

Let r denote the ratio of information bits to transmitted channel symbols. For binary codes, r is the code rate. For block codes with $m = \log_2 q$ information bits per symbol, $r = mk/n$. When coding is used but the information rate is preserved, the duration of a channel symbol is changed relative to that of an information bit. Thus, the energy per received channel symbol is

$$\mathcal{E}_s = r\mathcal{E}_b = \frac{mk}{n}\mathcal{E}_b \tag{1-28}$$

where $\mathcal{E}_b$ is the energy per information bit. When $r < 1$, a code is potentially beneficial if its error-control capability is sufficient to overcome the degradation due to the reduction in the energy per received symbol. For the AWGN channel and coherent binary phase-shift keying (PSK), the classical theory indicates that the symbol error probability at the demodulator output is

$$P_s = Q\left(\sqrt{\frac{2r\mathcal{E}_b}{N_0}}\right) \tag{1-29}$$

where

$$Q(x) = \frac{1}{\sqrt{2\pi}} \int_x^\infty \exp\left(-\frac{y^2}{2}\right) dy = \frac{1}{2}\operatorname{erfc}\left(\frac{x}{\sqrt{2}}\right) \tag{1-30}$$

and erfc( ) is the complementary error function. Consider the noncoherent detection of q-ary orthogonal signals over an AWGN channel. The channel symbols for multiple frequency-shift keying (MFSK) modulation are received as orthogonal signals. It is shown subsequently that P_s at the demodulator output is

$$P_s = \sum_{i=1}^{q-1} \frac{(-1)^{i+1}}{i+1} \binom{q-1}{i} \exp\left[-\frac{imr\mathcal{E}_b}{(i+1)N_0}\right] \tag{1-31}$$

which decreases as q increases for sufficiently large values of $\mathcal{E}_b/N_0$. The orthogonality of the signals ensures that at least the transmission channel is q-ary symmetric, and, hence, (1-27) is at least approximately correct.

If the alphabets of the code symbols and the transmitted channel symbols are the same, then the channel-symbol error probability P_{cs} equals the code-symbol error probability P_s. If not, then the q-ary code symbols may be mapped into q_1-ary channel symbols. If $q = 2^m$ and $q_1 = 2^{m_1}$, then choosing m/m_1 to be an integer is strongly preferred for implementation simplicity. Since any of the channel-symbol errors can cause an error in the corresponding code symbol, the independence of channel-symbol errors implies that

$$P_s = 1 - (1 - P_{cs})^{m/m_1} \tag{1-32}$$

A common application is to map nonbinary code symbols into binary channel symbols ($m_1 = 1$). In this case, (1-27) is no longer valid because the transmission channel plus the decoder is not necessarily q-ary symmetric. Since there is at least one bit error for every symbol error,

$$\frac{P_{is}}{m} \leq P_b \leq \frac{qP_{is}}{2(q-1)} \tag{1-33}$$

This lower bound is tight when P_{cs} is low because then there tends to be a single bit error per code-symbol error before decoding, and the decoder is unlikely to change an information symbol. For coherent binary PSK, (1-29) and (1-32) imply that

$$P_s = 1 - \left[1 - Q\left(\sqrt{\frac{2r\mathcal{E}_b}{N_0}}\right)\right]^m \tag{1-34}$$

Error Probabilities for Soft-Decision Decoding

A symbol is said to be erased when the demodulator, after deciding that a symbol is unreliable, instructs the decoder to ignore that symbol during the decoding. The simplest practical soft-decision decoding uses *erasures* to supplement hard-decision decoding. If a code has a minimum distance d_m and a received word is assigned ϵ erasures, then all codewords differ in at least $d_m - \epsilon$ of the unerased symbols. Hence, ν errors can be corrected if $2\nu + 1 \leq d_m - \epsilon$. If d_m or more erasures are assigned, a decoding failure occurs. Let P_ϵ denote the probability of an erasure. For independent symbol errors and erasures, the probability that a received sequence has i errors and j erasures is $P_s^i P_\epsilon^j (1 - P_s - P_\epsilon)^{n-i-j}$. Therefore, for a bounded-distance decoder,

$$P_w \leq \sum_{j=0}^{n} \sum_{i=i_0}^{n-j} \binom{n}{j} \binom{n-j}{i} P_s^i P_\epsilon^j (1 - P_s - P_\epsilon)^{n-i-j} ,$$

$$i_0 = \max(0, \lceil (d_m - j)/2 \rceil) \qquad (1\text{-}35)$$

where $\lceil x \rceil$ denotes the smallest integer greater than or equal to x. This inequality becomes an equality for an erasing decoder. For the AWGN channel, decoding with optimal erasures provides an insignificant performance improvement relative to hard-decision decoding, but erasures are often effective against fading or sporadic interference. Codes for which *errors-and-erasures decoding* is most attractive are those with relatively large minimum distances such as Reed-Solomon codes.

Soft decisions are made by associating a number called the *metric* with each possible codeword. The metric is a function of both the codeword and the demodulator output samples. A soft-decision decoder selects the codeword with the largest metric and then produces the corresponding information bits as its output. Let $\mathbf{y}$ denote the n-dimensional vector of noisy output samples $y_i, i = 1, 2, \ldots, n$, produced by a demodulator that receives a sequence of n symbols. Let $\mathbf{x}_l$ denote the lth codeword vector with symbols $x_{li}, i = 1, 2, \ldots, n$. Let $f(\mathbf{y}|\mathbf{x}_l)$ denote the *likelihood function*, which is the conditional probability density function of $\mathbf{y}$ given that $\mathbf{x}_l$ was transmitted. The maximum-likelihood decoder finds the value of l, $1 \leq l \leq q^k$, for which the likelihood function is largest. If this value is l_0, the decoder decides that codeword l_0 was transmitted. Any monotonically increasing function of $f(\mathbf{y}|\mathbf{x}_l)$ may serve as the metric of a maximum-likelihood decoder. A convenient choice is often proportional to the logarithm of $f(\mathbf{y}|\mathbf{x}_l)$, which is called the *log-likelihood function*. For statistically independent demodulator outputs, the log-likelihood function for each of the q^k possible codewords is

$$\ln f(\mathbf{y}|\mathbf{x}_l) = \sum_{i=1}^{n} \ln f(y_i|x_{li}) , \qquad l = 1, 2, \ldots, q^k \qquad (1\text{-}36)$$

where $f(y_i|x_{li})$ is the conditional probability density function of y_i given the value of x_{li}.

For coherent binary PSK communication over the AWGN channel, if codeword l is transmitted, then the received signal representing symbol i is

$$r_i(t) = \sqrt{2\mathcal{E}_s} x_{li} \psi(t) \cos 2\pi f_c t + n_i(t) , \quad 0 \leq t \leq T_s, \quad i = 1, 2, \ldots, n \qquad (1\text{-}37)$$

where $\mathcal{E}_s$ is the symbol energy, T_s is the symbol duration, f_c is the carrier frequency, $x_{li} = +1$ when binary symbol i is a 1 and $x_{li} = -1$ when binary symbol i is a 0, $\psi(t)$ is the unit-energy symbol waveform, and $n_i(t)$ is independent, zero-mean, white Gaussian noise. Since $\psi(t)$ has unit energy and vanishes outside $[0, T_s]$,

$$\int_0^{T_s} |\psi(t)|^2 dt = 1 \qquad (1\text{-}38)$$

For coherent demodulation, a frequency translation to baseband is provided by multiplying $r_i(t)$ by $\cos 2\pi f_c t$. After discarding a negligible integral, we find that the matched-filter demodulator, which is matched to $\psi(t)$, produces the output samples

$$y_i = \sqrt{\mathcal{E}_s/2}\, x_{li} + \int_0^{T_s} n_i(t)\psi(t)\cos 2\pi f_c t\, dt\ , \quad i = 1, 2, \ldots, n \qquad (1\text{-}39)$$

These outputs provide sufficient statistics because $\psi(t)\cos 2\pi f_c t$ is the sole basis function for the signal space. Since $n_i(t)$ is statistically independent of $n_k(t)$ when $i \neq k$, the $\{y_i\}$ are statistically independent.

The autocorrelation of each white noise process is

$$E[n_i(t)n_i(t+\tau)] = \frac{N_{0i}}{2}\delta(\tau)\ , \quad i = 1, 2, \ldots, n \qquad (1\text{-}40)$$

where $N_{0i}/2$ is the two-sided power spectral density of $n_i(t)$ and $\delta(\tau)$ is the Dirac delta function. A straightforward calculation using (1-40) and assuming that the spectrum of $\psi(t)$ is confined to $|f| < f_c$ indicates that the variance of the noise term of (1-39) is $N_{0i}/4$. Therefore, the conditional probability density function of y_i given that x_{li} was transmitted is

$$f(y_i|x_{li}) = \frac{1}{\sqrt{\pi N_{0i}/2}}\exp\left[-\frac{(y_i - \sqrt{\mathcal{E}_s/2}\, x_{li})^2}{N_{0i}/2}\right]\ , \quad i = 1, 2, \ldots, n \qquad (1\text{-}41)$$

Since y_i^2 and $x_{li}^2 = 1$ are independent of the codeword l, terms involving these quantities may be discarded in the log-likelihood function of (1-36). Therefore, the maximum-likelihood metric is

$$U(l) = \sum_{i=1}^{n} \frac{x_{li}y_i}{N_{0i}}\ , \quad l = 1, 2, \ldots, 2^k \qquad (1\text{-}42)$$

which requires knowledge of $N_{0i}, i = 1, 2, \ldots, n$.

If each $N_{0i} = N_0$, a constant, then this constant is irrelevant, and the maximum-likelihood metric is

$$U(l) = \sum_{i=1}^{n} x_{li}y_i\ , \quad l = 1, 2, \ldots, 2^k \qquad (1\text{-}43)$$

Let $P_2(\delta)$ denote the probability that the metric for an incorrect codeword at distance δ from the correct codeword exceeds the metric for the correct codeword. After reordering the samples $\{y_i\}$, the difference between the metrics for the correct codeword and the incorrect one may be expressed as

$$D(\delta) = \sum_{i=1}^{\delta}(x_{1i} - x_{2i})y_i = 2\sum_{i=1}^{\delta} x_{1i}y_i \qquad (1\text{-}44)$$

where the sum includes only the δ terms that differ, x_{1i} refers to the correct codeword, x_{2i} refers to the incorrect codeword, and $x_{2i} = -x_{1i}$. Then $P_2(\delta)$

is the probability that $D(\delta) < 0$. Since each of its terms is independent, $D(\delta)$ has a Gaussian distribution. A straightforward calculation using (1-41) and $\mathcal{E}_s = r\mathcal{E}_b$ yields

$$P_2(\delta) = Q\left(\sqrt{\frac{2\delta r\mathcal{E}_b}{N_0}}\right) \qquad (1\text{-}45)$$

which reduces to (1-29) when a single symbol is considered and $\delta = 1$.

A fundamental property of a probability, called *countable subadditivity*, is that the probability of a finite or countable union of events B_n, $n = 1, 2, \ldots$, satisfies

$$P[\cup_n B_n] \leq \sum_n P[B_n] \qquad (1\text{-}46)$$

In communication theory, a bound obtained from this inequality is called a *union bound*. To determine P_w for linear block codes, it suffices to assume that the all-zero codeword was transmitted. The union bound and the relation between weights and distances imply that P_w for soft-decision decoding satisfies

$$P_w \leq \sum_{l=d_m}^{n} A_l P_2(l) \qquad (1\text{-}47)$$

Let β_l denote the total information-symbol weight of the codewords of weight l. The union bound and (1-16) imply that

$$P_{is} \leq \sum_{l=d_m}^{n} \frac{\beta_l}{k} P_2(l) \qquad (1\text{-}48)$$

To determine β_l for any cyclic (n, k) code, consider the set S_l of A_l codewords of weight l. The total weight of all the codewords in S_l is $A_T = lA_l$. Let α and β denote any two fixed positions in the codewords. By definition, any cyclic shift of a codeword produces another codeword of the same weight. Therefore, for every codeword in S_l that has a zero in α, there is some codeword in S_l that results from a cyclic shift of that codeword and has a zero in β. Thus, among the codewords of S_l, the total weight of all the symbols in a fixed position is the same regardless of the position and is equal to A_T/n. The total weight of all the information symbols in S_l is $\beta_l = kA_T/n = klA_l/n$. Therefore,

$$P_{is} \leq \sum_{l=d_m}^{n} \frac{l}{n} A_l P_2(l) \qquad (1\text{-}49)$$

Optimal soft-decision decoding cannot be efficiently implemented except for very short block codes, primarily because the number of codewords for which the metrics must be computed is prohibitively large, but approximate maximum-likelihood decoding algorithms are available. The *Chase algorithm* [3] generates a small set of candidate codewords that will almost always include the codeword with the largest metric. Test patterns are generated by first making hard decisions on each of the received symbols and then altering the least reliable symbols, which are determined from the demodulator outputs given by (1-39). Hard-decision decoding of each test pattern and the discarding of decoding failures generate the candidate codewords. The decoder selects the candidate codeword with the largest metric.

The quantization of soft-decision information to more than two levels requires analog-to-digital conversion of the demodulator output samples. Since the optimal location of the levels is a function of the signal, thermal noise, and interference powers, automatic gain control is often necessary. For the AWGN channel, it is found that an eight-level quantization represented by three bits and a uniform spacing between threshold levels cause no more than a few tenths of a decibel loss relative to what could theoretically be achieved with unquantized analog voltages or infinitely fine quantization.

The *coding gain* of one code compared with a second one is the reduction in the signal power or value of E_b/N_0 required to produce a specified information-bit or information-symbol error probability. Calculations for specific communication systems and codes operating over the AWGN channel have shown that an optimal soft-decision decoder provides a coding gain of approximately 2 dB relative to a hard-decision decoder. However, soft-decision decoders are much more complex to implement and may be too slow for the processing of high information rates. For a given level of implementation complexity, hard-decision decoders can accommodate much longer block codes, thereby at least partially overcoming the inherent advantage of soft-decision decoders. In practice, soft-decision decoding other than erasures is seldom used with block codes of length greater than 50.

Performance Examples

Figure 1.2 depicts the information-bit error probability $P_b = P_{is}$ versus $\mathcal{E}_b/N_0$ for various binary block codes with coherent PSK over the AWGN channel. Equation (1-25) is used to compute P_b for the Golay (23,12) code with hard decisions. Since the packing density D_p is small for these codes, (1-26) is used for the BCH (63,36) code, which corrects $t = 5$ errors, and the BCH (127,64) code, which corrects $t = 10$ errors. Equation (1-29) is used for P_s. Inequality (1-49) and Table 1.2 are used to compute the upper bound on $P_b = P_{is}$ for the Golay (23,12) code with optimal soft decisions. The graphs illustrate the power of the soft-decision decoding. For the Golay (23,12) code, soft-decision decoding provides an approximately 2-dB coding gain for $P_b = 10^{-5}$ relative to hard-decision decoding. Only when $P_b < 10^{-5}$ does the BCH (127,64) begin to outperform the Golay (23,12) code with soft decisions. If $\mathcal{E}_b/N_0 \leq 3$ dB, an uncoded system with coherent PSK provides a lower P_b than a similar system that uses one of the block codes of the figure.

Figure 1.3 illustrates the performance of loosely packed Reed-Solomon codes with hard-decision decoding over the AWGN channel. The lower bound in (1-26) is used to compute the approximate information-bit error probabilities for binary channel symbols with coherent PSK and for nonbinary channel symbols with noncoherent MFSK. For the nonbinary channel symbols, (1-27) and (1-31)

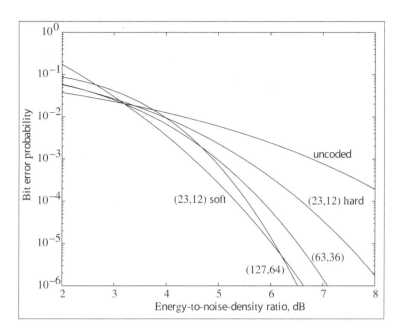

Figure 1.2: Information-bit error probability for binary block (n, k) codes and coherent PSK.

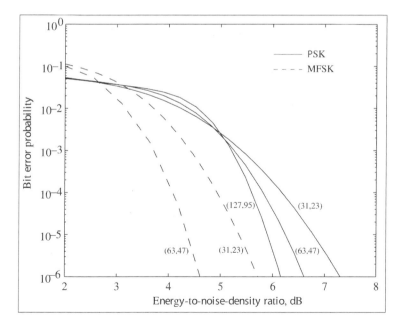

Figure 1.3: Information-bit error probability for Reed-Solomon (n, k) codes. Modulation is coherent PSK or noncoherent MFSK.

are used. For the binary channel symbols, (1-34) and the lower bound in (1-33) are used. For the chosen values of n, the best performance at $P_b = 10^{-5}$ is obtained if the code rate is $k/n \approx 3/4$. Further gains result from increasing n and hence the implementation complexity. Although the figure indicates the performance advantage of Reed-Solomon codes with MFSK, there is a major bandwidth penalty. Let B denote the bandwidth required for an uncoded binary PSK signal. If the same data rate is accommodated by using uncoded binary frequeny-shift keying (FSK), the required bandwidth for demodulation with envelope detectors is approximately $2B$. For uncoded MFSK using $q = 2^m$ frequencies, the required bandwidth is $2^m B/m$ because each symbol represents m bits. If a Reed-Solomon (n, k) code is used with MFSK, the required bandwidth becomes $2^m nB/mk$.

Code Metrics for Orthogonal Signals

For q-ary orthogonal symbol waveforms, $s_1(t), s_2(t), \ldots, s_q(t)$, q matched filters are needed, and the observation vector is $\mathbf{y} = [\mathbf{y}_1\ \mathbf{y}_2 \ldots\ \mathbf{y}_q]$, where each $\mathbf{y}_k$ is an n-dimensional row vector of matched-filter output samples for filter k with components $y_{ki}, i = 1, 2, \ldots, n$. Suppose that symbol i of codeword l uses unit-energy waveform $s_\nu(t)$, where the integer ν is a function of i and l. If codeword l is transmitted over the AWGN channel, the received signal for symbol i can be expressed in complex notation as

$$r_i(t) = \text{Re}\left[\sqrt{2\mathcal{E}_s}s_\nu(t)e^{j2\pi f_c t+\theta_i}\right] + n_i(t), \quad 0 \le t \le T_s, \quad i = 1, 2, \ldots, n \quad (1\text{-}50)$$

where $n_i(t)$ is independent, zero-mean, white Gaussian noise with two-sided power spectral density $N_{0i}/2$, f_c is the carrier frequency, and θ_i is the phase. Since the symbol energy for all the waveforms is unity,

$$\int_0^{T_s} |s_k(t)|^2 dt = 1, \quad k = 1, 2, \ldots, q \quad (1\text{-}51)$$

The orthogonality of symbol waveforms implies that

$$\int_0^{T_s} s_k(t)s_m^*(t)dt = 0, \quad k \ne m \quad (1\text{-}52)$$

A frequency translation or *downconversion* to baseband is followed by matched filtering. Matched-filter k, which is matched to $s_k(t)$, produces the output samples

$$y_{ki} = \int_0^{T_s} r_i(t)e^{-j2\pi f_c t}s_k^*(t)dt, \quad i = 1, 2, \ldots, n, \quad k = 1, 2, \ldots, q \quad (1\text{-}53)$$

The substitution of (1-50) into (1-53), (1-52), and the assumption that each of the $\{s_k(t)\}$ has a spectrum confined to $|f| < f_c$ yields

$$y_{ki} = \sqrt{\mathcal{E}_s/2}e^{j\theta_i}\delta_{k\nu} + n_{ki} \quad (1\text{-}54)$$

where $\delta_{k\nu} = 1$ if $k = \nu$ and $\delta_{k\nu} = 0$ otherwise, and

$$n_{ki} = \int_0^{T_s} n_i(t)e^{-j2\pi f_c t} s_k^*(t)dt \tag{1-55}$$

Since the real and imaginary components of n_{ki} are jointly Gaussian, this random process is a *complex-valued Gaussian random variable*. Straightforward calculations using (1-40) and the confined spectra of the $\{s_k(t)\}$ indicates that the real and are imaginary components of n_{ki} are uncorrelated and, hence, independent and have the same variance $N_{0i}/4$. Since the density of a complex-valued random variable is defined to be the joint density of its real and imaginary parts, the conditional probability density function of y_{ki} given θ_i is

$$f(y_{ki} \mid \theta_i) = \frac{1}{\pi N_{0i}/2} \exp\left(-\frac{\left| y_{ki} - \sqrt{\mathcal{E}_s/2}e^{j\theta_i}\delta_{k\nu} \right|^2}{N_{0i}/2}\right),$$

$$i = 1, 2, \ldots, n, \quad k = 1, 2, \ldots, q \tag{1-56}$$

The independence of the white Gaussian $\{n_i(t)\}$, the orthogonality condition (1-52), and the spectrally confined symbol waveforms ensure that both the real and imaginary parts of y_{ki} are independent of both the real and imaginary parts of y_{mp} unless $k=m$ and $i=p$. Thus, the likelihood function of .the observation vector $\mathbf{y}$ is the product of the qn densities specified by (1-56).

For *coherent* signals, the $\{\theta_i\}$ are tracked by the phase synchronization system and, thus, ideally may be set to zero. Forming the log-likelihood function with the $\{\theta_i\}$ set to zero, and eliminating irrelevant terms that are independent of l, we obtain the maximum-likelihood metric

$$U(l) = \sum_{i=1}^{n} \frac{\text{Re}(V_{li})}{N_{0i}} \tag{1-57}$$

where $V_{li} = y_{\nu i}$ is the sampled output i of the filter matched to $s_\nu(t)$, the signal representing symbol i of codeword l. If each $N_{0i} = N_0$, then the maximum-likelihood metric is

$$U(l) = \sum_{i=1}^{n} \text{Re}(V_{li}) \tag{1-58}$$

and the common value N_0 does not need to be known to apply this metric.

For *noncoherent* signals, it is assumed that each θ_i is independent and uniformly distributed over $[0, 2\pi)$, which preserves the independence of the $\{y_{ki}\}$. Expanding the argument of the exponential function in (1-56), expressing y_{ki} in polar form, and integrating over θ_i, we obtain the probability density function

$$f(y_{ki}) = \frac{1}{\pi N_{0i}/2} \exp\left[-\frac{|y_{ki}|^2 + \mathcal{E}_s\delta_{k\nu}/2}{N_{0i}/2}\right] I_0\left(\frac{\sqrt{8\mathcal{E}_s}\,|y_{ki}|\,\delta_{k\nu}}{N_{0i}}\right) \tag{1-59}$$

where $I_0(\ )$ is the modified Bessel function of the first kind and order zero, This function may be represented by

$$I_0(x) = \frac{1}{2\pi} \int_0^{2\pi} \exp(x\cos u)\,du$$

$$= \sum_0^\infty \frac{1}{i!i!} \left(\frac{x}{2}\right)^{2i} \tag{1-60}$$

Let $R_{li} = |y_{\nu i}|$ denote the sampled envelope produced by the filter matched to $s_\nu(t)$, the signal representing symbol i of codeword l. We form the log-likelihood function and eliminate terms and factors that do not depend on the codeword l, thereby obtaining the maximum-likelihood metric

$$U(l) = \sum_{i=1}^n \ln I_0 \left(\frac{\sqrt{8\mathcal{E}_s}\,R_{li}}{N_{0i}}\right) \tag{1-61}$$

If each $N_{0i} = N_0$, then the maximum-likelihood metric is

$$U(l) = \sum_{i=1}^n \ln I_0 \left(\frac{\sqrt{8\mathcal{E}_s}\,R_{li}}{N_0}\right) \tag{1-62}$$

and $\sqrt{\mathcal{E}_s}/N_0$ must be known to apply this metric.

From the series representation of $I_0(x)$, it follows that

$$I_0(x) \le \exp\left(\frac{x^2}{4}\right) \tag{1-63}$$

From the integral representation, we obtain

$$I_0(x) \le \exp(|\,x\,|) \tag{1-64}$$

The upper bound in (1-63) is tighter for $0 \le x < 2$, while the upper bound in (1-64) is tighter for $2 < x < \infty$. If we assume that R_{li}/N_{0i} is often less than 2, then the approximation of $I_0(x)$ by $\exp(x^2/4)$ is reasonable. Substitution into (1-61) and dropping an irrelevant constant gives the metric

$$U(l) = \sum_{i=1}^n \frac{R_{li}^2}{N_{0i}^2} \tag{1-65}$$

If each $N_{0i} = N_0$, then the value of N_0 is irrelevant, and we obtain the *Rayleigh metric*

$$U(l) = \sum_{i=1}^n R_{li}^2 \tag{1-66}$$

which is suboptimal for the AWGN channel but is the maximum-likelihood metric for the Rayleigh fading channel with identical statistics for each of the symbols (Section 5.6). Similarly, (1-64) can be used to obtain suboptimal metrics suitable for large values of R_{li}/N_{0i}.

To determine the maximum-likelihood metric for making a hard decision on each symbol, we set $n = 1$ and drop the subscript i in (1-57) and (1-61). We find that the maximum-likelihood symbol metric is $Re(V_l)$ for coherent MFSK and $\ln\left[I_0(\sqrt{8\mathcal{E}_s}R_l/N_0)\right]$ for noncoherent MFSK, where the index l ranges over the symbol alphabet. Since the latter function increases monotonically and $\sqrt{8\mathcal{E}_s}/N_0$ is a constant, optimal symbol metrics or decision variables for noncoherent MFSK are R_l or R_l^2 for $l = 1, 2, \ldots, q$.

Metrics and Error Probabilities for MFSK Symbols

For noncoherent MFSK, baseband matched-filter l is matched to the unit-energy waveform $s_l(t) = A\exp(j2\pi f_l t)$, $0 \le t \le T_s$, where $A = 1/\sqrt{T_s}$. If $r(t)$ is the received signal, a downconversion to baseband and a parallel set of matched filters and envelope detectors provide the decision variables

$$R_l^2 = A^2 \left| \int_0^{T_s} r(t) e^{-j2\pi f_c t} e^{-j2\pi f_l t} dt \right|^2 \tag{1-67}$$

The orthogonality condition (1-52) is satisfied if the adjacent frequencies are separated by k/T_s, where k is a nonzero integer. Expanding (1-67), we obtain

$$R_l^2 = R_{lc}^2 + R_{ls}^2 \tag{1-68}$$

$$R_{lc} = A\int_0^{T_s} r(t) \cos\left[2\pi(f_c + f_l)t\right] dt \tag{1-69}$$

$$R_{ls} = A\int_0^{T_s} r(t) \sin\left[2\pi(f_c + f_l)t\right] dt \tag{1-70}$$

These equations imply the correlator structure depicted in Figure 1.4, where the irrelevant constant A has been omitted. The comparator decides what symbol was transmitted by observing which comparator input is the largest.

To derive an alternative implementation, we observe that when the waveform is $s_l(t) = A\cos 2\pi(f_c + f_l)t$, $0 \le t \le T_s$, the impulse response of a filter matched to it is $A\cos 2\pi(f_c + f_l)(T_s - t)$, $0 \le t \le T_s$. Therefore, the matched-filter output at time t is

$$
\begin{aligned}
y_l(t) &= A\int_0^t r(\tau) \cos\left[2\pi(f_c + f_l)(\tau - t + T_s)\right] d\tau \\
&= A\left\{\int_0^t r(\tau) \cos\left[2\pi(f_c + f_l)\tau\right] d\tau\right\} \cos\left[2\pi(f_c + f_l)(t - T_s)\right] \\
&\quad + A\left\{\int_0^t r(\tau) \sin\left[2\pi(f_c + f_l)\tau\right] d\tau\right\} \sin\left[2\pi(f_c + f_l)(t - T_s)\right] \\
&= AR_l(t) \cos\left[2\pi(f_c + f_l)(t - T_s) + \phi(t)\right], \quad 0 \le t \le T_s \tag{1-71}
\end{aligned}
$$

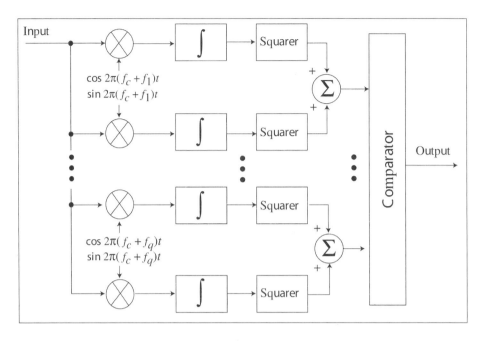

Figure 1.4: Noncoherent MFSK receiver using correlators.

where the envelope is

$$R_l(t) = A\left\{\left[\int_0^t r(\tau) \cos\left[2\pi(f_c + f_l)\tau\right] d\tau\right]^2\right.$$

$$\left. + \left[\int_0^t r(\tau) \sin\left[2\pi\left(f_c + f_l\right)\tau\right] d\tau\right]^2\right\}^{1/2} \tag{1-72}$$

Since $R_l(T_s) = R_l$ given by (1-68), we obtain the receiver structure depicted in Figure 1.5, where the irrelevant constant A has been omitted. A practical envelope detector consists of a peak detector followed by a lowpass filter.

To derive the symbol error probability for equally likely MFSK symbols, we assume that the signal $s_1(t)$ was transmitted over the AWGN channel. The received signal has the form $r(t) = \sqrt{2\mathcal{E}_s/T_s} \cos\left[2\pi\left(f_c + f_1\right)t + \theta\right] + n(t)$, $0 \leq t \leq T_s$. Since $n(t)$ is white,

$$E[n(t)n(t+\tau)] = \frac{N_0}{2}\delta(\tau) \tag{1-73}$$

Using the orthogonality of the symbol waveforms and (1-73) and assuming that

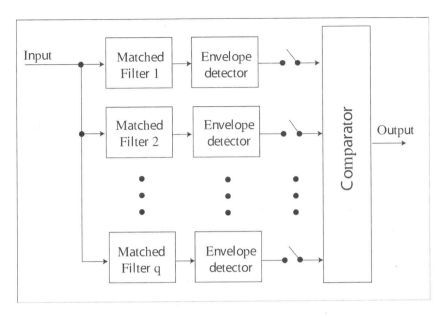

Figure 1.5: Noncoherent MFSK receiver with passband matched filters.

$f_c + f_l \gg 1/T_s$ in (1-69) and (1-70), we obtain

$$E[R_{1c}] = \sqrt{\mathcal{E}_s/2}\cos\theta , \quad E[R_{1s}] = \sqrt{\mathcal{E}_s/2}\sin\theta \qquad (1\text{-}74)$$

$$E[R_{lc}] = E[R_{ls}] = 0 , \quad l = 2,\ldots,q \qquad (1\text{-}75)$$

$$\mathrm{var}(R_{lc}) = \mathrm{var}(R_{ls}) = N_0/4 , \quad l = 1,2,\ldots,q \qquad (1\text{-}76)$$

Since $n(t)$ is Gaussian, R_{lc} and R_{ls} are jointly Gaussian. Since the covariance of R_{lc} and R_{ls} is zero, they are mutually statistically independent. Therefore, the joint probability density function of R_{lc} and R_{ls} is

$$g_1(r_{lc}, r_{ls}) = \frac{1}{\pi N_0/2} \exp\left[-\frac{(r_{lc} - m_{lc})^2 + (r_{ls} - m_{ls})^2}{N_0/2}\right] \qquad (1\text{-}77)$$

where $m_{lc} = E[R_{lc}]$ and $m_{ls} = E[R_{ls}]$.

Let R_l and Θ_l be implicitly defined by $R_{lc} = R_l \cos\Theta_l$ and $R_{ls} = R_l \sin\Theta_l$. Since the Jacobian of the transformation is r, we find that the joint density of R_l and Θ_l is

$$g_2(r, \theta) = \frac{2r}{\pi N_0} \exp\left[-\frac{r^2 - 2rm_{lc}\cos\theta - 2rm_{ls}\sin\theta + m_{lc}^2 + m_{ls}^2}{N_0/2}\right]$$

$$r \geq 0, \ |\theta| \leq \pi \qquad (1\text{-}78)$$

The density of the envelope R_l is obtained by integration of (1-78) over θ. Using trigonometry and the integral representation of the Bessel function, we obtain the density

$$g_3(r) = \frac{4r}{N_0} \exp\left(-\frac{r^2 + m_{lc}^2 + m_{ls}^2}{N_0/2}\right) I_0\left(4r\sqrt{m_{lc}^2 + m_{ls}^2}_0\right) u(r) \qquad (1\text{-}79)$$

where $u(r) = 1$ if $r \geq 0$, and $u(r) = 0$ if $r < 0$. Substituting (1-74), we obtain the densities for the R_l, $l = 1, 2, \ldots, q$:

$$f_1(r) = \frac{4r}{N_0} \exp\left(-\frac{r^2 + \mathcal{E}_s/2}{N_0/2}\right) I_0\left(\frac{\sqrt{8\mathcal{E}_s}r}{N_0}\right) u(r) \qquad (1\text{-}80)$$

$$f_l(r) = \frac{4r}{N_0} \exp\left(-r^2{}_0/2\right) u(r), \quad l = 2, \ldots, q \qquad (1\text{-}81)$$

The orthogonality of the symbol waveforms and (1-73) imply that the random variables $\{R_l\}$ are independent. A symbol error occurs when $s_1(t)$ was transmitted if R_1 is not the largest of the $\{R_l\}$. Since the $\{R_l\}$ are identically distributed for $l = 2, \cdots, q$, the probability of a symbol error when $s_1(t)$ was transmitted is

$$P_s = 1 - \int_0^\infty \left[\int_0^r f_2(y)dy\right]^{q-1} f_1(r)dr \qquad (1\text{-}82)$$

Substituting (1-81) into the inner integral gives

$$\int_0^r f_2(y)dy = 1 - \exp\left(-\frac{r^2}{N_0/2}\right) \qquad (1\text{-}83)$$

Expressing the $(q-1)$th power of this result as a binomial expansion and then substituting into (1-82), the remaining integration may be done by using the fact that for $\lambda > 0$,

$$\int_0^\infty r \exp\left(-\frac{r^2}{2b^2}\right) I_0\left(\frac{r\sqrt{\lambda}}{b^2}\right) dr = b^2 \exp\left(\frac{\lambda}{2b^2}\right) \qquad (1\text{-}84)$$

which follows from the fact that the density in (1-80) must integrate to unity. The final result is the symbol error probability for noncoherent MFSK over the AWGN channel:

$$P_s = \sum_{i=1}^{q-1} \frac{(-1)^{i+1}}{i+1} \binom{q-1}{i} \exp\left[-\frac{i\mathcal{E}_s}{(i+1)N_0}\right] \qquad (1\text{-}85)$$

When $q = 2$, this equation reduces to the classical formula for binary FSK:

$$P_s = \frac{1}{2} \exp\left(-\frac{\mathcal{E}_s}{2N_0}\right) \qquad (1\text{-}86)$$

Chernoff Bound

The Chernoff bound is an upper bound on the probability that a random variable equals or exceeds a constant. The usefulness of the Chernoff bound stems from the fact that it is often much more easily evaluated than the probability it bounds. The *moment generating function* of the random variable X with distribution function $F(x)$ is defined as

$$M(s) = E\left[e^{sX}\right] = \int_{-\infty}^{\infty} \exp(sx)dF(x) \tag{1-87}$$

for all real-valued s for which the integral is finite. For all nonnegative s, the probability that $X \geq 0$ is

$$P\left[X \geq 0\right] = \int_{0}^{\infty} dF(x) \leq \int_{0}^{\infty} \exp(sx)dF(x) \tag{1-88}$$

Thus,

$$P\left[X \geq 0\right] \leq M(s), \quad 0 \leq s < s_1 \tag{1-89}$$

where s_1 is the upper limit of an open interval in which $M(s)$ is defined. To make this bound as tight as possible, we choose the value of s that minimizes $M(s)$. Therefore,

$$P\left[X \geq 0\right] \leq \min_{0 \leq s < s_1} M(s) \tag{1-90}$$

which indicates the upper bound called the *Chernoff bound*. From (1-90) and (1-87), we obtain the generalization

$$P\left[X \geq b\right] \leq \min_{0 \leq s < s_1} M(s) \exp(-sb) \tag{1-91}$$

Since the moment generating function is finite in some neighborhood of $s = 0$, we may differentiate under the integral sign in (1-87) to obtain the derivative of $M(s)$. The result is

$$M'(s) = \int_{-\infty}^{\infty} x \exp(sx)dF(x) \tag{1-92}$$

which implies that $M'(0) = E[X]$. Differentiating (1-92) gives the second derivative

$$M''(s) = \int_{-\infty}^{\infty} x^2 \exp(sx)dF(x) \tag{1-93}$$

which implies that $M''(s) \geq 0$. Consequently, $M(s)$ is convex in its interval of definition. Consider a random variable is such that

$$E(X) < 0, \quad P[X > 0] > 0 \tag{1-94}$$

The first inequality implies that $M'(0) < 0$, and the second inequality implies that $M(s) \to \infty$ as $s \to \infty$. Thus, since $M(0) = 1$, the convex function $M(s)$ has a minimum value that is less than unity at some positive $s = s_0$. We conclude that (1-94) is sufficient to ensure that the Chernoff bound is less than unity and $s_0 > 0$.

The Chernoff bound can be tightened if X has a density function $f(x)$ such that

$$f(-x) \geq f(x), \quad x \geq 0 \tag{1-95}$$

For $s \in A$, where $A = (s_0, s_1)$ is the open interval over which $M(s)$ is defined, (1-87) implies that

$$
\begin{aligned}
M(s) &= \int_0^\infty \exp(sx)f(x)dx + \int_{-\infty}^0 \exp(sx)f(x)dx \\
&\geq \int_0^\infty [\exp(sx) + \exp(-sx)] f(x)dx = \int_0^\infty 2\cosh(sx)f(x)dx \\
&\geq 2 \int_0^\infty f(x)dx = 2P[X \geq 0]
\end{aligned}
\tag{1-96}
$$

Thus, we obtain the following version of the Chernoff bound:

$$P[X \geq 0] \leq \frac{1}{2} \min_{s \in A} M(s) \tag{1-97}$$

where the minimum value s_0 is not required to be nonnegative. However, if (1-94) holds, then the bound is less than $1/2$, $s_0 > 0$, and

$$P[X \geq 0] \leq \frac{1}{2} \min_{0 < s < s_1} M(s) \tag{1-98}$$

In soft-decision decoding, the encoded sequence or codeword with the largest associated metric is converted into the decoded output. Let $U(j)$ denote the value of the metric associated with sequence j of length L. Consider additive metrics having the form

$$U(j) = \sum_{i=1}^L m(j,i) \tag{1-99}$$

where $m(j,i)$ is the *symbol metric* associated with symbol i of the encoded sequence. Let $j = 1$ label the correct sequence and $j = 2$ label an incorrect one. Let $P_2(l)$ denote the probability that the metric for an incorrect codeword at distance l from the correct codeword exceeds the metric for the correct codeword. By suitably relabeling the l symbol metrics that may differ for the two sequences, we obtain

$$
\begin{aligned}
P_2(l) &\leq P\left[U(2) \geq U(1)\right] \\
&= P\left[\sum_{i=1}^l [m(2,i) - m(1,i)] \geq 0\right]
\end{aligned}
\tag{1-100}
$$

where the inequality results because $U(2) = U(1)$ does not necessarily cause an error if it occurs. In all practical cases, (1-94) is satisfied for the random variable $X = U(2) - U(1)$. Therefore, the Chernoff bound implies that

$$P_2(l) \leq \alpha \min_{0<s<s_1} E\left[\exp\left\{s\sum_{i=1}^{l}[m(2,i) - m(1,i)]\right\}\right] \qquad (1\text{-}101)$$

where s_1 is the upper limit of the interval over which the expected value is defined. Depending on which version of the Chernoff bound is valid, either $\alpha = 1$ or $\alpha = 1/2$. If $m(2,i) - m(1,i)$, $i = 1, 2, \ldots, l$, are independent, identically distributed random variables and we define

$$Z = \min_{0<s<s_1} E\left[\exp\left\{s[m(2,i) - m(1,i)]\right\}\right] \qquad (1\text{-}102)$$

then

$$P_2(l) \leq \alpha Z^l \qquad (1\text{-}103)$$

This bound is often much simpler to compute than the exact $P_2(l)$. As l increases, the central-limit theorem implies that the distribution of $X = U(2) - U(1)$ approximates the Gaussian distribution. Thus, for large enough l, (1-95) is satisfied when $E[X] < 0$, and we can set $\alpha = 1/2$ in (1-103). For small l, (1-95) may be difficult to establish mathematically, but is often intuitively clear; if not, setting $\alpha = 1$ in (1-103) is always valid.

These results can be applied to hard-decision decoding, which can be regarded as a special case of soft-decision decoding with the following symbol metric. If symbol i of a candidate binary sequence j agrees with the corresponding detected symbol at the demodulator output, then $m(j,i) = 1$; otherwise $m(j,i) = 0$. Therefore, $m(2,i) - m(1,i)$ in (1-102) is equal to $+1$ with probability P_s and -1 with probability $(1 - P_s)$. Thus,

$$Z = \min_{0<s}\left[(1 - P_s)e^{-s} + P_s e^s\right]$$
$$= [4P_s(1 - P_s)]^{1/2} \qquad (1\text{-}104)$$

for hard-decision decoding. Substituting this equation into (1-103) with $\alpha = 1$, we obtain

$$P_2(l) \leq [4P_s(1 - P_s)]^{l/2} \qquad (1\text{-}105)$$

This upper bound is not always tight but has great generality since no specific assumptions have been made about the modulation or coding.

1.2 Convolutional Codes and Trellis Codes

In contrast to a block codeword, a convolutional codeword represents an entire message of indefinite length. A *convolutional encoder* converts an input of k information bits into an output of n code bits that are Boolean functions of both the current k input bits and the preceding information bits. After k bits are shifted into a shift register and k bits are shifted out, n code bits are read out. Each code bit is a Boolean function of the outputs of selected shift-register

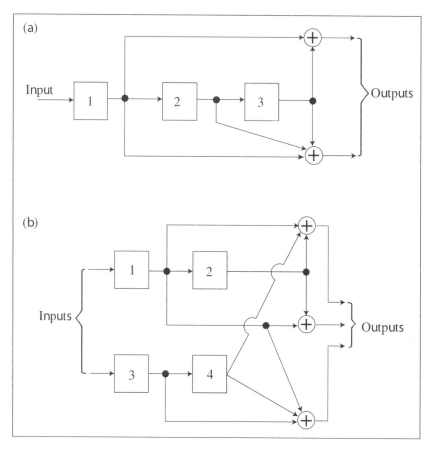

Figure 1.6: Encoders of nonsystematic convolutional codes with (a) $K = 3$ and rate $= 1/2$ and (b) $K = 2$ and rate $= 2/3$.

stages. A convolutional code is *linear* if each Boolean function is a modulo-2 sum because the superposition property applies to the input-output relations and the all-zero codeword is a member of the code. For a linear convolutional code, the minimum Hamming distance between codewords is equal to the minimum Hamming weight of a codeword. The *constraint length* K of a convolutional code is the maximum number of sets of n output bits that can be affected by an input bit. A convolutional code is *systematic* if the information bits appear unaltered in each codeword.

A nonsystematic linear convolutional encoder with $k = 1$, $n = 2$, and $K = 3$ is shown in Figure 1.6(a). The shift register consists of 3 stages, each of which is implemented as a bistable memory element. Information bits enter the shift register in response to clock pulses. After each clock pulse, the most recent information bit becomes the content and output of the first stage, the previous contents of the first two stages are shifted to the right, and the previous content

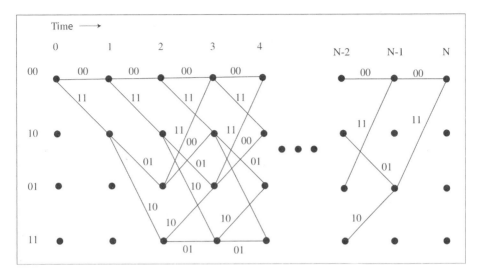

Figure 1.7: Trellis diagram corresponding to encoder of Figure 1.6(a).

of the third stage is shifted out of the register. The outputs of the modulo-2 adders (exclusive-OR gates) provide two code bits. The *generators* of the output bits are the sequences $\mathbf{g}_1 = [1\,0\,1]$ and $\mathbf{g}_2 = [1\,1\,1]$, which indicate the stages that are connected to the adders. In octal form, the two generator sequences are represented by (5, 7). The encoder of a nonsystematic convolutional code with $k = 2$, $n = 3$, and $K = 2$ is shown in Figure 1.6(b). In octal form(e.g., $1101 \rightarrow 13$), its generators are (13, 12, 11).

Since k bits exit from the shift register as k new bits enter it, only the contents of the first $(K-1)k$ stages prior to the arrival of new bits affect the subsequent output bits of a convolutional encoder. Therefore, the contents of these $(K-1)k$ stages define the *state* of the encoder. The initial state of the encoder is generally the all-zero state. After the message sequence has been encoded $(K-1)k$ zeros inust be inserted into the encoder to complete and terminate the codeword. If the number of message bits is much greater than $(K-1)k$, these terminal zeros have a negligible effect and the *code rate* is well approximated by $r = k/n$. However, the need for the terminal zeros renders the convolutional codes unsuitable for short messages. For example, if 12 information bits are to be transmitted, the Golay (23, 12) code provides a better performance than the same convolutional codes that are much more effective when 1000 or more bits are to be transmitted.

A *trellis diagram* corresponding to the encoder of Figure 1.6(a) is shown in Figure 1.7. Each of the nodes in a column of a trellis diagram represents the state of the encoder at a specific time prior to a clock pulse. The first bit of a state represents the content of stage 1, while the second bit represents the content of stage 2. Branches connecting nodes represent possible changes of state. Each branch is labeled with the output bits or symbols produced following a clock pulse and the formation of a new encoder state. In this example, the first bit of a branch label refers to the upper output of the encoder. The upper branch leaving a node corresponds to a 0 input bit, while the lower branch corresponds to a 1. Every path from left to right through the trellis represents a possible codeword. If the encoder begins in the all-zero state, not all of the other states can be reached until the initial contents have been shifted out. The trellis diagram then becomes identical from column to column until the final $(K-1)k$ input bits force the encoder back to the zero state.

Each branch of the trellis is associated with a branch metric, and the metric of a codeword is defined as the sum of the branch metrics for the path associated with the codeword. A maximum-likelihood decoder selects the codeword with the largest metric (or smallest metric, depending on how branch metrics are defined). The *Viterbi decoder* implements maximum-likelihood decoding efficiently by sequentially eliminating many of the possible paths. At any node, only the partial path reaching that node with the largest partial metric is retained, for any partial path stemming from the node will add the same branch metrics to all paths that merge at that node.

Since the decoding complexity grows exponentially with constraint length, Viterbi decoders are limited to use with convolutional codes of short constraint lengths. A Viterbi decoder for a rate-1/2, $K = 7$ convolutional code has approximately the same complexity as a Reed-Solomon (31,15) decoder. If the constraint length is increased to $K = 9$, the complexity of the Viterbi decoder increases by a factor of approximately 4.

The suboptimal *sequential decoding* of convolutional codes [2] does not invariably provide maximum-likelihood decisions, but its implementation complexity only weakly depends on the constraint length. Thus, very low error probabilities can be attained by using long constraint lengths. The number of computations needed to decode a frame of data is fixed for the Viterbi decoder, but is a random variable for the sequential decoder. When strong interference is present, the excessive computational demands and consequent memory overflows of sequential decoding usually result in a higher P_b than for Viterbi decoding and a much longer decoding delay. Thus, Viterbi decoding is preferable for most communication systems and is assumed in the subsequent performance analysis.

To bound P_b for the Viterbi decoder, we assume that the convolutional code is linear and that binary symbols are transmitted. With these assumptions, the distribution of either Hamming or Euclidean distances is invariant to the choice of a reference sequence. Consequently, whether the demodulator makes hard or soft decisions, the assumption that the all-zero sequence is transmitted entails no loss of generality in the derivation of the error probability. Let $a(l, i)$ denote the number of paths diverging at a node from the the correct path, each having Hamming weight l and i incorrect information symbols over the unmerged segment of the path before it merges with the correct path. Thus, the unmerged segment is at Hamming distance l from the correct all-zero segment. Let d_f denote the *minimum free distance*, which is the minimum distance between any two codewords. Although the encoder follows the all-zero path through the trellis, the decoder in the receiver essentially observes successive columns in the trellis, eliminating paths and thereby sometimes introducing errors at each node. The decoder may select an incorrect path that diverges at node j and introduces errors over its unmerged segment. Let $E[N_e(j)]$ denote the expected value of the number of errors introduced at node j. It is known from (1-16) that the P_b equals the *information-bit error rate*, which is defined as the ratio of the expected number of information-bit errors to the number of information bits applied to the convolutional encoder. Therefore, if there are N branches in a complete path,

$$P_b = \frac{1}{kN} \sum_{j=1}^{N} E[N_e(j)] \tag{1-106}$$

Let $B_j(l, i)$ denote the event that the path with the largest metric diverges at node j and has Hamming weight l and i incorrect information bits over its unmerged segment. Then,

$$E[N_e(j)] = \sum_{i=1}^{I_j} \sum_{l=d_f}^{D_j} E[N_e(j)|B_j(l, i)] \, P[B_j(l, i)] \tag{1-107}$$

when $E[N_e(j)|B_j(l, i)]$ is the conditional expectation of $N_e(j)$ given event $B_j(l, i)$, $P[B_j(l, i)]$ is the probability of this event, and I_j and D_j are the maximum values of i and l, respectively, that are consistent with the position of node j in the trellis. When $B_j(l, i)$ occurs, i bit errors are introduced into the decoded bits; thus,

$$E[N_e(j)|B_j(l, i)] = i \tag{1-108}$$

Since the decoder may already have departed from the correct path before node j, the union bound gives

$$P[B_j(l, i)] \leq a(l, i) P_2(l) \tag{1-109}$$

where $P_2(l)$ is the probability that the correct path segment has a smaller metric than an unmerged path segment that differs in l code symbols. Substituting (1-107) to (1-109) into (1-106) and extending the two summations to ∞, we obtain

$$P_b \leq \frac{1}{k} \sum_{i=1}^{\infty} \sum_{l=d_f}^{\infty} i a(l, i) P_2(l) \tag{1-110}$$

The *information-weight spectrum* or *distribution* is defined as

$$B(l) = \sum_{i=1}^{\infty} i a(l, i), \quad l \geq d_f \tag{1-111}$$

In terms of this distribution, (1-110) becomes

$$P_b \leq \frac{1}{k} \sum_{l=d_f}^{\infty} B(l) P_2(l) \tag{1-112}$$

For coherent PSK signals over an AWGN channel and soft decisions, (1-45)
indicates that

$$P_2(l) = Q\left(\sqrt{\frac{2lr\mathcal{E}_b}{N_0}}\right) \tag{1-113}$$

When the demodulator makes hard decisions and a correct path segment
is compared with an incorrect one, correct decoding results if the number of
symbol errors in the demodulator output is less than half the number of symbols
in which the two segments differ. If the number of symbol errors is exactly half
the number of differing symbols, then either of the two segments is chosen with
equal probability. Assuming the independence of symbol errors, it follows that
for hard-decision decoding

$$P_2(l) = \begin{cases} \sum_{i=(l+1)/2}^{l} \binom{l}{i} P_s^i (1 - P_s)^{l-i}, & l \text{ is odd} \\ \sum_{i=l/2+1}^{l} \binom{l}{i} P_s^i (1 - P_s)^{l-i} + \frac{1}{2} \binom{l}{l/2} [P_s (1 - P_s)]^{l/2}, & l \text{ is even} \end{cases}$$

$$(1\text{-}114)$$

Soft-decision decoding typically provides a 2 dB power savings at $P_b = 10^{-5}$
compared to hard-decision decoding for communications over the AWGN channel. Since the loss due to even three-bit quantization usually is 0.2 to 0.3 dB,
soft-decision decoding is highly preferable.

Among the convolutional codes of a given code rate and constraint length,
the one giving the smallest upper bound in (1-112) can sometimes be determined
by a complete computer search. The codes with the largest value of d_f are
selected, and the *catastrophic codes*, for which a finite number of demodulated
symbol errors can cause an infinite number of decoded information-bit errors,
are eliminated. All remaining codes that do not have the minimum value of
$B(d_f)$ are eliminated. If more than one code remains, codes are eliminated
on the basis of the minimal values of $B(d_f + 1)$, $B(d_f + 2)$, ..., until one code
remains. For binary codes of rates 1/2, 1/3, and 1/4, codes with these favorable
distance properties have been determined [6]. For these codes and constraint
lengths up to 12, Tables 1.4, 1.5, and 1.6 list the corresponding values of d_f
and $B(d_f + i)$, $i = 0, 1, ..., 7$. Also listed in octal form are the generator
sequences that determine which shift-register stages feed the modulo-two adders
associated with each code bit. For example, the best $K = 3$, rate-1/2 code
in Table 1.4 has generator sequences 5 and 7, which specify the connections
illustrated in Figure 1.6(a).

Approximate upper bounds on P_b for rate-1/2, rate-1/3, and rate-1/4 convolutional codes with coherent PSK, soft-decision decoding, and infinitely fine
quantization are depicted in Figures 1.8 to 1.10. The graphs are computed by
using (1-113), $k = 1$, and Tables 1.4 to 1.6 in (1-112) and then truncating the
series after seven terms. This truncation gives a tight upper bound in P_b for
$P_b \leq 10^{-2}$. However, the truncation may exclude significant contributions to
the upper bound when $P_b > 10^{-2}$, and the bound itself becomes looser as P_b increases. The figures indicate that the code performance improves with increases

Table 1.4: Parameter values of rate-1/2 convolutional codes with favorable distance properties.

K	d_f	Generators	$B(d_f + i)$ for $i = 0, 1, \ldots, 6$						
			0	1	2	3	4	5	6
3	5	5, 7	1	4	12	32	80	192	448
4	6	15, 17	2	7	18	49	130	333	836
5	7	23, 35	4	12	20	72	225	500	1324
6	8	53, 75	2	36	32	62	332	701	2342
7	10	133, 171	36	0	211	0	1404	0	11,633
8	10	247, 371	2	22	60	148	340	1008	2642
9	12	561, 763	33	0	281	0	2179	0	15,035
10	12	1131, 1537	2	21	100	186	474	1419	3542
11	14	2473, 3217	56	0	656	0	3708	0	27,518
12	15	4325, 6747	66	98	220	788	2083	5424	13,771

Table 1.5: Parameter values of rate-1/3 convolutional codes with favorable distance properties.

K	d_f	Generators	$B(d_f + i)$ for $i = 0, 1, \ldots, 6$						
			0	1	2	3	4	5	6
3	8	5, 7, 7	3	0	15	0	58	0	201
4	10	13, 15, 17	6	0	6	0	58	0	118
5	12	25, 33, 37	12	0	12	0	56	0	320
6	13	47, 53, 75	1	8	26	20	19	62	86
7	15	117, 127, 155	7	8	22	44	22	94	219
8	16	225, 331, 367	1	0	24	0	113	0	287
9	18	575, 673, 727	2	10	50	37	92	92	274
10	20	1167, 1375, 1545	6	16	72	68	170	162	340
11	22	2325, 2731, 3747	17	0	122	0	345	0	1102
12	24	5745, 6471, 7553	43	0	162	0	507	0	1420

Table 1.6: Parameter values of rate-1/4 convolutional codes with favorable distance properties.

K	d_f	Generators	$B(d_f + i)$ for $i = 0, 1, \ldots, 6$						
			0	1	2	3	4	5	6
3	10	5, 5, 7, 7	1	0	4	0	12	0	32
4	13	13, 13, 15, 17	4	2	0	10	3	16	34
5	16	25, 27, 33, 37	8	0	7	0	17	0	60
6	18	45, 53, 67, 77	5	0	19	0	14	0	70
7	20	117, 127, 155, 171	3	0	17	0	32	0	66
8	22	257, 311, 337, 355	2	4	4	24	22	33	44
9	24	533, 575, 647, 711	1	0	15	0	56	0	69
10	27	1173, 1325, 1467, 1751	7	10	0	28	54	58	54

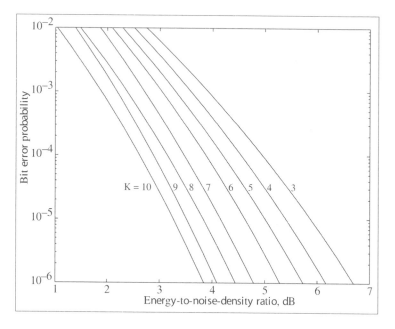

Figure 1.8: Information-bit error probability for rate $= 1/2$ convolutional codes with different constraint lengths and coherent PSK.

in the constraint length and as the code rate decreases if $K \geq 4$. The decoder complexity is almost exclusively dependent on K because there are 2^{K-1} encoder states. However, as the code rate decreases, more bandwidth and a more difficult bit synchronization are required.

For convolutional codes of rate $1/n$, two trellis branches enter each state. For higher-rate codes with k information bits per branch, 2^k trellis branches enter each state and the computational complexity may be large. This complexity can be avoided by using *punctured convolutional codes*. These codes are generated by periodically deleting bits from one or more output streams of an encoder for an unpunctured rate-$1/n$ code. For a period-p punctured code, p sets of n bits are written into a buffer from which $p + \nu$ bits are read out, where $1 \leq \nu \leq (n-1)p$. Thus, a punctured convolutional code has a rate of the form

$$r = \frac{p}{p + \nu} \;, \quad 1 \leq \nu \leq (n-1)p \tag{1-115}$$

The decoder of a punctured code uses the same decoder and trellis as the parent code, but uses only the metrics of the unpunctured bits as it proceeds through the trellis. The upper bound on P_b is given by (1-112) with $k = 1$. For most code rates, there are punctured codes with the largest minimum free distance of any convolutional code with that code rate. Punctured convolutional codes enable the efficient implementation of a variable-rate error-control system with a single encoder and decoder. However, the periodic character of the trellis of

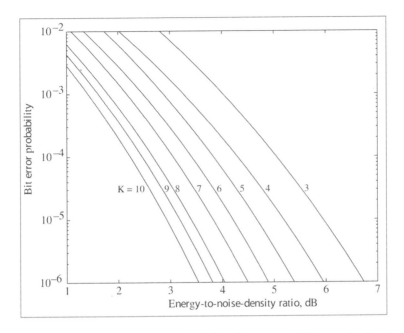

Figure 1.9: Information-bit error probability for rate $= 1/3$ convolutional codes with different constraint lengths and coherent PSK.

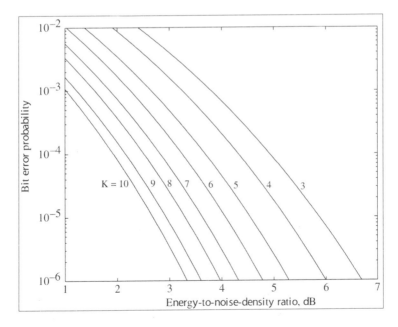

Figure 1.10: Information-bit error probability for rate $= 1/4$ convolutional codes with different constraint lengths and coherent PSK.

a punctured code requires that the decoder acquire frame synchronization.

Coded nonbinary sequences can be produced by converting the outputs of a binary convolutional encoder into a single nonbinary symbol, but this procedure does not optimize the nonbinary code's Hamming distance properties. Better nonbinary codes, such as the dual-k codes, are possible [3] but do not provide as good a performance as the nonbinary Reed-Solomon codes with the same transmission bandwidth.

In principle, $B(l)$ can be determined from the *generating function*, $T(D, I)$, which can be derived for some convolutional codes by treating the state diagram as a signal flow graph [1], [2]. The generating function is a polynomial in D and I of the form

$$T(D, I) = \sum_{i=1}^{\infty} \sum_{l=d_f}^{\infty} a(l, i) D^l I^i \tag{1-116}$$

where $a(l, i)$ represents the number of distinct unmerged segments characterized by l and i. The derivative at $I = 1$ is

$$\left.\frac{\partial T(D, I)}{\partial I}\right|_{I=1} = \sum_{i=1}^{\infty} \sum_{l=d_f}^{\infty} i a(l, i) D^l = \sum_{l=d_f}^{\infty} B(l) D^l \tag{1-117}$$

Thus, the bound on P_b given by (1-112), is determined by substituting $P_2(l)$ in place of D^l in the polynomial expansion of the derivative of $T(D, I)$ and multiplying the result by $1/k$. In many applications, it is possible to establish an inequality of the form

$$P_2(l) \leq \alpha Z^l \tag{1-118}$$

where α and Z are independent of l. It then follows from (1-112), (1-117), and (1-118) that

$$P_b \leq \frac{\alpha}{k} \left.\frac{\partial T(D, I)}{\partial I}\right|_{I=1, D=Z} \tag{1-119}$$

For soft-decision decoding and coherent PSK, $P_2(l)$ is given by (1-113). Using the definition of $Q(x)$ given by (1-30), changing variables, and comparing the two sides of the following inequality, we verify that

$$Q(\sqrt{\nu + \beta}) = \frac{1}{\sqrt{2\pi}} \int_0^{\infty} \exp\left[-\frac{1}{2}(y + \sqrt{\nu + \beta})^2\right] dy$$

$$\leq \frac{1}{\sqrt{2\pi}} \exp\left(-\frac{\beta}{2}\right) \int_0^{\infty} \exp\left[-\frac{1}{2}(y + \sqrt{\nu})^2\right] dy, \quad \nu \geq 0, \ \beta \geq 0 \tag{1-120}$$

A change of variables yields

$$Q(\sqrt{\nu + \beta}) \leq \exp\left(-\frac{\beta}{2}\right) Q(\sqrt{\nu}), \quad \nu \geq 0, \ \beta \geq 0 \tag{1-121}$$

Substituting this inequality into (1-113) with the appropriate choices for ν and β gives

$$P_2(l) \leq Q\left(\frac{\sqrt{2d_f r \mathcal{E}_b}}{N_0}\right) \exp\left[-(l - d_f) r \mathcal{E}_b / N_0\right] \tag{1-122}$$

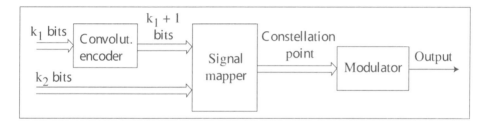

Figure 1.11: Encoder for trellis-coded modulation.

Thus, the upper bound on $P_2(l)$ may be expressed in the form given by (1-118) with

$$\alpha = Q\left(\frac{\sqrt{2d_f r \mathcal{E}_b}}{N_0}\right) \exp(d_f r \mathcal{E}_b/N_0) \tag{1-123}$$

$$Z = \exp(-r\mathcal{E}_b/N_0) \tag{1-124}$$

For other channels, codes, and modulations, an upper bound on $P_2(l)$ in the form given by (1-118) can often be derived from the Chernoff bound.

Trellis-Coded Modulation

To add an error-control code to a communication system while avoiding a bandwidth expansion, one may increase the number of signal constellation points. For example, if a rate-2/3 code is added to a system using quadriphase-shift keying (QPSK), then the bandwidth is preserved if the modulation is changed to eight-phase PSK (8-PSK). Since each symbol of the latter modulation represents 3/2 as many bits as a QPSK symbol, the channel-symbol rate is unchanged. The problem is that the change from QPSK to the more compact 8-PSK constellation causes an increase in the channel-symbol error probability that cancels most of the decrease due to the encoding. This problem is avoided by using *trellis-coded modulation,* which integrates the modulation and coding processes.

Trellis-coded modulation is produced by a system with the form shown in Figure 1.11. For $k > 1$, each input of k information bits is divided into two groups. One group of k_1 bits is applied to a convolutional encoder while the other group of $k_2 = k - k_1$ bits remains uncoded. The $k_1 + 1$ output bits of the convolutional encoder select one of 2^{k_1+1} possible subsets of the points in the constellation of the modulator. The k_2 uncoded bits select one of 2^{k_2} points in the chosen subset. If $k_2 = 0$, there are no uncoded bits and the convolutional encoder output bits select the constellation point. Each constellation point is a complex number representing an amplitude and phase.

For example, suppose that $k_1 = k_2 = 1$ and $n = 2$ in the encoder of Figure 1.11, and an 8-PSK modulator produces an output from a constellation of 8 points. Each of the four subsets that may be selected by the two convolutional-

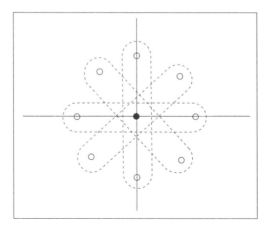

Figure 1.12: The constellation of 8-PSK symbols partitioned into 4 subsets.

code bits comprises two antipodal points in the 8-PSK constellation, as shown in Figure 1.12. If the convolutional encoder has the form of Figure 1.6(a), then the trellis of Figure 1.7 illustrates the state transitions of both the underlying convolutional code and the trellis code. The presence of the single uncoded bit implies that each transition between states in the trellis corresponds to two different transitions and two different phases of the transmitted 8-PSK waveform.

In general, there are 2^{k_2} parallel transitions between every pair of states in the trellis. Often, the dominant error events consist of mistaking one of these parallel transitions for the correct one. If the symbols corresponding to parallel transitions are separated by large Euclidean distances, and the constellation subsets associated with transitions are suitably chosen, then the trellis-coded modulation with soft-decision Viterbi decoding can yield a substantial coding gain [1], [2], [3]. This gain usually ranges from 4 to 6 dB, depending on the number of states and, hence, the implementation complexity. The minimum Euclidean distance between a correct trellis-code path and an incorrect one is called the *free Euclidean distance* and is denoted by $d_{fe}\sqrt{\mathcal{E}_s}$. Let B_{fe} denote the total number of information bit errors associated with erroneous paths that are at the free Euclidian distance from the correct path. The latter paths dominate the error events when the SNR is high. An analysis similar to the one for convolutional codes indicates that for the AWGN channel and a high SNR,

$$P_b \approx \frac{B_{fe}}{k} Q\left(\sqrt{\frac{d_{fe}^2 r \mathcal{E}_b}{2N_0}}\right) \qquad (1\text{-}125)$$

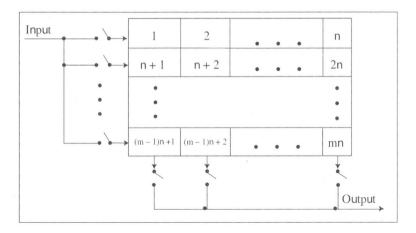

Figure 1.13: Block interleaver.

1.3 Interleaving

An *interleaver* is a device that permutes the order of a sequence of symbols. A *deinterleaver* is the corresponding device that restores the original order of the sequence. A major application is the interleaving of modulated symbols transmitted over a communication channel. After deinterleaving at the receiver, a burst of channel-symbol errors or corrupted symbols is dispersed over a number of codewords or constraint lengths, thereby facilitating the removal of the errors by the decoding. Ideally, the interleaving and deinterleaving ensures that the decoder encounters statistically independent symbol decisions or metrics, as it would if the channel were memoryless. Interleaving of channel symbols is useful when error bursts are caused by fast fading, interference, or even decision-directed equalization.

A *block interleaver* performs identical permutations on successive blocks of symbols. As illustrated in Figure 1.13, mn successive input symbols are stored in a random-access memory (RAM) as a matrix of m rows and n columns. The input sequence is written into the interleaver in successive rows, but successive columns are read to produce the interleaved sequence. Thus, if the input sequence is numbered $1, 2, \ldots, n, n + 1, \ldots, mn$, the interleaved sequence is $1, n + 1, 2n + 1, \ldots, 2, n + 2, \ldots, mn$. For continuous interleaving, two RAMs are needed. Symbols are written into one RAM matrix while previous symbols are read from the other. In the deinterleaver, symbols are stored by column in one matrix, while previous symbols are read by rows from another. Consequently, a delay of $2mnT_s$ must be accommodated and synchronization is required at the deinterleaver.

When channel symbols are interleaved, the parameter n equals or exceeds the block codeword length or a few constraint lengths of a convolutional code. Consequently, if a burst of m or fewer consecutive symbol errors occurs and there are no other errors, then each block codeword or constraint length, after deinterleaving, has at most one error, which can be eliminated by the error-correcting code. Similarly, a block code that can correct t errors is capable of correcting a single burst of errors spanning as many as mt symbols. Since fading can cause correlated errors, it is necessary that mT_s exceed the channel coherence time. Interleaving effectiveness can be thwarted by slow fading that cannot be accommodated without large buffers that cause an unacceptable delay.

Other types of interleavers that are closely related to the block interleaver include the *convolutional interleaver* and the *helical interleaver*. A helical interleaver reads symbols from its matrix diagonally instead of by column in such a way that consecutive interleaved symbols are never read from the same row or column. Both helical and convolutional interleavers and their corresponding deinterleavers confer advantages in certain applications, but do not possess the inherent simplicity and compatibility with block structures that block interleavers have.

A *pseudorandom interleaver* permutes each block of symbols pseudorandomly. Pseudorandom interleavers may be applied to channel symbols, but their main application is as critical elements in turbo encoders and encoders of serially concatenated codes that use iterative decoding (Section 1.4). The desired permutation may be stored in a read-only memory (ROM) as a sequence of addresses or permutation indices. Each block of symbols is written sequentially into a RAM matrix and then interleaved by reading them in the order dictated by the contents of the ROM.

If the interleaver is large, it is often preferable to generate the permutation indices by an algorithm rather than storing them in a ROM. If the interleaver size is $N = mn = 2^\nu - 1$, then a linear feedback shift register with ν stages that produces a maximal-length sequence can be used. The binary outputs of the shift-register stages constitute the *state* of the register. The state specifies the index from 1 to N that defines a specific interleaved symbol. The shift register generates all N states and indices periodically.

An *S-random interleaver* is a pseudorandom interleaver that constrains the minimum interleaving distance. A tentative permutation index is compared with the S previously selected indices, where $1 \leq S < N$. If the tentative index does not differ in absolute value from the S previous ones by at least S, then it is discarded and replaced by a new tentative index. If it does, then the tentative index becomes the next selected index. This procedure continues until all N pseudorandom indices are selected. The S-random interleaver is frequently used in turbo or serially concatenated encoders.

1.4 Concatenated and Turbo Codes

A *concatenated code* uses multiple levels of coding to achieve a large error-control capability with manageable implementation complexity by breaking the decoding process into stages. In practice, two levels of coding have been found to be effective. Figure 1.14 is a functional block diagram of a communication

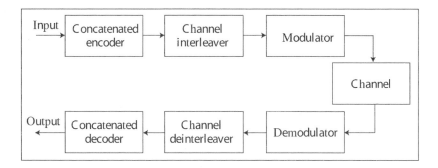

Figure 1.14: Concatenated coding in transmitter and receiver.

system incorporating a concatenated code. The channel interleaver permutes the code bits to ensure the random distribution of code-bit errors at the input of the concatenated decoder. Concatenated codes may be classified as classical concatenated codes, turbo codes, or serially concatenated turbo codes.

Classical Concatenated Codes

Classical concatenated codes are serially concatenated codes with the encoder and decoder forms shown in Figure 1.15. In the most common configuration for classical concatenated codes, an *inner code* uses binary symbols and a Reed-Solomon *outer code* uses nonbinary symbols. The outer-encoder output symbols are interleaved, and then these nonbinary symbols are converted into binary symbols that are encoded by the inner encoder. In the receiver, a grouping of the binary inner-decoder output symbols into nonbinary outer-code symbols is followed by symbol deinterleaving that disperses the outer-code symbol errors. Consequently, the outer decoder is able to correct most symbol errors originating in the inner-decoder output. The concatenated code has rate

$$r = r_1 r_0 \tag{1-126}$$

where r_1 is the inner-code rate and r_0 is the outer-code rate.

A variety of inner codes have been proposed. The dominant and most powerful concatenated code of this type comprises a binary convolutional inner code and a Reed-Solomon outer code. At the output of a convolutional inner decoder using the Viterbi algorithm, the bit errors occur over spans with an average length that depends on the signal-to-noise ratio (SNR). The deinterleaver is designed to ensure that Reed-Solomon symbols formed from bits in the same typical error span do not belong to the same Reed-Solomon codeword. Let $m = \log_2 q$ denote the number of bits in a Reed-Solomon code symbol. In the worst case, the inner decoder produces bit errors that are separated enough that each one causes a separate symbol error at the input to the Reed-Solomon decoder. Since there are m times as many bits as symbols, the symbol error probability P_{s1} is upper-bounded by m times the bit error probability at the

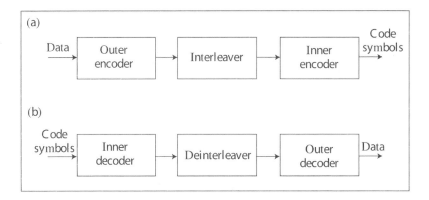

Figure 1.15: Structure of serially concatenated code: (a) encoder and (b) classical decoder.

inner-decoder output. Since P_{s1} is no smaller than it would be if each set of m bit errors caused a single symbol error, P_{s1} is lower-bounded by this bit error probability. Thus, for binary convolutional inner codes,

$$\frac{1}{k}\sum_{l=d_f}^{\infty} B(l)P_2(l) \le P_{s1} \le \frac{\log_2 q}{k}\sum_{l=d_f}^{\infty} B(l)P_2(l) \qquad (1\text{-}127)$$

where $P_2(l)$ is given by (1-103) and (1-102). Assuming that the deinterleaving ensures independent symbol errors at the outer-decoder input, and that the Reed-Solomon code is loosely packed, (1-26) and (1-27) imply that

$$P_b \approx \frac{q}{2(q-1)}\sum_{i=t+1}^{n}\binom{n-1}{i-1}P_{s1}^i(1-P_{s1})^{n-1} \qquad (1\text{-}128)$$

For coherent PSK modulation with soft decisions, $P_2(l)$ is given by (1-113); if hard decisions are made, (1-114) applies.

Figure 1.16 depicts examples of the approximate upper bound on the performance in white Gaussian noise of concatenated codes with coherent PSK, soft demodulator decisions, an inner binary convolutional code with $k = 1$, $K = 7$, and rate $= 1/2$, and various Reed-Solomon outer codes. Equation (1-128) and the upper bound in (1-127) are used. The bandwidth required by a concatenated code is B/r, where B is the uncoded PSK bandwidth. Since (1-126) gives $r < 1/3$, the codes of the figure require more bandwidth than rate-1/3 convolutional codes.

Turbo Codes

Turbo codes are parallel concatenated codes that use iterative decoding [1], [7], [8]. As shown in Figure 1.17, the encoder of a turbo code has two component encoders, one of which directly encodes the information bits while the other

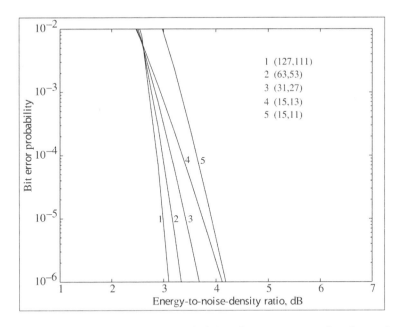

Figure 1.16: Information-bit error probability for concatenated codes with inner convolutional code ($K = 7$, rate $= 1/2$), various Reed-Solomon (n, k) outer codes, and coherent PSK.

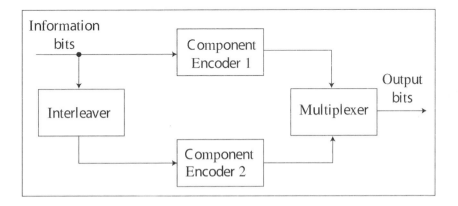

Figure 1.17: Encoder of turbo code.

encodes interleaved bits. The iterative decoding requires that both component codes be systematic and of the same type, that is, both convolutional or both block.

A *turbo convolutional code* uses two binary convolutional codes as its component codes. The multiplexer output comprises both the information and parity bits produced by encoder 1 but only the parity bits produced by encoder 2. Because of their superior distance properties, *recursive systematic convolutional encoders* are used in turbo encoders [1]. Each of these encoders has feedback that causes the shift-register state to depend on its previous outputs. Usually, identical rate-1/2 component codes are used, and a rate-1/3 turbo code is produced. However, if the multiplexer punctures the parity streams, a higher rate of 1/2 or 2/3 can be obtained. Although it requires frame synchronization in the decoder, the puncturing may serve as a convenient means of adapting the code rate to the channel conditions. The purpose of the interleaver, which may be a block or pseudorandom interleaver, is to permute the input bits of encoder 2 so that it is unlikely that both component codewords will have a low weight even if the input word has a low weight. Thus, a turbo code has very few low-weight codewords, whether or not its minimum distance is large.

Terminating tail bits are inserted into both component convolutional codes so that the turbo trellis terminates in the all-zero state and the turbo code can be treated as a block code. Recursive encoders require nonzero tail bits that are functions of the preceding nonsystematic output bits and, hence, the information bits.

To produce a rate-1/2 turbo code from rate-1/2 convolutional component codes, alternate puncturing of the even parity bits of encoder 1 and the odd parity bits of encoder 2 is done. Consequently, an odd information bit has its associated parity bit of code 1 transmitted. However, because of the interleaving that precedes encoder 2, an even information bit may have neither its associated parity bit of code 1 nor that of code 2 transmitted. Instead, some odd information bits may have both associated parity bits transmitted, although not successively because of the interleaving. Since some information bits have no associated parity bits transmitted, the decoder is less likely to be able to correct errors in those information bits. A convenient means of avoiding this problem, and ensuring that exactly one associated parity bit is transmitted for each information bit, is to use a block interleaver with an odd number of rows and an odd number of columns. If bits are written into the interleaver matrix in successive rows, but successive columns are read, then odd and even information bits alternate at the input of encoder 2, thereby ensuring that all information bits have an associated parity bit that is transmitted. This procedure, or any other that separates the odd and even information bits, is called *odd-even separation*. Simulation results confirm that odd-even separation improves the system performance when puncturing and block interleavers are used, but odd-even separation is not beneficial in the absence of puncturing [8]. In a system with a small interleaver size, block interleavers with odd-even separation usually give a better system performance than pseudorandom interleavers, but the latter are usually superior when the interleaver size is large.

The interleaver size is equal to the block length or frame length of the codes. The number of low-weight or minimum-distance codewords tends to be inversely proportional to the interleaver size. With a large interleaver and a sufficient number of decoder iterations, the performance of the turbo convolutional code can approach within less than 1 dB of the information-theoretic limit. However, as the block length increases, so does the *system latency*, which is the delay between the input and final output. As the symbol energy increases, the bit error rate of a turbo code decreases until it eventually falls to an *error floor* or bit error rate that continues to decrease very slowly. The potentially large system latency, the system complexity, and, rarely, the error floor are the primary disadvantages of turbo codes.

A maximum-likelihood decoder such as the Viterbi decoder minimizes the probability that a received codeword or an entire received sequence is in error. A turbo decoder is designed to minimize the error probability of each information bit. Under either criterion, an optimal decoder would use the sampled demodulator output streams for the information bits and the parity bits of both component codes. A turbo decoder comprises separate component decoders for each component code, which is theoretically suboptimal but crucial in reducing the decoder complexity. Each component decoder uses a version of the *maximum a posteriori* (MAP) or *BCJR algorithm* proposed by Bahl, Cocke, Jelinek, and Raviv [1], [8]. As shown in Figure 1.18, component decoder 1 of a turbo decoder is fed by demodulator outputs denoted by the vector $\mathbf{y}_1 = \begin{bmatrix} \mathbf{x}_0 & \mathbf{x}_1 \end{bmatrix}$, where the components of sequence $\mathbf{x}_0$ are the information bits and the components of sequence $\mathbf{x}_1$ are the parity bits of encoder 1. Similarly, component decoder 2 is fed by outputs denoted by $\mathbf{y}_2 = \begin{bmatrix} \mathbf{x}_0 & \mathbf{x}_2 \end{bmatrix}$, where the components of sequence $\mathbf{x}_2$ are the parity bits of encoder 2. For each information bit u_k, the MAP algorithm of decoder i computes estimates of the log-likelihood ratio (LLR) of the probabilities that this bit is +1 or −1 given the vector $\mathbf{y}_i$:

$$\Lambda_{ki} = \ln \left[\frac{P(u_k = +1 | \mathbf{y}_i)}{P(u_k = -1 | \mathbf{y}_i)} \right] , \quad i = 1, 2 \tag{1-129}$$

Since the *a posteriori* probabilities are related by $p(u_k = +1 \mid \mathbf{y}_i) = 1 - p(u_k = -1 \mid \mathbf{y}_i)$, Λ_{ki} completely characterizes the *a posteriori* probabilities. The LLRs of the information bits are iteratively updated in the two component decoders by passing information between them. Since it is interleaved or deinterleaved, arriving information is largely decorrelated from any other information in a decoder and thereby enables the decoder to improve its estimate of the LLR.

From the definition of a conditional probability, (1-129) may be expressed as

$$\Lambda_{ki} = \ln \left[\frac{P(u_k = +1, y_{sk}, \overline{\mathbf{y}}_i)}{P(u_k = -1, y_{sk}, \overline{\mathbf{y}}_i)} \right] , \quad i = 1, 2 \tag{1-130}$$

where y_{sk} is the demodulator output corresponding to the systematic or information bit u_k and $\overline{\mathbf{y}}_i$ is the sequence $\mathbf{y}_i$ excluding y_{sk}. Given u_k, y_{sk} is independent of $\mathbf{y}_i$. Therefore, for j = 1 or 2,

$$P(u_k = j, y_{sk}, \overline{\mathbf{y}}_i) = P(y_{sk} \mid u_k = j)P(\overline{\mathbf{y}}_i \mid u_k = j)P(u_k = j) \tag{1-131}$$

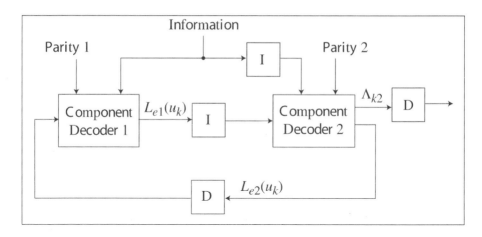

Figure 1.18: Decoder of turbo code.

Substitution of this equation into (1-130) and decomposing the results, we obtain

$$\Lambda_{ki} = L(u_k) + L(y_{sk}|u_k) + L_{ei}(u_k), \quad i = 1, 2 \tag{1-132}$$

where the *a priori* LLR is initially

$$L(u_k) = \ln\left[\frac{P(u_k = +1)}{P(u_k = -1)}\right] \tag{1-133}$$

and the *extrinsic information*

$$L_{ei}(u_k) = \ln\left[\frac{P(\overline{y}_i \mid u_k = +1)}{P(\overline{y}_i \mid u_k = -1)}\right], \quad i = 1, 2 \tag{1-134}$$

is a function of the parity bits processed by the component decoder i. The term $L(y_{sk}|u_k)$, which represents information about u_k provided by y_{sk}, is defined as

$$L(y_{sk}|u_k) = \ln\left[\frac{f(y_{sk}|u_k = +1)}{f(y_{sk}|u_k = -1)}\right] \tag{1-135}$$

where $f(y_{sk}|u_k = j)$ is the conditional density of y_{sk} given that $u_k = j$. Let N_{0k} denote the noise power spectral density associated with u_k. For coherent PSK, (1-41) with $y_i \to y_{sk}$, $N_{0i} \to N_{0k}$, and $x_{li} \to \alpha u_k$, where α accounts for the fading attenuation, gives the conditional density

$$f(y_{sk}|u_k = \pm 1) = \frac{1}{\sqrt{\pi N_{0k}/2}} \exp\left[-\frac{(y_{sk} \mp \sqrt{\mathcal{E}_s/2}\alpha)^2}{N_{0k}/2}\right] \tag{1-136}$$

Substitution into (1-135) yields

$$L(y_{sk}|u_k) = L_c y_{sk}, \quad L_c = 4\alpha\frac{\sqrt{2\mathcal{E}_s}}{N_{0k}} \tag{1-137}$$

The *channel reliability factor* L_c must be known or estimated to compute Λ_{ki}.

Since almost always no *a priori* knowledge of the likely value of the bit u_k is available, $P(u_k) = 0.5$ is assumed, and $L(u_k)$ is set to zero for the first iteration of component decoder 1. However, for subsequent iterations of either component decoder, $L(u_k)$ for one decoder is set equal to the extrinsic information calculated by the other decoder at the end of its previous iteration. As indicated by (1-132), $L_{ei}(u_k)$ can be calculated by subtracting $L(u_k)$ and $L_c y_{sk}$ from Λ_{ki}, which is computed by the MAP algorithm. Since the extrinsic information depends primarily on the constraints imposed by the code used, it provides additional information to the decoder to which it is transferred. As indicated in Figure 1.18, appropriate interleaving or deinterleaving is required to ensure that the extrinsic information $L_{e1}(u_k)$ or $L_{e2}(u_k)$ is applied to each component decoder in the correct sequence. Let $B\{\ \}$ denote the function calculated by the MAP algorithm during a single iteration, $I[\ ]$ denote the interleave operation, $D[\ ]$ denote the deinterleave operation, and a numerical superscript (n) denote the nth iteration. The turbo decoder calculates the following functions for $n \geq 1$:

$$\Lambda_{k1}^{(n)} = B\{\mathbf{x}_0, \mathbf{x}_1, D[L_{e2}^{(n-1)}(u_k)]\} \tag{1-138}$$

$$L_{e1}^{(n)}(u_k) = \Lambda_{k1}^{(n)} - L_c y_{sk} - D[L_{e2}^{(n-1)}(u_k)] \tag{1-139}$$

$$\Lambda_{k2}^{(n)} = B\{I[\mathbf{x}_0], \mathbf{x}_2, I[L_{e1}^{(n)}(u_k)]\} \tag{1-140}$$

$$L_{e2}^{(n)}(u_k) = \Lambda_{k2}^{(n)} - L_c y_{sk} - I[L_{e1}^{(n)}(u_k)] \tag{1-141}$$

where $D[L_{e2}^{(0)}] = L(u_k)$. When the iterative process terminates after N iterations, the LLR $\Lambda_{k2}^{(N)}$ from component decoder 2 is deinterleaved and then applied to a device that makes a hard decision. Thus, the decision for bit k is

$$\widehat{u_k} = sgn\{D[\Lambda_{k2}^{(N)}(u_k)]\} \tag{1-142}$$

Performance improves with the number of iterations, but simulation results indicate that typically little is gained beyond roughly 4 to 12 iterations.

The generic name for a version of the MAP algorithm or an approximation of it is *soft-in soft-out (SISO) algorithm*. The *log-MAP algorithm* is an SISO algorithm that transforms the MAP algorithm into the logarithmic domain, thereby simplifying operations and reducing numerical problems while causing no performance degradation. The *max-log-MAP algorithm* and the *soft-output Viterbi algorithm (SOVA)* are SISO algorithms that reduce the complexity of the log-MAP algorithm at the cost of some performance degradation [1], [8]. The max-log-MAP algorithm is roughly 2/3 as complex as the log-MAP algorithm and typically degrade the performance by 0.1 dB to 0.2 dB at $P_b = 10^{-4}$. The SOVA algorithm is roughly 1/3 as complex as the log-MAP algorithm and typically degrades the performance by 0.5 dB to 1.0 dB at $P_b = 10^{-4}$. The MAP, log MAP, max-log-MAP, and SOVA algorithms have complexities that increase linearly with the number of states of the component codes.

The log-MAP algorithm requires both a forward and a backward recursion through the code trellis. Since the log-MAP algorithm also requires additional

memory and calculations, it is roughly 4 times as complex as the standard
Viterbi algorithm [8]. For 2 identical component decoders and typically 8 algorithm iterations, the overall complexity of a turbo decoder is roughly 64 times
that of a Viterbi decoder for one of the component codes. The complexity of
the decoder increases while the performance improves as the constraint length
K of each component code increases. The complexity of a turbo decoder using
8 iterations and component convolutional codes with $K = 3$ is approximately
the same as that of a Viterbi decoder for a convolutional code with $K = 9$.

If N_{0k} is unknown and may be significantly different from symbol to symbol,
a standard procedure is to replace the LLR of (1-135) with the *generalized log-
likelihood ratio*

$$L(y_{sk}|u_k) = \ln \left[\frac{f(y_{sk}|u_k = +1, N_1)}{f(y_{sk}|u_k = -1, N_2)} \right] \tag{1-143}$$

where N_1 and N_2 are maximum-likelihood estimates of N_{0k} obtained from (1-
136) with $u_k = +1$ and $u_k = -1$, respectively. Calculations yield the estimates

$$N_1 = 4(y_{sk} - \sqrt{\mathcal{E}_s/2}\alpha)^2 \ , \quad N_2 = 4(y_{sk} + \sqrt{\mathcal{E}_s/2}\alpha)^2 \tag{1-144}$$

Substituting these estimates into (1-136) and then substituting the results into
(1-143), we obtain

$$L(y_{sk}|u_k) = \ln \left[\frac{|y_{sk} + \sqrt{\mathcal{E}_s/2}\alpha|}{|y_{sk} - \sqrt{\mathcal{E}_s/2}\alpha|} \right] \tag{1-145}$$

This equation replaces (1-137).

A *turbo block code* uses two linear block codes as its component codes. To
limit the decoding complexity, high-rate binary BCH codes are generally used
as the component codes, and the turbo code is called a *turbo BCH code*. The
encoder of a turbo block code has the form of Figure 1.17. Puncturing is
generally not used as it causes a significant performance degradation. Suppose
that the component block codes are binary systematic (n_1, k_1) and (n_2, k_2)
codes, respectively. Encoder 1 converts k_1 information bits into k_2 codeword
bits.   Each block of $k_1 k_2$ information bits are written successively into the
interleaver as k_1 columns and k_2 rows. Encoder 2 converts each column of k_2
interleaver bits into a codeword of n_2 bits. The multiplexer passes the n_1 bits of
each of k_2 encoder-1 codewords, but only the $n_2 - k_2$ parity bits of k_1 encoder-2
codewords so that information bits are transmitted only once. Consequently,
the code rate of the turbo block code is

$$r = \frac{k_1 k_2}{k_2 n_1 + (n_2 - k_2)k_1} \tag{1-146}$$

If the two block codes are identical, then $r = k/(2n - k)$. If the minimum
Hamming distances of the component codes are d_{m1} and d_{m2}, respectively,
then the minimum distance of the concatenated code is

$$d_m = d_{m1} + d_{m2} - 1 \tag{1-147}$$

The decoder of a turbo block code has the form of Figure 1.18, and only slight modifications of the SISO decoding algorithms are required. Long, high-rate turbo BCH codes approach the Shannon limit in performance, but their complexities are higher then those of turbo convolutional codes of comparable performance [8].

Approximate upper bounds on the bit error probability for turbo codes have been derived [1], [8]. Since these bounds are difficult to evaluate except for short codewords, simulation results are generally used to predict the performance of a turbo code.

Serially Concatenated Turbo Codes

Serially concatenated turbo codes differ from classical concatenated codes in their use of large interleavers and iterative decoding. The interchange of information between the inner and outer decoders gives the serially concatenated codes a major performance advantage. Both the inner and outer codes must be amenable to efficient decoding by an SISO algorithm and, hence, are either binary systematic block codes or binary systematic convolutional codes. The encoder for a serially concatenated turbo code has the form of Figure 1.15(a). The outer encoder generates n_1 bits for every k_1 information bits. After the interleaving, each set of n_1 bits is converted by the inner encoder into n_2 bits. Thus, the overall code rate of the serially concatenated code is k_1/n_2. If the component codes are block codes, then an outer (n_1, k_1) code and an inner (n_2, n_1) code are used. A functional block diagram of an iterative decoder for a serially concatenated code is illustrated in Figure 1.19. For each inner codeword, the input comprises the demodulator outputs corresponding to the n_2 bits. For each iteration, the inner decoder computes the LLRs for the n_1 systematic bits. After a deinterleaving, these LLRs provide extrinsic information about the n_1 code bits of the outer code. The outer decoder then computes the LLRs for all its code bits. After an interleaving, these LLRs provide extrinsic information about the n_1 systematic bits of the inner code. The final output of the iterative decoder comprises the k_1 information bits of the concatenated code. Simulation results indicate that a serially concatenated code with convolutional codes tends to outperform a comparable turbo convolutional code for the AWGN channel when low bit error probabilities are required [1].

Turbo Product Codes

A *product code* is a special type of serially concatenated code that is constructed from multidimensional arrays and linear block codes. An encoder for a two-dimensional turbo product code has the form of Figure 1.15(a). The outer encoder produces codewords of an (n_1, k_1) code. For an inner (n_2, k_2) code, k_2 codewords are placed in a $k_2 \times n_1$ interleaver array of k_2 rows and n_1 columns. The block interleaver columns are read by the inner encoder to produce n_1 codewords of length n_2 that are transmitted. The resulting product code has

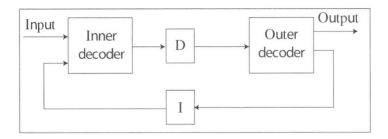

Figure 1.19: Iterative decoder for serially concatenated code. D = deinterleaver; I = interleaver.

$n = n_1 n_2$ code symbols, $k = k_1 k_2$ information symbols, and code rate

$$r = \frac{k_1 k_2}{n_1 n_2} \tag{1-148}$$

If the minimum Hamming distances of the outer and inner codes are d_{m1} and d_{m2}, respectively, then a straightforward analysis indicates that the minimum Hamming distance of the product code is

$$d_m = d_{m1} d_{m2} \tag{1-149}$$

Hard-decision decoding is done sequentially on an $n_2 \times n_1$ array of received code symbols. The inner codewords are decoded and code-symbol errors are corrected. Any residual errors are then corrected during the decoding of the outer codewords. Let t_1 and t_2 denote the error-correcting capability of the outer and inner codes, respectively. Incorrect decoding of the inner codewords requires that there are at least $t_2 + 1$ errors in at least one inner codeword or array column. For the outer decoder to fail to correct the residual errors, there must be at least $t_1 + 1$ inner codewords that have $t_2 + 1$ or more errors, and the errors must occur in certain array positions. Thus, the number of errors that is always correctable is

$$t = (t_1 + 1)(t_2 + 1) - 1 \tag{1-150}$$

which is roughly half of what (1-1) guarantees for classical block codes. However, although not all patterns with more than t errors are correctable, most of them are.

When iterative decoding is used, a product code is called a *turbo product code*. A comparison of (1-149) with (1-147) indicates that d_m for a turbo product code is generally larger than d_m for a turbo block code with the same component codes. The decoder for a turbo product code has the form shown in Figure 1.20. The demodulator outputs are applied to both the inner decoder, and after deinterleaving, the outer decoder. The LLRs of both the information and parity bits of the corresponding code are computed by each decoder. These LLRs are then exchanged between the decoders after the appropriate deinterleaving or interleaving converts the LLRs into extrinsic information. A large

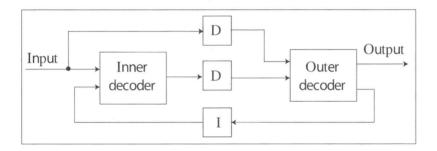

Figure 1.20: Decoder of turbo product code. D = deinterleaver; I = interleaver.

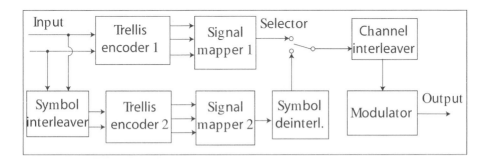

Figure 1.21: Encoder for turbo trellis-coded modulation.

reduction in the complexity of a turbo product code in exchange for a relatively small performance loss is obtained by using the Chase algorithm (Section 1.5) in the SISO algorithm of the component decoders [9]. For a given complexity, the performance of turbo product codes and turbo block codes are similar [8].

Turbo Trellis-Coded Modulation

Turbo trellis-coded modulation (TTCM), which produces a nonbinary bandwidth-efficient modulation, is obtained by using identical trellis codes as the component codes in a turbo code [10]. The encoder has the form illustrated in Figure 1.21. The code rate and, hence, the required bandwidth of the component trellis code is preserved by the TTCM encoder because it alternately selects constellation points or complex symbols generated by the two parallel component encoders. To ensure that all information bits, which constitute the encoder input, are transmitted only once and that the parity bits are provided alternately by the two component encoders, the symbol interleaver transfers symbols in odd positions to odd positions and symbols in even positions to even positions, where each symbol is a group of bits. After the complex symbols are produced by signal mapper 2, the symbol deinterleaver restores the original ordering. The selector passes the odd-numbered complex symbols from mapper 1 and the even-numbered complex symbols from mapper 2. The channel interleaver permutes the selected complex symbols prior to the modulation. The TTCM decoder uses a symbol-based SISO algorithm analogous to the SISO algorithm used by turbo-code decoders. TTCM can provide a performance close to the theoretical limit for the AWGN channel, but its implementation complexity is much greater than that of conventional trellis-coded modulation [8].

The iterative decoding principle of turbo codes can be applied to equalization, demodulation, and even other codes, notably the *low-density parity-check codes* [11]. Recently, these codes have been shown to be competitive with turbo codes in both performance and complexity.

1.5   Problems

1. Verify that both Golay perfect codes satisfy the Hamming bound with equality.

2. (a) Use (1-12) to show that $N(d_m, d_m - t) = \binom{d_m}{t}$. Can the same result be derived directly? (b) Use (1-13) to derive $N(l, i)$ for Hamming codes. Consider the cases $l = i$, $i + 1$, $i - 1$, and $i - 2$ separately.

3. (a) Use (1-21) to derive an upper bound on A_{d_m}.(b) Explain why this upper bound becomes an equality for perfect codes. (c) Show that $A_3 = \frac{n(n-1)}{6}$ for Hamming codes. (d) Show that for perfect codes as $P_s \to 0$, both the exact equation (1-22) and the approximation (1-25) give the same expression for P_b.

4. Evaluate P_b for the Hamming (7,4) code using both the exact equation and the approximate one. Use the result of problem 2(b) and the weight distribution given in the text. Compare the two results.

5. Use erasures to show that a Reed-Solomon codeword can be recovered from any k correct symbols.

6. Suppose that a binary Hamming (7,4) code is used for coherent PSK communications with a constant noise-power spectral density.   A codeword has $x_{ji} = +1$ if symbol i in candidate codeword j is a 1, and $x_{ji} = -1$ if it is a 0. The received output samples are -0.4, 1.0, 1.0, 1.0, 1.0, 1.0, 0.4. Use the table of Hamming (7,4) codewords to find the decision made when the maximum-likelihood metric is used.

7. Prove that the word error probability for a block code with soft-decision decoding satisfies $P_w \leq (q^k - 1)Q(d)$.

8. Use (1-49) and (1-45) to show that the coding gain of a block code is roughly $d_m r$ relative to no code when P_{is} is low.

9. Derive a generalization of the symbol error probability for binary FSK. Let $N_{01}/2$ and $N_{02}/2$ denote the two-sided power spectral densities of the

white Gaussian noise in the filter matched to the transmitted signal and the other matched filter, respectively. Change (1-81) and (1-82) appropriately and then derive P_s.

10. (a) Show that $P[X \geq b] \geq 1 - \min_{0 \leq s \leq s_1} [M(-s)e^{sb}]$. (b) Derive the Chernoff bound for a Gaussian random variable with mean μ and variance σ^2.

11. Consider the convolutional code defined in Figures 1.6(a) and 1.7. The input of a Viterbi decoder is 1000100000. Show the surviving paths and their partial metrics.

12. Consider a system that uses coherent PSK and a convolutional code in the presence of white Gaussian noise. (a) What is the coding gain of a binary system with soft decisions, $K = 7$, and $r = 1/2$ relative to an uncoded system for large E_b/N_0 ? (b) Use the approximation

$$Q(x) \approx \frac{1}{\sqrt{2\pi x}} \exp(-\frac{x^2}{2}), \quad x > 0$$

to show that as $E_b/N_0 \to \infty$, soft-decision decoding of a binary convolutional code has a 3 dB coding gain relative to hard-decision decoding.

13. A concatenated code comprises an inner binary block $(2^m, m)$ code, which is called a *Hadamard code*, and an outer Reed-Solomon (n, k) code. The outer encoder maps every m bits into one Reed-Solomon symbol, and every k symbols are encoded as an n-symbol codeword. After the symbol interleaving, the inner encoder maps every Reed-Solomon symbol into 2^m bits. After the interleaving of these bits, they are transmitted using a binary modulation. (a) Describe the removal of the encoding by the inner and outer decoders. (b) What is the value of n as a function of m ? (c) What are the block length and code rate of the concatenated code?

14. Derive (1-144) and (1-145) using the steps outlined in the text.

15. Show that the minimum Hamming distance of a product code is equal to the product of the minimum Hamming distances of the outer and inner codes, respectively.

1.6 References

1. S. Benedetto and E. Biglieri, *Principles of Digital Transmission.* New York: Kluwer Academic, 1999.

2. S. B. Wicker, *Error Control Systems for Digital Communication and Storage.* Upper Saddle River, NJ: Prentice-Hall, 1995.

3. S. G. Wilson, *Digital Modulation and Coding.* Upper Saddle River, NJ: Prentice-Hall, 1996.

4. J. G. Proakis, *Digital Communications, 4th ed.* New York: McGraw-Hill, 2001.

5. D. Torrieri, "Information-Bit, Information-Symbol, and Decoded-Symbol Error Rates for Linear Block Codes," *IEEE Trans. Commun.*, vol. 36, pp. 613-617, May 1988.

6. J.-J. Chang, D.-J. Hwang, and M.-C. Lin, "Some Extended Results on the Search for Good Convolutional Codes," *IEEE Trans. Inform. Theory*, vol. 43, pp. 1682–1697, September 1997.

7. C. Berrou and A. Glavieux, "Near Optimum Error-Correcting Coding and Decoding: Turbo Codes," *IEEE Trans. Commun.*, vol. 44, pp. 1261–1271, October 1996.

8. L. Hanzo, T. H. Liew, and B. L. Yeap, *Turbo Coding, Turbo Equalisation and Space-Time Coding.* Chichester, England: Wiley, 2002.

9. R. Pyndiah, "Near-Optimum Decoding of Product Codes: Block Turbo Codes," *IEEE Trans. Commun.*, vol. 46, pp. 1003–1010, August 1998.

10. P. Robertson and T. Worz, "Bandwidth Efficient Turbo Trellis-Coded Modulation Using Punctured Component Codes," *IEEE J. Selected Areas Commun.*, vol. 16, pp. 206–218, February 1998.

11. D. R. Barry, E. A. Lee, and D. G. Messerschmitt, *Digital Communication, 3rd ed.* Boston: Kluwer Academic, 2004.

Chapter 2

Direct-Sequence Systems

2.1 Definitions and Concepts

A *spread-spectrum signal* is a signal that has an extra modulation that expands the signal bandwidth beyond what is required by the underlying data modulation. Spread-spectrum communication systems [1], [2], [3] are useful for suppressing interference, making interception difficult, accommodating fading and multipath channels, and providing a multiple-access capability. The most practical and dominant methods of spread-spectrum communications are direct-sequence modulation and frequency hopping of digital communications.

At first it might seem that a spread-spectrum signal is counterproductive insofar as the receive filter will require an increased bandwidth and, hence, will pass more noise power to the demodulator. However, when any signal and white Gaussian noise are applied to a filter matched to the signal, the sampled filter output has a signal-to-noise ratio (SNR) that is inversely proportional to the noise-power spectral density. The remarkable aspect of this result is that the filter bandwidth and, hence, the output noise power are irrelevant. Thus, we observe that there is no fundamental barrier to the use of spread-spectrum communications.

A *direct-sequence signal* is a spread-spectrum signal generated by the direct mixing of the data with a spreading waveform before the final carrier modulation. Ideally, a direct-sequence signal with binary phase-shift keying (PSK) or differential PSK (DPSK) data modulation can be represented by

$$s(t) \;=\; Ad(t)p(t)\cos(2\pi f_c t + \theta) \qquad (2\text{-}1)$$

where A is the signal amplitude, $d(t)$ is the data modulation, $p(t)$ is the spreading waveform, f_c is the carrier frequency, and θ is the phase at $t = 0$. The data modulation is a sequence of nonoverlapping rectangular pulses of duration T_s, each of which has an amplitude $d_i = +1$ if the associated data symbol is a 1 and $d_i = -1$ if it is a 0 (alternatively, the mapping could be $1 \rightarrow -1$ and $0 \rightarrow +1$). Equation (2-1) implies that $s(t) = Ap(t)cos[2\pi f_c t + \theta + \pi d(t)]$, which explicitly exhibits the phase-shift keying by the data modulation. The

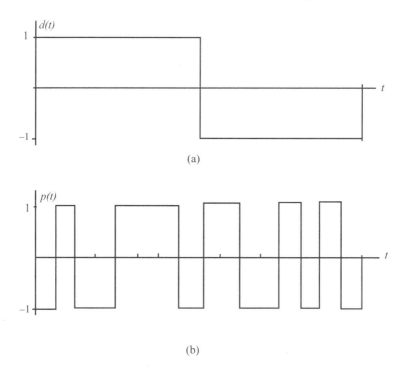

Figure 2.1: Examples of (a) data modulation and (b) spreading waveform.

spreading waveform has the form

$$p\left(t\right) = \sum_{i=-\infty}^{\infty} p_i \psi\left(t - iT_c\right) \tag{2-2}$$

where each p_i equals $+1$ or -1 and represents one *chip* of the *spreading sequence*. The *chip waveform* $\psi(t)$ is ideally confined to the interval $[0, T_c]$ to prevent interchip interference in the receiver. A rectangular chip waveform has $\psi(t) = w(t, T_c)$, where

$$w\left(t, T\right) = \begin{cases} 1, & 0 \leq t < T \\ 0, & \text{otherwise} \end{cases} \tag{2-3}$$

Figure 2.1 depicts an example of $d(t)$ and $p(t)$ for a rectangular chip waveform.

Message privacy is provided by a direct-sequence system if a transmitted message cannot be recovered without knowledge of the spreading sequence. To ensure message privacy, which is assumed henceforth, the data-symbol transitions must coincide with the chip transitions. Since the transitions coincide, the *processing gain* $G = T_s/T_c$ is an integer equal to the number of chips in a symbol interval. If W is the bandwidth of $p(t)$ and B is the bandwidth of $d(t)$, the spreading due to $p(t)$ ensures that $s(t)$ has a bandwidth $W >> B$.

Figure 2.2 is a functional or conceptual block diagram of the basic operation

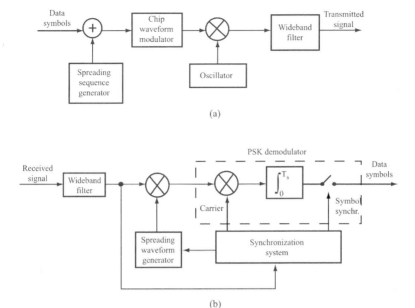

(a)

(b)

Figure 2.2: Functional block diagram of direct-sequence systemn with PSK or DPSK: (a) transmitter and (b) receiver.

of a direct-sequence system with PSK. To provide message privacy, data symbols and chips, which are represented by digital sequences of 0's and 1's, are synchronized by the same clock and then modulo-2 added in the transmitter. The adder output is converted according to $0 \to -1$ and $1 \to +1$ before the chip and carrier modulations. Assuming that chip and symbol synchronization has been established, the received signal passes through the wideband filter and is multiplied by a synchronized local replica of $p(t)$. If $\psi(t)$ is rectangular, then $p(t) = \pm 1$ and $p^2(t) = 1$. Therefore, if the filtered signal is given by (1-1), the multiplication yields the *despread signal*

$$s_1(t) = p(t)s(t) = Ad(t) \cos\left(2\pi f_c t + \theta\right) \qquad (2\text{-}4)$$

at the input of the PSK demodulator. Since the despread signal is a PSK signal, a standard coherent demodulator extracts the data symbols.

Figure 2.3(a) is a qualitative depiction of the relative spectra of the desired signal and narrowband interference at the output of the wideband filter. Multiplication of the received signal by the spreading waveform, which is called *despreading*, produces the spectra of Figure 2.3(b) at the demodulator input. The signal bandwidth is reduced to B, while the interference energy is spread over a bandwidth exceeding W. Since the filtering action of the demodulator then removes most of the interference spectrum that does not overlap the signal spectrum, most of the original interference energy is eliminated. An approximate measure of the interference rejection capability is given by the ratio W/B.

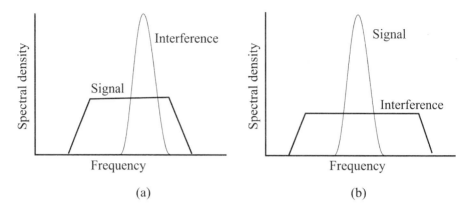

Figure 2.3: Spectra of desired signal and interference: (a) wideband-filter output and (b) demodulator input.

Whatever the precise definition of a bandwidth, W and B are proportional to $1/T_c$ and $1/T_s$, respectively, with the same proportionality constant. Therefore,

$$G = \frac{T_s}{T_c} = \frac{W}{B} \tag{2-5}$$

which links the processing gain with the interference rejection illustrated in the figure. Since its spectrum is unchanged by the despreading, white Gaussian noise is not suppressed by a direct-sequence system.

In practical systems, the wideband filter in the transmitter is used to limit the out-of-band radiation. This filter and the propagation channel disperse the chip waveform so that it is no longer confined to $[0, T_c]$. To avoid interchip interference in the receiver, the filter might be designed to generate a pulse that satisfies the Nyquist criterion for no intersymbol interference. A convenient representation of a direct-sequence signal when the chip waveform may extend beyond $[0, T_c]$ is

$$s(t) = A \sum_{i=-\infty}^{\infty} d_{\lfloor i/G \rfloor} p_i \psi\left(t - iT_c\right) \cos\left(2\pi f_c t + \theta\right) \tag{2-6}$$

where $\lfloor x \rfloor$ denotes the integer part of x. When the chip waveform is assumed to be confined to $[0, T_c]$, then (2-6) can be expressed by (2-1) and (2-2).

2.2   Spreading Sequences and Waveforms

Random Binary Sequence

A *random binary sequence* $x(t)$ is a stochastic process that consists of independent, identically distributed symbols, each of duration T. Each symbol takes

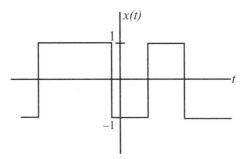

Figure 2.4: Sample function of a random binary sequence.

the value $+1$ with probability $\frac{1}{2}$ or the value -1 with probability $\frac{1}{2}$. Therefore, $E[x(t)] = 0$ for all t, and

$$P\left[x\left(t\right) = i\right] = \frac{1}{2}, \quad i = +1, -1 \tag{2-7}$$

The process is wide-sense stationary if the location of the first symbol transition or start of a new symbol after $t = 0$ is a random variable uniformly distributed over the half-open interval $(0, T]$. A sample function of a *wide-sense-stationary random binary sequence* $x(t)$ is illustrated in Figure 2.4.

The *autocorrelation* of a stochastic process $x(t)$ is defined as

$$R_x(t, \tau) = E\left[x(t)x(t + \tau)\right] \tag{2-8}$$

If $x(t)$ is a wide-sense stationary process, then $R_x(t, \tau)$ is a function of τ alone, and the autocorrelation is denoted by $R_x(\tau)$. From (2-7) and the definitions of an expected value and a conditional probability, it follows that the autocorrelation of a random binary sequence is

$$R_x(t, \tau) = \frac{1}{2}P[x(t + \tau) = 1 | x(t) = 1] - \frac{1}{2}P[x(t + \tau) = -1 | x(t) = 1]$$

$$+ \frac{1}{2}P[x(t + \tau) = -1 | x(t) = -1] - \frac{1}{2}P[x(t + \tau) = 1 | x(t) = -1] \tag{2-9}$$

where $P[A|B]$ denotes the conditional probability of event A given the occurrence of event B. From the theorem of total probability, it follows that

$$P[x(t + \tau) = i | x(t) = i] + P[x(t + \tau) = -i | x(t) = i] = 1,$$
$$i = +1, -1 \tag{2-10}$$

Since both of the following probabilities are equal to the probability that $x(t)$ and $x(t + \tau)$ differ,

$$P[x(t + \tau) = 1 | x(t) = -1] = P[x(t + \tau) = -1 | x(t) = 1] \tag{2-11}$$

Substitution of (2-10) and (2-11) into (2-9) yields

$$R_x(t, \tau) = 1 - 2P[x(t + \tau) = 1 | x(t) = -1] \tag{2-12}$$

If $|\tau| \geq T$, then $x(t)$ and $x(t+\tau)$ are independent random variables because t and $t+\tau$ are in different symbol intervals. Therefore,

$$P[x(t+\tau) = 1 | x(t) = -1] = P[x(t+\tau) = 1] = \frac{1}{2}$$

and (2-6) implies that $R_x(t, \tau) = 0$ for $|\tau| \geq T$. If $|\tau| < T$, then $x(t)$ and $x(t+\tau)$ are independent only if a symbol transition occurs in the half-open interval $I_0 = (t, t+\tau]$. Consider any half-open interval I_1 of length T that includes I_0. Exactly one transition occurs in I_1. Since the first transition for $t > 0$ is assumed to be uniformly distributed over $(0, T]$, the probability that a transition in I_1 occurs in I_0 is $|\tau|/T$. If a transition occurs in I_0, then $x(t)$ and $x(t+\tau)$ are independent and differ with probability $\frac{1}{2}$; otherwise, $x(t) = x(t+\tau)$. Consequently, $P[x(t+\tau) = 1 | x(t) = -1] = |\tau|/2T$ if $|\tau| < T$. Substitution of the preceding results into (2-12) confirms the wide-sense stationarity of $x(t)$ and gives the *autocorrelation of the random binary sequence*:

$$R_x(t, \tau) = R_x(\tau) = \Lambda\left(\frac{\tau}{T}\right) \tag{2-13}$$

where the *triangular function* is defined by

$$\Lambda(t) = \begin{cases} 1 - |t|, & |t| \leq 1 \\ 0, & |t| > 1 \end{cases} \tag{2-14}$$

Shift-Register Sequences

Ideally, one would prefer a random binary sequence as the spreading sequence. However, practical synchronization requirements in the receiver force one to use periodic binary sequences. A *shift-register sequence* is a periodic binary sequence generated by combining the outputs of feedback shift registers. A *feedback shift register*, which is diagrammed in Figure 2.5, consists of consecutive two-state memory or storage stages and feedback logic. Binary sequences drawn from the alphabet $\{0,1\}$ are shifted through the shift register in response to clock pulses. The *contents* of the stages, which are identical to their outputs, are logically combined to produce the input to the first stage. The initial contents of the stages and the feedback logic determine the successive contents of the stages. If the feedback logic consists entirely of modulo-2 adders (exclusive-OR gates), a feedback shift register and its generated sequence are called *linear*.

Figure 2.6(a) illustrates a linear feedback shift register with three stages and an output sequence extracted from the final stage. The input to the first stage is the modulo-2 sum of the contents of the second and third stages. After each clock pulse, the contents of the first two stages are shifted to the right, and the input to the first stage becomes its content. If the initial contents of the shift-register stages are 0 0 1, the subsequent contents after successive shifts are listed in Figure 2.6(b). Since the shift register returns to its initial state after 7 shifts, the periodic output sequence extracted from the final stage has a period of 7 bits.

The *state* of the shift register after clock pulse i is the vector

$$\mathbf{S}(i) = [s_1(i) \ s_2(i) \ldots s_m(i)], \quad i \geq 0 \tag{2-15}$$

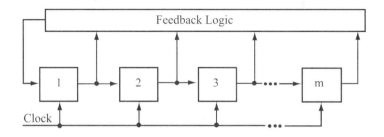

Figure 2.5: General feedback shift register with m stages.

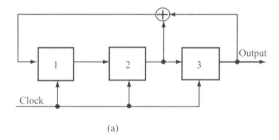

(a)

Shift	Contents		
	Stage 1	Stage 2	Stage 3
Initial	0	0	1
1	1	0	0
2	0	1	0
3	1	0	1
4	1	1	0
5	1	1	1
6	0	1	1
7	0	0	1

(b)

Figure 2.6: (a) Three-stage linear feedback shift register and (b) contents after successive shifts.

where $s_j(i)$ denotes the content of stage j after clock pulse i and $\mathbf{S}(0)$ is the initial state. The definition of a shift register implies that

$$s_j(i) = s_{j-k}(i - k), \quad i \geq k \geq 0, \quad k \leq j \leq m \qquad (2\text{-}16)$$

where $s_0(i)$ denotes the input to stage 1 after clock pulse i. If a_i denotes the state of bit i of the output sequence, then $a_i = s_m(i)$. The state of a feedback shift register uniquely determines the subsequent sequence of states and the shift-register sequence. The period N of a periodic sequence $\{a_i\}$ is defined as the smallest positive integer for which $a_{i+N} = a_i, i \geq 0$. Since the number of distinct states of an m-stage shift register is 2^m, the sequence of states and the shift-register sequence have period $N \leq 2^m$.

The *Galois field* of two elements, which is denoted by $GF(2)$, consists of the symbols 0 and 1 and the operations of modulo-2 addition and modulo-2 multiplication. These binary operations are defined by

$$0 \oplus 0 = 0, \qquad 0 \oplus 1 = 1, \qquad 1 \oplus 0 = 1, \qquad 1 \oplus 1 = 0$$
$$0 \cdot 0 = 0, \qquad 0 \cdot 1 = 0, \qquad 1 \cdot 0 = 0, \qquad 1 \cdot 1 = 1 \qquad (2\text{-}17)$$

where $\oplus$ denotes modulo-2 addition. From these equations, it is easy to verify that the field is closed under both modulo-2 addition and modulo-2 multiplication and that both operations are *associative* and *commutative*. Since -1 is defined as that element which when added to 1 yields 0, we have $-1 = 1$, and subtraction is the same as addition. From (2-11), it follows that the additive identity element is 0, the multiplicative identity is 1, and the multiplicative inverse of 1 is $1^{-1} = 1$. The substitutions of all possible symbol combinations verify the *distributive laws*:

$$a(b \oplus c) = ab \oplus ac, \quad (b \oplus c)a = ba \oplus ca \qquad (2\text{-}18)$$

where $a, b,$ and c can each equal 0 or 1. The equality of subtraction and addition implies that if $a \oplus b = c$, then $a = b \oplus c$.

The input to stage 1 of a linear feedback shift register is

$$s_0(i) = \sum_{k=1}^{m} c_k s_k(i), \quad i \geq 0 \qquad (2\text{-}19)$$

where the operations are modulo-2 and the feedback coefficient c_k equals either 0 or 1, depending on whether the output of stage k feeds a modulo-2 adder. An m-stage shift register is defined to have $c_m = 1$; otherwise, the final state would not contribute to the generation of the output sequence, but would only provide a one-shift delay. For example, Figure 2.6 gives $c_1 = 0, c_2 = c_3 = 1$, and $s_0(i) = s_2(i) \oplus s_3(i)$. A general representation of a linear feedback shift register is shown in Figure 2.7(a). If $c_k = 1$, the corresponding switch is closed; if $c_k = 0$, it is open.

Since the output bit $a_i = s_m(i)$, (2-16) and (2-19) imply that for $i \geq m$,

$$a_i = s_0(i - m) = \sum_{l=1}^{m} c_k s_k(i - m) = \sum_{l=1}^{m} c_k s_m(i - k)$$

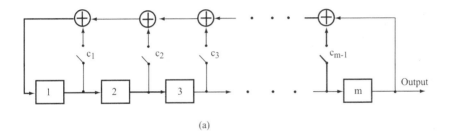

(a)

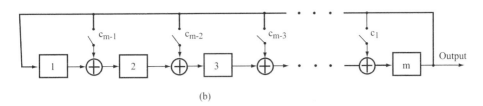

(b)

Figure 2.7: Linear feedback shift register: (a) standard representation and (b) high-speed form.

which indicates that each output bit satisfies the *linear recurrence relation*:

$$a_i = \sum_{k=1}^{m} c_k a_{i-k}, \quad i \geq m \tag{2-20}$$

The first m output bits are determined solely by the initial state:

$$a_i = s_{m-i}(0), \quad 0 \leq i \leq m-1 \tag{2-21}$$

Figure 2.7(a) is not necessarily the best way to generate a particular shift-register sequence. Figure 2.7(b) illustrates an implementation that allows higher-speed operation. From this diagram, it follows that

$$s_j(i) = s_{j-1}(i-1) \oplus c_{m-j+1} s_m(i-1), \quad i \geq 1, \quad 2 \leq j \leq m \tag{2-22}$$

$$s_1(i) = s_m(i-1) \qquad\qquad\qquad\qquad\qquad i \geq 1 \tag{2-23}$$

Repeated application of (2-22) implies that

$$s_m(i) = s_{m-1}(i-1) \oplus c_1 s_m(i-1) \;, \qquad i \geq 1$$

$$s_{m-1}(i-1) = s_{m-2}(i-2) \oplus c_2 s_m(i-2) \;, \qquad i \geq 2$$

$$\vdots \tag{2-24}$$

$$s_2(i-m+2) = s_1(i-m+1) \oplus c_{m-1} s_m(i-m+1) \;, \qquad i \geq m-1$$

Addition of these m equations yields

$$s_m(i) = s_1(i - m + 1) \oplus \sum_{k=1}^{m-1} c_k s_m(i - k), \quad i \geq m - 1 \tag{2-25}$$

Substituting (2-23) and then $a_i = s_m(i)$ into (2-25), we obtain

$$a_i = a_{i-m} \oplus \sum_{k=1}^{m-1} c_k a_{i-k}, \quad i \geq m \tag{2-26}$$

Since $c_m = 1$, (2-26) is the same as (2-20). Thus, the two implementations can produce the same output sequence indefinitely if the first m output bits coincide. However, they require different initial states and have different sequences of states. Successive substitutions into the first equation of sequence (2-24) yields

$$s_m(i) = s_{m-i}(0) \oplus \sum_{k=1}^{i} c_k s_m(i - k), \quad 1 \leq i \leq m - 1 \tag{2-27}$$

Substituting $a_i = s_m(i), a_{i-k} = s_m(i - k)$, and $j = m - i$ into (2-27) and then using binary arithmetic, we obtain

$$s_j(0) = a_{m-j} \oplus \sum_{k=1}^{m-j} c_k a_{m-j-k}, \quad 1 \leq j \leq m \tag{2-28}$$

If $a_0, a_1, \ldots a_{m-1}$ are specified, then (2-28) gives the corresponding initial state of the high-speed shift register.

The sum of binary sequence $\mathbf{a} = (a_0, a_1, \cdots)$ and binary sequence $\mathbf{b} = (b_0, b_1, \cdots)$ is defined to be the binary sequence $\mathbf{a} \oplus \mathbf{b}$, each bit of which is the modulo-2 sum of the corresponding bits of $\mathbf{a}$ and $\mathbf{b}$. Thus, if $\mathbf{d} = \mathbf{a} \oplus \mathbf{b}$ we can write

$$d_i = a_i \oplus b_i, \quad i \geq 0 \tag{2-29}$$

Consider sequences $\mathbf{a}$ and $\mathbf{b}$ that are generated by the same linear feedback shift register but may differ because the initial states may be different. For the sequence $\mathbf{d} = \mathbf{a} \oplus \mathbf{b}$, (2-29) and the associative and distributive laws of binary fields imply that

$$d_i = \sum_{k=1}^{m} c_k a_{i-k} \oplus \sum_{k=1}^{m} c_k b_{i-k} = \sum_{k=1}^{m} (c_k a_{i-k} \oplus c_k b_{i-k})$$

$$= \sum_{k=1}^{m} c_k (a_{i-k} \oplus b_{i-k}) = \sum_{k=1}^{m} c_k d_{i-k} \tag{2-30}$$

Since the linear recurrence relation is identical, $\mathbf{d}$ can be generated by the same linear feedback logic as $\mathbf{a}$ and $\mathbf{b}$. Thus, if $\mathbf{a}$ and $\mathbf{b}$ are two output sequences of a linear feedback shift register, then $\mathbf{a} \oplus \mathbf{b}$ is also. If $\mathbf{a} = \mathbf{b}$, then $\mathbf{a} \oplus \mathbf{b}$ is the sequence of all 0's, which can be generated by any linear feedback shift register.

If a linear feedback shift register reached the zero state with all its contents equal to 0 at some time, it would always remain in the zero state, and the output sequence would subsequently be all 0's. Since a linear m-stage feedback shift register has exactly $2^m - 1$ nonzero states, the period of its output sequence cannot exceed $2^m - 1$. A sequence of period $2^m - 1$ generated by a linear feedback shift register is called a *maximal* or *maximal-length sequence*. If a linear feedback shift register generates a maximal sequence, then all of its nonzero output sequences are maximal, regardless of the initial states.

Out of 2^m possible states, the content of the last stage, which is the same as the output bit, is a 0 in 2^{m-1} states. Among the nonzero states, the output bit is a 0 in $2^{m-1} - 1$ states. Therefore, in one period of a maximal sequence, the number of 0's is exactly $2^{m-1} - 1$, while the number of 1's is exactly 2^{m-1}.

Given the binary sequence $\mathbf{a}$, let $\mathbf{a}(j) = (a_j, a_{j+1}, \ldots)$ denote a shifted binary sequence. If $\mathbf{a}$ is a maximal sequence and $j \neq 0$, modulo $2^m - 1$, then $\mathbf{a} \oplus \mathbf{a}(j)$ is not the sequence of all 0's. Since $\mathbf{a} \oplus \mathbf{a}(j)$ is generated by the same shift register as $\mathbf{a}$, it must be a maximal sequence and, hence, some cyclic shift of $\mathbf{a}$. We conclude that the modulo-2 sum of a maximal sequence and a cyclic shift of itself by j digits, where $j \neq 0$, modulo $2^m - 1$, produces another cyclic shift of the original sequence; that is,

$$\mathbf{a} \oplus \mathbf{a}(j) = \mathbf{a}(k), \quad j \neq 0 \ (\text{modulo } 2^m - 1) \qquad (2\text{-}31)$$

In contrast, a non-maximal linear sequence $\mathbf{a} \oplus \mathbf{a}(j)$ is not necessarily a cyclic shift of $\mathbf{a}$ and may not even have the same period. As an example, consider the linear feedback shift register depicted in Figure 2.8. The possible state transitions depend on the initial state. Thus, if the initial state is 0 1 0, then the second state diagram indicates that there are two possible states, and, hence, the output sequence has a period of two. The output sequence is $\mathbf{a} = (0, 1, 0, 1, 0, 1, \ldots)$, which implies that $\mathbf{a}(1) = (1, 0, 1, 0, 1, 0, \ldots)$ and $\mathbf{a} \oplus \mathbf{a}(1) = (1, 1, 1, 1, 1, 1, \ldots)$; this result indicates that there is no value of k for which (2-31) is satisfied.

Periodic Autocorrelations

A binary sequence $\mathbf{a}$ with components $a_i \in GF(2)$, can be mapped into a binary antipodal sequence $\mathbf{p}$ with components $p_i \in \{-1, +1\}$ by means of the transformation

$$p_i = (-1)^{a_i+1}, \quad i \geq 0 \qquad (2\text{-}32)$$

or, alternatively, $p_i = (-1)^{a_i}$. The *periodic autocorrelation* of a periodic binary sequence $\mathbf{a}$ with period N is defined as

$$\theta_p(j) = \frac{1}{N} \sum_{i=0}^{N} p_i p_{i+j} \qquad (2\text{-}33)$$

Substitution of (2-32) into (2-33) yields

$$\theta_p(j) = \frac{1}{N} \sum_{i=0}^{N-1} (-1)^{a_i + a_{i+j}} = \frac{1}{N} \sum_{i=0}^{N-1} (-1)^{a_i \oplus a_{i+j}} = \frac{A_j - D_j}{N} \qquad (2\text{-}34)$$

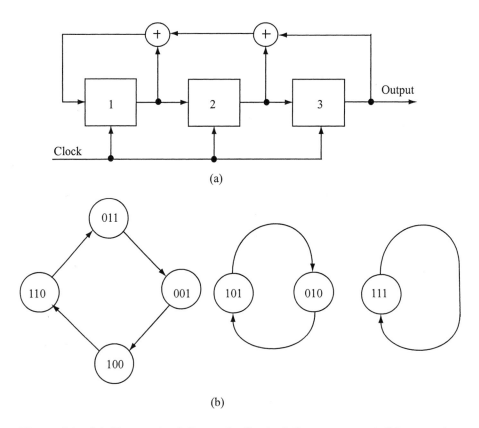

Figure 2.8: (a) Nonmaximal linear feedback shift register and (b) state diagrams.

where A_j denotes the number of agreements in the corresponding bits of $\mathbf{a}$ and $\mathbf{a}(j)$, and D_j denotes the number of disagreements. Equivalently, A_j is the number of 0's in one period of $\mathbf{a} \oplus \mathbf{a}(j)$, and $D_j = N - A_j$ is the number of 1's.

Consider a maximal sequence. From (2-31), it follows that A_j equals the number of 0's in a maximal sequence if $j \neq 0$, modulo N. Thus, $A_j = (N-1)/2$ and, similarly, $D_j = (N+1)/2$ if $j \neq 0$, modulo N. Therefore,

$$\theta_p(j) = \begin{cases} 1, & j = 0(\mathrm{mod}\ N) \\ -\frac{1}{N}, & j \neq 0(\mathrm{mod}\ N) \end{cases} \tag{2-35}$$

The *periodic autocorrelation* of a periodic function $x(t)$ with period T is defined as

$$R_x(\tau) = \frac{1}{T} \int_c^{c+T} x(t)x(t+\tau)dt \tag{2-36}$$

where τ is the relative delay variable and c is an arbitrary constant. It follows that $R_x(\tau)$ has period T. We derive the periodic autocorrelation of $p(t)$ assuming an ideal periodic spreading waveform of infinite extent and a rectangular chip waveform. If the spreading sequence has period N, then $p(t)$ has period $T = NT_c$. Equations (2-2) and (2-36) with $c = 0$ yield the autocorrelation of $p(t)$:

$$R_p(\tau) = \frac{1}{NT_c} \sum_{i=0}^{N-1} p_i \sum_{l=0}^{N-1} p_l \int_0^{NT_c} \psi(t - iT_c)\psi(t - lT_c + \tau)\,dt \qquad (2\text{-}37)$$

If $\tau = jT_c$, where j is an integer, then $\psi(t) = w(t, T_c)$, (2-3), and (2-37) yield

$$R_p(jT_c) = \frac{1}{N} \sum_{i=0}^{N-1} p_i p_{i+j} = \theta_p(j) \qquad (2\text{-}38)$$

Any delay can be expressed in the form $\tau = jT_c + \epsilon$, where j is an integer and $0 \le \epsilon < T_c$. Therefore, (2-37) and $\psi(t) = w(t, T_c)$ give

$$R_p(jT_c + \epsilon) = \frac{1}{NT_c} \sum_{i=0}^{N-1} p_i p_{i+j} \int_0^{NT_c} w(t - iT_c, T_c)w(t - iT_c + \epsilon, T_c)\,dt$$

$$+ \frac{1}{NT_c} \sum_{i=0}^{N-1} p_i p_{i+j+1} \int_0^{NT_c} w(t - iT_c, T_c)w(t - iT_c + \epsilon - T_c, T_c)\,dt$$

$$(2\text{-}39)$$

Using (2-38) and (2-3) in (2-39), we obtain

$$R_p(jT_c + \epsilon) = \left(1 - \frac{\epsilon}{T_c}\right)\theta_p(j) + \frac{\epsilon}{T_c}\theta_p(j+1) \qquad (2\text{-}40)$$

For a maximal sequence, the substitution of (2-35) into (2-40) yields $R_p(\tau)$ over one period:

$$R_p(\tau) = \frac{N+1}{N}\Lambda\left(\frac{\tau}{T_c}\right) - \frac{1}{N}, \qquad |\tau| \le NT_c/2 \qquad (2\text{-}41)$$

where $\Lambda(\ )$ is the triangular function defined by (2-14). Since it has period NT_c, the autocorrelation can be compactly expressed as

$$R_p(\tau) = -\frac{1}{N} + \frac{N+1}{N}\sum_{i=-\infty}^{\infty}\Lambda\left(\frac{\tau - iNT_c}{T_c}\right) \qquad (2\text{-}42)$$

Over one period, this autocorrelation resembles that of a random binary sequence, which is given by (2-13) with $T = T_c$. Both autocorrelations are shown in Figure 2.9.

A straightforward calculation or the use of tables gives the Fourier transform of the triangular function:

$$\mathcal{F}\left\{\Lambda\left(\frac{t}{T}\right)\right\} = \int_{-\infty}^{\infty}\Lambda\left(\frac{t}{T}\right)\exp\left(-j2\pi ft\right)\,dt$$

$$= T\mathrm{sinc}^2 fT \qquad (2\text{-}43)$$

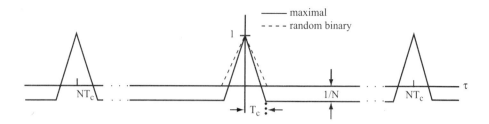

Figure 2.9: Autocorrelations of maximal sequence and random binary sequence.

where $j = \sqrt{-1}$ and sinc $x = (\sin \pi x)/\pi x$. Since the infinite series in (2-42) is a periodic function of τ, it can be expressed as a complex exponential Fourier series. From (2-43) and the fact that the Fourier transform of a complex exponential is a delta function, we obtain

$$\mathcal{F}\left\{ \sum_{i=-\infty}^{\infty} \Lambda\left(\frac{t - iNT_c}{T_c}\right)\right\} = \frac{1}{N} \sum_{i=-\infty}^{\infty} \text{sinc}^2\left(\frac{i}{N}\right) \delta\left(f - \frac{i}{NT_c}\right) \qquad (2\text{-}44)$$

where $\delta(\ )$ is the Dirac delta function. Applying this identity to (2-42), we determine $S_p(f)$, the *power spectral density* of $p(t)$, which is defined as the Fourier transform of $R_p(\tau)$:

$$S_p(f) = \frac{N+1}{N^2} \sum_{\substack{i=-\infty \\ i \neq 0}}^{\infty} \text{sinc}^2\left(\frac{i}{N}\right) \delta\left(f - \frac{i}{NT_c}\right) + \frac{1}{N^2}\delta(f) \qquad (2\text{-}45)$$

This function, which consists of an infinite series of delta functions, is depicted in Figure 2.10.

A *pseudonoise* or *pseudorandom sequence* is a periodic binary sequence with a nearly even balance of 0's and 1's and an autocorrelation that roughly resembles, over one period, the autocorrelation of a random binary sequence. Pseudonoise sequences, which include the maximal sequences, provide practical spreading sequences because their autocorrelations facilitate code synchronization in the receiver (Chapter 4). Other sequences have peaks that hinder synchronization.

To derive the power spectral density of a direct-sequence signal with a periodic spreading sequence, it is necessary to define the *average autocorrelation* of $x(t)$:

$$\bar{R}_x(\tau) = \lim_{T \to \infty} \frac{1}{2T} \int_{-T}^{T} R_x(t, \tau)dt \qquad (2\text{-}46)$$

The limit exists and may be nonzero if $x(t)$ has finite power and infinite duration. If $x(t)$ is stationary, $\bar{R}_x(\tau) = R_x(\tau)$. The *average power spectral density* $\bar{S}_x(f)$ is defined as the Fourier transform of the average autocorrelation.

For the direct-sequence signal of (2-1), $d(t)$ is modeled as a random binary sequence with autocorrelation given by (2-13), and θ is modeled as a random

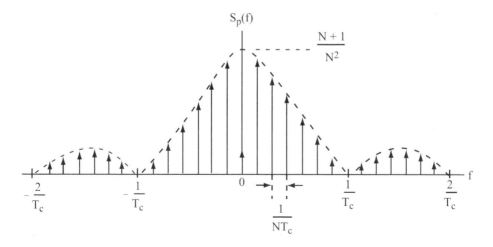

Figure 2.10: Power spectral density of maximal sequence.

variable uniformly distributed over $[0, 2\pi)$ and statistically independent of $d(t)$. Neglecting the constraint that the bit transitions must coincide with chip transitions, we obtain the autocorrelation of the direct-sequence signal $s(t)$:

$$R_s(t, \tau) = \frac{A^2}{2} p(t) p(t + \tau) \Lambda \left(\frac{\tau}{T_s} \right) \cos 2\pi f_c \tau \qquad (2\text{-}47)$$

where $p(t)$ is the periodic spreading waveform. Substituting this equation into (2-46) and using (2-36), we obtain

$$\bar{R}_s(\tau) = \frac{A^2}{2} R_p(\tau) \Lambda \left(\frac{\tau}{T_s} \right) \cos 2\pi f_c \tau \qquad (2\text{-}48)$$

where $R_p(\tau)$ is the periodic autocorrelation of $p(t)$. For a maximal spreading sequence, the convolution theorem, (2-48), (2-43), and (2-45) provide the average power spectral density of $s(t)$:

$$\bar{S}_s(f) = \frac{A^2}{4} [S_{s1}(f - f_c) + S_{s1}(f + f_c)] \qquad (2\text{-}49)$$

where the lowpass equivalent density is

$$S_{s1}(f) = \frac{T_s}{N^2} \operatorname{sinc}^2 f T_s + \frac{N + 1}{N^2} T_s \sum_{\substack{i=-\infty \\ i \neq 0}}^{\infty} \operatorname{sinc}^2 \left(\frac{i}{N} \right) \operatorname{sinc}^2 \left(f T_s - \frac{i T_s}{N T_c} \right)$$

$$(2\text{-}50)$$

For a random binary sequence, $S_s(f) = \bar{S}_s(f)$ is given by (2-49) with $S_{s1}(f) = T_c \operatorname{sinc}^2 f T_c$.

Polynomials over the Binary Field

Polynomials allow a compact description of the dependence of the output sequence of a linear feedback shift register on its feedback coefficients and initial state. A *polynomial* over the binary field $GF(2)$ has the form

$$f(x) = f_0 + f_1 x + f_2 x^2 + \cdots + f_n x^n \tag{2-51}$$

where the coefficients $f_0, f_1, \cdots, f_n$ are elements of $GF(2)$ and the symbol x is an indeterminate introduced for convenience in calculations. The *degree* of a polynomial is the largest power of x with a nonzero coefficient. The *sum* of a polynomial $f(x)$ of degree n_1 and a polynomial $g(x)$ of degree n_2 is another polynomial over $GF(2)$ defined as

$$f(x) + g(x) = \sum_{i=0}^{\max(n_1,n_2)} (f_i \oplus g_i) x^i \tag{2-52}$$

where $\max(n_1, n_2)$ denotes the larger of n_1 and n_2. An example is

$$(1 + x^2 + x^3) + (1 + x^2 + x^4) = x^3 + x^4 \tag{2-53}$$

The *product* of two polynomials over $GF(2)$ is another polynomial over $GF(2)$ defined as

$$f(x)g(x) = \sum_{i=0}^{n_1+n_2} \left(\sum_{j=0}^{i} f_j g_{i-j} \right) x^i \tag{2-54}$$

where the inner addition is modulo 2. For example,

$$(1 + x^2 + x^3)(1 + x^2 + x^4) = 1 + x^3 + x^5 + x^6 + x^7 \tag{2-55}$$

It is easily verified that associative, commutative, and distributive laws apply to polynomial addition and multiplication.

The *characteristic polynomial* associated with a linear feedback shift register of m stages is defined as

$$f(x) = 1 + \sum_{i=1}^{m} c_i x^i \tag{2-56}$$

where $c_m = 1$ assuming that stage m contributes to the generation of the output sequence. The *generating function* associated with the output sequence is defined as

$$G(x) = \sum_{i=0}^{\infty} a_i x^i \tag{2-57}$$

Substitution of (2-20) into this equation yields

$$G(x) = \sum_{i=0}^{m-1} a_i x^i + \sum_{i=m}^{\infty} \sum_{k=1}^{m} c_k a_{i-k} x^i$$

$$= \sum_{i=0}^{m-1} a_i x^i + \sum_{k=1}^{m} c_k x^k \sum_{i=m}^{\infty} a_{i-k} x^{i-k}$$

$$= \sum_{i=0}^{m-1} a_i x^i + \sum_{k=1}^{m} c_k x^k \left[G(x) + \sum_{i=0}^{m-k-1} a_i x^i \right] \qquad (2\text{-}58)$$

Combining this equation with (2-56), and defining $c_0 = 1$, we obtain

$$G(x) f(x) = \sum_{i=0}^{m-1} a_i x^i + \sum_{k=1}^{m} c_k x^k \left(\sum_{i=0}^{m-k-1} a_i x^i \right)$$

$$= \sum_{k=0}^{m-1} c_k x^k \left(\sum_{i=0}^{m-k-1} a_i x^i \right) = \sum_{k=0}^{m-1} \sum_{i=0}^{m-k-1} c_k a_i x^{k+i}$$

$$= \sum_{k=0}^{m-1} \sum_{j=k}^{m-1} c_k a_{j-k} x^j = \sum_{j=0}^{m-1} \sum_{k=0}^{j} c_k a_{j-k} x^j \qquad (2\text{-}59)$$

which implies that

$$G(x) = \frac{\displaystyle\sum_{i=0}^{m-1} x^i \left(\sum_{k=0}^{i} c_k a_{i-k} \right)}{f(x)} \,, \qquad c_0 = 1 \qquad (2\text{-}60)$$

Thus, the generating function of the output sequence generated by a linear feedback shift register with characteristic polynomial $f(x)$ may be expressed in the form $G(x) = \phi(x)/f(x)$, where the degree of $\phi(x)$ is less than the degree of $f(x)$. The output sequence is said to be *generated* by $f(x)$. Equation (2-60) explicitly shows that the output sequence is completely determined by the feedback coefficients $c_k, k = 1, 2, \ldots, m$, and the initial state $a_i = s_{m-i}(0), i = 0, 1, \ldots, m-1$.

In Figure 2.6, the feedback coefficients are $c_1 = 0, c_2 = 1$, and $c_3 = 1$, and the initial state gives $a_0 = 1, a_1 = 0$, and $a_2 = 0$. Therefore,

$$G(x) = \frac{1 + x^2}{1 + x^2 + x^3} \qquad (2\text{-}61)$$

Performing the long polynomial division according to the rules of binary arithmetic yields $1 + x^3 + x^5 + x^6 + x^7 + x^{10} + \ldots$, which implies the output sequence listed in the figure.

The polynomial $p(x)$ is said to *divide* the polynomial $b(x)$ if there is a polynomial $h(x)$ such that $b(x) = h(x)p(x)$. A polynomial $p(x)$ over $GF(2)$ of degree m is called *irreducible* if $p(x)$ is not divisible by any polynomial over $GF(2)$ of degree less than m but greater than zero. If $p(x)$ is irreducible over $GF(2)$, then $p(0) \neq 0$, for otherwise x would divide $p(x)$. If $p(x)$ has an even number of terms, then $p(1) = 0$ and the fundamental theorem of algebra implies that $x + 1$ divides $p(x)$. Therefore, an irreducible polynomial over $GF(2)$ must have an odd number of terms, but this condition is not sufficient for irreducibility. For example, $1 + x + x^2$ is irreducible, but $1 + x + x^5 = (1 + x^2 + x^3)(1 + x + x^2)$ is not.

If a shift-register sequence $\{a_i\}$ is periodic with period n, then its generating function $G(x) = \phi(x)/f(x)$ may be expressed as

$$G(x) = g(x) + x^n g(x) + x^{2n} g(x) + \cdots$$

$$= g(x) \sum_{i=0}^{\infty} x^{in}$$

$$= \frac{g(x)}{1 + x^n} \tag{2-62}$$

where $g(x)$ is a polynomial of degree $n - 1$. Therefore,

$$g(x) = \frac{\phi(x)(1 + x^n)}{f(x)} \tag{2-63}$$

Suppose that $f(x)$ and $\phi(x)$ have no common factors, which is true if $f(x)$ is irreducible since $\phi(x)$ is of lower degree than $f(x)$. Then $f(x)$ must divide $1 + x^n$. Conversely, if the characteristic polynomial $f(x)$ divides $1 + x^n$, then $f(x)h(x) = 1 + x^n$ for some polynomial $h(x)$, and

$$G(x) = \frac{\phi(x)}{f(x)} = \frac{\phi(x)h(x)}{1 + x^n} \tag{2-64}$$

which has the form of (2-62). Thus, $f(x)$ generates a sequence of period n for all $\phi(x)$ and, hence, all initial states.

A polynomial over $GF(2)$ of degree m is called *primitive* if the smallest positive integer n for which the polynomial divides $1 + x^n$ is $n = 2^m - 1$. Thus, a primitive characteristic polynomial of degree m can generate a sequence of period $2^m - 1$, which is the period of a maximal sequence generated by a characteristic polynomial of degree m. Suppose that a primitive characteristic polynomial of positive degree m could be factored so that $f(x) = f_1(x)f_2(x)$, where $f_1(x)$ is of positive degree m_1 and $f_2(x)$ is of positive degree $m - m_1$. A partial-fraction expansion yields

$$\frac{1}{f(x)} = \frac{a(x)}{f_1(x)} + \frac{b(x)}{f_2(x)} \tag{2-65}$$

Since $f_1(x)$ and $f_2(x)$ can serve as characteristic polynomials, the period of the first term in the expansion cannot exceed $2^{m_1} - 1$ while the period of the second term cannot exceed $2^{m-m_1} - 1$. Therefore, the period of $1/f(x)$ cannot exceed

Table 2.1: Primitive polynomials.

Degree	Primitive	Degree	Primitive	Degree	Primitive
2	7	7	103	8	534
3	51		122	9	1201
	31		163	10	1102
4	13		112	11	5004
	32		172	12	32101
5	15		543	13	33002
	54		523	14	30214
	57		532	15	300001
	37		573	16	310012
	76		302	17	110004
	75		323	18	1020001
6	141		313	19	7400002
	551		352	20	1100004
	301		742	21	50000001
	361		763	22	30000002
	331		712	23	14000004
	741		753	24	702000001
			772	25	110000002

$(2^{m_1} - 1)(2^{m-m_1} - 1) \leq 2^m - 3$, which contradicts the assumption that $f(x)$ is primitive. Thus, a *primitive characteristic polynomial must be irreducible.*

Theorem. *A characteristic polynomial of degree m generates a maximal sequence of period $2^m - 1$ if and only if it is a primitive polynomial.*

Proof: To prove sufficiency, we observe that if $f(x)$ is a primitive characteristic polynomial, it divides $1 + x^n$ for $n = 2^m - 1$ so a maximal sequence of period $2^m - 1$ is generated. If a sequence of smaller period could be generated, then the irreducible $f(x)$ would have to divide $1 + x^{n_1}$ for $n_1 < n$, which contradicts the assumption of a primitive polynomial. To prove necessity, we observe that if the characteristic polynomial $f(x)$ generates a maximal sequence with period $n = 2^m - 1$, then $f(x)$ cannot divide $1 + x^{n_1}, n_1 < n$, because a sequence with a smaller period would result, and such a sequence cannot be generated by a maximal sequence generator. Since $f(x)$ does divide $1 + x^n$, it must be a primitive polynomial. $\square$

Primitive polynomials are difficult to find, but many have been tabulated (e.g., [4]). Those for which $m \leq 7$ and one of those of minimal coefficient weight for $8 \leq m \leq 25$ are listed in Table 2.1 as octal numbers in increasing order (e.g., $51 \leftrightarrow 1\,0\,1\,1\,0\,0 \leftrightarrow 1 + x^2 + x^3$). For any positive integer m, the number of different primitive polynomials of degree m over $GF(2)$ is

$$\lambda(m) = \frac{\phi_e (2^m - 1)}{m} \tag{2-66}$$

where the *Euler function* $\phi_e(n)$ is the number of positive integers that are less than and relatively prime to the positive integer n. If n is a prime number,

$\phi_e(n) = n - 1$. In general,

$$\phi_e(n) = n \prod_{i=1}^{k} \frac{\nu_i - 1}{\nu_i} \leq n - 1 \qquad (2\text{-}67)$$

where $\nu_1, \nu_2, \ldots, \nu_k$ are the prime integers that divide n. Thus, $\lambda(6) = \phi_e(63)/6 = 6$ and $\lambda(13) = \phi_e(8191)/13 = 630$.

Long Nonlinear Sequences

A *long sequence* or *long code* is a spreading sequence with a period that is much longer than the data-symbol duration and may even exceed the message duration. A *short sequence* or *short code* is a spreading sequence with a period that is equal to or less than the data-symbol duration. Since short sequences are susceptible to interception and linear sequences are inherently susceptible to mathematical cryptanalysis [1], long nonlinear pseudonoise sequences and programmable code generators are needed for communications with a high level of security. However, if a modest level of security is acceptable, short or moderate-length pseudonoise sequences are preferable for rapid acquisition, burst communications, and multiuser detection.

The algebraic structure of linear feedback shift registers makes them susceptible to cryptanalysis. Let

$$\mathbf{c} = [c_1 \ c_2 \ldots c_m]^T \qquad (2\text{-}68)$$

denote the column vector of the m feedback coefficients of an m-stage linear feedback shift register, where T denotes the transpose. The column vector of m successive sequence bits produced by the shift register starting at bit i is

$$\mathbf{a}_i = [a_i \ a_{i+1} \ \ldots a_{i+m-1}]^T \qquad (2\text{-}69)$$

Let $\mathbf{A}(i)$ denote the $m \times m$ matrix with columns consisting of the $\mathbf{a}_j$ vectors for $i \leq j \leq i + m - 1$:

$$\mathbf{A}(i) = \begin{bmatrix} a_{i+m-1} & a_{i+m-2} & \cdots & a_i \\ a_{i+m} & a_{i-m-1} & \cdots & a_{i+1} \\ \vdots & \vdots & & \vdots \\ a_{i+2m-2} & a_{i+2m-3} & \cdots & a_{i+m-1} \end{bmatrix} \qquad (2\text{-}70)$$

The linear recurrence relation (2-14) indicates that the output sequence and feedback coefficients are related by

$$\mathbf{a}_{i+m} = \mathbf{A}(i)\mathbf{c}, \quad i \geq 0 \qquad (2\text{-}71)$$

If $2m$ consecutive sequence bits are known, then $\mathbf{A}(i)$ and $\mathbf{a}_{i+m}$ are completely known for some i. If $\mathbf{A}(i)$ is invertible, then the feedback coefficients can be computed from

$$\mathbf{c} = \mathbf{A}^{-1}(i)\mathbf{a}_{i+m} \ , \quad i \geq 0 \qquad (2\text{-}72)$$

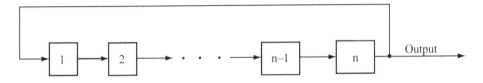

Figure 2.11: Linear generator of binary sequence with period n.

A shift-register sequence is completely determined by the feedback coefficients and any state vector. Since any m successive sequence bits determine a state vector, $2m$ successive bits provide enough information to reproduce the output sequence unless $\mathbf{A}(i)$ is not invertible. In that case, one or more additional bits are required.

If a binary sequence has period n, it can always be generated by a n-stage linear feedback shift register by connecting the output of the last stage to the input of the first stage and inserting n consecutive bits of the sequence into the output sequence, as illustrated in Figure 2.11. The polynomial associated with one period of the binary sequence is

$$g(x) = \sum_{i=0}^{n-1} a_i x^i \tag{2-73}$$

Let $gcd(g(x), 1 + x^n)$ denote the greatest common polynomial divisor of the polynomials $g(x)$ and $1 + x^n$. Then (2-62) implies that the generating function of the sequence may be expressed as

$$G(x) = \frac{g(x)/gcd\,(g(x), 1 + x^n)}{(1 + x^n)\,/gcd\,(g(x), 1 + x^n)} \tag{2-74}$$

If $gcd(g(x), 1 + x^n) \neq 1$, the degree of the denominator of $G(x)$ is less than n. Therefore, the sequence represented by $G(x)$ can be generated by a linear feedback shift register with fewer stages than n and with the characteristic function given by the denominator. The appropriate initial state can be determined from the coefficients of the numerator.

The *linear equivalent* of the generator of a sequence is the linear shift register with the fewest stages that produces the sequence. The number of stages in the linear equivalent is called the *linear complexity* of the sequence. If the linear complexity is equal to m, then (2-72) determines the linear equivalent after the observation of $2m$ consecutive sequence bits. Security improves as the period of a sequence increases, but there are practical limits to the number of shift-register stages. To produce sequences with a long enough period for high security, the feedback logic in Figure 2.5 must be nonlinear. Alternatively, one or more shift-register sequences or several outputs of shift-register stages may be applied to a nonlinear device to produce the sequence [5]. Nonlinear generators with relatively few shift-register stages can produce sequences of enormous linear complexity. As an example, Figure 2.12(a) depicts a nonlinear generator in

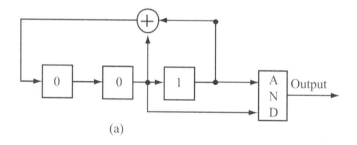

(a)

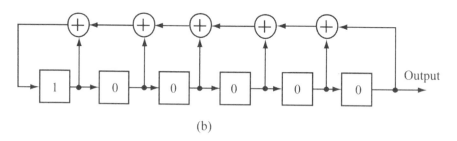

(b)

Figure 2.12: (a) Nonlinear generator and (b) its linear equivalent.

which two stages of a linear feedback shift register have their outputs applied to an AND gate to produce the output sequence. The initial contents of the shift-register stages are indicated by the enclosed binary numbers. Since the linear generator produces a maximal sequence of length 7, the output sequence has period 7. The first period of the sequence is (0 0 0 0 0 1 1), from which the linear equivalent with the initial contents shown in Figure 2.12(b) is derived by evaluating (2-74).

While a large linear complexity is necessary for the cryptographic integrity of a sequence, it is not necessarily sufficient because other statistical characteristics, such as a nearly even distribution of 1's and 0's, are required. For example, a long sequence of many 0's followed by a single 1 has a linear complexity equal to the length of the sequence, but the sequence is very weak. The generator of Figure 2.12(a) produces a relatively large number of 0's because the AND gate produces a 1 only if both of its inputs are 1's.

As another example, a nonlinear generator that uses a multiplexer is shown in Figure 2.13. The outputs of various stages of feedback shift register 1 are applied to the multiplexer, which interprets the binary number determined by these outputs as an address. The multiplexer uses this address to select one of the stages of feedback shift register 2. The selected stage provides the multiplexer output and, hence, one bit of the output sequence. Suppose that register 1 has m stages and register 2 has n stages. If h stages of register 1, where $h < m$, are applied to the multiplexer, then the address is one of the numbers $0, 1, ..., 2^h - 1$. Therefore, if $n \geq 2^h$, each address specifies a distinct stage of

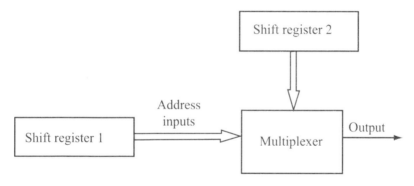

Figure 2.13: Nonlinear generator that uses a multiplexer.

register 2. The initial states of the two registers, the feedback connections, and which stages are used for addressing may be parts of a variable *key* that provides security. The security of the nonlinear generator is further enhanced if nonlinear feedback is used in both shift registers.

2.3 Systems with PSK Modulation

A received direct-sequence signal with coherent PSK modulation and ideal carrier synchronization can be represented by (2-1) or (2-6) with $\theta = 0$ to reflect the absence of phase uncertainty. Assuming that the chip waveform is well approximated by a waveform $\psi(t)$ of duration T_c, the received signal is

$$s(t) = \sqrt{2S}d(t)p(t)\cos 2\pi f_c t \qquad (2\text{-}75)$$

where S is the average power, $d(t)$ is the data modulation, $p(t)$ is the spreading waveform, and f_c is the carrier frequency. The data modulation is a sequence of nonoverlapping rectangular chip waveforms, each of which has an amplitude equal to $+1$ or -1. Each pulse of $d(t)$ represents a data symbol and has a duration of T_s. The spreading waveform has the form

$$p(t) = \sum_{i=-\infty}^{\infty} p_i \psi(t - iT_c) \qquad (2\text{-}76)$$

where p_i is equal to $+1$ or -1 and represents one chip of a spreading sequence $\{p_i\}$. It is convenient, and entails no loss of generality, to normalize the energy content of the chip waveform according to

$$\frac{1}{T_c} \int_0^{T_c} \psi^2(t)dt = 1 \qquad (2\text{-}77)$$

Because the transitions of a data symbol and the chips coincide on both sides of a symbol, the *processing gain*, defined as

$$G = \frac{T_s}{T_c} \qquad (2\text{-}78)$$

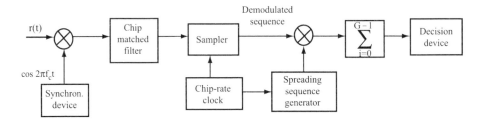

Figure 2.14: Basic elements of correlator for direct-sequence signal with coherent PSK.

is an integer equal to the number of chips in a symbol interval.

A practical direct-sequence system differs from the functional diagram of Figure 2.2. The transmitter needs practical devices, such as a power amplifier and a filter, to limit out-of-band radiation. In the receiver, the radio-frequency front end includes devices for wideband filtering and automatic gain control. These devices are assumed to have a negligible effect on the operation of the demodulator, at least for the purposes of this analysis. Thus, the front-end circuitry is omitted from Figure 2.14, which shows the optimum demodulator in the form of a correlator for the detection of a single symbol in the presence of white Gaussian noise. This correlator is more practical and flexible for digital processing than the alternative one shown in Figure 2.2. It is a suboptimal but reasonable approach against non-Gaussian interference. An equivalent matched-filter demodulator is implemented with a transversal filter or tapped delay line and a stored spreading sequence. However, the matched-filter implementation is not practical for a long sequence that extends over many data symbols. If the chip-rate synchronization in Figure 2.14 is accurate, then the demodulated sequence and the receiver-generated spreading sequence are multiplied together, and G successive products are added in an accumulator to produce the decision variable. The effective sampling rate of the decision variable is the symbol rate. The sequence generator, multiplier, and summer function as a discrete-time filter matched to the spreading sequence.

In the subsequent analysis, perfect phase, sequence, and symbol synchronization are assumed. The received signal is

$$r(t) = s(t) + i(t) + n(t) \qquad (2\text{-}79)$$

where $i(t)$ is the interference, and $n(t)$ denotes the zero-mean white Gaussian noise. The chip matched filter has impulse response $\psi(-t)$. Its output is sampled at the chip rate to provide G samples per data symbol. If $d(t) = d_0$ over $[0, T_s]$, then (2-75) to (2-79) indicate that the demodulated sequence corresponding to this data symbol is

$$Z_i = \int_{iT_c}^{(i+1)T_c} r(t)\psi\left(t - iT_c\right)\cos 2\pi f_c t\, dt = S_i + J_i + N_i, \qquad 0 \le i \le G - 1$$

$$(2\text{-}80)$$

where

$$S_i = \int_{iT_c}^{(i+1)T_c} s(t)\psi(t - iT_c)\cos 2\pi f_c t\, dt = p_i d_0 \sqrt{\frac{S}{2}}\, T_c \qquad (2\text{-}81)$$

$$J_i = \int_{iT_c}^{(i+1)T_c} i(t)\psi(t - iT_c)\cos 2\pi f_c t\, dt \qquad (2\text{-}82)$$

$$N_i = \int_{iT_c}^{(i+1)T_c} n(t)\psi\,(t - iT_c)\cos 2\pi f_c t\, dt \qquad (2\text{-}83)$$

and it is assumed that $f_c \gg 1/T_c$ so that the integral over a double-frequency term in (2-81) is negligible. The input to the decision device is

$$V = \sum_{i=0}^{G-1} p_i Z_i = d_0 \sqrt{\frac{S}{2}}\, T_s + V_1 + V_2 \qquad (2\text{-}84)$$

where

$$V_1 = \sum_{\nu=0}^{G-1} p_i J_i \qquad (2\text{-}85)$$

$$V_2 = \sum_{\nu=0}^{G-1} p_i N_i \qquad (2\text{-}86)$$

Suppose that $d_0 = +1$ represents the logic symbol 1 and $d_0 = -1$ represents the logic symbol 0. The decision device produces the symbol 1 if $V > 0$ and the symbol 0 if $V < 0$. An error occurs if $V < 0$ when $d_0 = +1$ or if $V > 0$ when $d_0 = -1$. The probability that $V = 0$ is zero.

The white Gaussian noise has autocorrelation

$$R_n(\tau) = \frac{N_0}{2}\delta(t - \tau) \qquad (2\text{-}87)$$

where $N_0/2$ is the two-sided noise power spectral density. Since $E[n(t)] = 0$, (2-86) implies that $E[V_2] = 0$. A straightforward calculation using (2-83), (2-86), (2-87), the limited duration of $\psi(t)$, and $f_c \gg 1/T_c$ yields

$$var\,(V_2) = \frac{1}{4}N_0 T_s \qquad (2\text{-}88)$$

It is natural and analytically desirable to model a long spreading sequence as a random binary sequence. The random-binary-sequence model does not seem to obscure important exploitable characteristics of long sequences and is a reasonable approximation even for short sequences in networks with asynchronous communications. A random binary sequence consists of statistically independent symbols, each of which takes the value $+1$ with probability $1/2$ or the value -1 with probability $1/2$. Thus, $E[p_i] = E[p(t)] = 0$. It then follows from (2-84) to (2-86) that $E[V_1] = E[V_2] = 0$, and the mean value of the decision variable is

$$E[V] = d_0 \sqrt{\frac{S}{2}}T_s \qquad (2\text{-}89)$$

for the direct-sequence system with coherent PSK. Since p_i and p_j are independent for $i \neq j$;

$$E\left[p_i p_j\right] = 0, \quad i \neq j \tag{2-90}$$

Therefore, the independence of p_i and J_j for all i and j implies that $E[p_i J_i p_j J_j] = 0, i \neq j$, and hence

$$var\left(V_1\right) = \sum_{i=0}^{G-1} E\left[J_i^2\right] \tag{2-91}$$

Tone Interference at Carrier Frequency

For tone interference with the same carrier frequency as the desired signal, a nearly exact, closed-form equation for the symbol error probability can be derived. The tone interference has the form

$$i\left(t\right) = \sqrt{2I} \cos\left(2\pi f_c t + \phi\right) \tag{2-92}$$

where I is the average power and ϕ is the phase relative to the desired signal. Assuming that $f_c \gg 1/T_c$, (2-82), (2-85), (2-92) and a change of variables give

$$V_1 = \sqrt{\frac{I}{2}} \cos \phi \sum_{i=0}^{G-1} p_i \int_0^{T_c} \psi\left(t\right) dt \tag{2-93}$$

A rectangular chip waveform has $\psi(t) = w(t, T_c)$, which is given by (2-3). For sinusoidal chips in the spreading waveform, $\psi(t) = \psi_s(t, T_c)$, where

$$\psi_s(t, T) = \begin{cases} \sqrt{2} \sin\left(\frac{\pi}{T} t\right), & 0 \leq t \leq T \\ 0, & \text{otherwise} \end{cases} \tag{2-94}$$

Let k_1 denote the number of chips in $[0, T_s]$ for which $p_i = +1$; the number for which $p_i = -1$ is $G - k_1$. Equations (2-93), (2-3), and (2-94) yield

$$V_1 = \sqrt{\frac{I\kappa}{2}} T_c\left(2k_1 - G\right) \cos \phi \tag{2-95}$$

where κ depends on the chip waveform, and

$$\kappa = \begin{cases} 1, & \textit{rectangular chip} \\ \frac{8}{\pi^2}, & \textit{sinusoidal chip} \end{cases} \tag{2-96}$$

These equations indicate that the use of sinusoidal chip waveforms instead of rectangular ones effectively reduces the interference power by a factor $8/\pi^2$ if $V_1 \neq 0$. Thus, the advantage of sinusoidal chip waveforms is 0.91 dB against tone interference at the carrier frequency. Equation (2-95) indicates that tone interference at the carrier frequency would be completely rejected if $k_1 = G/2$ in every symbol interval.

In the random-binary-sequence model, p_i is equally likely to be $+1$ or -1. Therefore, the conditional symbol error probability given the value of ϕ is

$$P_s(\phi) = \sum_{k_1=0}^{G} \binom{G}{k_1} \left(\frac{1}{2}\right)^G \left[\frac{1}{2}P_s(\phi, k_1, +1) + \frac{1}{2}P_s(\phi, k_1, -1)\right] \qquad (2\text{-}97)$$

where $P_s(\phi, k_1, d_0)$ is the conditional symbol error probability given the values of ϕ, k_1 and d_0. Under these conditions, V_1 is a constant, and V has a Gaussian distribution. Equations (2-84) and (2-95) imply that the conditional expected value of V is

$$E[V|\phi, k_1, d_0] = d_0 \sqrt{\frac{S}{2}} T_s + \sqrt{\frac{I\kappa}{2}} T_c (2k_1 - G) \cos \phi \qquad (2\text{-}98)$$

The conditional variance of V is equal to the variance of V_2, which is given by (2-88). Using the Gaussian density to evaluate $P_s(\phi, k_1, +1)$ and $P_s(\phi, k_1, -1)$ separately and then consolidating the results yields

$$P_s(\phi, k_1, d_0) = Q\left[\sqrt{\frac{2\mathcal{E}_s}{N_0}} + d_0\sqrt{\frac{2IT_c\kappa}{GN_0}}(2k_1 - G)\cos\phi\right] \qquad (2\text{-}99)$$

where $\mathcal{E}_s = ST_s$ is the energy per symbol and (1-30) defines

$$Q(x) = \frac{1}{\sqrt{2\pi}} \int_x^\infty \exp\left(-\frac{y^2}{2}\right) dy \qquad (2\text{-}100)$$

Assuming that ϕ is uniformly distributed over $[0, 2\pi)$ and exploiting the periodicity of $\cos\phi$, we obtain the symbol error probability

$$P_s = \frac{1}{\pi} \int_0^\pi P_s(\phi) \, d\phi \qquad (2\text{-}101)$$

where $P_s(\phi)$ is given by (2-97) and (2-99).

General Tone Interference

To simplify the preceding equations for P_s and to examine the effects of tone interference with a carrier frequency different from the desired frequency, a Gaussian approximation is used. Consider interference due to a single tone of the form

$$i(t) = \sqrt{2I} \cos(2\pi f_1 t + \theta_1) \qquad (2\text{-}102)$$

where I, f_1, and θ_1 are the average power, frequency, and phase angle of the interference signal at the receiver. The frequency f_1 is assumed to be close enough to the desired frequency f_c that the tone is undisturbed by the initial wideband filtering that precedes the correlator. If $f_1 + f_c \gg f_d = f_1 - f_c$ so that a term involving $f_1 + f_c$ is negligible, (2-102) and (2-82) and a change of variable yield

$$J_i = \sqrt{\frac{I}{2}} \int_0^{T_c} \psi(t) \cos(2\pi f_d t + \theta_1 + i2\pi f_d T_c) dt \qquad (2\text{-}103)$$

For a rectangular chip waveform, evaluation of the integral and trigonometry yield

$$J_i = \sqrt{\frac{I}{2}} T_c \, \text{sinc} \, (f_d T_c) \cos (i2\pi f_d T_c + \theta_2) \tag{2-104}$$

where

$$\theta_2 = \theta_1 + \pi f_d T_c \tag{2-105}$$

Substituting (2-104) into (2-91) and expanding the squared cosine, we obtain

$$var \, (V_1) = \frac{1}{4} I T_c^2 \text{sinc}^2 \, (f_d T_c) \left[G + \sum_{i=0}^{G-1} \cos (i4\pi f_d T_c + 2\theta_2) \right] \tag{2-106}$$

To evaluate the inner summation, we use the identity

$$\sum_{\nu=0}^{n-1} \cos (a + \nu b) = \cos \left(a + \frac{n-1}{2} b \right) \frac{\sin (nb/2)}{\sin (b/2)} \tag{2-107}$$

which is proved by using mathematical induction and trigonometric identities. Evaluation and simplification yield

$$var(V_1) = \frac{1}{4} I T_s T_c \text{sinc}^2 \, (f_d T_c) \left[1 + \frac{\text{sinc} \, (2f_d T_s)}{\text{sinc} \, (2f_d T_c)} \cos 2\phi \right] \tag{2-108}$$

where

$$\phi = \theta_2 + \pi f_d \, (T_s - T_c) = \theta_1 + \pi f_d T_s \tag{2-109}$$

Given the value of ϕ, the J_i in (2-104) are uniformly bounded constants, and, hence, the terms of V_1 in (2-85) are independent and uniformly bounded. Since $var(V_1) \to \infty$ as $G \to \infty$, the central limit theorem [6] implies that when G is large, the conditional distribution of V_1 is approximately Gaussian. Thus, V is nearly Gaussian with mean given by (2-89) and $var(V) = var(V_1) + var(V_2)$. Because of the symmetry of the model, the conditional symbol error probability may be calculated by assuming $d_0 = 1$ and evaluating the probability that $V < 0$. A straightforward derivation using (2-108) indicates that the conditional symbol error probability is well approximated by

$$P_s \, (\phi) = Q \left[\sqrt{\frac{2\mathcal{E}_s}{N_{0e}(\phi)}} \right] \tag{2-110}$$

where

$$N_{0e}(\phi) = N_0 + I T_c \text{sinc}^2 \, (f_d T_c) \left[1 + \frac{\text{sinc} \, (2f_d T_s)}{\text{sinc} \, (2f_d T_c)} \cos 2\phi \right] \tag{2-111}$$

and $N_{0e}(\phi)/2$ can be interpreted as the *equivalent two-sided power spectral density* of the interference plus noise, given the value of ϕ. For sinusoidal chip waveforms, a similar derivation yields (2-110) with

$$N_{0e}(\phi) = N_0 + I T_c \left(\frac{8}{\pi^2} \right) \left(\frac{\cos \pi f_d T_c}{1 - 4f_d^2 T_c^2} \right)^2 \left[1 + \frac{\text{sinc} \, (2f_d T_s)}{\text{sinc} \, (2f_d T_c)} \cos 2\phi \right] \tag{2-112}$$

To explicitly exhibit the reduction of the interference power by the factor G, we may substitute $T_c = T_s/G$ in (2-111) or (2-112). A comparison of these two equations confirms that sinusoidal chip waveforms provide a $\pi^2/8 = 0.91$ dB advantage when $f_d = 0$, but this advantage decreases as $|f_d|$ increases and ultimately disappears. The preceding analysis can easily be extended to multiple tones, but the resulting equations are complicated.

If θ_1 in (2-109) is modeled as a random variable that is uniformly distributed over $[0, 2\pi)$, then the modulo-2π character of $\cos 2\phi$ in (2-111) implies that its distribution is the same as it would be if ϕ were uniformly distributed over $[0, 2\pi)$. Therefore, we can henceforth assign a uniform distribution for ϕ. The symbol error probability, which is obtained by averaging $P_s(\phi)$ over the range of ϕ, is

$$P_s = \frac{2}{\pi} \int_0^{\pi/2} Q\left[\sqrt{\frac{2\mathcal{E}_s}{N_{0e}(\phi)}}\right] d\phi \qquad (2\text{-}113)$$

where the fact that $\cos 2\phi$ takes all its possible values over $[0, \pi/2]$ has been used to shorten the integration interval.

Figure 2.15 depicts the symbol error probability as a function of the despread signal-to-interference ratio, GS/I, for one tone-interference signal, rectangular chip waveforms, $f_d = 0, G = 50 = 17$ dB, and $\mathcal{E}_s/N_0 = 14$ dB and 20 dB. One pair of graphs are computed using the approximate model of (2-111) and (2-113), while the other pair are derived from the nearly exact model of (2-97), (2-99), and (2-101) with $\kappa = 1$. For the nearly exact model, P_s depends not only on GS/I, but also on G. A comparison of the two graphs indicates that the error introduced by the Gaussian approximation is on the order of or less than 0.1 dB when $P_s \geq 10^{-6}$. This example and others provide evidence that the Gaussian approximation introduces insignificant error if $G \geq 50$ and practical values for the other parameters are assumed.

Figure 2.16 uses the approximate model to plot P_s versus the normalized frequency offset $f_d T_c$ for rectangular and sinusoidal chip waveforms, $G = 17$ dB, $\mathcal{E}_s/N_0 = 14$ dB, and $GS/I = 10$ dB. The performance advantage of sinusoidal chip waveforms is apparent, but their realization or that of Nyquist chip waveforms in a transmitted PSK waveform is difficult because of the distortion introduced by a nonlinear power amplifier in the transmitter when the signal does not have a constant envelope.

Gaussian Interference

Gaussian interference is interference that approximates a zero-mean, stationary Gaussian process. If $i(t)$ is modeled as Gaussian interference and $f_c \gg 1/T_c$, then (2-82), a trigonometric expansion, the dropping of a negligible double integral, and a change of variables give

$$E\left[J_i^2\right] = \frac{1}{2} \int_0^{T_c} \int_0^{T_c} R_j(t_1 - t_2)\, \psi(t_1)\, \psi(t_2) \cos\left[2\pi f_c (t_1 - t_2)\right] dt_1 dt_2$$

$$(2\text{-}114)$$

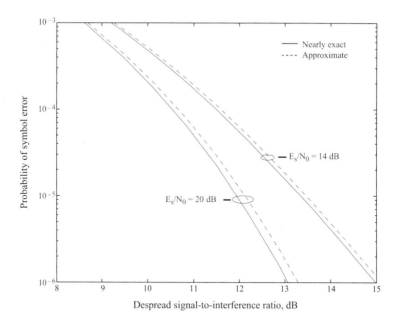

Figure 2.15: Symbol error probability of binary direct-sequence system with tone interference at carrier frequency and $G = 17$ dB.

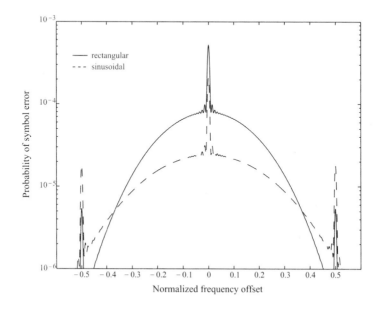

Figure 2.16: Symbol error probability for direct-sequence system with PSK, rectangular and sinusoidal chip waveforms, $G = 17$ dB, $\mathcal{E}_s/N_0 = 14$ dB, and $GS/I = 10$ dB in the presence of tone interference.

where $R_j(t)$ is the autocorrelation of $i(t)$. Since $E[J_i^2]$ does not depend on the index i, (2-91) gives

$$var\,(V_1) = GE\left[J_i^2\right] \qquad (2\text{-}115)$$

Assuming that $\psi(t)$ is rectangular, we change variables in (2-114) by using $\tau = t_1 - t_2$ and $s = t_1 + t_2$. The Jacobian of this transformation is 2. Evaluating one of the resulting integrals and substituting the result into (2-115) yields

$$var\,(V_1) = \frac{1}{2}T_s \int_{-T_c}^{T_c} R_j\,(\tau)\Lambda\left(\frac{\tau}{T_c}\right)\cos 2\pi f_c \tau\; d\tau \qquad (2\text{-}116)$$

The limits in this equation can be extended to $\pm\infty$ because the integrand is truncated. Since $R_j(\tau)\Lambda\left(\frac{\tau}{T_c}\right)$ is an even function, the cosine function may be replaced by a complex exponential. Then the convolution theorem and the known Fourier transform of $\Lambda(t)$ yield the alternative form

$$var\,(V_1) = \frac{1}{2}T_s T_c \int_{-\infty}^{\infty} S_j\,(f)\,\mathrm{sinc}^2\left[(f - f_c)\,T_c\right]df \qquad (2\text{-}117)$$

where $S_j(f)$ is the power spectral density of the interference after passage through the initial wideband filter of the receiver.

Since $i(t)$ is a zero-mean Gaussian process, the $\{J_i\}$ are zero-mean and jointly Gaussian. Therefore, if the $\{p_i\}$ are given, then (V_1) is conditionally zero-mean and Gaussian. Since $var(V_1)$ does not depend on the $\{p_i\}$, V_1 without conditioning is a zero-mean Gaussian random variable. The independence of the thermal noise and the interference imply that $V = V_1 + V_2$ is a zero-mean Gaussian random variable. Thus, a standard derivation yields the symbol error probability:

$$P_s = Q\left(\sqrt{\frac{2\mathcal{E}_s}{N_{0e}}}\right) \qquad (2\text{-}118)$$

where

$$N_{0e} = N_0 + 2T_c \int_{-\infty}^{\infty} S_j\,(f)\,\mathrm{sinc}^2\left[(f - f_c)\,T_c\right]df \qquad (2\text{-}119)$$

If $S_j'(f)$ is the interference power spectral density at the input and $H(f)$ is the transfer function of the initial wideband filter, then $S_j(f) = S_j'(f)|H(f)|^2$. Suppose that the interference has a flat spectrum over a band within the passband of the wideband filter so that

$$S_j\,(f) = \begin{cases} \frac{I}{2W_1}, & |f - f_1| \le \frac{W_1}{2}, \quad |f + f_1| \le \frac{W_1}{2} \\ 0, & \text{otherwise} \end{cases} \qquad (2\text{-}120)$$

If $f_c \gg 1/T_c$, the integration over negative frequencies in (2-119) is negligible and

$$N_{0e} = N_0 + \frac{IT_c}{W_1}\int_{f_1-W_1/2}^{f_1+W_1/2} \mathrm{sinc}^2\left[(f - f_c)\,T_c\right]df \qquad (2\text{-}121)$$

This equation shows that $f_1 = f_c$ or $f_d = 0$ coupled with a narrow bandwidth increases the impact of the interference power. Since the integrand is upper-bounded by unity, $N_{0e} \leq N_0 + IT_c$. This upper bound is intuitively reasonable because $IT_c \approx I/B = I_0$, where $B \approx 1/T_c$ is the bandwidth of narrowband interference after the despreading, and I_0 is its power spectral density. Equation (2-118) yields

$$P_s \leq Q\left(\sqrt{\frac{2\mathcal{E}_s}{N_0 + IT_c}}\right) \qquad (2\text{-}122)$$

This upper bound is tight if $f_d \approx 0$ and the Gaussian interference is narrowband. A plot of (2-122) with the parameter values of Figure 2.15 indicates that roughly 2 dB more interference power is required for worst-case Gaussian interference to degrade P_s as much as tone interference at the carrier frequency.

2.4   Quaternary Systems

A received *quaternary direct-sequence signal* with ideal carrier synchronization and a chip waveform of duration T_c can be represented by

$$s(t) = \sqrt{S}d_1(t)p_1(t)\cos 2\pi f_c t + \sqrt{S}d_2(t+t_0)p_2(t+t_0)\sin 2\pi f_c t \qquad (2\text{-}123)$$

where two spreading waveforms, $p_1(t)$ and $p_2(t)$, and two data signals, $d_1(t)$ and $d_2(t)$, are used with two quadrature carriers, and t_0 is the relative delay between the in-phase and quadrature components of the signal. For a *quadriphase direct-sequence system*, which uses QPSK, $t_0 = 0$. For a direct-sequence system with offset QPSK (OQPSK) or minimum-shift keying (MSK), $t_0 = T_c/2$. For OQPSK, the chip waveforms are rectangular; for MSK, they are sinusoidal. One might use MSK to limit the spectral sidelobes of the direct-sequence signal, which may interfere with other signals.

Consider the *classical* or *dual* quaternary system in which $d_1(t)$ and $d_2(t)$ are independent. Let T_s denote the duration of the data symbols before the generation of (2-123), and let $T_{s1} = 2T_s$ denote the duration of the channel symbols, which are transmitted in pairs. Let T_c denote the common chip duration of $p_1(t)$ and $p_2(t)$. The number of chips per channel symbol is $2G$, where $G = T_s/T_c$. It is assumed that the synchronization is perfect in the receiver, which is shown in Figure 2.17. Consequently, if the received signal is given by (2-123), then the upper decision variable applied to the decision device at the end of a symbol interval during which $d_1(t) = d_{10}$ is

$$V = d_{10}\sqrt{S}\,T_s + \sum_{i=0}^{2G-1} p_{1i}J_i + \sum_{i=0}^{2G-1} p_{1i}N_i \qquad (2\text{-}124)$$

where J_i and N_i are given by (2-82) and (2-83), respectively. The term representing crosstalk,

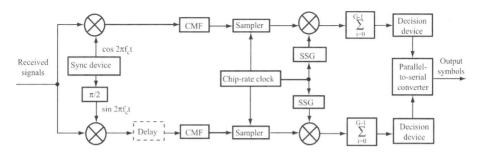

Figure 2.17: Receiver for direct-sequence signal with dual quaternary modulation; CMF = chip-matched filter; SSG = spreading sequence generator. Delay = 0 for QPSK; delay = $T_c/2$ for OQPSK and MSK.

$$V_c = \sum_{i=0}^{2G-1} p_{1i} \frac{\sqrt{S}}{2} \int_{iT_c}^{(i+1)T_c} d_2 \left(t + t_0\right) p_2 \left(t + t_0\right) \psi \left(t - iT_c\right) \sin 4\pi f_c t \; dt$$

$$(2\text{-}125)$$

is negligible if $f_c \gg 1/T_c$ so that the sinusoid in (2-125) varies much more rapidly than the other factors. Similarly, the lower decision variable at the end of a channel-symbol interval during which $d_2(t) = d_{20}$ is

$$U = d_{20}\sqrt{S}T_s + \sum_{i=0}^{2G-1} p_{2i} J_i' + \sum_{i=0}^{2G-1} p_{2i} N_i' \qquad (2\text{-}126)$$

where

$$J_i' = \int_{iT_c}^{(i+1)T_c} i(t)\psi \left(t - iT_c\right) \sin 2\pi f_c t \; dt \qquad (2\text{-}127)$$

$$N_i' = \int_{iT_c}^{(i+1)T_c} n(t)\psi \left(t - iT_c\right) \sin 2\pi f_c t \; dt \qquad (2\text{-}128)$$

Of the available desired-signal power S, half is in each of the two components of (2-123). Since $T_{s1} = 2T_s$, the energy per channel symbol is $\mathcal{E}_s = ST_s$, the same as for a direct-sequence system with PSK, and

$$E[V] = d_{10}\sqrt{S}\,T_s, \quad E(U) = d_{20}\sqrt{S}\,T_s \qquad (2\text{-}129)$$

A derivation similar to the one leading to (2-88) gives the variances of the noise terms V_2 and U_2 in (2-124) and (2-126):

$$var\left(V_2\right) = var\left(U_2\right) = \frac{1}{2}N_0 T_s \qquad (2\text{-}130)$$

Using the tone-interference model of Section 2.3, and averaging the error probabilities for the two parallel symbol streams, we obtain the conditional symbol error probability:

$$P_s\left(\phi\right) = \frac{1}{2}Q\left[\sqrt{\frac{2\mathcal{E}_s}{N_{0e}^{(0)}\left(\phi\right)}}\right] + \frac{1}{2}Q\left[\sqrt{\frac{2\mathcal{E}_s}{N_{0e}^{(1)}\left(\phi\right)}}\right]   \qquad (2\text{-}131)$$

where $N_{0e}^{(0)}(\phi)$ and $N_{0e}^{(1)}(\phi)$ arise from the upper and lower branches of Figure 2.17, respectively. For rectangular chip waveforms (QPSK and OQPSK signals),

$$N_{0e}^{(l)}\left(\phi\right) = N_0 + IT_c\text{sinc}^2\left(f_dT_c\right)\left[1 + \frac{\text{sinc}\left(4f_dT_s\right)}{\text{sinc}\left(2f_dT_c\right)}\cos(2\phi + l\pi)\right]   \qquad (2\text{-}132)$$

and for sinusoidal chip waveforms,

$$N_{oe}^{(l)}\left(\phi\right) = N_0 + IT_c\left(\frac{8}{\pi^2}\right)\left(\frac{\cos\pi f_dT_c}{1 - 4f_d^2T_c^2}\right)^2\left[1 + \frac{\text{sinc}\left(4f_dT_s\right)}{\text{sinc}\left(2f_dT_c\right)}\cos(2\phi + l\pi)\right]$$
$$\qquad (2\text{-}133)$$

where $l = 0, 1$, and we have used $T_{s1} = 2T_s$ and

$$\phi = \theta_1 + 2\pi f_dT_s   \qquad (2\text{-}134)$$

These equations indicate that $P_s(\phi)$ for a quaternary direct-sequence system and the worst value of ϕ is usually lower than $P_s(\phi)$ for a binary direct-sequence system with the same chip waveform and the worst value of ϕ. The symbol error probability is determined by integrating $P_s(\phi)$ over the distribution of ϕ. For a uniform distribution, the two integrals are equal. Using the periodicity of $\cos 2\phi$ to shorten the integration interval, we obtain

$$P_s = \frac{2}{\pi}\int_0^{\pi/2}Q\left[\sqrt{\frac{2E_s}{N_{0e}^{(0)}\left(\phi\right)}}\right]d\phi   \qquad (2\text{-}135)$$

The quaternary system provides a slight advantage relative to the binary system against tone interference. Both systems provide the same P_s when $f_d = 0$ and nearly the same P_s when $f_d > 1/T_s$. Figure 2.18 illustrates P_s versus the normalized frequency offset f_dT_c for quaternary and binary systems, $G = 17$ dB, $\mathcal{E}_s/N_0 = 14$ dB, and $GS/I = 10$ dB.

In a *balanced quaternary system,* the same data symbols are carried by both the in-phase and quadrature components, which implies that the received direct-sequence signal has the form given by (2-123) with $d_1(t) = d_2(t) = d(t)$. Thus, although the spreading is done by quadrature carriers, the data modulation may be regarded as binary PSK. A receiver for this system is shown in Figure 2.19. The synchronization system is assumed to operate perfectly in the subsequent analysis. If $f_c \gg 1/T_c$, the crosstalk terms similar to (2-125) are negligible. If the transmitted symbol is $d_{10} = d_{20} = d_0$, then the input to the decision device is

$$V = d_0\sqrt{S}\,T_s + \sum_{i=0}^{G-1}p_{1i}J_i + \sum_{i=0}^{G-1}p_{2i}J_i' + \sum_{i=0}^{G-1}p_{1i}N_i + \sum_{i=0}^{G-1}p_{2i}N_i'   \qquad (2\text{-}136)$$

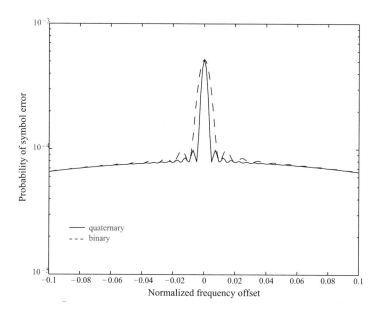

Figure 2.18: Symbol error probability for quaternary and binary direct-sequence systems with $G = 17$ dB, $\mathcal{E}_s/N_0 = 14$ dB, and $GS/I = 10$ dB in the presence of tone interference.

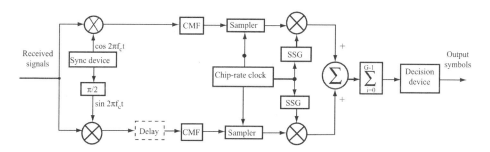

Figure 2.19: Receiver for direct-sequence signal with balanced quaternary modulation (delay $= 0$ for QPSK and delay $= T_c/2$ for OQPSK and MSK); CMF $=$ chip-matched filter; SSG $=$ spreading sequence generator.

where T_s is the duration of both a data symbol and a channel symbol. If $p_1(t)$ and $p_2(t)$, are approximated by independent random binary sequences, then the last four terms of (2-136) are zero-mean uncorrelated random variables. Therefore, the variance of V is equal to the sum of the variances of these four random variables, and

$$E[V] = d_0\sqrt{S}T_s \tag{2-137}$$

Straightforward evaluations verify that both types of quaternary signals provide the same performance against Gaussian interference as direct-sequence signals with PSK.

Consider a *balanced QPSK system*, for which $t_0 = 0$. If $i(t)$ is a tone, then a straightforward extension of the preceding analysis for general tone interference (Section 2.3) yields a $P_s(\phi)$ that is independent of ϕ. Therefore,

$$P_s = P_s(\phi) = Q\left(\sqrt{\frac{2\mathcal{E}_s}{N_{0e}}}\right) \tag{2-138}$$

where for rectangular chip waveforms,

$$N_{0e} = N_0 + IT_c\mathrm{sinc}^2\left(f_dT_c\right) \tag{2-139}$$

and for sinusoidal chip waveforms,

$$N_{0e} = N_0 + IT_c\left(\frac{8}{\pi^2}\right)\left(\frac{\cos\pi f_dT_c}{1 - 4f_d^2T_c^2}\right)^2 \tag{2-140}$$

If $f_d = 0$, a nearly exact model similar to the one in Section 2.3 implies that the conditional symbol error probability is

$$P_s(\phi) = \sum_{k_1=0}^{G}\sum_{k_2=0}^{G}\binom{G}{k_1}\binom{G}{k_2}\left(\frac{1}{2}\right)^{2G}\left[\frac{1}{2}P_s(\phi, k_1, k_2, +1) + \frac{1}{2}P_s(\phi, k_1, k_2, -1)\right] \tag{2-141}$$

where k_1 and k_2 are the number of chips in a symbol for which $p_1(t) = +1$ and $p_2(t) = +1$, respectively, and $P_s(\phi, k_1, k_2, d_0)$ is the conditional symbol error probability given the values of $\phi, k_1,$ and k_2 and that $d(t) = d_0$. A derivation analogous to that of (2-99) yields

$$P_s(\phi, k_1, k_2, d_0) = Q\left\{\sqrt{\frac{2\mathcal{E}_s}{N_0}} + d_0\sqrt{\frac{IT_c\kappa}{GN_0}}\left[(2k_1 - G)\cos\phi - (2k_2 - G)\sin\phi\right]\right\} \tag{2-142}$$

If ϕ is uniformly distributed over $[0, 2\pi)$, then

$$P_s = \frac{1}{2\pi}\int_0^{2\pi}P_s(\phi)d\phi \tag{2-143}$$

Numerical comparisons of the nearly exact model with the approximate results given by (2-138) for $f_d = 0$ indicate that the approximate results typically introduce an insignificant error if $G \geq 50$.

If $g(x)$ is a convex function over an interval containing the range of a random variable X, then Jensen's inequality (Appendix A) is

$$g\left(E\left[X\right]\right) \le E\left[g(X)\right] \tag{2-144}$$

provided that the indicated expected values exist. Consider the function

$$g(x) = Q\left(\sqrt{\frac{1}{a+bx}}\right) \tag{2-145}$$

Since the second derivative of $g(x)$ is nonnegative over the interval such that $0 < a + bx \le 1/3$, $g(x)$ is a convex function over that interval, and Jensen's inequality is applicable.

The application of this result to (2-135) with $X = \cos 2\phi$ and the fact that $E[\cos 2\phi] = 0$ yields a lower bound identical to the right-hand side of (2-138). Thus, the *balanced QPSK* system, for which $d_1(t) = d_2(t)$, provides a lower symbol error probability against tone interference than the dual quaternary or QPSK system for which $d_1(t) \ne d_2(t)$. A sufficient convexity condition for all f_d is

$$\mathcal{E}_s \ge \frac{3}{2}\left(N_0 + 2IT_c\right) \tag{2-146}$$

Figure 2.20 illustrates the performance advantage of the balanced QPSK system of Figure 2.19 against tone interference when $f_d < 1/T_s$. Equations (2-131) to (2-135) and (2-138) to (2-140) are used for the dual quaternary and the balanced QPSK systems, respectively, and $G = 17$ dB, $\mathcal{E}_s/N_0 = 14$ dB, and $GS/I = 10$ dB. The normalized frequency offset is $f_d T_c$. The advantage of the balanced QPSK system when f_d is small exists because a tone at the carrier frequency cannot have a phase that causes desired-signal cancellation simultaneously in both receiver branches.

2.5 Pulsed Interference

Pulsed interference is interference that occurs periodically or sporadically for brief durations. Whether it is generated unintentionally or by an opponent, pulsed interference can cause a substantial increase in the bit error rate of a communication system relative to the rate caused by continuous interference with the same average power. Pulsed interference may be produced in a receiver by a signal with a variable center frequency that sweeps over a frequency range that intersects or includes the receiver passband.

Consider a direct-sequence system with binary PSK that operates in the presence of pulsed interference. Let μ denote either the pulse duty cycle, which is the ratio of the pulse duration to the repetition period, or the probability of pulse occurrence if the pulses occur randomly. During a pulse, the interference is modeled as Gaussian interference with power I/μ, where I is the average interference power. According to (2-121), the equivalent noise-power spectral density may be decomposed as

$$N_{0e} = N_0 + I_0 \tag{2-147}$$

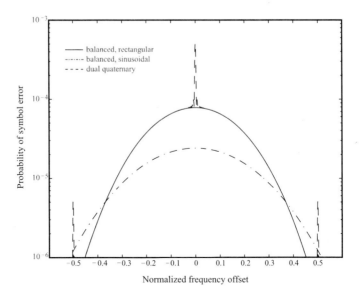

Figure 2.20: Symbol error probability for direct-sequence systems with balanced QPSK and dual quaternary modulations, rectangular and sinusoidal chip waveforms, $G = 17$ dB, $\mathcal{E}_s/N_0 = 14$ dB, and $GS/I = 10$ dB in the presence of tone interference.

where the power spectral density of continuous interference ($\mu = 1$) is

$$I_0 = \frac{IT_c}{W_1} \int_{f_1 - W_1/2}^{f_1 + W_1/2} \operatorname{sinc}^2[\,(f - f_c)\,T_c]\,df \qquad (2\text{-}148)$$

In the absence of a pulse, $N_{0e} = N_0$, whereas $N_{0e} = N_0 + I_0/\mu$ in the presence of a pulse. If the interference pulse duration approximately equals or exceeds the channel-symbol duration, then (2-118) implies that

$$P_s \cong \mu Q\left(\sqrt{\frac{2\mathcal{E}_s}{N_0 + I_0/\mu}}\right) + (1 - \mu)\,Q\left(\sqrt{\frac{2\mathcal{E}_s}{N_0}}\right), \qquad 0 \le \mu \le 1 \qquad (2\text{-}149)$$

If μ is treated as a continuous variable over $[0, 1]$ and $I_0 \gg N_0$, calculus gives the value of μ that maximizes P_s:

$$\mu_0 \cong \begin{cases} 0.7\left(\dfrac{\mathcal{E}_s}{I_0}\right)^{-1}, & \dfrac{\mathcal{E}_s}{I_0} > 0.7 \\[2ex] 1, & \dfrac{\mathcal{E}_s}{I_0} \le 0.7 \end{cases} \qquad (2\text{-}150)$$

Thus, worst-case pulsed interference is more damaging than continuous interference if $\mathcal{E}_s/I_0 > 0.7$.

By substituting $\mu = \mu_0$ into (2-149), we obtain an approximate expression for the worst-case P_s when $I_0 \gg N_0$:

$$
P_s \cong \begin{cases} 0.083 \left(\frac{\mathcal{E}_s}{I_0} \right)^{-1} , & \frac{\mathcal{E}_s}{I_0} > 0.7 \\ Q \left(\sqrt{\frac{2\mathcal{E}_s}{I_0}} \right) , & \frac{\mathcal{E}_s}{I_0} \leq 0.7 \end{cases}
\qquad (2\text{-}151)
$$

This equation indicates that the worst-case P_s varies inversely, rather than exponentially, with $\mathcal{E}_s/I_0$ if this ratio is sufficiently large. To restore a nearly exponential dependence on $\mathcal{E}_s/I_0$, a channel code and symbol interleaving are necessary.

Decoding metrics that are effective against white Gaussian noise are not necessarily effective against worst-case pulsed interference. We examine the performance of five different metrics against pulsed interference when the direct-sequence system uses PSK, ideal symbol interleaving, a binary convolutional code, and Viterbi decoding [7]. The results are the same when either dual or balanced QPSK is the modulation.

Let $B(l)$ denote the total information weight of the paths at Hamming distance l from the correct path over an unmerged segment in the trellis diagram of the convolutional code. Let $P_2(l)$ denote the probability of an error in comparing the correct path segment with a path segment that differs in l symbols. According to (1-112) with $k = 1$, the information-bit error rate is upper-bounded by

$$
P_b \leq \sum_{l=d_f}^{\infty} B(l) P_2(l) \qquad (2\text{-}152)
$$

where d_f is the minimum free distance. If r is the code rate, $\mathcal{E}_b$ is the energy per information bit, T_b is the bit duration, and G_u is the processing gain of the uncoded system, then

$$
\mathcal{E}_s = r\mathcal{E}_b, \quad T_s = rT_b, \quad G = rG_u. \qquad (2\text{-}153)
$$

The decrease in the processing gain is compensated by the coding gain. An upper bound on P_b for worst-case pulsed interference is obtained by maximizing the right-hand side of (2-152) with respect to μ, where $0 \leq \mu \leq 1$. The maximizing value of μ, which depends on the decoding metric, is not necessarily equal to the actual worst-case μ because a bound rather than an equality is maximized. However, the discrepancy is small when the bound is tight.

The simplest practical metric to implement is provided by hard-decision decoding. Assuming that the deinterleaving ensures the independence of symbol errors, (1-114) indicates that

$$
P_2(l) = \begin{cases} \displaystyle\sum_{i=(l+1)/2}^{l} \binom{l}{i} P_s^i (1 - P_s)^{l-i} , & l \text{ is odd} \\ \displaystyle\sum_{i=l/2+1}^{l} \binom{l}{i} P_s^i (1 - P_s)^{l-i} + \frac{1}{2} \binom{l}{l/2} \left[P_s (1 - P_s) \right]^{l/2} , & l \text{ is even} \end{cases}
$$

$$(2\text{-}154)$$

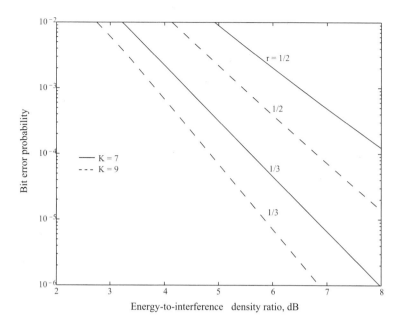

Figure 2.21: Worst-case performance against pulsed interference for convolutional codes of constraint length K, rate r, $\mathcal{E}_b/N_0 = 20$ dB, and hard decisions.

Since $\mu = \mu_0$ approximately maximizes P_s, it also approximately maximizes the upper bound on P_b for hard-decision decoding given by (2-152) to (2-154).

Figure 2.21 depicts the upper bound on P_b as a function of $\mathcal{E}_b/I_0$ for worst-case pulsed interference, $\mathcal{E}_b/N_0 = 20$ dB, and binary convolutional codes with several constraint lengths and rates. Tables 1.4 and 1.5 for $B(l)$ are used, and the series in (2-152) is truncated after the first 7 terms. This truncation gives reliable results only if $P_b \leq 10^{-3}$ because the series converges very slowly. However, the truncation error is partially offset by the error incurred by the use of the union bound because the latter error is in the opposite direction. Figure 2.21 indicates the significant advantage of raising the constraint length K and reducing r at the cost of increased implementation complexity and synchronization requirements, respectively.

Let N_{0i} denote the equivalent one-sided noise-power spectral density in output sample y_i of a coherent PSK demodulator. For convenience, y_i is assumed to have the form of the right-hand side of (2-84) normalized by multiplying the latter by $\sqrt{2/T_s}$. Thus, y_i has variance $N_{0i}/2$. Given that code symbol i of sequence j has value x_{ji}, the conditional probability density function of y_i is determined from the Gaussian character of the interference and noise. For a sequence of L code symbols, the density is

$$f\left(y_i | x_{ji}\right) = \frac{1}{\sqrt{\pi N_{0i}}} \exp\left[-\frac{(y_i - x_{ji})^2}{N_{0i}}\right], \quad i = 1, 2, \ldots, L \qquad (2\text{-}155)$$

From the log-likelihood function and the statistical independence of the samples, it follows that when the values of $N_{01}, N_{02}, \ldots, N_{0L}$ are known, the *maximum-likelihood metric* for optimal soft-decision decoding of the sequence is

$$U(j) = \sum_{i=1}^{L} \frac{x_{ji} y_i}{N_{0i}} \qquad (2\text{-}156)$$

This metric weights each output sample y_i according to the level of the equivalent noise. Since each y_i is assumed to be an independent Gaussian random variable, $U(j)$ is a Gaussian random variable.

Without loss of generality, let $j = 1$ label the correct sequence and $j = 2$ label an incorrect one at distance l. We assume that there is no quantization of the sample values or that the quantization is infinitely fine. Therefore, the probability that $U(2) = U(1)$ is zero, and the probability of an error in comparing a correct sequence with an incorrect one that differs in l symbols, $P_2(l)$, is equal to probability that $M_0 = U(2) - U(1) > 0$. The symbols that are the same in both sequences are irrelevant to the calculation of $P_2(l)$ and are ignored subsequently. Let $P_2(l|\nu)$ denote the conditional probability that $M_0 > 0$ given that an interference pulse occurs during ν out of l differing symbols and does not occur during $l - \nu$ symbols. Because of the interleaving, the probability that a symbol is interfered is statistically independent of the rest of the sequence and equals μ. Thus, (2-152) yields

$$P_b \leq \sum_{l=d_f}^{\infty} B(l) \sum_{\nu=0}^{l} \binom{l}{\nu} \mu^{\nu} (1-\mu)^{l-\nu} P_2(l/\nu) \qquad (2\text{-}157)$$

Since M_0 is a Gaussian random variable, $P_2(l|\nu)$ is determined from the conditional mean and variance. A straightforward calculation gives

$$P_2(l|\nu) = Q\left(\frac{-E\left[M_0|\nu\right]}{\sqrt{var\left[M_0|\nu\right]}} \right) \qquad (2\text{-}158)$$

where $E[M_0|\nu]$ is the conditional mean and $var[M_0|\nu]$ is the conditional variance. When an interference pulse occurs, $N_{0i} = N_0 + I_0$; otherwise, $N_{0i} = N_0$. Reordering the symbols for calculative simplicity and observing that $x_{2i} = -x_{1i}, x_{1i}^2 = \mathcal{E}_s$, and $E[y_i] = x_{1i}$, we obtain

$$
\begin{aligned}
E[M_0|\nu] &= \sum_{i=1}^{\nu} \frac{(x_{2i} - x_{1i}) E\left[y_i\right]}{N_0 + I_0/\mu} + \sum_{i=\nu+1}^{l} \frac{(x_{2i} - x_{1i}) E\left[y_i\right]}{N_0} \\
&= \sum_{i=1}^{\nu} \frac{-2\mathcal{E}_s}{N_0 + I_0/\mu} + \sum_{i=\nu+1}^{l} \frac{-2\mathcal{E}_s}{N_0} \\
&= -2\mathcal{E}_s \left[\frac{\nu}{N_0 + I_0/\mu} + \frac{l-\nu}{N_0} \right] \qquad (2\text{-}159)
\end{aligned}
$$

Using the statistical independence of the samples and observing that $var[y_i] = N_{0i}/2$, we find similarly that

$$var\,[M_0|\nu] = 2\mathcal{E}_s \left[\frac{\nu}{N_0 + I_0/\mu} + \frac{l - \nu}{N_0} \right] \tag{2-160}$$

Substituting (2-159) and (2-160) into (2-158), we obtain

$$P_2[l|\nu] = Q\left\{ \sqrt{\frac{2\mathcal{E}_s}{N_0}} \left[l - \nu \left(1 + \frac{\mu N_0}{I_0} \right)^{-1} \right]^{1/2} \right\} \tag{2-161}$$

The substitution of this equation into (2-157) gives the upper bound on P_b for the maximum-likelihood metric.

The upper bound on P_b versus $\mathcal{E}_b/I_0$ for worst-case pulsed interference, $\mathcal{E}_b/N_0 = 20$ dB, and several binary convolutional codes is shown in Figure 2.22. Although the worst value of μ varies with $\mathcal{E}_b/I_0$, it is found that worst-case pulsed interference causes very little degradation relative to continuous interference. When $K = 9$ and $r = 1/2$, the *maximum-likelihood metric* provides a performance that is more than 4 dB superior at $P_b = 10^{-5}$ to that provided by hard-decision decoding; when $K = 9$ and $r = 1/3$, the advantage is approximately 2.5 dB. However, the implementation of the maximum-likelihood metric entails knowledge of not only the presence of interference, but also its density level. Estimates of the N_{0i} might be based on power measurements in adjacent frequency bands only if the interference spectral density is fairly uniform over the desired-signal and adjacent bands. Any measurement of the power within the desired-signal band is contaminated by the presence of the desired signal, the average power of which is usually unknown *a priori* because of the fading. Since iterative estimation of the N_{0i} and decoding is costly in terms of system latency and complexity, we examine another approach.

Consider an *automatic gain control* (AGC) device that measures the average power at the demodulator output before sampling and then weights the sampled demodulator output y_i in proportion to the inverse of the measured power to form the *AGC metric*. The average power during channel-symbol i is $N_{0i}B + \mathcal{E}_s/T_s$, where B is the equivalent bandwidth of the demodulator and T_s is the channel-symbol duration. If the power measurement is perfect and $BT_s \approx 1$, then the AGC metric is

$$U(j) = \sum_{i=1}^{L} \frac{x_{ji}y_i}{N_{0i} + \mathcal{E}_s} \tag{2-162}$$

which is a Gaussian random variable. This metric and (2-158) yield

$$P_2(l/\nu) = Q\left\{ \sqrt{\frac{2\mathcal{E}_s}{N_0}} \frac{l\,(N_0 + \mathcal{E}s + I_0/\mu) - \nu I_0/\mu}{\left[l\,(N_0 + \mathcal{E}_s + I_0/\mu)^2 - \nu\,(N_0 + I_0/\mu - \mathcal{E}_s^2/N_0)\,I_0/\mu \right]^{1/2}} \right\} \tag{2-163}$$

This equation and (2-157) give the upper bound on P_b for the AGC metric.

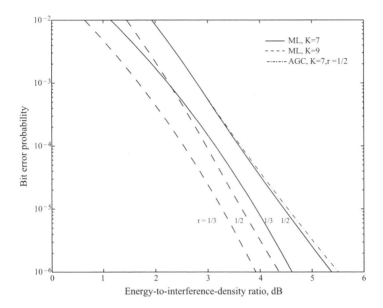

Figure 2.22: Worst-case performance against pulsed interference for convolutional codes of constraint length K, rate r, $\mathcal{E}_b/N_0 = 20$ dB and maximum-likelihood (ML) and AGC metrics.

The upper bound on P_b versus $\mathcal{E}_b/I_0$ for worst-case pulsed interference, the AGC metric, the rate-1/2 binary convolutional code with $K = 7$, and $\mathcal{E}_b/N_0 = 20$ dB is plotted in Figure 2.22. The figure indicates that the potential performance of the AGC metric is nearly as good as that of the maximum-likelihood metric.

The measurement of $N_{0i} BT_s + \mathcal{E}_s$ may be performed by a *radiometer*, which is a device that measures the energy at its input. An ideal radiometer (Chapter 7) provides an unbiased estimate of the energy received during a symbol interval. The radiometer outputs are accurate estimates only if the standard deviation of the output is much less than its expected value. This criterion and theoretical results for $BT_s = 1$ indicate that the energy measurements over a symbol interval will be unreliable if $\mathcal{E}_s/N_{0i} \leq 10$ during interference pulses. Thus, the potential performance of the AGC metric is expected to be significantly degraded in practice unless each interference pulse extends over many channel symbols and its energy is measured over the corresponding interval.

The maximum-likelihood metric for continuous interference (N_{0i} is constant for all i) is the *white-noise metric*:

$$U(j) = \sum_{i=1}^{L} x_{ji} y_i \qquad (2\text{-}164)$$

which is much simpler to implement than the AGC metric. For the white-noise

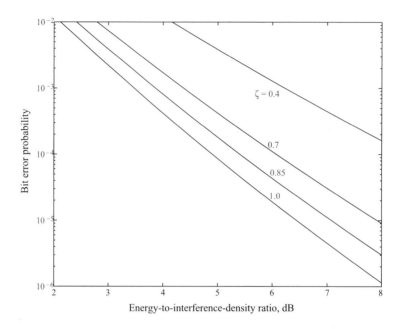

Figure 2.23: Performance against pulsed interference for convolutional code with white-noise metric, $K = 7$, $r = 1/2$, and $\mathcal{E}_b/N_0 = 20$ dB.

metric, calculations similar to the preceding ones yield

$$P_2(l|\nu) = Q\left\{\sqrt{\frac{2\mathcal{E}_s}{N_0}}\, l\left(1 + \nu\frac{I_0}{\mu N_0}\right)^{-1/2}\right\} \qquad (2\text{-}165)$$

This equation and (2-157) give the upper bound on P_b for the white-noise metric. Figure 2.23 illustrates the upper bound on P_b versus $\mathcal{E}_b/I_0$ for $K = 7$, $r = 1/2, \mathcal{E}_b/N_0 = 20$ dB, and several values of $\zeta = \mu/\mu_0$. The figure demonstrates the vulnerability of soft-decision decoding with the white-noise metric to short high-power pulses if interference power is conserved. The high values of P_b for $\zeta < 1$ are due to the domination of the metric by a few degraded symbol metrics.

Consider a coherent PSK demodulator that erases its output and, hence, a received symbol whenever an interference pulse occurs. The presence of the pulse might be detected by examining a sequence of the demodulator outputs and determining which ones have inordinately large magnitudes compared to the others. Alternatively, the demodulator might decide that a pulse has occurred if an output has a magnitude that exceeds a known upper bound for the desired signal. Consider an ideal demodulator that unerringly detects the pulses and erases the corresponding received symbols. Following the deinterleaving of the demodulated symbols, the decoder processes symbols that have a probability of being erased equal to μ. The unerased symbols are decoded by using the

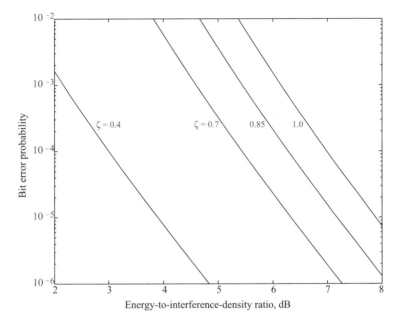

Figure 2.24: Performance against pulsed interference for convolutional code with erasures, $K = 7, r = 1/2$, and $\mathcal{E}_b/N_0 = 20$ dB.

white-noise metric. The erasing of ν symbols causes two sequences that differ in l symbols to be compared on the basis of $l - \nu$ symbols where $0 \le \nu \le l$. As a result,

$$P_2(l|\nu) = Q\left[\sqrt{\frac{2\mathcal{E}_s}{N_0}(l - \nu)}\right] \qquad (2\text{-}166)$$

The substitution of this equation into (2-157) give the upper bound on P_b for errors-and-erasures decoding.

The upper bound on P_b is illustrated in Figure 2.24 for $K = 7, r = 1/2, \mathcal{E}_b/N_0$ 20 dB, and several values of $\zeta = \mu/\mu_0$. In this example, erasures provide no advantage over the white-noise metric in reducing the required $\mathcal{E}_b/I_0$ for $P_b = 10^{-5}$ if $\zeta > 0.85$, but are increasingly useful as ζ decreases. Consider an ideal demodulator that activates erasures only when μ is small enough that the erasures are more effective than the white-noise metric. When this condition does not occur, the white-noise metric is used. The upper bound on P_b for this *ideal erasure decoding*, worst-case pulsed interference, $\mathcal{E}_b/N_0 = 20$ dB, and several binary convolutional codes is illustrated in Figure 2.25. The required $\mathcal{E}_b/I_0$ at $P_b = 10^{-5}$ is roughly 2 dB less than for worst-case hard-decision decoding. However, a practical demodulator will sometimes erroneously make erasures or fail to erase, and its performance advantage may be much more modest.

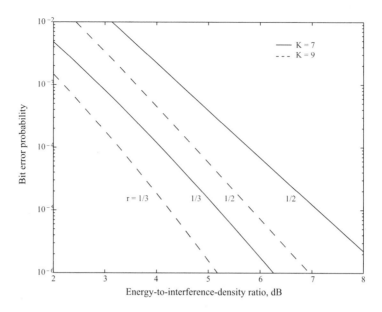

Figure 2.25: Worst-case performance against pulsed interference for convolutional codes with ideal erasure decoding, constraint length K, rate r, and $\mathcal{E}_b/N_0 = 20$ dB.

2.6   Despreading with Matched Filters

Despreading short spreading sequences with matched filters provides inherent code synchronization. The spreading waveform for a short sequence may be expressed as

$$p(t) = \sum_{i=-\infty}^{\infty} p_1(t - iT) \tag{2-167}$$

where $p_1(t)$ is one period of the spreading waveform and T is its period. If the short spreading sequence has length N, then

$$p_1(t) = \begin{cases} \sum_{i=0}^{N-1} p_i \psi\left(t - iT_c\right), & 0 \le t \le T \\ 0, & \text{otherwise} \end{cases} \tag{2-168}$$

where $p_i = \pm 1$, and $T = NT_c$.

Consider a signal $x(t)$ that is zero outside the interval $[0, T]$. A filter is said to be *matched* to this signal if the impulse response of the filter is $h(t) = x(T - t)$.

When $x(t)$ is applied to a filter matched to it, the filter output is

$$y(t) = \int_{-\infty}^{\infty} x(u)h(t-u)du = \int_{-\infty}^{\infty} x(u)x(u+T-t)du$$

$$= \int_{\max(t-T,0)}^{\min(t,T)} x(u)x(u+T-t)du \qquad (2\text{-}169)$$

The *aperiodic autocorrelation* of a deterministic signal with finite energy is defined as

$$R_x(\tau) = \int_{-\infty}^{\infty} x(u)x(u+\tau)du = \int_{-\infty}^{\infty} x(u)x(u-\tau)du \qquad (2\text{-}170)$$

Therefore, the response of a matched filter to the matched signal is

$$y(t) = R_x(t-T) \qquad (2\text{-}171)$$

If this output is sampled at $t=T$, then $y(T) = R_x(0)$, the signal energy.
  Consider a *bandpass matched filter* that is matched to

$$x(t) = \begin{cases} p_1(t)\cos\left(2\pi f_c t + \theta_1\right), & 0 \le t \le T \\ 0, & \text{otherwise} \end{cases} \qquad (2\text{-}172)$$

where $p_1(t)$ is one period of a spreading waveform and f_c is the desired carrier frequency. We evaluate the filter response to the received signal corresponding to a single data symbol:

$$s(t) = \begin{cases} 2Ap_1(t-t_0)\cos\left(2\pi f_1 t + \theta\right), & t_0 \le t \le t_0+T \\ 0, & \text{otherwise} \end{cases} \qquad (2\text{-}173)$$

where t_0 is a measure of the unknown arrival time, the polarity of A is determined by the data symbol, and f_1 is the received carrier frequency, which differs from f_c because of oscillator instabilities and the Doppler shift. The matched-filter output is

$$y_s(t) = \int_{t-T}^{t} s(u)p_1(u+T-t)\cos\left[2\pi f_c(u+T-t)+\theta_1\right]du \qquad (2\text{-}174)$$

If $f_c \gg 1/T$, then substituting (2-173) into (2-174) yields

$$y_s(t) = A\int_{\max(t-T,t_0)}^{\min(t,t_0+T)} p_1\left(u-t_0\right)p_1\left(u-t+T\right)\cos\left(2\pi f_d u + 2\pi f_c t + \theta_2\right)du \qquad (2\text{-}175)$$

where $\theta_2 = \theta-\theta_1-2\pi f_c T$ is the phase mismatch and $f_d = f_1-f_c$. If $f_d \ll 1/T$, the carrier-frequency error is inconsequential, and

$$y_s(t) \approx A_s(t)\cos\left(2\pi f_c t + \theta_2\right) \qquad (2\text{-}176)$$

where

$$A_s(t) = A\int_{\max(t-T,t_0)}^{\min(t,t_0+T)} p_1\left(u-t_0\right)p_1\left(u-t+T\right)du \qquad (2\text{-}177)$$

In the absence of noise, the matched-filter output is a sinusoidal spike with a polarity determined by A. Assuming that (2-77) is applicable, the peak magnitude, which occurs at $t = t_0 + T$, equals $|A|T$. However, if $f_d > 0.1/T$, then (2-175) is not well-approximated by (2-176), and the matched-filter output is significantly degraded.

The response of the matched filter to the interference plus noise, denoted by $N(t) = i(t) + n(t)$, may be expressed as

$$y_n(t) = \int_{t-T}^{t} N(u)p_1(u + T - t) \cos\left[2\pi f_c(u + T - t) + \theta_1\right] du$$

$$= N_1(t) \cos\left(2\pi f_c t + \theta_2\right) + N_2(t) \sin\left(2\pi f_c t + \theta_2\right) \quad (2\text{-}178)$$

where

$$N_1(t) = \int_{t-T}^{t} N(u)p_1(u + T - t) \cos\left(2\pi f_c u + \theta\right) du \quad (2\text{-}179)$$

$$N_2(t) = \int_{t-T}^{t} N(u)p_1(u + T - t) \sin\left(2\pi f_c u + \theta\right) du \quad (2\text{-}180)$$

These equations exhibit the spreading of the interference spectrum.

The envelope of the matched-filter output $y(t) = y_s(t) + y_n(t)$ is

$$E(t) = \left\{[A_s(t) + N_1(t)]^2 + N_2^2(t)\right\}^{1/2} \quad (2\text{-}181)$$

Define ϵ such that $2\pi f_c(t_0 + \epsilon) + \theta - \theta_1$ is an integer times 2π. If f_c is sufficiently large that $\epsilon << t_0 + T$, then (2-176) and (2-178) imply that if $y(t)$ is sampled at $t = t_0 + T + \epsilon$,

$$\begin{aligned} y(t_0 + T + \epsilon) &= y_s(t_0 + T + \epsilon) + y_n(t_0 + T + \epsilon) \\ &= A_s(t_0 + T + \epsilon) + N_1(t_0 + T + \epsilon) \\ &\approx AT + N_1(t_0 + T + \epsilon) \end{aligned} \quad (2\text{-}182)$$

where $A_s(t_0 + T) = AT$. If

$$|AT + N_1(t_0 + T)| >> |N_2(t_0 + T)| \quad (2\text{-}183)$$

then (2-181) implies that

$$E(t_0 + T) \approx |AT + N_1(t_0 + T)| \quad (2\text{-}184)$$

A comparison of this equation with (2-182) indicates that there is relatively little degradation in using an envelope detector after the matched filter rather than directly detecting the peak magnitude of the matched-filter output, which is much more difficult.

Figure 2.26 illustrates the basic form of a *surface-acoustic-wave* (SAW) *transversal filter*, which is a passive matched filter that essentially stores a replica of the underlying spreading sequence and waits for the received sequence to align itself with the replica. The SAW delay line consists primarily

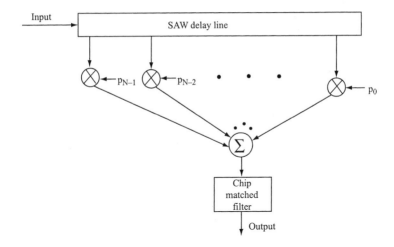

Figure 2.26: Matched filter that uses a SAW transversal filter.

of a piezoelectric substrate, which serves as the acoustic propagation medium, and interdigital transducers, which serve as the taps and the input transducer. The transversal filter is matched to one period of the spreading waveform, the propagation delay between taps is T_c, and $f_c T_c$ is an integer. The chip matched filter following the summer is matched to $\psi(t) \cos(2\pi f_c t + \theta)$. It is easily verified that the impulse response of the transversal filter is that of a filter matched to $p_1(t) \cos(2\pi f_c t + \theta)$.

A *convolver* is an *active matched filter* that produces the convolution of the received signal with a local reference [8]. When used as a direct-sequence matched filter, a convolver uses a recirculating, time-reversed replica of the spreading waveform as a reference waveform. In a *SAW elastic convolver*, which is depicted in Figure 2.27, the received signal and the reference are applied to interdigital transducers that generate acoustic waves at opposite ends of the substrate. The acoustic waves travel in opposite directions with speed v, and

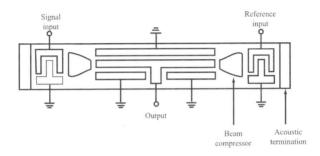

Figure 2.27: SAW elastic convolver.

the acoustic terminations suppress reflections. The signal wave is launched at position $x = 0$ and the reference wave at $x = L$. The signal wave travels to the right in the substrate and has the form

$$F(t, x) = f\left(t - \frac{x}{v}\right) \cos\left[2\pi f_c\left(t - \frac{x}{v}\right) + \theta\right] \qquad (2\text{-}185)$$

where $f(t)$ is the modulation at position $x = 0$. The reference wave travels to the left and has the form

$$G(t, x) = g\left(t + \frac{x}{v} - \frac{L}{v}\right) \cos\left[2\pi f_c\left(t + \frac{x}{v} - \frac{L}{v}\right) + \theta_1\right] \qquad (2\text{-}186)$$

where $g(t)$ is the modulation at position $x = L$. Both $f(t)$ and $g(t)$ are assumed to have bandwidths much smaller than f_c. The beam compressors, which consist of thin metallic strips, focus the acoustic energy to increase the convolver's efficiency. When the acoustic waves overlap beneath the central electrode, a nonlinear piezoelectric effect causes a surface charge distribution that is spatially integrated by the electrode. The primary component of the convolver output is proportional to

$$y(t) = \int_0^L [F(t, x) + G(t, x)]^2 dx \qquad (2\text{-}187)$$

Substituting (2-185) and (2-186) into (2-187) and using trigonometry, we find that $y(t)$ is the sum of a number of terms, some of which are negligible if $f_c L/v \gg 1$. Others are slowly varying and are easily blocked by a filter. The most useful component of the convolver output is

$$y_s(t) = \left[\int_0^L f\left(t - \frac{x}{v}\right) g\left(t + \frac{x}{v} - \frac{L}{v}\right) dx\right] \cos\left(4\pi f_c t + \theta_2\right) \qquad (2\text{-}188)$$

where $\theta_2 = \theta + \theta_1 - 2\pi f_c L/v$. Changing variables, we find that the amplitude of the output is

$$A_s(t) = \int_{t-L/v}^t f(y)g(2t - y - L/v)dy \qquad (2\text{-}189)$$

where the factor $2t$ results from the counterpropagation of the two acoustic waves.

Suppose that an acquisition pulse is a single period of the spreading waveform. Then $f(t) = A p_1(t - t_0)$ and $g(t) = p(T - t)$, where t_0 is the uncertainty in the arrival time of an acquisition pulse relative to the launching of the reference signal at $x = L$. The periodicity of $g(t)$ allows the time origin to be selected so that $0 \leq t_0 \leq T$. Equations (2-189) and (2-167) and a change of variables yield

$$A_s(t) = A \sum_{i=-\infty}^{\infty} \int_{t-t_0-L/v}^{t-t_0} p_1(y)p_1(y + iT + t_0 - 2t + L/v) dy \qquad (2\text{-}190)$$

Since $p_1(t) = 0$ unless $0 \le t < T$, $A_s(t) = 0$ unless $t_0 < t < t_0 + T + L/v$. For every positive integer k, let

$$\tau_k = \frac{kT + t_0 + L/v}{2}, \qquad k = 1, 2, \ldots \tag{2-191}$$

Only one term in (2-190) can be nonzero when $t = \tau_k$, and

$$A_s(\tau_k) = A \int_{\tau_k - t_0 - L/v}^{\tau_k - t_0} p_1^2(y) dy \tag{2-192}$$

The maximum possible magnitude of $A_s(\tau_k)$ is produced if $\tau_k - t_0 \ge T$ and $\tau_k - t_0 - L/v \le 0$; that is, if

$$t_0 + T \le \tau_k \le t_0 + \frac{L}{v} \tag{2-193}$$

Since (2-191) indicates that $\tau_{k+1} - \tau_k = T/2$, there is some τ_k that satisfies (2-193) if

$$L \ge \frac{3}{2} vT \tag{2-194}$$

Thus, if L is large enough, then there is some k such that $A_s(\tau_k) = AT$, and the envelope of the convolver output at $t = \tau_k$ has the maximum possible magnitude $|A|T$. If $L = 3vT/2$ and $t_0 \ne T/2$, only one peak value occurs in response to the single received pulse.

As an example, let $t_0 = 0$, $L/v = 6T_c$, and $T = 4T_c$. The chips propagating in the convolver for three separate time instants $t = 4T_c$, $5T_c$, and $6T_c$ are illustrated in Figure 2.28. The top diagrams refer to the counterpropagating periodic reference signal, whereas the bottom diagrams refer to the single received pulse of four chips. The chips are numbered consecutively. The received pulse is completely contained within the convolver during $4T_c \le t \le 6T_c$. The maximum magnitude of the output occurs at time $t = 5T_c$, which is the instant of perfect alignment of the reference signal and the received chips.

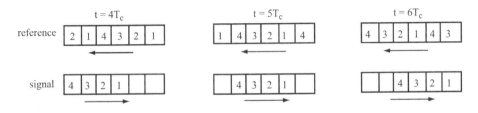

Figure 2.28: Chip configurations within convolver at time instants $t = 4T_c$, $5T_c$, and $6T_c$ when $t_0 = 0$, $L/v = T_c$, and $T = 4T_c$.

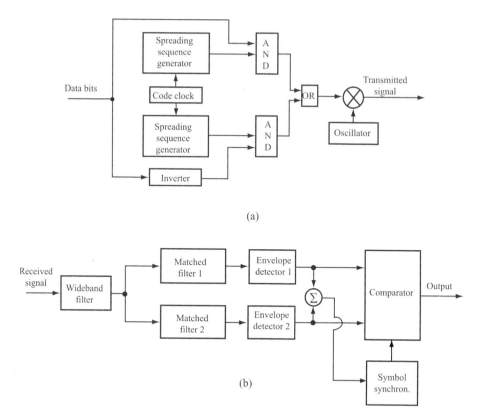

Figure 2.29: Direct-sequence system with binary code-shift keying: (a) transmitter and (b) receiver.

Noncoherent Systems

In a noncoherent direct-sequence system with binary *code-shift keying* (CSK), one of two orthogonal spreading sequences is transmitted, as shown in Figure 2.29(a). One sequence represents the symbol 1, and the other represents the symbol 0. The receiver uses two matched filters, each matched to a different sequence and followed by an envelope detector, as shown in Figure 2.29(b). In the absence of noise and interference, each sequence causes only one envelope detector to produce a significant output. The data is recovered by comparing the two detector outputs every symbol period.

Since each of the two orthogonal sequences has a period equal to the symbol duration, symbol or bit synchronization is identical to code synchronization. The *symbol synchronizer*, which provides timing pulses to the comparator or decision device, must lock onto the autocorrelation spikes appearing in the envelope-detector outputs. Ideally, these spikes have a triangular shape. The symbol synchronizer must be impervious to the autocorrelation sidelobe peaks and any cross-correlation peaks. A simple implementation with a single threshold detector would result in an unacceptable number of false alarms, premature detections, or missed detections when the received signal amplitude is unknown and has a wide dynamic range. Limiting or automatic gain control only exacerbates the problem when the signal power level is below that of the interference plus noise. More than one threshold detector with precedence given to the highest threshold crossed will improve the accuracy of the decision timing or sampling instants produced by the symbol synchronizer [9]. Another approach is to use peak detection based on a differentiator and a zero-crossing detector. Finally, a phase-locked or feedback loop of some type could be used in the symbol synchronizer. A preamble may be transmitted to initiate accurate synchronization so that symbols are not incorrectly detected while synchronization is being established.

Consider the detection of a symbol represented by (2-173), where $p_1(t)$ is the CSK waveform to which filter 1 is matched. Assuming perfect symbol synchronization, the channel symbol is received during the interval $0 \leq t \leq T_s$. From (2-176) to (2-181) with $T = T_s$ and $t_0 = 0$, we find that the output of envelope detector 1 at $t = T_s$ is

$$R_1 = \left(Z_1^2 + Z_2^2\right)^{1/2} \tag{2-195}$$

where

$$Z_1 = AT_s + \int_0^{T_s} N(u)p_1(u) \cos\left(2\pi f_c u + \theta\right) du \tag{2-196}$$

$$Z_2 = \int_0^{T_s} N(u)p_1(u) \sin\left(2\pi f_c u + \theta\right) du \tag{2-197}$$

Similarly, if filter 2 is matched to sequence $p_2(t)$, then the output of envelope detector 2 at $t = T_s$ is

$$R_2 = \left(Z_3^2 + Z_4^2\right)^{1/2} \tag{2-198}$$

where

$$Z_3 = \int_0^{T_s} N(u)p_2(u) \cos\left(2\pi f_c u + \theta\right) du \tag{2-199}$$

$$Z_4 = \int_0^{T_s} N(u)p_2(u) \sin\left(2\pi f_c u + \theta\right) du \tag{2-200}$$

and the response to the transmitted symbol at $t = T_s$ is zero because of the orthogonality of the sequences.

Suppose that the interference plus noise $N(t)$ is modeled as zero-mean, Gaussian interference, and the spreading sequences are modeled as deterministic and orthogonal. Then $E[Z_1] = AT_s$ and $E[Z_i] = 0, i = 2, 3, 4$. If $N(t)$ is assumed to be wideband enough that its autocorrelation is approximated by (2-87), then straightforward calculations using $f_c T_s \gg 1$ and the orthogonality of $p_1(t)$ and $p_2(t)$ indicate that Z_1, Z_2, Z_3, and Z_4 are all uncorrelated with each other. The jointly Gaussian character of the random variables then implies that they are statistically independent of each other, and hence R_1 and R_2 are independent. Analogous results can be obtained when the transmitted symbol is represented by CSK waveform $p_2(t)$. A straightforward derivation similar to the classical one for orthogonal signals then yields the symbol error probability

$$P_s = \frac{1}{2} \exp\left(-\frac{\mathcal{E}_s}{2N_{0e}}\right) \qquad (2\text{-}201)$$

where N_{0e} is given by (2-121). A comparison of (2-201) with (2-118) indicates that the performance of the direct-sequence system with noncoherent binary CSK in the presence of wideband Gaussian interference is approximately 4 dB worse than that of a direct-sequence system with coherent binary PSK. This difference arises because binary CSK uses orthogonal rather than antipodal signaling. A much more complicated coherent version of Figure 2.29 would only recover roughly 1 dB of the disparity.

A direct-sequence system with q-ary CSK encodes each group of m binary symbols as one of $q = 2^m$ sequences chosen to have negligible cross correlations. Suppose that bandwidth constraints limit the chip rate of a binary CSK system to G chips per data bit. For a fixed data-bit rate, the q-ary CSK system produces mG chips to represent each group of m bits, which may be regarded as a single q-ary symbol. Thus, the processing gain relative to a data symbol is mG, which indicates an enhanced ability to suppress interference. In the presence of wideband Gaussian interference, the performance improvement of quaternary CSK is more than 2 dB relative to binary CSK, but four filters matched to four double-length sequences are required. When the chip rate is fixed, q-ary CSK provides a means of increasing the data-bit or code-symbol rate without sacrificing the processing gain.

Elimination of the lower branch in Figure 2.29(b) leaves a system that uses a single CSK sequence and a minimum amount of hardware. The symbol 1 is signified by the transmission of the sequence, whereas the symbol 0 is signified by the absence of a transmission. Decisions are made after comparing the envelope-detector output with a threshold. One problem with this system is that the optimal threshold is a function of the amplitude of the received signal, which must somehow be estimated. Another problem is the degraded performance of the symbol synchronizer when many consecutive zeros are transmitted. Thus, a system with binary CSK is much more practical.

A direct-sequence system with DPSK signifies the symbol 1 by the transmission of a spreading sequence without any change in the carrier phase; the symbol 0 is signified by the transmission of the same sequence after a phase shift of π radians in the carrier phase or multiplication of the signal by -1. A matched filter despreads the received direct-sequence signal, as illustrated in Figure 2.30. The filter output is applied to a standard DPSK demodulator that makes symbol decisions. An analysis of this system in the presence of wideband Gaussian interference indicates that it is more than 2 dB superior to the system with binary CSK. However, the system with DPSK is more sensitive to Doppler shifts and is more than 1 dB inferior to a system with coherent binary PSK.

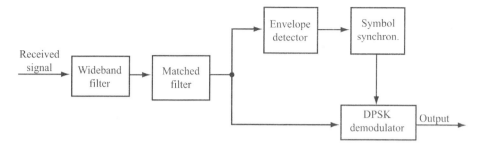

Figure 2.30: Receiver for direct-sequence system with differential phase-shift keying.

Multipath-Resistant Coherent System

Carrier synchronization is essential for the coherent demodulation of a direct-sequence signal. Prior to despreading, the signal-to-interference-plus-noise ratio (SINR) may be too low for the received signal to serve as the input to a phase-locked loop that produces a phase-coherent carrier. Although the despread matched-filter output has a large SINR near the autocorrelation peak, the average SINR may be insufficient for a phase-locked loop. An alternative approach is to use a recirculation loop to produce a synchronized carrier during the main lobe of the matched-filter output.

A *recirculation loop*, is designed to reinforce a periodic input signal by positive feedback. As illustrated in Figure 2.31, the feedback elements are an attenuator of gain K and a delay line with delay $\hat{T}_s$ approximating a symbol duration T_s. The basic concept behind this architecture is that successive signal pulses are coherently added while the interference and noise are noncoherently added, thereby producing an output pulse with an improved SINR. The periodic input

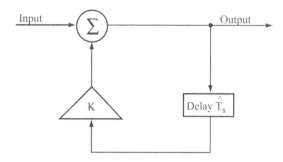

Figure 2.31: Recirculation loop.

consists of N symbol pulses such that

$$s_0(t) = \sum_{i=0}^{N} g(t - iT_s) \tag{2-202}$$

where $g(t) = 0$ for $t < 0$ or $t \geq T_s$. The figure indicates that the loop output is

$$s_1(t) = s_0(t) + K s_1\left(t - \hat{T}_s\right) \tag{2-203}$$

Substitution of this equation into itself yields

$$s_1(t) = s_0(t) + K s_0\left(t - \hat{T}_s\right) + K^2 s_1\left(t - 2\hat{T}_s\right) \tag{2-204}$$

Repeating this substitution process n times leads to

$$s_1(t) = \sum_{m=0}^{n} K^m s_0\left(t - m\hat{T}_s\right) + K^{n+1} s_1\left[t - (n+1)\hat{T}_s\right] \tag{2-205}$$

which indicates that $s_1(t)$ increases with n if $K \geq 1$ and enough input pulses are available. To prevent an eventual loop malfunction, $K < 1$ is a design requirement that is assumed henceforth.

During the interval $[n\hat{T}_s, (n+1)\hat{T}_s], n$ or fewer recirculations of the symbols have occurred. Since $s_1(t) = 0$ for $t < 0$, the substitution of (2-202) into (2-205) yields

$$s_1(t) = \sum_{m=0}^{n} \sum_{i=0}^{N} K^m g\left(t - m\hat{T}_s - iT_s\right), \qquad n\hat{T}_s \leq t < (n+1)\hat{T}_s \tag{2-206}$$

This equation indicates that if $\hat{T}_s$ is not exactly equal to T_s, then the pulses do not add coherently, and may combine destructively. However, since $K < 1$, the effect of a particular pulse decreases as m increases and will eventually be negligible. The delay $\hat{T}_s$ is designed to match T_s. Suppose that the design error is small enough that

$$N\left|\hat{T}_s - T_s\right| << \hat{T}_s \tag{2-207}$$

Since $t - m\hat{T}_s - iT_s = t - (m+i)T_s - m(\hat{T}_s - T_s)$ and $g(t)$ is time-limited, (2-207) and $n \leq N$ imply that only the term in (2-206) with $i = n - m$ contributes appreciably to the output. Therefore,

$$s_1(t) \approx \sum_{m=0}^{n} K^m g\left[t - nT_s - m\left(\hat{T}_s - T_s\right)\right], \qquad nT_s \leq t < (n+1)T_s \tag{2-208}$$

Let ν denote a positive integer such that K^m is negligible if $m > \nu$. Consider an input pulse of the form

$$g(t) = A(t) \cos 2\pi f_c t, \qquad 0 \leq t < \min\left(T_s, \hat{T}_s\right) \tag{2-209}$$

which implies that each of the N pulses in (2-202) has the same initial phase. Assume that the amplitude $A(t)$ varies slowly enough that

$$A\left[t - nT_s - m\left(\hat{T}_s - T_s\right)\right] \approx A\left(t - nT_s\right), \qquad 0 \le m \le \nu \qquad (2\text{-}210)$$

and that the design error is small enough that

$$\nu f_c \left|\hat{T}_s - T_s\right| << 1 \qquad (2\text{-}211)$$

Then (2-208) to (2-211) yield

$$s_1(t) \approx g\left(t - nT_s\right) \sum_{m=0}^{n} K^m$$

$$= g\left(t - nT_s\right) \left(\frac{1 - K^{n+1}}{1 - K}\right), \qquad nT_s \le t \le (n+1)T_s \qquad (2\text{-}212)$$

If S is the average power in an input pulse, then (2-212) indicates that the average power in an output pulse during the interval $nT_s \le t < (n+1)T_s$ is approximately

$$S_n = \left(\frac{1 - K^{n+1}}{1 - K}\right)^2 S, \qquad K < 1 \qquad (2\text{-}213)$$

If $\hat{T}_s$ is large enough that the recirculated noise is uncorrelated with the input noise, which has average power σ^2, then the output noise power after n recirculations is

$$\sigma_n^2 = \sigma^2 \sum_{m=0}^{n} \left(K^2\right)^m$$

$$= \sigma^2 \left(\frac{1 - K^{2n+2}}{1 - K^2}\right), \qquad K < 1 \qquad (2\text{-}214)$$

The improvement in the SNR due to the presence of the recirculation loop is

$$I\left(n, K\right) = \frac{S_n/\sigma_n^2}{S/\sigma^2} = \frac{\left(1 - K^{n+1}\right)\left(1 + K\right)}{\left(1 + K^{n+1}\right)\left(1 - K\right)}$$

$$\le \frac{1 + K}{1 - K}, \qquad K < 1 \qquad (2\text{-}215)$$

Since it was assumed that K^m is negligibly small when $m > \nu$, the maximum improvement is nearly attained when $n \ge \nu$. However, the upper bound on ν for the validity of (2-211) decreases as the loop phase error $2\pi f_c |\hat{T}_s - T_s|$ increases. Thus, K must be decreased as the phase error increases. The phase error of a practical SAW recirculation loop may be caused by a temperature fluctuation, a Doppler shift, oscillator instability, or an imprecise delay-line length. Various other loop imperfections limit the achievable value of K and, hence, the improvement that the loop can provide [10].

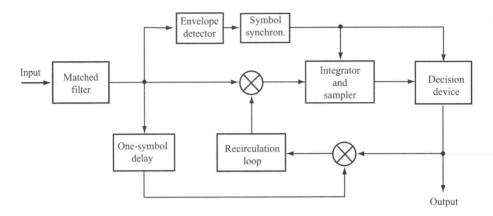

Figure 2.32: Coherent decision-directed demodulator.

Figure 2.32 illustrates a *coherent decision-directed demodulator* for a direct-sequence signal with binary PSK and the same carrier phase at the beginning of each symbol. The bandpass matched filter removes the spreading waveform and produces compressed sinusoidal pulses, as indicated by (2-176) and (2-177) when A is bipolar. A compressed pulse due to a direct-path signal may be followed by one or more compressed pulses due to multipath signals, as illustrated conceptually in Figure 2.33(a) for pulses corresponding to the transmitted symbols 101. Each compressed pulse is delayed by one symbol and then mixed with the demodulator's output symbol. If this symbol is correct, it coincides with the same data symbol that is modulated onto the compressed pulse. Consequently, the mixer removes the data modulation and produces a phase-coherent reference pulse that is independent of the data symbol, as illustrated in Figure 2.33(b), where the middle pulses are inverted in phase relative to the corresponding pulses in Figure 2.33(a). The reference pulses are amplified by a recirculation loop. The loop output and the matched-filter output are applied to a mixer that produces the baseband integrator input illustrated in Figure 2.33(c). The length of the integration interval is equal to a symbol duration. The integrator output is sampled and applied to a decision device that produces the data output. Since multipath components are coherently integrated, the demodulator provides an improved performance in a fading environment.

Even if the desired-signal multipath components are absent, the coherent decision-directed receiver potentially suppresses interference approximately as much as the correlator of Figure 2.14. The decision-directed receiver is much simpler to implement because code acquisition and tracking systems are unnecessary, but it requires a short spreading sequence and an accurate recirculation loop. More efficient exploitation of multipath components is possible with rake combining (Chapter 5).

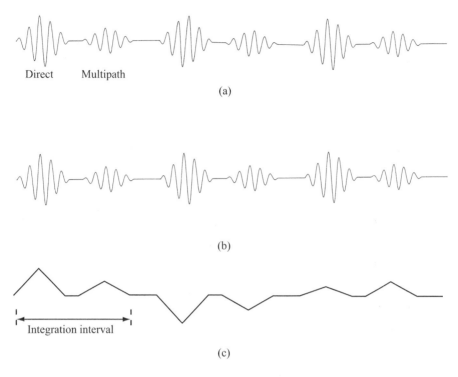

Figure 2.33: Conceptual waveforms of demodulator: (a) matched-filter output, (b) recirculation loop input or output, and (c) baseband integrator input.

2.7 Rejection of Narrowband Interference

Narrowband interference presents a crucial problem for *spread-spectrum overlay systems*, which are systems that have been assigned a spectral band already occupied by narrowband communication systems. Jamming against tactical spread-spectrum communications is another instance of narrowband interference that may exceed the natural resistance of a practical spread-spectrum system, which has a limited processing gain. There are a wide variety of techniques that supplement the inherent ability of a direct-sequence system to reject narrowband interference [11], [12]. All of the techniques directly or indirectly exploit the spectral disparity between the narrowband interference and the wideband direct-sequence signal. The most useful methods can be classified as time-domain adaptive filtering, transform-domain processing, nonlinear filtering, or code-aided techniques. The general form of a receiver that rejects narrowband interference and demodulates a direct-sequence signal with binary PSK is shown in Figure 2.34. The processor, which follows the chip-rate sampling of the baseband signal, implements one of the rejection methods.

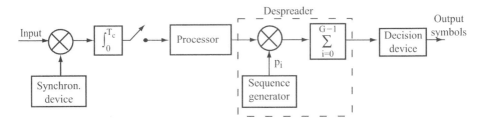

Figure 2.34: Direct-sequence receiver with processor for rejecting narrowband interference.

Time-Domain Adaptive Filtering

A time-domain *adaptive filter* [13] for interference suppression processes the baseband sample values of a received signal to adaptively estimate the interference. This estimate is subtracted from the sample values, thereby canceling the interference. The adaptive filter is primarily a predictive system that exploits the inherent predictability of a narrowband signal to form an accurate replica of it for the subtraction. Since the wideband desired signal is largely unpredictable, it does not significantly impede the prediction of a narrowband signal. When adaptive filtering is used, the processor in Figure 2.34 has the form of Figure 2.35(a). The adaptive filter may be a one-sided or two-sided transversal filter.

The *two-sided adaptive transversal filter* multiplies each tap output by a weight except for the central tap output, as diagrammed in Figure 2.35(b). This filter is an *interpolator* in that it uses *both* past and future samples to estimate the value to be subtracted. The two-sided filter provides a better performance than the one-sided filter, which is a *predictor*. The adaptive algorithm of the weight-control mechanism is designed to adjust the weights so that the power in the filter output is minimized. The direct-sequence components of the tap outputs, which are delayed by integer multiples of a chip duration, are largely uncorrelated with each other, but the narrowband interference components are strongly correlated. As a result, the adaptive algorithm causes the interference cancellation in the filter output, but the direct-sequence signal is largely unaffected.

An adaptive filter with $2N + 1$ taps and $2N$ weights, as shown in Figure 2.35(b), has input vector at iteration k given by

$$\mathbf{x}(k) = [x_1(k) \ x_2(k) \dots x_{2N}(k)]^T \qquad (2\text{-}216)$$

and weight vector

$$\mathbf{W}(k) = [W_{-N}(k) \ W_{-N+1}(k) \dots W_{-1}(k) \ W_1(k) \dots W_N(k)]^T \qquad (2\text{-}217)$$

where T denotes the transpose and the central tap output, which is denoted by d, has been excluded from $\mathbf{x}$. Since coherent demodulation produces real-

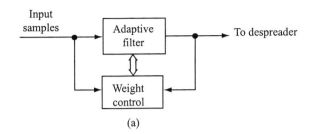

(a)

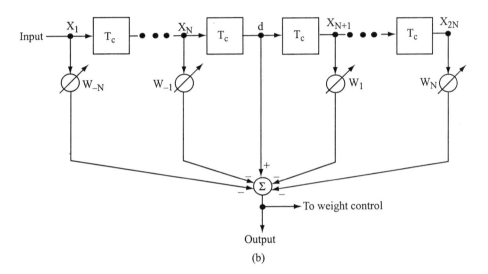

Output

(b)

Figure 2.35: (a) Processor using adaptive filter and (b) two-sided adaptive transversal filter.

valued inputs to the adaptive filter, $\mathbf{x}(k)$ and $\mathbf{W}(k)$ are assumed to have real-valued components. The symmetric *correlation matrix* of $\mathbf{x}$ is defined as $\mathbf{R}_{xx} = E[\mathbf{xx}^T]$. The *cross-correlation vector* is defined as $\mathbf{R}_{xd} = E[\mathbf{x}d]$. According to the Wiener-Hopf equation (Appendix B), the optimal weight vector is

$$\mathbf{W}_0 = \mathbf{R}_{xx}^{-1}\mathbf{R}_{xd} \qquad (2\text{-}218)$$

The least-mean-square (LMS) algorithm (Appendix B) computes the weight vector at iteration k as

$$\mathbf{W}(k) = \mathbf{W}(k-1) + \mu\epsilon_k\mathbf{x}(k) \qquad (2\text{-}119)$$

where $\epsilon_k = d - y_k$ is the estimation error, $y_k = \mathbf{W}^T(k)\mathbf{x}(k)$ is the filter output, and μ is the adaptation constant, which controls the rate of convergence of the algorithm. The output of the adaptive filter is ϵ_k, which is applied to the despreader. Under certain conditions, the mean weight vector converges to $\mathbf{W}_0$ after a number of iterations of the adaptive algorithm. If it is assumed

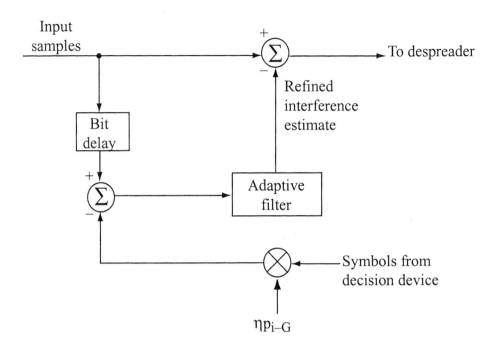

Figure 2.36: Processor with decision-directed adaptive filter.

that $\mathbf{W} = \mathbf{W}_0$, then a straightforward analysis indicates that the adaptive transversal filter provides a substantial suppression of narrowband interference [11]. Although the interference suppression increases with the number of taps, it is always incomplete if the interference has a nonzero bandwidth because a finite-impulse-response filter can only place a finite number of zeros in the frequency domain.

The adaptive transversal filter is inhibited by the presence of direct-sequence components in the filter input vector $\mathbf{x}(k)$. These components can be suppressed by using decision-directed feedback, as shown in Figure 2.36. Previously detected symbols remodulate the spreading sequence delayed by G chips (long sequence) or one period of the spreading sequence (short sequence). After an amplitude compensation by a factor η, the resulting sequence provides estimates of the direct-sequence components of previous input samples. A subtraction then provides estimated sample values of the interference plus noise that are largely free of direct-sequence contamination. These samples are then applied to an adaptive transversal filter that has the form of Figure 2.35 except that it has no central tap. The transversal filter output consists of refined interference estimates that are subtracted from the input samples to produce samples that have relatively small interference components. An erroneous symbol from the decision device causes an enhanced direct-sequence component in samples applied to the transversal filter, and error propagation is possible. However, for moderate values of the signal-to-interference ratio at the input, the performance is not degraded significantly.

Adaptive filtering is only effective after the convergence of the adaptive algorithm, which may not be able to track time-varying interference. In contrast, transform-domain processing suppresses interference almost instantaneously.

Transform-Domain Processing

The input of a *transform-domain processor* could be a continuous-time received signal that feeds a real-time Fourier transformer implemented as a chirp transform processor [1]. In a more versatile implementation, which is depicted in Figure 2.37 and assumed henceforth, the input consists of the output samples of a chip-matched filter. Blocks of these samples feed a discrete-time Fourier or wavelet transformer. The transform is selected so that the transform-domain forms of the desired signal and interference are easily distinguished. Ideally, the transform produces interference components that are confined to a few transform bins while the desired-signal components have nearly the same magnitude in all the transform bins. A simple exciser can then suppress the interference with little impact on the desired signal by setting to zero the components in bins containing the interference. The decision as to which bins contain interference can be based on the comparison of each component to a threshold. After the excision operation, the desired signal is largely restored by the inverse transformer.

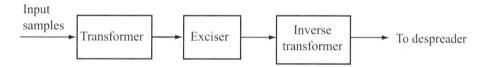

Figure 2.37: Transform-domain processor.

Much better performance against stationary narrowband interference may be obtained by using a transform-domain adaptive filter as the exciser [14]. This filter adjusts a single nonbinary weight at each transform-bin output. The adaptive algorithm is designed to minimize the difference between the weighted transform and a desired signal that is the transform of the spreading sequence used by the input block of the processor. If the direct-sequence signal uses the same short spreading sequence for each data symbol and each processor input block includes the chips for a single data symbol, then the desired-signal transform may be stored in a read-only memory. However, if a long spreading sequence is used, then the desired-signal transform must be continuously produced from the output of the receiver's code generator. The main disadvantage of the adaptive filter is that its convergence rate may be insufficient to track rapidly time-varying interference.

A transform that operates on disjoint blocks of N input samples may be defined in terms of N orthonormal, N-component basis vectors:

$$\boldsymbol{\phi}_i = [\phi_{i1}\ \phi_{i2}\ldots\phi_{iN}]^T, \qquad i = 1, 2, \ldots, N \tag{2-220}$$

which span a linear vector space of dimension N. Since the components may be complex numbers, the orthonormality implies that

$$\boldsymbol{\phi}_i^H \boldsymbol{\phi}_k = \begin{cases} 0, & i \neq k \\ 1, & i = k \end{cases} \tag{2-221}$$

where H denotes the complex conjugate of the transpose. The input block

$$\mathbf{x} = [x_1\ x_2\ldots x_N]^T \tag{2-222}$$

may be expressed in terms of the basis as

$$\mathbf{x} = \sum_{i=1}^{N} c_i \boldsymbol{\phi}_i \tag{2-223}$$

where

$$c_i = \boldsymbol{\phi}_i^H \mathbf{x}, \quad i = 1, 2, \ldots N \tag{2-224}$$

If the discrete Fourier transform is used, then $\phi_{ik} = \exp(j2\pi ik/N)$, where $j = \sqrt{-1}$.

The transformer extracts the vector

$$\mathbf{c} = [c_1\ c_2\ \ldots\ c_N]^T \tag{2-225}$$

by computing

$$\mathbf{c} = \mathbf{B}^H \mathbf{x} \tag{2-226}$$

where $\mathbf{B}$ is the unitary matrix of basis vectors:

$$\mathbf{B} = [\boldsymbol{\phi}_1\ \boldsymbol{\phi}_2\ \ldots\ \boldsymbol{\phi}_N] \tag{2-227}$$

The exciser weights each component of the transform $\mathbf{c}$ by computing

$$\mathbf{e} = \mathbf{W}_d \mathbf{c} \tag{2-228}$$

where $\mathbf{W}_d$ is the $N \times N$ diagonal weight matrix with diagonal elements W_1, $W_2, \ldots, W_N$. The inverse transformer then produces the excised block that is applied to the despreader:

$$\mathbf{z} = [z_1\ z_2\ldots z_N]^T = \mathbf{B}\,\mathbf{e} = \mathbf{B}\,\mathbf{W}_d\mathbf{c} = \mathbf{B}\,\mathbf{W}_d\mathbf{B}^H\mathbf{x} \tag{2-229}$$

If there were no weighting, then $\mathbf{W}_d = \mathbf{I}$. Since $\mathbf{BB}^H = \mathbf{I}$, $\mathbf{z} = \mathbf{x}$ would result, as expected when the transformer and inverse transformer are in tandem. In general, the diagonal elements of $\mathbf{W}_d$ are either set by a threshold device fed by $\mathbf{c}$ or they are the outputs of the weight-control mechanism of an adaptive filter. When N equals the processing gain G and the input comprises the unmodulated spreading sequence, the despreader correlates its input block with the appropriate segment of the spreading sequence to form the decision variable:

$$V = \sum_{i=1}^{G} p_i z_i \qquad (2\text{-}230)$$

The filtering and despreading can be simultaneously performed in the transform domain. Let

$$\mathbf{p} = [p_1 \; p_2 \; \cdots \; p_G]^T \qquad (2\text{-}231)$$

denote a synchronous replica of the spreading sequence, which is generated by the receiver code generator. Then (2-229) to (2-231) give

$$V = \mathbf{p}^T \mathbf{z} = \mathbf{p}^T \mathbf{B} \mathbf{W}_d \mathbf{c} \qquad (2\text{-}232)$$

Thus, if the spreading sequence is used to produce the matrix $\mathbf{p}^T \mathbf{B} \mathbf{W}_d$, then the product of this matrix and the transform $\mathbf{c}$ gives V without the need for an inverse transformer and a separate despreader.

Nonlinear Filtering

By modeling the narrowband interference as part of a dynamic linear system, one can use the Kalman-Bucy filter [13] to extract an optimal linear estimate of the interference. A subtraction of this estimate from the filter input then removes a large part of the interference from the despreader input. However, a superior nonlinear filter can be designed by approximating an extension of the Kalman-Bucy filter.

Consider the estimation of an $n \times 1$ state vector $\mathbf{x}_k$ of a dynamic system based on the $r \times 1$ observation vector $\mathbf{z}_k$. Let ϕ_k denote the $n \times n$ state transition matrix, $\mathbf{H}_k$ an $r \times n$ observation matrix, and $\mathbf{u}_k$ and $\mathbf{v}_k$ disturbance vectors of dimensions $n \times 1$ and $r \times 1$, respectively. According to the linear dynamic system model, the state and observation vectors satisfy

$$\mathbf{x}_{k+1} = \phi_k \mathbf{x}_k + \mathbf{u}_k, \qquad 0 \le k < \infty \qquad (2\text{-}233)$$

$$\mathbf{z}_k = \mathbf{H}_k \mathbf{x}_k + \mathbf{v}_k, \qquad 0 \le k < \infty \qquad (2\text{-}234)$$

It is assumed that the sequences $\{\mathbf{u}_k\}, \{\mathbf{v}_k\}$ are independent sequences of independent, zero-mean random vectors that are also independent of the initial state $\mathbf{x}_0$. The covariance of $\mathbf{u}_k$ is $E\left[\mathbf{u}_k \mathbf{u}_k^T\right] = \mathbf{Q}_k$. Let $Z^k = (\mathbf{z}_1, \; \mathbf{z}_2, \; \cdots \; \mathbf{z}_k)$ denote the first k observation vectors. Let $f(\mathbf{z}_k | Z^{k-1})$ and $f(\mathbf{x}_k | Z^{k-1})$ denote the probability density functions of $\mathbf{z}_k$ and $\mathbf{x}_k$, respectively, conditioned on Z^{k-1}. A fundamental result of estimation theory is that the estimate $\hat{\mathbf{x}}_k$ that minimizes the mean-norm-squared error $E\left[\|\mathbf{x}_k - \hat{\mathbf{x}}_k\|^2\right]$ is the expectation conditioned on Z^k:

$$\hat{\mathbf{x}}_k = E\left[\mathbf{x}_k | Z^k\right] \qquad (2\text{-}235)$$

The corresponding conditional covariance is denoted by

$$\mathbf{P}_k = E\left[(\mathbf{x}_k - \hat{\mathbf{x}}_k)(\mathbf{x}_k - \hat{\mathbf{x}}_k)^T | Z^k\right] \qquad (2\text{-}236)$$

From (2-233), it follows that the expectation of $\mathbf{x}_k$ conditioned on Z^{k-1} is

$$\bar{\mathbf{x}}_k = E\left[\mathbf{x}_k \,|\, Z^{k-1}\right] = \phi_{k-1}\hat{\mathbf{x}}_{k-1} \tag{2-237}$$

The covariance of $\mathbf{x}_k$ conditioned on Z^{k-1} is defined as

$$\mathbf{M}_k = E\left[\left(\mathbf{x}_k - \bar{\mathbf{x}}_k\right)\left(\mathbf{x}_k - \bar{\mathbf{x}}_k\right)^T \,\big|\, Z^{k-1}\right] \tag{2-238}$$

The following theorem due to Masreliez [15] extends the Kalman-Bucy filter.

Theorem. Assume that $f(\mathbf{x}_k|Z^{k-1})$ is a Gaussian density with mean $\bar{\mathbf{x}}_k$ and $n \times n$ covariance matrix $\mathbf{M}_k$, and that $f(\mathbf{z}_k|Z^{k-1})$ is twice differentiable with respect to the components of $\mathbf{z}_k$. Then the conditional expectation $\hat{\mathbf{x}}_k$ and the conditional covariance $\mathbf{P}_k$ satisfy

$$\hat{\mathbf{x}}_k = \bar{\mathbf{x}}_k + \mathbf{M}_k\mathbf{H}_k^T\mathbf{g}_k\left(\mathbf{z}_k\right) \tag{2-239}$$

$$\mathbf{P}_k = \mathbf{M}_k - \mathbf{M}_k\mathbf{H}_k^T\mathbf{G}_k\left(\mathbf{z}_k\right)\mathbf{H}_k\mathbf{M}_k \tag{2-240}$$

$$\mathbf{M}_{k+1} = \phi_k\mathbf{P}_k\phi_k^T + \mathbf{Q}_k \tag{2-241}$$

$$\bar{\mathbf{x}}_{k+1} = \phi_k\hat{\mathbf{x}}_k \tag{2-242}$$

where $\mathbf{g}_k(\mathbf{z}_k)$ is an $r \times 1$ vector with components

$$\{\mathbf{g}_k\left(\mathbf{z}_k\right)\}_i = -\frac{1}{f\left(\mathbf{z}_k|Z^{k-1}\right)}\frac{\partial f\left(\mathbf{z}_k|Z^{k-1}\right)}{\partial z_{ki}} \tag{2-243}$$

$\mathbf{G}_k(\mathbf{z}_k)$ is an $r \times r$ matrix with elements

$$\{\mathbf{G}_k\left(\mathbf{z}_k\right)\}_{ij} = \frac{\partial\{\mathbf{g}_k\left(\mathbf{z}_k\right)\}_i}{\partial z_{kj}} \tag{2-244}$$

and z_{kj} is the jth component of $\mathbf{z}_k$.

Proof: When $\mathbf{x}_k$ is given, (2-234) indicates that $\mathbf{z}_k$ is independent of Z^{k-1}. Therefore, Bayes' rule gives

$$f\left(\mathbf{x}_k\,|\,Z^k\right) = \frac{f\left(\mathbf{x}_k\,|\,Z^{k-1}\right)f\left(\mathbf{z}_k\,|\,\mathbf{x}_k\right)}{f\left(\mathbf{z}_k\,|\,Z^{k-1}\right)} \tag{2-245}$$

With the concise notation $b = [f(\mathbf{z}_k|Z^{k-1})]^{-1}$, (2-235) and the fact that a density is a scalar function yield

$$\begin{aligned}
\hat{\mathbf{x}}_k - \bar{\mathbf{x}}_k &= b\int_{R^n}\left(\mathbf{x}_k - \bar{\mathbf{x}}_k\right)f\left(\mathbf{z}_k\,|\,\mathbf{x}_k\right)f\left(\mathbf{x}_k\,|\,Z^{k-1}\right)d\mathbf{x}_k \\
&= b\mathbf{M}_k\int_{R^n}f\left(\mathbf{z}_k\,|\,\mathbf{x}_k\right)\mathbf{M}_k^{-1}\left(\mathbf{x}_k - \bar{\mathbf{x}}_k\right)f\left(\mathbf{x}_k\,|\,Z^{k-1}\right)d\mathbf{x}_k
\end{aligned}$$

Using the Gaussian density $f(\mathbf{x}_k|Z^{k-1})$, (2-237), and (2-238), and then integrating by parts, we obtain

$$\begin{aligned}
\hat{\mathbf{x}}_k - \bar{\mathbf{x}}_k &= -b\mathbf{M}_k\int_{R^n}f\left(\mathbf{z}_k\,|\,\mathbf{x}_k\right)\frac{\partial}{\partial\mathbf{x}_k}f\left(\mathbf{x}_k\,|\,Z^{k-1}\right)d\mathbf{x}_k \\
&= b\mathbf{M}_k\int_{R^n}f\left(\mathbf{x}_k\,|\,Z^{k-1}\right)\frac{\partial}{\partial\mathbf{x}_k}f\left(\mathbf{z}_k\,|\,\mathbf{x}_k\right)d\mathbf{x}_k
\end{aligned}$$

where the $n \times 1$ gradient vector $\partial/\partial \mathbf{x}_k$ has $\partial/\partial x_{ki}$ as its ith component. Equation (2-234) implies that

$$
\begin{aligned}
\frac{\partial}{\partial \mathbf{x}_k} f\left(\mathbf{z}_k \,|\mathbf{x}_k\right) &= \frac{\partial}{\partial \mathbf{x}_k} f_v\left(\mathbf{z}_k - \mathbf{H}_k \mathbf{x}_k\right) = -\mathbf{H}_k^T \frac{\partial}{\partial \mathbf{z}_k} f_v\left(\mathbf{z}_k - \mathbf{H}_k \mathbf{x}_k\right) \\
&= -\mathbf{H}_k^T \frac{\partial}{\partial \mathbf{z}_k} f\left(\mathbf{z}_k \,|\mathbf{x}_k\right)
\end{aligned}
$$

where $f_v(\ )$ is the density of $\mathbf{v}_k$. Substitution of this equation into the preceding one gives

$$
\begin{aligned}
\hat{\mathbf{x}}_k - \bar{\mathbf{x}}_k &= -b\mathbf{M}_k \mathbf{H}_k^T \int_{R^n} f\left(\mathbf{x}_k \,|Z^{k-1}\right) \frac{\partial}{\partial \mathbf{z}_k} f\left(\mathbf{z}_k \,|\mathbf{x}_k\right) d\mathbf{x}_k \\
&= -b\mathbf{M}_k \mathbf{H}_k^T \frac{\partial}{\partial \mathbf{z}_k} \int_{R^n} f\left(\mathbf{x}_k \,|Z^{k-1}\right) f\left(\mathbf{z}_k \,|\mathbf{x}_k\right) d\mathbf{x}_k
\end{aligned}
$$

where the second equality results because $f(\mathbf{x}_k|Z^{k-1})$ is not a function of $\mathbf{z}_k$. Substituting (2-245) into this equation and evaluating the integral, we obtain (2-239).

To derive (2-240), we add and subtract $\bar{\mathbf{x}}_k$ in (2-236) and simplify, which gives

$$
\mathbf{P}_k = E\left[\left(\mathbf{x}_k - \bar{\mathbf{x}}_k\right)\left(\mathbf{x}_k - \bar{\mathbf{x}}_k\right)^T \,\big|Z^k\right] - \left(\hat{\mathbf{x}}_k - \bar{\mathbf{x}}_k\right)\left(\hat{\mathbf{x}}_k - \bar{\mathbf{x}}_k\right)^T
$$

The second term of this equation may be evaluated by substituting (2-239). The first term may be evaluated in a similar manner as the derivation of (2-239) except that an integration by parts must be done twice. After a tedious calculation, we obtain (2-240). Equation (2-241) is derived by using the definition of $\mathbf{M}_{k+1}$ given by (2-238) and then substituting (2-233), (2-237), and (2-236). Equation (2-242) follows from (2-237). □

The filter defined by this theorem is the Kalman-Bucy filter if $f(\mathbf{z}_k|Z^{k-1})$ is a Gaussian density. Since (2-234) and (2-238) indicate that the covariance of $\mathbf{z}_k$ conditioned on Z^{k-1} is $\mathbf{H}_k \mathbf{M}_k \mathbf{H}_k^T + \mathbf{R}_k$, where $\mathbf{R}_k = E\left[\mathbf{v}_k \mathbf{v}_k^T\right]$, a Gaussian density implies that

$$
\mathbf{g}_k\left(\mathbf{z}_k\right) = \left(\mathbf{H}_k \mathbf{M}_k \mathbf{H}_k^T + \mathbf{R}_k\right)^{-1}\left(\mathbf{z}_k - \mathbf{H}_k \bar{\mathbf{x}}_k\right) \tag{2-246}
$$

$$
\mathbf{G}_k\left(\mathbf{z}_k\right) = \left(\mathbf{H}_k \mathbf{M}_k \mathbf{H}_k^T + \mathbf{R}_k\right)^{-1} \tag{2-247}
$$

Substitution of these two equations into (2-239) and (2-240) yields the usual Kalman-Bucy equations.

To apply this theorem to the interference suppression problem, the narrowband interference sequence $\{i_k\}$ at the filter input is modeled as an autoregressive process that satisfies

$$
i_k = \sum_{l=1}^{q} \phi_l i_{k-l} + e_k \tag{2-248}
$$

where e_k is a white Gaussian process with variance σ_i^2 and the $\{\phi_l\}$ are known to the receiver. The state-space representation of the system is

$$\mathbf{x}_k = \boldsymbol{\phi}\mathbf{x}_{k-1} + \mathbf{u}_k \qquad (2\text{-}249)$$

$$z_k = \mathbf{H}\mathbf{x}_k + v_k \qquad (2\text{-}250)$$

where

$$\mathbf{x}_k = \begin{bmatrix} i_k & i_{k-1} & \cdots & i_{k-q+1} \end{bmatrix}^T \qquad (2\text{-}251)$$

$$\boldsymbol{\phi} = \begin{bmatrix} \phi_1 & \phi_2 & \cdots & \phi_{q-1} & \phi_q \\ 1 & 0 & \cdots & 0 & 0 \\ 0 & 1 & \cdots & 0 & 0 \\ \vdots & \vdots & \cdots & \vdots & \vdots \\ 0 & 0 & \cdots & 1 & 0 \end{bmatrix} \qquad (2\text{-}252)$$

$$\mathbf{u}_k = \begin{bmatrix} e_k & 0 & \cdots & 0 \end{bmatrix}^T \qquad (2\text{-}253)$$

$$\mathbf{H} = \begin{bmatrix} 1 & 0 & \cdots & 0 \end{bmatrix} \qquad (2\text{-}254)$$

The observation noise v_k is the sum of the direct-sequence signal s_k and the white Gaussian noise n_k:

$$v_k = s_k + n_k \qquad (2\text{-}255)$$

Since the first component of the state vector $\mathbf{x}_k$ is the interference i_k, the state estimate $\mathbf{H}\hat{\mathbf{x}}_k$ provides an interference estimate that can be subtracted from the received signal to cancel the interference.

For a random spreading sequence, $s_k = +c$ or $-c$ with equal probability. If n_k is zero-mean and Gaussian with variance σ_n^2, then v_k has the density

$$f_v(v) = \frac{1}{2}N_{\sigma_n^2}(v - c) + \frac{1}{2}N_{\sigma_n^2}(v + c) \qquad (2\text{-}256)$$

where

$$N_{\sigma^2}(x) = \frac{1}{\sqrt{2\pi}\sigma} \exp\left(-\frac{x^2}{2\sigma^2}\right) \qquad (2\text{-}257)$$

For this non-Gaussian density, the optimal filter that computes the exact conditional mean given by (2-235) is nonlinear with exponentially increasing complexity and, thus, is impractical. The density $f(\mathbf{x}_k|Z^{k-1})$ is not Gaussian as required by Masreliez's theorem. However, by assuming that this density is approximately Gaussian, we can use results of the theorem to derive the *approximate conditional mean* (ACM) filter [16].

Conditioned on Z^{k-1} and s_k, the expected value of z_k is $\mathbf{H}\bar{\mathbf{x}}_k + s_k$ since $\mathbf{x}_k$ and n_k are independent of s_k. From the definition of $\mathbf{M}_k$ and (2-250), it follows that the conditional variance of z_k is

$$\sigma_z^2 = \mathbf{H}\,\mathbf{M}_k\,\mathbf{H}^T + \sigma_n^2 \qquad (2\text{-}258)$$

Since $f(\mathbf{x}_k|Z^{k-1})$ is approximated by a Gaussian density, we obtain

$$f\left(z_k|Z^{k-1}\right) = \frac{1}{2}N_{\sigma_z^2}(z_k - \mathbf{H}\bar{\mathbf{x}}_k - c) + \frac{1}{2}N_{\sigma_z^2}(z_k - \mathbf{H}\bar{\mathbf{x}}_k + c) \qquad (2\text{-}259)$$

Substitution of this equation into (2-243) and (2-244) yields

$$g_k(z_k) = \frac{1}{\sigma_z^2}\left[\epsilon_k - c\tanh\left(\frac{c\epsilon_k}{\sigma_z^2}\right)\right] \tag{2-260}$$

$$G_k(z_k) = \frac{1}{\sigma_z^2}\left[1 - \frac{c^2}{\sigma_z^2}\operatorname{sech}^2\left(\frac{c\epsilon_k}{\sigma_z^2}\right)\right] \tag{2-261}$$

where the *innovation* or prediction residual is

$$\epsilon_k = z_k - \mathbf{H}\bar{\mathbf{x}}_k = z_k - \bar{z}_k \tag{2-262}$$

and

$$\bar{z}_k = \mathbf{H}\bar{\mathbf{x}}_k \tag{2-263}$$

is the predicted observation based on Z^{k-1}. The update equations of the ACM filter are given by (2-239) to (2-242) and (2-260) to (2-263). The difference between the ACM filter and the Kalman-Bucy filter is the presence of the nonlinear *tanh* and *sech* functions in (2-260) and (2-261).

Adaptive ACM filter

In practical applications, the elements of the matrix ϕ in (2-252) are unknown and may vary with time. To cope with these problems, an adaptive algorithm that can track the interference is desirable. The *adaptive ACM filter* receives $z_k = i_k + s_k + n_k$ and produces the interference estimate denoted by $\bar{z}_k$. The output of the filter is denoted by $\epsilon_k = z_k - \bar{z}_k$ and ideally is $s_k + n_k$ plus a small residual of i_k. An adaptive transversal filter is embedded in the adaptive ACM filter. To use the structure of the nonlinear ACM filter, we observe that the second term inside the brackets in (2-260) would be absent if s_k were absent. Therefore, $c\tanh(c\epsilon_k/\sigma_z^2)$ may be interpreted as a soft decision on the direct-sequence signal s_k. The input to the adaptive transversal filter at time k is taken to be the difference between the observation z_k and the soft decision:

$$\tilde{z}_k = z_k - c\tanh\left(\frac{c\epsilon_k}{\sigma_z^2}\right) = \bar{z}_k + \rho(\epsilon_k) \tag{2-264}$$

where

$$\rho(\epsilon_k) = \epsilon_k - c\tanh\left(\frac{c\epsilon_k}{\sigma_z^2}\right) \tag{2-265}$$

The input $\tilde{z}_k$ is a reasonable estimate of the interference that is improved by the adaptive filter. The architecture of the one-sided adaptive ACM filter [18] is shown in Figure 2.38. The output of the N-tap transversal filter provides the interference estimate

$$\bar{z}_k = \mathbf{W}^T(k)\tilde{\mathbf{z}}_k \tag{2-266}$$

where $\mathbf{W}(k)$ is the weight vector and

$$\tilde{\mathbf{z}}_k = \left[\tilde{z}_{k-1}\ \tilde{z}_{k-2}\ \ldots\ \tilde{z}_{k-N}\right]^T \tag{2-267}$$

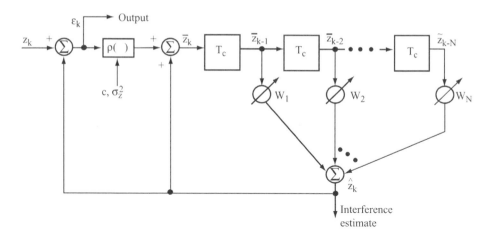

Figure 2.38: Adaptive ACM filter.

which is extracted from the filter taps. When $\tilde{z}_k$ has only a small component due to s_k, the filter can effectively track the interference, and $\bar{z}_k$ is a good estimate of this interference.

A normalized version of the LMS algorithm for the adaptive ACM filter is given by the weight-update equation:

$$\mathbf{W}(k) = \mathbf{W}(k-1) + \frac{\mu_0}{r_k}\left(\tilde{z}_k - \bar{z}_k\right)\bar{\mathbf{z}}_k \qquad (2\text{-}268)$$

where μ_0 is the adaptation constant and r_k is an estimate of the input power iteratively determined by

$$r_k = r_{k-1} + \mu_0\left[\left|\tilde{\mathbf{z}}_k\right|^2 - r_{k-1}\right] \qquad (2\text{-}269)$$

The division by r_k in (2-268) normalizes the algorithm by making the choice of an appropriate μ_0 for fast convergence and good performance much less dependent on the input power level.

The calculation of $\rho(\epsilon_k)$ requires the estimation of σ_z^2. If the $\bar{z}_k$ produced by the adaptive filter approximates the prediction residual of (2-263), then (2-262), (2-260), (2-255), and (2-238) imply that $var(\epsilon_k^2) \approx \sigma_z^2 + c^2$. Therefore, if $var(\epsilon_k^2)$ is estimated by computing the sample variance of the filter output, then the subtraction of c^2 from the sample variance gives an estimate of σ_z^2.

A figure of merit for filters is the *SINR improvement*, which is the ratio of the output SINR to the input SINR. Since the filters of concern do not change the signal power, the SINR improvement is

$$R = \frac{E\left\{\left|z_k - s_k\right|^2\right\}}{E\left\{\left|\epsilon_k - s_k\right|^2\right\}} \qquad (2\text{-}270)$$

In terms of this performance measure, the nonlinear adaptive ACM filter has been found to provide much better suppression of narrowband interference than the linear Kalman-Bucy filter if the noise power in n_k is less than the direct-sequence signal power in s_k. If the latter condition is not satisfied, the advantage is small or absent. Disadvantages apparent from (2-265) are the requirements to estimate the parameters c and σ_z^2 and to compute or store the *tanh* function. At the cost of additional complexity and delay, a nonlinear *adaptive interpolator* [17] gives a slight performance gain.

The preceding linear and nonlinear methods are primarily predictive methods that exploit the inherent predictability of narrowband interference. Further improvements in interference suppression are theoretically possible by using *code-aided methods*, which exploit the predictability of the spread-spectrum signal itself [18]. Most of these methods are based on methods that were originally developed for multiuser detection (Chapter 6). Some of them can potentially be used to simultaneously suppress both narrowband interference and multiple-access interference. However, code-aided methods require even more computation and parameter estimation than the ACM filter, and the most powerful of the adaptive methods are practical only for short spreading sequences.

2.8   Problems

1. Consider a linear feedback shift register with characteristic polynomial $f(x) = 1 + x^3$. Find all possible state sequences.

2. Derive (2-44) using the steps specified in the text.

3. The characteristic polynomial associated with a linear feedback shift register is $f(x) = 1 + x^2 + x^3 + x^5 + x^6$. The initial state is $a_0 = a_1 = 0, a_2 = a_3 = a_4 = a_5 = 1$. Use polynomial long division to determine the first nine bits of the output sequence.

4. If the characteristic polynomial associated with a linear feedback shift register is $1 + x^m$, what is the linear recurrence relation?   Write the generating function associated with the output sequence.   What is the period of the output sequence?   Derive it by polynomial long division.

5. Prove by exhaustive search that the polynomial $f(x) = 1 + x^2 + x^3$ is primitive.

6. Derive the characteristic function of the linear equivalent of Figure 2.12(a). Verify the structure of Figure 2.12(b) and derive the initial contents indicated in the figure.

7. This problem illustrates the limitations of an approximate model in an extreme case.   Suppose that tone interference at the carrier frequency is coherent with a PSK direct-sequence signal so that $\phi = 0$ in (2-92). Assume that $N_0 \to 0$ and $E_s > \kappa I T_c$.   Show that $P_s = 0$.   Show

that the general tone-interference model of Section 2.2 leads to a nonzero approximate expression for P_s .

8. Derive (2-116) using the steps specified in the text.

9. To assess the effect of wideband filtering on the thermal noise, we may substitute bN_0 in place of N_0, where b is the factor that accounts for the presence of the filter. Show that for an ideal rectangular bandpass filter of bandwidth W,

$$b = 2 \int_0^{WT_c/2} sinc^2(x)dx$$

If $WT_c \geq 2$, then $0.9 \leq b \leq 1.0$, and the impact of the wideband filtering is modest or small.

10. Derive (2-131) and (2-132) using the results of Section 2.2.

11. Derive (2-138) and (2-139) using the results of Section 2.2.

12. Derive the expression for $E[V \mid \phi, k_1, k_2, d_0]$ that leads to (2-142).

13. Use the general interference model to plot P_s versus GS/I for dual and balanced quadriphase direct-sequence systems with tone interference at the carrier frequency and $E_s/N_0 = 20$ dB. Observe that the balanced system has more than a 2 dB advantage at $P_s = 10^{-6}$.

14. Consider a direct-sequence system with binary PSK, a required $P_s = 10^{-5}$, and $N_0 = 0$. How much additional power is required against worst-case pulsed interference beyond that required against continuous interference. Use $Q(\sqrt{20}) = 10^{-5}$.

15. For a direct-sequence system with binary DPSK, $P_s = \frac{1}{2}\exp(E_s/N_0)$ in the presence of white Gaussian noise. Derive the worst-case duty cycle and P_s for strong pulsed interference when the power spectral density of continuous interference is I_0. Show that DPSK has a more than 3 dB disadvantage relative to PSK against worst-case pulsed interference when E_s/I_0 is large.

16. What are the values of $E[M_0|\nu]$and $var[M_0|\nu]$ for the white noise metric and for the AGC metric?

17. Expand (2-175) to determine the degradation in $A_s(t_0 + T)$ when $f_d \neq 0$ and the chip waveform is rectangular.

18. Evaluate the impulse response of a transversal filter with the form of Figure 2.26. Show that this impulse response is equal to that of a filter matched to $p_1(t)cos(2\pi f_c t + \theta)$.

19. Consider an elastic convolver for which $L/v = nT$ for some positive integer n and $g(t) = p(T - t)$, where p(t) is the periodic spreading waveform. The received signal is $f(t) = Ap(t-t_0)$, where A is a positive constant. Express $A_s(t)$ as a function of $R_p(\ )$, the periodic autocorrelation of the spreading waveform. How might this result be applied to acquisition?

20. Consider the soft-decision term in (2-264). What are its values as $\sigma_z \to \infty$ and as $\sigma_z \to 0$? Give an engineering interpretation of these results.

2.9 References

1. D. Torrieri, *Principles of Secure Communication Systems*, 2nd ed. Boston: Artech House, 1992.

2. R. L. Peterson, R. E. Ziemer, and D. E. Borth, *Introduction to Spread Spectrum Communications.* Upper Saddle River, NJ: Prentice Hall, 1995.

3. M. K. Simon et al., *Spread-Spectrum Communications Handbook.* New York: McGraw-Hill, 1994.

4. S. B. Wicker, *Error Control Systems for Digital Communication and Storage.* Upper Saddle River, NJ: Prentice Hall, 1995.

5. G. J. Simmons, ed., *Contemporary Cryptology: The Science of Information Integrity.* New York: IEEE Press, 1992.

6. R. B. Ash and C. A. Doleans-Dade, *Probability and Measure Theory, 2nd ed.* San Diego: Academic Press, 2000.

7. D. J. Torrieri, "The Performance of Five Different Metrics Against Pulsed Jamming," *IEEE Trans. Commun.*, vol. 34, pp. 200–207, February 1986.

8. C. Campbell, *Surface Acoustic Wave Devices for Mobile and Wireless Communications.* New York: Academic Press, 1998.

9. M. Kowatsch, "Application of Surface-Acoustic-Wave Technology to Burst-Format Spread-Spectrum Communications," *IEE Proc.*, vol. 131, pt. F, pp. 734–741, December 1984.

10. D. P. Morgan and J. M. Hannah, "Surface Wave Recirculation Loops for Signal Processing," *IEEE Trans. Sonics and Ultrason.*, vol. 25, pp. 30–38, January 1978.

11. L. B. Milstein, "Interference Rejection Techniques in Spread Spectrum Communications," *Proc. IEEE*, vol. 76, pp. 657–671, June 1988.

12. H. V. Poor, "Active Interference Suppression in CDMA Overlay Systems," *IEEE J. Select. Areas Commun.*, vol. 19, pp. 4–20, January 2001.

13. S. Haykin, *Adaptive Filter Theory, 4th ed.* Upper Saddle River, NJ: Prentice-Hall, 2002.

14. M. Medley, G. Saulnier, and P. Das, "The Application of Wavelet-Domain Adaptive Filtering to Spread-Spectrum Communications," *Proc. SPIE Wavelet Applications for Dual-Use*, vol. 2491, pp. 233–247, April 1995.

15. C. J. Masreliez, "Approximate Non-Gaussian Filtering with Linear State and Observation Relations," *IEEE Trans. Automat. Contr.*, vol. 20, pp. 107–110, February 1975.

16. R. Vijayan and H. V. Poor, "Nonlinear Techniques for Interference Suppression in Spread-Spectrum Systems," *IEEE Trans. Commun.*, vol. 37, pp. 1060–1065, July 1990.

17. L. A. Rusch and H. V. Poor, "Narrowband Interference Suppression in CDMA Spread Spectrum Communications," *IEEE Trans. Commun.*, vol. 42, pp. 1969–1979, Feb./March/April 1994.

18. S. Buzzi, M. Lops, and H. V. Poor, "Code-Aided Interference Suppression for DS/CDMA Overlay Systems," *Proc. IEEE*, vol. 90, pp. 394–435, March 2002.

Chapter 3

Frequency-Hopping Systems

3.1 Concepts and Characteristics

Frequency hopping is the periodic changing of the carrier frequency of a transmitted signal. The sequence of carrier frequencies is called the *frequency-hopping pattern*. The set of M possible carrier frequencies $\{f_1, f_2, \ldots, f_M\}$ is called the *hopset*. The rate at which the carrier frequency changes is called the *hop rate*. Hopping occurs over a frequency band called the *hopping band* that includes M *frequency channels*. Each frequency channel is defined as a spectral region that includes a single carrier frequency of the hopset as its center frequency and has a bandwidth B large enough to include most of the power in a signal pulse with a specific carrier frequency. Figure 3.1 illustrates the frequency channels associated with a particular frequency-hopping pattern. The time interval between hops is called the *hop interval*. Its duration is called the *hop duration* and is denoted by T_h. The hopping band has bandwidth $W \geq MB$.

Figure 3.2 depicts the general form of a frequency-hopping system. The frequency synthesizers (Section 3.4) produce frequency-hopping patterns determined by the time-varying multilevel sequence specified by the output bits of the code generators. In the transmitter, the data-modulated signal is mixed with the synthesizer output pattern to produce the frequency-hopping signal. If the data modulation is some form of angle modulation $\phi(t)$, then the received signal for the ith hop is

$$s(t) = \sqrt{2S} \cos\left[2\pi f_j t + \phi(t) + \phi_i\right] , \quad (i-1)T_h \leq t \leq iT_h \qquad \text{(3-1)}$$

where S is the average power, f_j is the carrier frequency for this hop, and ϕ_i is a random phase angle for the ith hop. The frequency-hopping pattern produced by the receiver synthesizer is synchronized with the pattern produced by the transmitter, but is offset by a fixed intermediate frequency, which may be zero. The mixing operation removes the frequency-hopping pattern from the

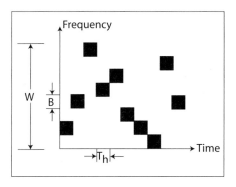

Figure 3.1: Frequency-hopping patterns.

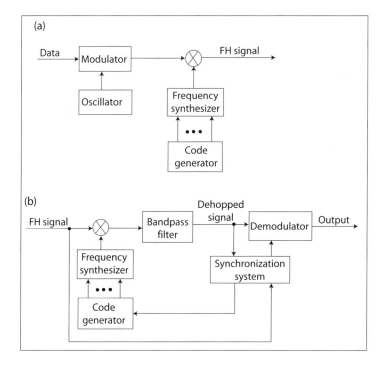

Figure 3.2: General form of frequency-hopping system: (a) transmitter and (b) receiver.

received signal and, hence, is called *dehopping*. The mixer output is applied to a bandpass filter that excludes double-frequency components and power that originated outside the appropriate frequency channel and produces the data-modulated *dehopped signal*, which has the form of (3-1) with f_j replaced by the intermediate frequency.

Although it provides no advantage against white noise, frequency hopping enables signals to hop out of frequency channels with interference or slow frequency-selective fading. To fully exploit this capability against narrowband interference signals, disjoint frequency channels are necessary. The disjoint channels may be contiguous or have unused spectral regions between them. Some spectral regions with steady interference or a susceptibility to fading may be omitted from the hopset, a process called *spectral notching*. Multiple frequency-shift keying (MFSK) differs fundamentally from frequency hopping in that all the MFSK subchannels affect each receiver decision. No escape from or avoidance of a subchannel with interference is possible.

To ensure the secrecy and unpredictably of the frequency-hopping pattern, the pattern should be a pseudorandom sequence of frequencies. The sequence should have a large period and a uniform distribution over the frequency channels and should be generated by a multilevel sequence with a large linear span. The large period prevents the capture and storage of a period of the pattern by an opponent. The *linear span* of a multilevel sequence is the smallest degree of any linear recursion that the sequence satisfies. A large linear span inhibits the reconstruction of the pattern from a short segment of it. The set of *control bits* produced by the code generator usually constitutes a symbol drawn from a finite field with the necessary properties. A frequency-hopping pattern is obtained by associating a distinct frequency with each symbol. A number of methods have been found to ensure a large linear span [1], [2].

An architecture that enhances the *transmission security* by encrypting the control bits is shown in Figure 3.3. The specific algorithm for generating the control bits is determined by the key and the time-of-day (TOD). The *key*, which is the ultimate source of security, is a set of bits that are changed infrequently and must be kept secret. The TOD is a set of bits that are derived from the stages of the TOD counter and change with every transition of the TOD clock. For example, the key might change daily while the TOD might change every second. The purpose of the TOD is to vary the generator algorithm without constantly changing the key. In effect, the generator algorithm is controlled by a time-varying key. The code clock, which regulates the changes of state in the code generator and thereby controls the hop rate, operates at a much higher rate than the TOD clock. In a receiver, the code clock is produced by the synchronization system. In both the transmitter and the receiver, the TOD clock may be derived from the code clock.

A frequency-hopping pulse with a fixed carrier frequency occurs during a portion of the hop interval called the *dwell interval*. As illustrated in Figure 3.4, the *dwell time* is the duration of the dwell interval during which the channel symbols are transmitted. The hop duration T_h is equal to the sum of the dwell time T_d and the switching time T_{sw}. The *switching time* is equal to the

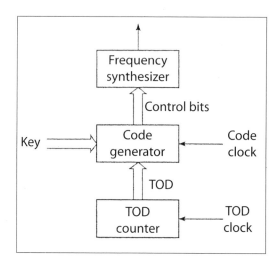

Figure 3.3: Secure method of synthesizer control.

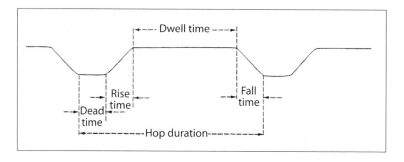

Figure 3.4: Time durations of a frequency-hopping pulse.

dead time, which is the duration of the interval when no signal is present, plus the rise and fall times of a pulse. Even if the switching time is absent in the transmitted signal, it will be present in the dehopped signal in the receiver because of the imperfect synchronization of received and receiver-generated waveforms. The nonzero switching time, which may include an intentional *guard time*, decreases the transmitted symbol duration T_s. If T_{so} is the symbol duration in the absence of frequency hopping, then $T_s = T_{so}(T_d/T_h)$. The reduction in symbol duration expands the transmitted spectrum and thereby reduces the number of frequency channels within a fixed hopping band. Since the receiver filtering will ensure that rise and fall times of pulses have durations on the order of a symbol duration, $T_{sw} > T_s$ in practical systems. Implementing a short switching time becomes an obstacle as the hop rate decreases.

Frequency hopping may be classified as fast or slow. *Fast frequency hopping* occurs if there is more than one hop for each information symbol. *Slow frequency hopping* occurs if one or more information symbols are transmitted in the time interval between frequency hops. Although these definitions do not refer to the hop rate, fast frequency hopping is an option only if a hop rate that exceeds the information-symbol rate can be implemented. Slow frequency hopping is usually preferable because the transmitted waveform is much more spectrally compact (cf. Table 3.1, Section 3.2) and the overhead cost of the switching time is reduced.

Let M denote the hopset size, B denote the bandwidth of frequency channels, and F_s denote the minimum separation between adjacent carriers in a hopset. For full protection against stationary narrowband interference and jamming, it is desirable that $F_s \geq B$ so that the frequency channels are nearly spectrally disjoint. A hop then enables the transmitted signal to escape the interference in a frequency channel.

To obtain the full advantage of block or convolutional channel codes in a slow frequency-hopping system, it is important to interleave the code symbols in such a way that the symbol errors in a code word or constraint length are independent(for hard-decision decoding) or that the symbols are degraded independently (for soft-decision decoding). In frequency-hopping systems operating over a frequency-selective fading channel (Chapter 5), the realization of this independence requires certain constraints among the system parameter values. Symbol errors are independent if the fading is independent in each frequency channel and each symbol is transmitted in a different frequency channel. If each of the interleaved code symbols is transmitted at the same location in each hop dwell interval, then adjacent symbols are separated by T_h after the interleaving. Thus, a sufficient condition for nearly independent symbol errors is

$$T_h \geq T_{coh} \tag{3-2}$$

where T_{coh} is the coherence time of the fading channel. Another sufficient condition for nearly independent symbol errors is

$$F_s \geq B_{coh} \tag{3-3}$$

where B_{coh} is the coherence bandwidth of the fading channel. For practical mobile communication networks with hop rates exceeding 100 hops/s, (3-2) is rarely satisfied. For a hopping band with bandwidth W, and a hopset with a uniform carrier separation, $F_s = W/M \geq B$. Thus, (3-3) implies that the number of frequency channels is constrained by

$$M \leq \frac{W}{\max(B, B_{coh})} \tag{3-4}$$

if nearly independent symbol errors are to be ensured. If (3-4) is not satisfied, there will be a performance loss due to the correlated symbol errors. If $B < B_{coh}$, equalization will not be necessary because the channel transfer function is nearly flat over each frequency channel. If $B \geq B_{coh}$, either equalization may be used to prevent intersymbol interference or a multicarrier modulation may be combined with the frequency hopping.

Let n denote the length of a block codeword or the constraint length of a convolutional code. Let T_{del} denote the maximum tolerable processing delay. Since the delay caused by coding and ideal interleaving over n hops is $(n - 1)T_h + T_s$, and n distinct frequencies are desired,

$$n \leq \min\left(M, 1 + \frac{T_{\text{del}} - T_s}{T_h}\right) \qquad (3\text{-}5)$$

is required. If this inequality is not satisfied, then nonideal interleaving is necessary, and some performance degradation results.

Frequency-selective fading and Doppler shifts make it difficult to maintain phase coherence from hop to hop between frequency synthesizers in the transmitter and the receiver. Furthermore, the time-varying delay between the frequency changes of the received signal and those of the synthesizer output in the receiver causes the phase shift in the dehopped signal to differ for each hop interval. Thus, practical frequency-hopping systems use noncoherent or differentially coherent demodulators unless a pilot signal is available, the hop duration is very long, or elaborate iterative phase estimation (perhaps as part of turbo decoding) is used.

In military applications, the ability of frequency-hopping systems to avoid interference is potentially neutralized by a *repeater jammer* (also known as a *follower jammer*), which is a device that intercepts a signal, processes it, and then transmits jamming at the same center frequency. To be effective against a frequency-hopping system, the jamming energy must reach the victim receiver before it hops to a new set of frequency channels. Thus, the hop rate is the critical factor in protecting a system against a repeater jammer. Required hop rates and the limitations of repeater jamming are analyzed in reference [3].

3.2 Modulations

MFSK

An FH/MFSK system uses MFSK as its data modulation. One of q frequencies is selected as the carrier or center frequency for each transmitted symbol, and the set of q possible frequencies changes with each hop. The general transmitter of Figure 3.2(a) can be simplified for an FH/MFSK system, as illustrated in Figure 3.5(a), where the code generator output bits and the digital input are combined to determine the frequency generated by the synthesizer. An FH/MFSK signal has the form

$$s(t) = \sqrt{2S} \sum_{i=-\infty}^{\infty} \sum_{l=0}^{N_h-1} w[t - i(N_hT_s + T_{sw}) - lT_s]\cos[2\pi(f_i + f_{il})t + \phi_i + \phi_{il}] \quad (3\text{-}6)$$

where $S = \mathcal{E}_s/T_s$ is the average signal power during a dwell interval, $w(t)$ is a unit-amplitude rectangular pulse of duration T_s, N_h is the number of symbols per dwell interval, f_i is the carrier frequency during dwell interval i, $f_i + f_{il}$ is

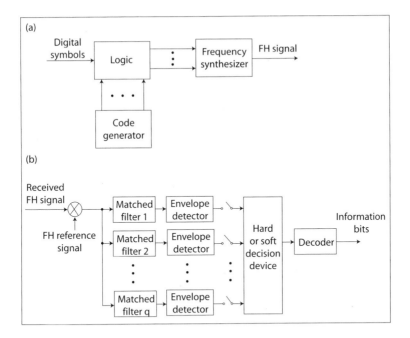

Figure 3.5: FH/MFSK (a) transmitter and (b) receiver.

the MFSK frequency used for symbol l of dwell interval i, ϕ_i is the phase at the beginning of dwell interval i, and ϕ_{il} is the phase associated with MFSK symbol l during dwell interval i. If the MFSK is phase continuous from symbol to symbol, then $\phi_{il} = 0$; otherwise, it may be modeled as a random variable uniformly distributed over $[0,2\pi)$. The implementation of phase continuity is highly desirable to prevent excessive spectral splatter outside a frequency channel (Section 3.3).

In an FH/MFSK system, each of the q frequencies or tones in an MFSK set can be considered as the center frequency of an MFSK subchannel. Therefore, the *effective number of frequency channels* is

$$M_e = qM \tag{3-7}$$

where M is the hopset size. In the standard implementation, the q subchannels of each MFSK set are contiguous, and each set constitutes a frequency channel within the hopping band. For noncoherent orthogonal signals, the MFSK tones must be separated enough that a received signal produces negligible responses in the incorrect subchannels. As shown subsequently, the frequency separation must be $f_d = k/T_s$, where k is a nonzero integer, and T_s denotes the symbol duration. To maximize the hopset size when the MFSK subchannels are contiguous, $k = 1$ is selected. Consequently, the bandwidth of a frequency channel for slow frequency hopping with many symbols per dwell interval is

$$B = \frac{q}{T_s} = \frac{q}{T_b \log_2 q} \tag{3-8}$$

where T_b is the duration of a bit, and the factor $\log_2 q$ accounts for the increase in symbol duration when a nonbinary modulation is used. If the hopping band has bandwidth W, the hopset size is

$$M = \left\lfloor \frac{W}{B} \right\rfloor \tag{3-9}$$

where $\lfloor x \rfloor$ denotes the largest integer in x. Figure 3.5(b) depicts the main elements of a noncoherent FH/MFSK receiver. Each matched filter corresponds to an MFSK subchannel. In practical FH/MFSK systems, the orthogonality of the q MFSK tones is imperfect because of transients that occur after every hop in the receiver.

Soft-Decision Decoding

To illustrate some basic issues of frequency-hopping communications and the effectiveness of soft-decision decoding, we consider an FH/MFSK system that uses a repetition code and the receiver of Figure 3.5(b). Each information symbol, which is transmitted as L code symbols, may be regarded as a codeword or as an uncoded symbol that uses diversity combining. The interference is modeled as wideband Gaussian noise uniformly distributed over part of the hopping band. Slow frequency hopping with a fixed hop rate and ideal interleaving or variable-rate fast frequency hopping is assumed. Both ensure the independence of code-symbol errors. The optimal metric for the Rayleigh-fading channel (Chapter 5) and a good metric for the additive-white-Gaussian-noise (AWGN) channel without fading is the Rayleigh metric defined by (1-66), which is

$$U(l) = \sum_{i=1}^{L} R_{li}^2 , \quad l = 1, 2, \ldots, q \tag{3-10}$$

where R_{li} is the sample value of the envelope-detector output that is associated with code symbol i of candidate information-symbol l, and L is the number of repetitions or code symbols. The diversity combining required by the Rayleigh metric is often called *linear square-law combining*. This metric has the advantage that no *side information*, which is specific information about the reliability of symbols, is required for its implementation. A performance analysis of a frequency-hopping system with binary FSK and soft-decision decoding with the Rayleigh metric indicates that the system performs poorly against worst-case partial-band jamming [6] primarily because a single jammed frequency can corrupt the metrics. Furthermore, the repetition code is counterproductive because the *noncoherent combining loss* resulting from the fragmentation of the symbol energy is greater than any coding or diversity gain.

The difficulty of implementing the maximum-likelihood metric (1-61) leads to consideration of the approximation (1-65), which requires *nonlinear square-law combining:*

$$U(l) = \sum_{i=1}^{L} \frac{R_{li}^2}{N_{0i}^2} \ , \quad l = 1, 2, \ldots, q \tag{3-11}$$

where $N_{0i}/2$ is the two-sided power spectral density of the interference and noise over all the MFSK subchannels during code symbol i. A plausible simplification [8] that is much easier to analyze is the *variable-gain metric:*

$$U(l) = \sum_{i=1}^{L} \frac{R_{li}^2}{N_{0i}} \ , \quad l = 1, 2, \ldots, q \tag{3-12}$$

The advantage of both metrics is that they incorporate *side information* contained in the $\{N_{0i}\}$, which must be known. The subsequent analysis is for the the variable-gain metric.

The union bound (1-46) implies that the information-symbol error probability satisfies

$$P_{is} \le (q-1)P_2 \tag{3-13}$$

where P_2 is the probability of an error in comparing the metric associated with the transmitted information symbol with the metric associated with an alternative one. It is assumed that there are enough frequency channels that L distinct carrier frequencies are used for the L code symbols. Since the MFSK tones are orthogonal, the symbol metrics $\{R_{li}^2/N_{0i}\}$ are independent and identically distributed for all values of l and i (Chapter 1). Therefore, the Chernoff bound given by (1-103) and (1-102) with $\alpha = 1/2$ yields

$$P_2 \le \frac{1}{2} Z^L \tag{3-14}$$

$$Z = \min_{0 < s < s_1} E\left[\exp\left\{ \frac{s}{N_1}(R_2^2 - R_1^2) \right\} \right] \tag{3-15}$$

where R_1 is the sampled output of an envelope detector when the desired signal is present at the input of the associated matched filter, R_2 is the output when the desired signal is absent, and $N_1/2$ is the two-sided power-spectral density of the interference and noise over all the MFSK subchannels during a code symbol. For the q-ary symmetric channel, (1-27), (3-13), and (3-14) give an upper bound on the information-bit error probability:

$$P_b \le \frac{q}{4} Z^L \tag{3-16}$$

For a Gaussian random variable X with mean m and variance σ^2, a direct calculation yields

$$E[\exp(aX^2)] = \frac{1}{\sqrt{1 - 2a\sigma^2}} \exp\left(\frac{am^2}{1 - 2a\sigma^2} \right), \quad a < \frac{1}{2\sigma^2} \tag{3-17}$$

From the analysis of Chapter 1 leading to (1-78), it follows that

$$R_l^2 = x_l^2 + y_l^2 , \quad l = 1, 2 \tag{3-18}$$

where x_l and y_l are the real and imaginary parts of R_l, respectively, and are independent Gaussian random variables with the moments

$$E[x_1] = \sqrt{\mathcal{E}_s/2} \cos\theta , \quad E[y_1] = \sqrt{\mathcal{E}_s/2} \sin\theta \tag{3-19}$$

$$E[x_2] = E[y_2] = 0 \tag{3-20}$$

$$\mathrm{var}[x_l] = \mathrm{var}[y_l] = N_1/4 , \quad l = 1, 2 \tag{3-21}$$

where $\mathcal{E}_s$ is the energy per symbol. By conditioning on N_1, the expectation in (3-15) can be partially evaluated. Equations (3-17) to (3-21) and the substitution of $\lambda = s/2$ give

$$Z = \min_{0 < \lambda < 1} E\left[\frac{1}{1 - \lambda^2} \exp\left(-\frac{\lambda \mathcal{E}_s/N_1}{1 + \lambda} \right) \right] \tag{3-22}$$

where the remaining expectation is over the statistics of N_1.

To simplify the analysis, it is assumed that the thermal noise is negligible. When a repetition symbol encounters no interference, $N_1 = 0$; when it does, $N_1 = I_{t0}/\mu$, where μ is the fraction of the hopping band with interference, and I_{t0} is the spectral density that would exist if the interference power were uniformly spread over the entire hopping band. Since μ is the probability that interference is encountered, (3-22) becomes

$$Z = \min_{0 < \lambda < 1} \left[\frac{\mu}{1 - \lambda^2} \exp\left(-\frac{\lambda \mu \gamma}{1 + \lambda} \right) \right] \tag{3-23}$$

where

$$\gamma = \frac{\mathcal{E}_s}{I_{t0}} = \left(\frac{m}{L} \right) \frac{\mathcal{E}_b}{I_{t0}} \tag{3-24}$$

and $m = \log_2 q$ is the number of bits per information symbol, and $\mathcal{E}_b$ is the energy per information bit. Using calculus, we find that

$$Z = \frac{\mu}{1 - \lambda_0^2} \exp\left(-\frac{\lambda_0 \mu \gamma}{1 + \lambda_0} \right) \tag{3-25}$$

where

$$\lambda_0 = -\left(\frac{1}{2} + \frac{\mu\gamma}{4} \right) + \left[\left(\frac{1}{2} + \frac{\mu\gamma}{4} \right)^2 + \frac{\mu\gamma}{2} \right]^{1/2} \tag{3-26}$$

Substituting (3-25) and (3-24) into (3-16), we obtain

$$P_b \leq 2^{m-2} \left(\frac{\mu}{1 - \lambda_0^2} \right)^L \exp\left[-\left(\frac{\lambda_0 \mu m}{1 + \lambda_0} \right) \frac{\mathcal{E}_b}{I_{t0}} \right] \tag{3-27}$$

Suppose that the interference is worst-case partial-band jamming. An upper bound on P_b is obtained by maximizing the right-hand side of (3-27) with respect to μ, where $0 \le \mu \le 1$. Calculus yields the maximizing value of μ:

$$\mu_0 = \min \left[\frac{3L}{m} \left(\frac{\mathcal{E}_b}{I_{t0}} \right)^{-1}, 1 \right] \tag{3-28}$$

Substituting (3-28) into (3-27), we obtain an upper bound on P_b for worst-case partial-band jamming:

$$P_b \le \begin{cases} 2^{m-2} \left[\dfrac{4L}{me} \left(\dfrac{\mathcal{E}_b}{I_{t0}} \right)^{-1} \right]^L & , \quad L \le \dfrac{m}{3} \left(\dfrac{\mathcal{E}_b}{I_{t0}} \right) \\[3mm] 2^{m-2}(1-\lambda_0^2)^{-L} \exp\left[-\left(\dfrac{m\lambda_0}{1+\lambda_0} \right) \dfrac{\mathcal{E}_b}{I_{t0}} \right] & , \quad L > \dfrac{m}{3} \left(\dfrac{\mathcal{E}_b}{I_{t0}} \right) \end{cases} \tag{3-29}$$

Since μ_0 is obtained by maximizing a bound rather than an equality, it is not necessarily equal to the actual worst-case μ, which would provide a tighter bound than the one in (3-29).

If $\mathcal{E}_b/I_{t0}$ is known, then the number of repetitions can be chosen to minimize the upper bound on P_b for worst-case partial-band jamming. We treat L as a continuous variable such that $L \ge 1$ and let L_0 denote the minimizing value of L. A calculation indicates that the derivative with respect to L of the second line on the right-hand side of (3-29) is positive. Therefore, if $\mathcal{E}_b/I_{t0} < 3/m$ so that the second line is applicable for $L \ge 1$, then $L_0 = 1$. If $\mathcal{E}_b/I_{t0} \ge 3/m$, the continuity of (3-29) as a function of L implies that L_0 is determined by the first line in (3-29). Further calculation yields

$$L_0 = \max \left(\frac{m\mathcal{E}_b}{4I_{t0}}, 1 \right) \tag{3-30}$$

Since L must be an integer, its minimizing value is approximately $\lfloor L_0 \rfloor$.

The upper bound on P_b for worst-case partial-band jamming when $L = L_0$ is given by

$$P_b \le \begin{cases} 2^{m-2} \exp\left(-\dfrac{m\mathcal{E}_b}{4I_{t0}} \right) & , \quad \dfrac{\mathcal{E}_b}{I_{t0}} \ge \dfrac{4}{m} \\[3mm] \dfrac{2^m}{me} \left(\dfrac{\mathcal{E}_b}{I_{t0}} \right)^{-1} & , \quad \dfrac{3}{m} \le \dfrac{\mathcal{E}_b}{I_{t0}} < \dfrac{4}{m} \\[3mm] 2^{m-2}(1-\lambda_0^2)^{-1} \exp\left[-\left(\dfrac{m\lambda_0}{1+\lambda_0} \right) \dfrac{\mathcal{E}_b}{I_{t0}} \right] & , \quad \dfrac{\mathcal{E}_b}{I_{t0}} < \dfrac{3}{m} \end{cases} \tag{3-31}$$

This upper bound indicates that P_b decreases exponentially as $\mathcal{E}_b/I_{t0}$ increases if the appropriate number of repetitions is chosen and $\mathcal{E}_b/I_{t0}$ is large enough. Thus, the nonlinear diversity combining with the variable-gain metric sharply limits the performance degradation caused by worst-case partial-band jamming relative to full-band jamming. Setting $N_0 \to I_{t0}$ in (1-86) and $m = 1$ in (3-31) and then comparing the equations, we find that this degradation is approximately 3 dB for binary FSK. Substituting (3-30) into (3-28), we obtain

$$
\mu_0 = \begin{cases}
\dfrac{3}{4} & , \quad \dfrac{\mathcal{E}_b}{I_{t0}} \geq \dfrac{4}{m} \\[2ex]
\dfrac{3}{m}\left(\dfrac{\mathcal{E}_b}{I_{t0}}\right)^{-1} & , \quad \dfrac{3}{m} < \dfrac{\mathcal{E}_b}{I_{t0}} < \dfrac{4}{m} \\[2ex]
1 & , \quad \dfrac{\mathcal{E}_b}{I_{t0}} \leq \dfrac{3}{m}
\end{cases}
\tag{3-32}
$$

This result shows that the appropriate choice of L implies that worst-case jamming must cover three-fourths or more of the hopping band, a task that may not be a practical possibility for a jammer.

For slow frequency hopping with a fixed hop rate, the suppression of partial-band interference is improved by decreasing the data rate so that $\mathcal{E}_b$ is increased. If $\mathcal{E}_b$ is increased enough, then (3-30) indicates that the optimal amount of diversity combining is proportional to $\mathcal{E}_b$.

For frequency hopping with binary FSK and the variable-gain metric, a more precise derivation [8] that does not use the Chernoff bound and allows $N_0 > 0$ confirms that (3-31) provides an approximate upper bound on the information-bit error rate caused by worst-case partial-band jamming when N_0 is small, although the optimal number of repetitions is much smaller than is indicated by (3-30). Thus, the appropriate weighting of terms in nonlinear square-law combining prevents the domination by a single corrupted term and limits the inherent noncoherent combining loss.

The implementation of the variable-gain metric requires the measurement of the interference power. One might attempt to measure this power in frequency channels immediately before the hopping of the signal into those channels, but this method will not be reliable if the interference is frequency-hopping or non-stationary. Another approach is to clip (soft-limit) each envelope-detector output R_{li} to prevent a single erroneous sample from undermining the metric. This method is potentially effective, but its implementation requires an accurate measurement of the signal power for properly setting the clipping level. A sufficiently accurate measurement is often impractical because of fading or power variations across the hopping band. A metric that requires no side information is the *self-normalization metric* defined for binary FSK as [9]

$$
U(l) = \sum_{i=1}^{n} \frac{R_{li}^2}{R_{1i}^2 + R_{2i}^2} \, , \quad l = 1, 2
\tag{3-33}
$$

Although it does not provide as good a performance against partial-band jamming as the variable-gain metric, the self-normalization metric is far more practical and is generally superior to hard-decision decoding.

The assumption was made that either all or none of the subchannels in an MFSK set are jammed. However, this assumption ignores the threat of narrowband jamming signals that are randomly distributed over the frequency channels. Although (3-31) indicates that it is advantageous to use nonbinary signaling when $\mathcal{E}_b/I_{t0} \geq 4/m$, this advantage is completely undermined when distributed, narrowband jamming signals are a threat. A fundamental problem, which also limits the applicability of FH/MFSK in networks, is the reduced hopset size for nonbinary MFSK indicated by (3-9) and (3-8).

Narrowband Jamming Signals

When the MFSK subchannels are contiguous, it is not advantageous to a jammer to transmit the jamming in all the subchannels of an MFSK set because only a single subchannel needs to be jammed to cause a symbol error. A sophisticated jammer with knowledge of the spectral locations of the MFSK sets can cause increased system degradation by placing one jamming tone or narrowband jamming signal in every MFSK set.

To assess the impact of this sophisticated multitone jamming on hard-decision decoding in the receiver of Figure 3.5(b), it is assumed that thermal noise is absent and that each jamming tone coincides with one MFSK tone in a frequency channel encompassing q MFSK tones [4], [5]. Whether a jamming tone coincides with the transmitted MFSK tone or an incorrect one, there will be no symbol error if the desired-signal power S exceeds the jamming power. Thus, if I_t is the total available jamming power, then the jammer can maximize symbol errors by placing tones with power levels slightly above S whenever possible in approximately J frequency channels such that

$$J = \begin{cases} 1 & , & I_t < S \\ \left\lfloor \dfrac{I_t}{S} \right\rfloor & , & S \leq I_t \\ M & , & MS < I_t \end{cases} \tag{3-34}$$

If a transmitted tone enters a jammed frequency channel and $I_t \geq S$, then with probability $(q-1)/q$ the jamming tone will not coincide with the transmitted tone and will cause a symbol error after hard-decision decoding. If the jamming tone does coincide with the correct tone, it may cause a symbol error in the absence of thermal noise only if its power level is exactly S and it has exactly a 180° phase shift relative to the desired signal, an event with zero probability. Since J/M is the probability that a frequency channel is jammed, and no error occurs if $I_t < S$, the symbol error probability is

$$P_s = \begin{cases} 0 & , & I_t < S \\ \dfrac{J}{M}\left(\dfrac{q-1}{q}\right) & , & I_t \geq S \end{cases} \tag{3-35}$$

Substitution of (3-8), (3-9), and (3-34) into (3-35) and the approximation $\lfloor x \rfloor \approx x$ yields

$$P_s = \begin{cases} \dfrac{q-1}{q} & , & \dfrac{\mathcal{E}_b}{I_{t0}} < \dfrac{q}{\log_2 q} \\ \left(\dfrac{q-1}{\log_2 q}\right)\left(\dfrac{\mathcal{E}_b}{I_{t0}}\right)^{-1} & , & \dfrac{q}{\log_2 q} \leq \dfrac{\mathcal{E}_b}{I_{t0}} \leq WT_b \\ 0 & , & \dfrac{\mathcal{E}_b}{I_{t0}} > WT_b \end{cases} \tag{3-36}$$

where $\mathcal{E}_b = ST_b$ denotes the energy per bit and $I_{t0} = I_t/W$ denotes the spectral density of the interference power that would exist if it were uniformly spread over the hopping band. This equation exhibits an inverse linear dependence of P_s on $\mathcal{E}_b/I_{t0}$, which indicates that the jamming has an impact qualitatively similar to that of Rayleigh fading. It is observed that P_s increases with q, which is the opposite of what is observed over the AWGN channel. Thus, binary FSK is advantageous against this sophisticated multitone jamming.

To preclude this jamming, each MFSK tone in an MFSK set may be independently hopped. However, this approach demands a large increase in the amount of hardware, and uniformly distributed, narrowband jamming signals are almost as damaging as the worst-case multitone jamming. Thus, contiguous MFSK subchannels are usually preferable, and the FH/MFSK receiver has the form of Figure 3.5(b). An analysis of FH/MFSK systems with hard-decision decoding in the presence of uniformly distributed, narrowband jamming signals confirms the superior robustness of binary FSK relative to nonbinary MFSK whether the MFSK tones hop independently or not [6].

Other Modulations

In a network of frequency-hopping systems, it is highly desirable to choose a spectrally compact modulation so that the number of frequency channels is large and, hence, the number of collisions between frequency-hopping signals is kept small. Binary orthogonal FSK allows more frequency channels than MFSK and, hence, is advantageous against narrowband interference distributed throughout the hopping band. A spectrally compact modulation helps ensure that $B < B_{coh}$ so that equalization in the receiver is not necessary. This section considers spectrally compact alternatives to orthogonal FSK.

The demodulator transfer function following the dehopping in Figure 3.2 is assumed to have a bandwidth approximately equal to B, the bandwidth of a frequency channel. The bandwidth is determined primarily by the percentage of the signal power that must be processed by the demodulator if the demodulated signal distortion and the intersymbol interference are to be negligible. In practice, this percentage must be at least 90 percent and is often more than 95 percent. The relation between B and the symbol duration may be expressed as

$$B = \frac{\zeta}{T_s} \tag{3-37}$$

where ζ is a constant determined by the signal modulation. For example, if minimum-shift keying is used, the transfer function is rectangular, and many symbols are transmitted during a dwell interval, then $\zeta = 0.8$ if 90 percent of the signal power is included in a frequency channel, and $\zeta = 1.2$ if 99 percent is included.

Spectral splatter is the interference produced in frequency channels other than the one being used by a frequency-hopping pulse. It is caused by the time-limited nature of transmitted pulses. The degree to which spectral splatter may cause errors depends primarily on F_s (see Section 3.1) and the percentage of the signal power included in a frequency channel. Usually, only pulses in adjacent channels produce a significant amount of spectral splatter in a frequency channel.

The *adjacent splatter ratio* K_s is the ratio of the power due to spectral splatter from an adjacent channel to the corresponding power that arrives at the receiver in that channel. For example, if B is the bandwidth of a frequency channel that includes 97 percent of the signal power and $F_s \geq B$, then no more than 1.5 percent of the power from a transmitted pulse can enter an adjacent channel on one side of the frequency channel used by the pulse; therefore, $K_s \leq 0.015$. A given maximum value of K_s can be reduced by an increase in F_s, but eventually the value of M must be reduced if W is fixed. As a result, the rate at which users hop into the same channel increases. This increase may cancel any improvement due to the reduction of the spectral splatter. The opposite procedure (reducing F_s and B so that more frequency channels become available) increases not only the spectral splatter but also signal distortion and intersymbol interference, so the amount of useful reduction is limited.

To avoid spectral spreading due to amplifier nonlinearity, it is desirable for the signal modulation to have a constant envelope, as it is often impossible to implement a filter with the appropriate bandwidth and center frequency for spectral shaping of a signal after it emerges from the final power amplifier. Noncoherent demodulation is nearly always a practical necessity in frequency-hopping systems unless the dwell interval is large. Accordingly, good modulation candidates are DPSK and MSK or some other form of spectrally compact continuous-phase modulation (CPM).

The general form of a CPM signal is

$$s(t) = A \cos[2\pi f_c t + \phi(t, \boldsymbol{\alpha})] \tag{3-38}$$

where A is the amplitude, f_c is the carrier frequency, and $\phi(t, \boldsymbol{\alpha})$ is the phase function that carries the message. The phase function has the ideal form

$$\phi(t, \boldsymbol{\alpha}) = 2\pi h \int_{-\infty}^{t} \left[\sum_{i=-\infty}^{\infty} \alpha_i g(x - iT_s) \right] dx \tag{3-39}$$

where h is a constant called the *deviation ratio* or *modulation index*, T_s is the symbol duration, and the vector $\boldsymbol{\alpha}$ is a sequence of q-ary channel symbols. Each symbol α_i takes one of q values; if q is even,

$$\alpha_i = \pm 1, \pm 3, \ldots, \pm(q - 1), \quad i = 0, 1, 2, \ldots \tag{3-40}$$

Equation (3-39) exhibits the phase continuity and indicates that the phase in any specified symbol interval depends on the previous symbols.

It is assumed that the integrand in (3-39) is piecewise continuous so that $\phi(t, \boldsymbol{\alpha})$ is differentiable. The frequency function of the CPM signal, which is proportional to the derivative of $\phi(t, \boldsymbol{\alpha})$, is

$$\frac{1}{2\pi}\phi'(t, \boldsymbol{\alpha}) = h \sum_{i=-\infty}^{\infty} \alpha_i g(t - iT_s) \tag{3-41}$$

The frequency pulse $g(t)$ is assumed to vanish outside an interval; that is,

$$g(t) = 0 , \quad t < 0 , \quad t > LT_s \tag{3-42}$$

where L is a positive integer and may be infinite. The presence of h as a multiplicative factor in the pulse function makes it convenient to normalize $g(t)$ by assuming that

$$\int_0^{LT_s} g(x)dx = \frac{1}{2} \tag{3-43}$$

If $L = 1$, the continuous-phase modulation is called a *full-response modulation*; if $L > 1$, it is called a *partial-response modulation*, and each frequency pulse extends over two or more symbol intervals. The normalization condition for a full-response modulation implies that the phase change over a symbol interval is equal to $h\pi\alpha_i$.

Continuous-phase frequency-shift keying (CPFSK) is a subclass of CPM for which the instantaneous frequency is constant over each symbol interval. Because of the normalization, a CPFSK frequency pulse is given by

$$g(t) = \begin{cases} \dfrac{1}{2LT_s} , & 0 \leq t \leq LT_s \\ 0 , & \text{otherwise} \end{cases} \tag{3-44}$$

A binary CPFSK signal shifts between two frequencies separated by $f_d = h/T_s$. *Minimum-shift keying(MSK)* is defined as binary CPFSK with $h = 1/2$ and, hence, the frequencies are separated by $f_d = 1/2T_s$. The main difference between CPFSK and MFSK is that h can have any positive value for CPFSK but is relegated to integer values for MFSK so that the tones are orthogonal to each other. A second difference is that MFSK is detected with matched filters and envelope detectors, whereas CPFSK with $h < 1$ is usually detected with a frequency discriminator. Although CPFSK explicitly requires phase continuity and MFSK does not, MFSK is usually implemented with phase continuity to avoid the generation of spectral splatter.

A measure of the spectral compactness of signals is provided by the *fractional out-of-band* power defined as

$$P_{ob}(f) = 1 - \frac{\int_{-f}^{f} S(f')df'}{\int_{-\infty}^{\infty} S(f')df'} , \quad f \geq 0 \tag{3-45}$$

where f is the frequency variable and $S(f)$ is the two-sided power spectral density of the complex envelope of the signal, which is often called the *equivalent lowpass waveform*. The closed-form expressions for the power spectral densities of QPSK and binary MSK (Appendix C.2) can be used to generate Figure 3.6. The graphs depict $P_{ob}(f)$ in decibels as a function of f in units of $1/T_b$, where $T_b = T_s/\log_2 q$ for a q-ary modulation. The *fractional power* within a transmission channel of one-sided bandwidth B is given by

$$K_0 = 1 - P_{ob}(B/2) \tag{3-46}$$

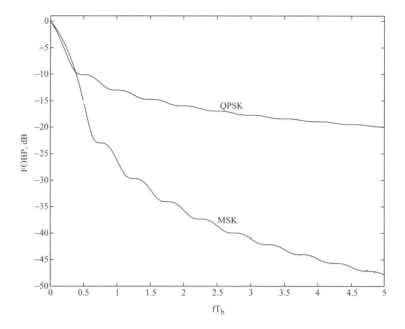

Figure 3.6: Fractional out-of-band power (FOBP) for equivalent lowpass wave-forms of QPSK and MSK.

Usually, the fractional power K_0 must exceed at least 0.9 to prevent significant performance degradation in communications over a bandlimited channel. The transmission bandwidth for which $K_0 = 0.99$ is approximately $1.2/T_b$ for binary MSK, but approximately $8/T_b$ for PSK or QPSK. The adjacent splatter ratio, which is due to out-of-band power on one side of the center frequency, has the upper bound given by

$$K_s < \frac{1}{2}P_{ob}(B/2) \qquad (3\text{-}47)$$

An even more compact spectrum than MSK is obtained by passing the MSK frequency pulses through a Gaussian filter with transfer function

$$H(f) = \exp\left[-\frac{(\ln 2)}{B^2}f^2\right] \qquad (3\text{-}48)$$

where B is the one-sided 3-dB bandwidth. The filter response to an MSK frequency pulse is the *Gaussian MSK (GMSK)* pulse:

$$g(t) = Q\left[\frac{2\pi B}{\sqrt{\ln 2}}(t - \frac{T_s}{2})\right] - Q\left[\frac{2\pi B}{\sqrt{\ln 2}}(t + \frac{T_s}{2})\right] \qquad (3\text{-}49)$$

where $T_s = T_b$. As B decreases, the spectrum of a GMSK signal becomes more compact. However, each pulse has a longer duration and, hence, there is more intersymbol interference. If $BT_b = 0.3$, which is specified in the Global System for Mobile (GSM) cellular communication system, the bandwidth for which $K_0 = 0.99$ is approximately $0.92/T_b$. Each pulse may be truncated for $| t | > 1.5T_s$ with little loss. The performance loss relative to MSK is approximately 0.46 dB for coherent demodulation and presumably also for discriminator demodulation.

An FH/CPM signal has a continuous phase over each dwell interval with N symbols but has a phase discontinuity every $T_h = NT_s + T_{sw}$ seconds at the beginning of another dwell interval. The signal may be expressed as

$$s(t) = \sqrt{2S} \sum_{i=-\infty}^{\infty} w(t - iT_h, T_d) \cos\left[2\pi f_i t + \phi(t, \boldsymbol{\alpha}) + \theta_i\right] \qquad (3\text{-}50)$$

where $S = \mathcal{E}_s/T_s$ is the average signal power during a dwell interval, $w(t, T_d)$ is a unit-amplitude rectangular pulse of duration $T_d = NT_s$, f_i is the carrier frequency during hop-interval i, and θ_i is the phase at the beginning of dwell-interval i.

Consider multitone jamming of an FH/CPM or FH/CPFSK system in which the thermal noise is absent and each jamming tone is randomly placed within a single frequency channel. It is reasonable to assume that a symbol error occurs with probability $(q-1)/q$ when the frequency channel contains a jamming tone with power exceeding S. Thus, (3-34), (3-35), and (3-9) are applicable to FH/CPM or FH/CPFSK, but (3-8) is not. The substitution of (3-9), (3-34), $\mathcal{E}_b = ST_b$, and $I_{t0} = I_t/W$ into (3-35) yield

$$P_s = \begin{cases} \dfrac{q-1}{q} & , & \dfrac{\mathcal{E}_b}{I_{t0}} < BT_b \\[2ex] \left(\dfrac{q-1}{q}\right) BT_b \left(\dfrac{\mathcal{E}_b}{I_{t0}}\right)^{-1} & , & BT_b \le \dfrac{\mathcal{E}_b}{I_{t0}} \le WT_b \\[2ex] 0 & , & \dfrac{\mathcal{E}_b}{I_{t0}} > WT_b \end{cases} \qquad (3\text{-}51)$$

for sophisticated multitone jamming. Since the orthogonality of the MFSK tones is not a requirement for CPM or CPFSK, the bandwidth B for FH/CPM or FH/CPFSK may be much smaller than the bandwidth for FH/MFSK given by (3-8). Thus, P_s may be much lower.

Consider multitone jamming of an FH/DPSK system with negligible thermal noise. Each tone is assumed to have a frequency identical to the center frequency of one of the frequency channels. A DPSK demodulator compares the phases of two successive received symbols. If the magnitude of the phase difference is less then $\pi/2$, then the demodulator decides that a 1 was transmitted; otherwise, it decides that a 0 was transmitted. The composite signal, consisting of the transmitted signal plus the jamming tone, has a constant phase over two successive received symbols in the same dwell interval, if a 1 was transmitted and the thermal noise is absent; thus, the demodulator will correctly detect the 1.

Suppose that a 0 was transmitted. Then the desired signal is $\sqrt{2S} \cos 2\pi f_c t$ during the first symbol and $-\sqrt{2S} \cos 2\pi f_c t$ during the second symbol, respectively, where f_c is the carrier frequency of the frequency-hopping signal during the dwell interval. When a jamming tone is present, trigonometric identities indicate that the composite signal during the first symbol may be expressed as

$$\sqrt{2S}\cos 2\pi f_c t + \sqrt{2I}\cos\left(2\pi_c t + \theta\right) = \sqrt{2S + 2I_t + 4\sqrt{SI}\cos\theta}\cos\left(2\pi f_c t + \phi_1\right)$$

(3-52)

where I is the average power of the tone, θ is the phase of the tone relative to the phase of the transmitted signal, and ϕ_1 is the phase of the composite signal:

$$\phi_1 = \tan^{-1}\left(\frac{\sqrt{I}\sin\theta}{\sqrt{S}+\sqrt{I}\cos\theta}\right)$$

(3-53)

Since the desired signal during the second symbol is $-\sqrt{2S}\cos 2\pi f_c t$, the phase of the composite signal during the second symbol is

$$\phi_2 = \tan^{-1}\left(\frac{\sqrt{I}\sin\theta}{-\sqrt{S}+\sqrt{I}\cos\theta}\right)$$

(3-54)

Using trigonometry, it is found that

$$\cos\left(\phi_2 - \phi_1\right) = \frac{I - S}{\sqrt{S^2 + I^2 + 2SI(1 - 2\cos^2\theta)}}$$

(3-55)

If $I > S$, $|\phi_2 - \phi_1| < \pi/2$ so the demodulator incorrectly decides that a 1 was transmitted. If $I < S$, no mistake is made. Thus, multitone jamming with total power I_t is most damaging when J frequency channels given by (3-34) are jammed and each tone has power $I = I_t/J$. If the information bits 0 and 1 are equally likely, then the symbol error probability given that a frequency channel is jammed with $I > S$ is $P_s = 1/2$, the probability that a 0 was transmitted. Therefore, $P_s = J/2M$ if $I_t \geq S$, and $P_s = 0$, otherwise. Using (3-9) and (3-34) with $S = E_b/T_b$, $I_t = I_{t0}W$, and $\lfloor x \rfloor \approx x$, we obtain the symbol error probability for DPSK and multitone jamming:

$$P_s = \begin{cases} \dfrac{1}{2} & , & \dfrac{\mathcal{E}_b}{I_{t0}} < BT_b \\[2ex] BT_b\left(\dfrac{\mathcal{E}_b}{I_{t0}}\right)^{-1} & , & BT_b \leq \dfrac{\mathcal{E}_b}{I_{t0}} \\[2ex] 0 & , & \dfrac{\mathcal{E}_b}{I_{t0}} > WT_b \end{cases}$$

(3-56)

The same result holds for binary CPFSK.

As implied by Figure 3.6, the bandwidth requirement of DPSK with $K_0 > 0.9$, which is the same as that of PSK or QPSK and less than that of orthogonal FSK, exceeds that of MSK. Thus, if the hopping bandwidth W is fixed, the number of frequency channels available for FH/DPSK is smaller than it is for noncoherent FH/MSK. This increase in B and reduction in frequency channels offsets the intrinsic performance advantage of DPSK and implies that noncoherent FH/MSK will give a lower P_s than FH/DPSK in the presence of worst-case multitone jamming, as indicated in (3-56). Alternatively, if the bandwidth of a frequency channel is fixed, an FH/DPSK signal will experience more distortion and spectral splatter than an FH/MSK signal. Any pulse shaping of the DPSK symbols will alter their constant envelope. An FH/DPSK system is more sensitive to Doppler shifts and frequency instabilities than an FH/MSK system. Another disadvantage of FH/DPSK is due to the usual lack of phase coherence from hop to hop, which necessitates an extra phase-reference symbol at the start of every dwell interval. This extra symbol reduces $\mathcal{E}_s$ by a factor $(N_h - 1)/N_h$, where N_h is the number of symbols per hop or dwell interval and $N_h \geq 2$. Thus, DPSK does not appear to be as suitable a means of modulation as noncoherent MSK for most applications of frequency-hopping communications, and the main competition for MSK comes from other forms of CPM.

The *cross-correlation parameter* for two signals $s_1(t)$ and $s_2(t)$, each with energy $\mathcal{E}_s$, is defined as

$$C = \frac{1}{\mathcal{E}_s} \int_0^{T_s} s_1(t)s_2(t)dt \qquad (3\text{-}57)$$

For CPFSK, two possible transmitted signals, each representing a different channel symbol, are

$$s_1(t) = \sqrt{2\mathcal{E}_s/T_s}\cos(2\pi f_1 t + \phi_1)\,, \quad s_2(t) = \sqrt{2\mathcal{E}_s/T_s}\cos(2\pi f_2 t + \phi_2) \quad (3\text{-}58)$$

The substitution of these equations into (3-57), a trigonometric expansion and discarding of an integral that is negligible if $(f_1+f_2)T_s \gg 1$, and the evaluation of the remaining integral give

$$C = \frac{1}{2\pi f_d T_s}[\sin(2\pi f_d T_s + \phi_d) - \sin\phi_d]\,, \quad f_d \neq 0 \qquad (3\text{-}59)$$

where $f_d = f_1 - f_2$ and $\phi_d = \phi_1 - \phi_2$. Because of the phase synchronization in a coherent demodulator, we may take $\phi_d = 0$. Therefore, the orthogonality condition $C = 0$ is satisfied if $h = f_d T_s = k/2$, where k is any nonzero integer. The smallest value of h for which $C = 0$ is $h = 1/2$, which corresponds to MSK.

In a noncoherent demodulator, ϕ_d is a random variable that is assumed to be uniformly distributed over $[0, 2\pi)$. Equation (3-59) indicates that $E[C] = 0$ for all values of h. The variance of C is

$$var(C) = \left(\frac{1}{2\pi f_d T_s}\right)^2 E\left[\sin^2(2\pi f_d T_s + \phi_d) + \sin^2\phi_d - 2\sin\phi_d\sin(2\pi f_d + \phi_d)\right]$$

$$= \left(\frac{1}{2\pi f_d T_s}\right)^2 (1 - \cos 2\pi f_d T_s)$$

$$= \frac{1}{2}\left(\frac{\sin \pi h}{\pi h}\right)^2 \qquad (3\text{-}60)$$

Table 3.1: Bandwidth(99 percent) for FH/CPFSK.

Symbols/dwell	Deviation ratio	
	$h = 0.5$	$h = 0.7$
1	18.844	18.688
2	9.9375	9.9688
4	5.1875	5.2656
16	1.8906	2.1250
64	1.2813	1.8750
256	1.2031	1.8125
1024	1.1875	1.7969
No hopping	1.1875	1.7813

Since var(C) $\neq 0$ for $h = 1/2$, MSK does not provide orthogonal signals for noncoherent demodulation. If h is any nonzero integer, then both (3-60) and (3-59) indicate that the two CPFSK signals are orthogonal for any ϕ_d. This result justifies the previous assertion that MFSK tones must be separated by $f_d = k/T_s$ to provide noncoherent orthogonal signals.

A noncoherent FH/CPFSK signal can be represented by (3-50). The power spectral density of the complex envelope of this signal, which is the same as the dehopped power spectral density, depends on the number of symbols per dwell interval, N_h, because of the random phases $\{\theta_i\}$. The power spectral density has been calculated [10] for binary CPFSK, assuming that each θ_i is an independent random variable uniformly distributed over $[0,2\pi)$ and the information symbols are ± 1 with equal probability. The 99-percent bandwidths of FH/CPFSK with deviation ratios $h = 0.5$ and $h = 0.7$ are listed in Table 3.1 for different values of N_h. As N_h increases, the power spectral density becomes more compact and approaches that of coherent CPFSK without frequency hopping. For $N_h \geq 64$, the frequency hopping causes little spectral spreading. However, fast frequency hopping, which corresponds to $N_h = 1$, entails a very large 99-percent bandwidth. This fact is the main reason why slow frequency hopping is usually preferable to fast frequency hopping.

With multisymbol noncoherent detection, full-response CPFSK systems can provide a better symbol error probability than coherent PSK systems [11]. For r-symbol detection, where r is odd, the optimal receiver correlates the received waveform over all possible r-symbol patterns before making a decision about the middle symbol. The drawback is the considerable implementation complexity of multisymbol detection, even for three-symbol detection. An additional problem for FH/CPFSK with multisymbol detection is that the first and last $(r-1)/2$ symbols during a dwell interval cannot use the multisymbol detection without accessing other dwell intervals, which may cause practical difficulties.

Symbol-by-symbol noncoherent detection after the dehopping of an FH/CPFSK signal can be inexpensively implemented by using a limiter and frequency discriminator, as illustrated in Figure 3.7. Analysis of the *limiter-discriminator* or *frequency discriminator* [12] provides complicated expressions for the symbol error probability in the presence of white Gaussian noise. However, the

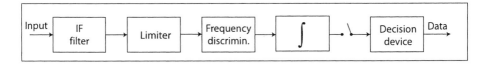

Figure 3.7: Frequency discriminator for CPFSK.

theoretical P_s can be approximated to within a few tenths of a decibel by

$$P_s = \frac{1}{2} \exp\left(-\xi \frac{\mathcal{E}_s}{N_0} \right) \tag{3-61}$$

where the parameter ξ depends on h and the product BT_s, and $N_0/2$ is the two-sided power spectral density of the noise. If the frequency discriminator has a Gaussian IF filter, an integrate-and-dump postdetection filter, and $BT_s = 1$, then it is found that P_s is minimized when $h \approx 0.7$. For CPFSK with $h = 0.7$ and $BT_s = 1$, setting $\xi = 0.7$ in (3-61) provides an approximate least-squares fit to the theoretical curve for P_s over the range $10^{-6} \le P_s \le 10^{-2}$. If $BT_s = 1$, then $\xi = 0.5$ provides a close fit over the same range for orthogonal CPFSK with $h = 1$ and a fairly close fit for MSK($h = 0.5$). Thus, the discriminator demodulation of MSK or orthogonal CPFSK provides approximately the same performance as optimal noncoherent detection of orthogonal FSK. The favorable performance of the frequency discriminator is due to its ability to exploit the phase continuity from symbol-to-symbol of a CPFSK signal. In view of the known 0.46 dB loss of GMSK relative to MSK when coherent demodulation is used, it is expected that P_s for GMSK and discriminator demodulation is well approximated by (3-61) with $\xi = 0.45$.

The practical advantage of noncoherent MSK is that it requires roughly half the bandwidth of orthogonal FSK for specified levels of spectral splatter and intersymbol interference. The increased number of frequency channels due to the decreased value of B does not give FH/MSK an advantage over the AWGN channel. However, the increase is advantageous against a fixed number of interference tones, optimized jamming, and multiple-access interference in a network of frequency-hopping systems, as discussed in the next section. A further increase in the number of frequency channels is possible with FH/GMSK.

Since $\xi = 0.7$ for an FH/CPFSK system with $h = 0.7$, this system has a potential 1.46 dB advantage in $\mathcal{E}_s$ relative to an FH/MSK system with $BT_s = 1$. However, since CPFSK with $h = 0.7$ does not have as compact a spectrum as MSK, the FH/CPFSK system will have increased intersymbol interference due to bandlimiting and spectral splatter relative to the FH/MSK system. Only if these effects are negligible can the potential 1.46 dB advantage be realized. When $N_s \ge 64$, reducing the spectral splatter of the FH/CPFSK to the same level that it is for FH/MSK with $B = 1/T_s$ requires that $B = 1.4/T_s$. The increased bandwidth lowers ξ and decreases the number of frequency channels.

Hybrid Systems

Frequency-hopping systems reject interference by avoiding it, whereas direct-sequence systems reject interference by spreading it. Channel codes are more essential for frequency-hopping systems than for direct-sequence systems because partial-band interference is a more pervasive threat than high-power pulsed interference. When frequency-hopping and direct-sequence systems are constrained to use the same fixed bandwidth, then direct-sequence systems have an inherent advantage because they can use coherent PSK rather than a noncoherent modulation. Coherent PSK has an approximately 4 dB advantage relative to noncoherent MSK over the AWGN channel and an even larger advantage over fading channels. However, the potential performance advantage of direct-sequence systems is often illusory for practical reasons. A major advantage of frequency-hopping systems relative to direct-sequence systems is that it is possible to hop into noncontiguous frequency channels over a much wider band than can be occupied by a direct-sequence signal. This advantage more than compensates for the relatively inefficient noncoherent demodulation that is usually required for frequency-hopping systems. Other major advantages of frequency hopping are the possibility of excluding frequency channels with steady or frequent interference, the reduced susceptibility to the near-far problem (Chapter 6), and the relatively rapid acquisition.

A *hybrid frequency-hopping direct-sequence system* is a frequency-hopping system that uses direct-sequence spreading during each dwell interval or, equivalently, a direct-sequence system in which the carrier frequency changes periodically. In the transmitter of the hybrid system of Figure 3.8, a single code generator controls both the spreading and the hopping pattern. The spreading sequence is added modulo-2 to the data sequence. Hops occur periodically after a fixed number of sequence chips. In the receiver, the frequency hopping and the spreading sequence are removed in succession to produce a carrier with the message modulation. Because of the phase changes due to the frequency hopping, noncoherent modulation, such as DPSK, is usually required unless the hop rate is very low. Serial-search acquisition occurs in two stages. The first stage provides alignment of the hopping patterns, whereas the second stage over the phase of the pseudonoise sequence finishes acquisition rapidly because the timing uncertainty has been reduced by the first stage to less than a hop duration.

A hybrid system curtails partial-band interference in two ways. The hopping allows the avoidance of the interference spectrum part of the time. When the system hops into the interference, the interference is spread and filtered as in a direct-sequence system. However, during a hop interval, interference that would be avoided by an ordinary frequency-hopping receiver is passed by the bandpass filter of a hybrid receiver because the bandwidth must be large enough to accommodate the direct-sequence signal that remains after the dehopping. This large bandwidth also limits the number of available frequency channels, which increases the susceptibility to narrowband interference and the near-far problem. Thus, hybrid systems are seldom used except perhaps in specialized military applications because the additional direct-sequence spreading weakens

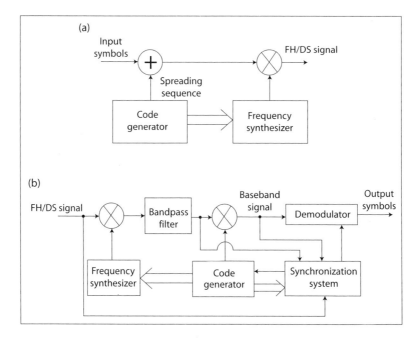

Figure 3.8: Hybrid frequency-hopping direct-sequence system: (a) transmitter and (b) receiver.

the major strengths of frequency hopping.

3.3   Codes for Partial-Band Interference

When partial-band interference is present, let I_{t0} denote the one-sided interference power spectral density that would exist if the power were uniformly distributed over the hopping band. If a fixed amount of interference power is uniformly distributed over J frequency channels out of M in the hopping band, then the fraction of the hopping band with interference is

$$\mu = \frac{J}{M} \tag{3-62}$$

and the interference power spectral density in each of the interfered channels is I_{t0}/μ. When the frequency-hopping signal uses a carrier frequency that lies within the spectral region occupied by the partial-band interference, this interference is modeled as additional white Gaussian noise that increases the noise-power spectral density from N_0 to $N_0 + I_{t0}/\mu$. Therefore, for hard-decision decoding, the symbol error probability is

$$P_s = \mu F \left(\frac{\mathcal{E}_s}{N_0 + I_{t0}/\mu} \right) + (1 - \mu) F \left(\frac{\mathcal{E}_s}{N_0} \right) \tag{3-63}$$

where the conditional symbol error probability $F(\ )$ depends on the modulation and fading. For noncoherent FH/MFSK and the AWGN channel, (1-85) indicates that

$$F(x) = \sum_{i=1}^{q-1} \frac{(-1)^{i+1}}{i+1} \binom{q-1}{i} \exp\left[-\frac{ix}{(i+1)}\right] \qquad (3\text{-}64)$$

where q is the alphabet size of the MFSK symbols. When frequency-nonselective or flat fading (Chapter 5) occurs, the symbol energy may be expressed as $\mathcal{E}_s \alpha^2$, where $\mathcal{E}_s$ represents the average energy and α is a random variable with $E[\alpha^2] = 1$. For Ricean fading, the probability density function of α is

$$f_\alpha(r) = 2(\kappa+1)r \exp\{-\kappa - (\kappa+1)r^2\} I_0(\sqrt{\kappa(\kappa+1)}2r)u(r) \qquad (3\text{-}65)$$

where κ is the Rice factor. Replacing x by $x\alpha^2$ in (3-62), an integration over the density (3-65) and the use of (1-84) yield

$$F(x) = \sum_{i=1}^{q-1}(-1)^{i+1}\binom{q-1}{i}\frac{\kappa+1}{\kappa+1+(\kappa+1+x)i}\exp\left[-\frac{\kappa x i}{\kappa+1+(\kappa+1+x)i}\right]$$
$$(3\text{-}66)$$

When there is no fading and the modulation is binary CPFSK, then (3-61) implies that

$$F(x) = \frac{1}{2}\exp(-\xi x) \qquad (3\text{-}67)$$

For the AWGN channel and no fading, classical communication theory indicates that $F(x)$ for DPSK is given by (3-67) with $\xi = 1$. However, $\mathcal{E}_s$ in (3-63) must be reduced by the factor $N_h/(N_h+1)$ because of the reference symbol that must be included in each dwell interval. When Ricean fading is present, (3-67) and (3-65) yield

$$F(x) = \frac{\kappa+1}{2(\kappa+1)+2\xi x}\exp\left[-\frac{2\xi\kappa x}{2(\kappa+1)+2\xi x}\right] \qquad (3\text{-}68)$$

If μ is treated as a continuous variable over $[0, 1]$ and $I_{t0} \gg N_0$, then straightforward calculations using (3-63) and (3-67) indicate that the worst-case value of μ is

$$\mu_0 = \min\left[\left(\frac{\xi\mathcal{E}_s}{I_{t0}}\right)^{-1}, 1\right] \qquad (3\text{-}69)$$

The corresponding worst-case symbol error probability is

$$P_s = \begin{cases} \dfrac{1}{2e}\left(\dfrac{\xi\mathcal{E}_s}{I_{t0}}\right)^{-1}, & \dfrac{\xi\mathcal{E}_s}{I_{t0}} \geq 1 \\[3mm] \dfrac{1}{2e}\exp\left(-\dfrac{\xi\mathcal{E}_s}{I_{t0}}\right), & \dfrac{\xi\mathcal{E}_s}{I_{t0}} < 1 \end{cases} \qquad (3\text{-}70)$$

which does not depend on M because of the assumption that μ is a continuous variable. For Rayleigh fading and binary FSK, similar calculations using (3-68) with $\kappa = 0$ yield $\mu_0 = 1$. Thus, in the presence of Rayleigh fading, interference spread uniformly over the entire hopping band hinders communications more than interference concentrated over part of the band.

Consider a frequency-hopping system with a fixed hop interval and negligible switching time. For FH/MFSK with a channel code, the bandwidth of a frequency channel must be increased to $B = qB_u/2(\log_2 q)r$, where $r = k/n$ is the code rate and B_u is the bandwidth for binary FSK in the absence of coding. If the bandwidth W of the hopping band is fixed, then the number of disjoint frequency channels available for hopping is reduced to

$$M = \left\lfloor \frac{2(\log_2 q)rW}{qB_u} \right\rfloor \tag{3-71}$$

The energy per channel symbol is

$$\mathcal{E}_s = r(\log_2 q)\mathcal{E}_b \tag{3-72}$$

When the interference is partial-band jamming, J and, hence, μ are parameters that may be varied by a jammer. It is assumed henceforth that M is large enough that μ in (3-63) may be treated as a continuous variable over $[0, 1]$. With this assumption, the error probabilities do not explicitly depend on M.

If a large amount of interference power is received over a small portion of the hopping band, then soft-decision decoding metrics for the AWGN channel will be ineffective because of the possible dominance of a path or code metric by a single symbol metric (cf. Section 2.5 on pulsed interference). Thus, in choosing a suitable code for FH/MFSK in the presence of partial-band interference, we seek one that gives a strong performance when the decoder uses hard decisions and/or erasures.

Reed-Solomon Codes

The use of a Reed-Solomon code with MFSK is advantageous against partial-band interference for two principal reasons. First, a Reed-Solomon code is maximum-distance-separable (Chapter 1) and, hence, accommodates many erasures. Second, the use of nonbinary MFSK symbols to represent code symbols allows a relatively large symbol energy, as indicated by (3-72).

Consider an FH/MFSK system that uses a Reed-Solomon code with no erasures in the presence of partial-band interference and Ricean fading. The demodulator comprises a parallel bank of noncoherent detectors and a device that makes hard decisions. In a slow frequency-hopping system, symbol interleaving among different dwell intervals and subsequent deinterleaving in the receiver may be needed to disperse errors due to the fading or interference and thereby facilitate their removal by the decoder. In a fast frequency-hopping system, symbol errors may be independent so that interleaving is unnecessary. The MFSK modulation implies a q-ary symmetric channel. Therefore, for ideal symbol interleaving and hard-decision decoding of loosely packed codes, (1-26)

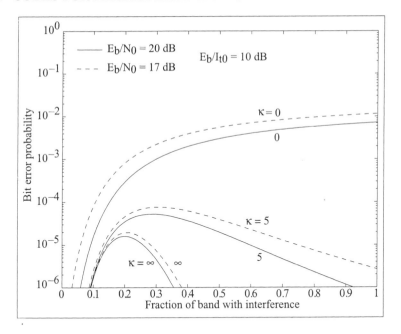

Figure 3.9: Performance of FH/MFSK with Reed-Solomon (32,12) code, non-binary channel symbols, no erasures, and Ricean factor κ.

and (1-27) indicate that

$$P_b \approx \frac{q}{2(q-1)} \sum_{i=t+1}^{n} \binom{n-1}{i-1} P_s^i (1-P_s)^{n-i} \qquad (3\text{-}73)$$

Figure 3.9 shows P_b for FH/MFSK with $q = 32$ and an extended Reed-Solomon (32,12) code in the presence of Ricean fading. The frequency channels are assumed to be separated enough that fading events are independent. Thus, (3-63), (3-64), and (3-73) are applicable. For $\kappa > 0$, the graphs exhibit peaks as the fraction of the band with interference varies. These peaks indicate that for a specific value of $\mathcal{E}_b/I_{t0}$, the concentration of the interference power over part of the hopping band (perhaps intentionally by a jammer) is more damaging than uniformly distributed interference. The peaks become sharper and occur at smaller values of μ as $\mathcal{E}_b/I_{t0}$ increases. For Rayleigh fading, which corresponds to $\kappa = 0$, peaks are absent in the figure, and full-band interference is the most damaging. As κ increases, the peaks appear and become more pronounced.

Much better performance against partial-band interference can be obtained by inserting erasures (Chapter 1) among the demodulator output symbols before the symbol deinterleaving and hard-decision decoding. The decision to erase, which is made independently for each code symbol, is based on *side information*, which indicates which codeword symbols have a high probability of being incorrectly demodulated. The side information must be reliable so that only degraded symbols are erased, not correctly demodulated ones.

Side information may be obtained from known *test symbols* that are transmitted along with the data symbols in each dwell interval of a slow frequency-hopping signal [13]. A dwell interval during which the signal is in partial-band interference is said to be *hit*. If one or more of the N_t test symbols are incorrectly demodulated, then the receiver decides that a hit has occurred, and all codeword symbols in the same dwell interval are erased. Only one symbol of each codeword is erased if the interleaving ensures that only a single symbol of a codeword is in any particular dwell interval. Test symbols decrease the information rate, but this loss is negligible if $N_t \ll N_h$, which is assumed henceforth.

The probability of the erasure of a code symbol is

$$P_\epsilon = \mu P_{\epsilon 1} + (1 - \mu) P_{\epsilon 0} \tag{3-74}$$

where $P_{\epsilon 1}$ is the erasure probability given that a hit occurred, and $P_{\epsilon 0}$ is the erasure probability given that no hit occurred. If δ or more errors among the N_t known test symbols causes an erasure, then

$$P_{\epsilon i} = \sum_{j=\delta}^{N_t} \binom{N_t}{j} P_{si}^j (1 - P_{si})^{N_t - j} , \quad i = 0, 1 \tag{3-75}$$

where P_{s1} is the conditional channel-symbol error probability given that a hit occurred and P_{s0} is the conditional channel-symbol error probability given that no hit occurred.

A codeword symbol error can only occur if there is no erasure. Since test and codeword symbol errors are statistically independent when the partial-band interference is modeled as a white Gaussian process, the probability of a codeword symbol error is

$$P_s = \mu(1 - P_{\epsilon 1}) P_{s1} + (1 - \mu)(1 - P_{\epsilon 0}) P_{s0} \tag{3-76}$$

and the conditional channel-symbol error probabilities are

$$P_{s1} = F\left(\frac{\mathcal{E}_s}{N_0 + I_{t0}/\mu}\right) , \quad P_{s0} = F\left(\frac{\mathcal{E}_s}{N_0}\right) \tag{3-77}$$

where (3-64) is applicable for MFSK symbols. To account for Ricean fading, one must integrate (3-76) and (3-74) over the Ricean density (3-65). In the remainder of this section, we assume the absence of fading.

The word error probability for errors-and-erasures decoding is upper-bounded in (1-35). Since most word errors result from decoding failures, it is reasonable to assume that $P_b \lesssim P_w/2$. Therefore, the information-bit error probability is given by

$$P_b \approx \frac{1}{2} \sum_{j=0}^{n} \sum_{i=i_0}^{n-j} \binom{n}{j} \binom{n-j}{i} P_s^i P_\epsilon^j (1 - P_s - P_\epsilon)^{n-i-j} \tag{3-78}$$

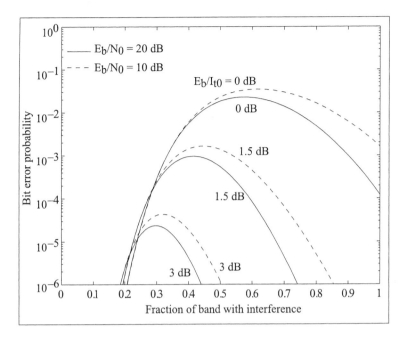

Figure 3.10: Performance of FH/MFSK with Reed-Solomon (32,12) code, non-binary channel symbols, erasures, $N_t = 2$, and no fading.

where $i_0 = \max(0, \lceil (d_m - j)/2 \rceil)$ and $\lceil x \rceil$ denotes the smallest integer greater than or equal to x.

The P_b for FH/MFSK with $q = 32$, an extended Reed-Solomon (32,12) code, and errors-and-erasures decoding with $N_t = 2$ and $\delta = 0$ is shown in Figure 3.10. Fading is absent, and (3-74) to (3-78) are used. A comparison of this figure with the $\kappa = \infty$ graphs of Figure 3.9 indicates that when $\mathcal{E}_b/N_0 = 20$ dB, erasures provide nearly a 7 dB improvement in the required $\mathcal{E}_b/I_{t0}$ for $P_b = 10^{-5}$. The erasures also confer immunity to partial-band interference that is concentrated in a small fraction of the hopping band and decrease the sensitivity to $\mathcal{E}_b/N_0$.

There are other options for generating side information and, hence, erasure insertion in addition to demodulating test symbols. One might use a radiometer to measure the energy in the current frequency channel, a future channel, or an adjacent channel. Erasures are inserted if the energy is inordinately large. This method does not have the overhead cost in information rate that is associated with the use of test symbols. Other methods without overhead cost use iterative decoding [14], the soft information provided by the inner decoder of a concatenated code, or the outputs of the parallel MFSK envelope detectors.

Consider the decision variables applied to the MFSK decision device of Figure 3.5(b). The *output threshold test* (OTT) compares the largest decision variable to a threshold to determine whether the corresponding demodulated

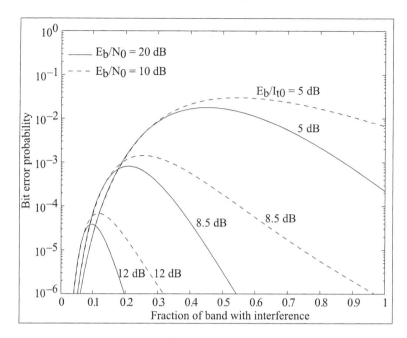

Figure 3.11: Performance of FH/MFSK with Reed-Solomon (8,3) code, nonbinary channel symbols, erasures, $N_t = 4$, and no fading.

symbol should be erased. The *ratio threshold test* (RTT) computes the ratio of the largest decision variable to the second largest one. This ratio is then compared to a threshold to determine an erasure. If the values of both $\mathcal{E}_b/N_0$ and $\mathcal{E}_b/I_{t0}$ are known, then optimum thresholds for the OTT, the RTT, or a hybrid method can be calculated [15]. It is found that the OTT tends to outperform the RTT when $\mathcal{E}_b/I_{t0}$ is sufficiently low, but the opposite is true when $\mathcal{E}_b/I_{t0}$ is sufficiently high. If side information concerning the presence or absence of the partial-band interference is available at the receiver and if the interference power is high, then a threshold determined by $\mathcal{E}_b/N_0$ only and a separate threshold determined by $\mathcal{E}_b/(N_0 + I_{t0})$ can be used to further improve the performance of the errors and erasures decoding. The main disadvantage of the OTT and the RTT relative to the test-symbol method is the need to estimate $\mathcal{E}_b/N_0$ and either $\mathcal{E}_b/I_{t0}$ or $\mathcal{E}_b/(N_0 + I_{t0})$.

Proposed erasure methods are based on the use of MFSK symbols, and their performances against partial-band interference improve as the alphabet size q increases. For a fixed hopping band, the number of frequency channels decreases as q increases, thereby making an FH/MFSK system more vulnerable to narrowband jamming signals (Section 3.2) or multiple-access interference (Chapter 6). Thus, we examine alternatives that give less protection against partial-band interference in exchange for enhanced protection against multiple-access interference.

Figure 3.11 depicts P_b for FH/MFSK with $q = 8$, an extended Reed-Solomon

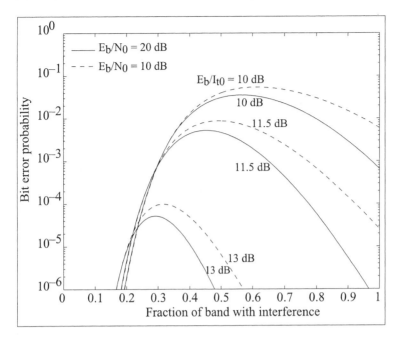

Figure 3.12: Performance of FH/DPSK with Reed-Solomon (32,12) code, binary channel symbols, erasures, $N_t = 10$, and no fading.

(8,3) code, $N_t = 4$, and $\delta = 0$. A comparison of Figures 3.11 and 3.10 indicates that reducing the alphabet size while preserving the code rate has increased the system sensitivity to $\mathcal{E}_b/N_0$, increased the susceptibility to interference concentrated in a small fraction of the hopping band, and raised the required $\mathcal{E}_b/I_{t0}$ for a specified P_b by 5 to 9 dB.

Another approach is to represent each nonbinary code symbol by a sequence of $m = \log_2 q$ consecutive binary channel symbols. Then an FH/MSK or FH/DPSK system can be implemented to provide a large number of frequency channels and, hence, better protection against multiple-access interference. Equations (3-74), (3-75), and (3-77) are still valid. However, since a code-symbol error occurs if any of its m component channel symbols is incorrect, (3-76) is replaced by

$$P_s = 1 - [1 - \mu(1 - P_{\epsilon1})P_{s1} - (1 - \mu)(1 - P_{\epsilon0})P_{s0}]^m \qquad (3\text{-}79)$$

and (3-64) is replaced by (3-67), where $\xi = 1/2$ for MSK and $\xi = 1$ for DPSK. The results for an FH/DPSK system with an extended Reed-Solomon (32,12) code, $N_t = 10$ binary test symbols, and $\delta = 0$ are shown in Figure 3.12. It is assumed that $N_h \gg 1$ so that the loss due to the reference symbol in each dwell interval is negligible. The graphs in Figure 3.12 are similar in form to those of Figure 3.10, but the transmission of binary rather than nonbinary symbols has caused approximately a 10 dB increase in the required $\mathcal{E}_b/I_{t0}$ for a specified P_b.

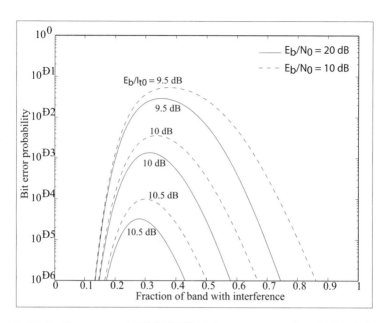

Figure 3.13: Performance of FH/DPSK with concatenated code, hard decisions, and no fading. Inner code is convolutional (rate = 1/2, $K = 7$) code and outer code is Reed-Solomon (31,21) code.

Figure 3.12 is applicable to orthogonal FSK and MSK if $\mathcal{E}_b/I_{t0}$ and $\mathcal{E}_b/N_0$ are both increased by 3 dB to compensate for the lower value of ξ.

An alternative to erasures that uses binary channel symbols is an FH/DPSK system with concatenated coding, which has the form illustrated in Figures 1.14 and 1.15. Although generally unnecessary in a fast frequency-hopping system, the channel interleaver and deinterleaver may be required in a slow frequency-hopping system to ensure independent symbol errors at the decoder input. Consider a concatenated code comprising a Reed-Solomon (n, k) outer code and a binary convolutional inner code. The inner Viterbi decoder performs hard-decision decoding to limit the impact of individual symbol metrics. Assuming that $N_h >> 1$, the symbol error probability is given by (3-63) and (3-67) with $\xi = 1$. The probability of a Reed-Solomon symbol error, P_{s1}, at the output of the Viterbi decoder is upper-bounded by (1-127) and (1-114). Setting $P_s = P_{s1}$ in (3-73) then provides an upper bound on P_b. Figure 3.13 depicts this bound for an outer Reed-Solomon (31,21) code and an inner rate-1/2, $K = 7$ convolutional code. This concatenated code provides a better performance than the Reed-Solomon (32,12) code with binary channel symbols, but a much worse performance than the latter code with nonbinary channel symbols. Figures 3.10 through 3.13 indicate that a reduction in the alphabet size for channel symbols increases the system susceptibility to partial-band interference. The primary reason is the reduced energy per channel symbol.

Trellis-Coded Modulation

Trellis-coded modulation is a combined coding and modulation method that is usually applied to coherent digital communications over bandlimited channels (Chapter 1). Multilevel and multiphase modulations are used to enlarge the signal constellation while not expanding the bandwidth beyond what is required for the uncoded signals. Since the signal constellation is more compact, there is some modulation loss that detracts from the coding gain, but the overall gain can be substantial. Since a noncoherent demodulator is usually required for frequency-hopping communications, the usual coherent trellis-coded modulations are not suitable. Instead, the trellis coding may be implemented by expanding the signal set for $M/2$-ary MFSK to M-ary MFSK [16]. Although the frequency tones are uniformly spaced, they are allowed to be nonorthogonal to limit or avoid bandwidth expansion.

Trellis-coded 4-ary MFSK is illustrated in Figure 3.14 for a rate-1/2 code with four states. The signal set partitioning, shown in Figure 3.14(a), partitions the set of four signals or tones into two subsets, each with two tones. The partitioning doubles the frequency separation between tones from Δ Hz to 2Δ Hz. The mapping of code bits into signals is indicated. In Figure 3.14(b), the numerical labels denote the signal assignments associated with the state transitions in the trellis for a four-state encoder. The bandwidth of the frequency channel that accommodates the four tones is approximately $B = 4\Delta$.

There is a trade-off in the choice of Δ because a small Δ allows more frequency channels and thereby limits the effect of multiple-access interference or multitone jamming, whereas a large Δ tends to improve the system performance against partial-band interference. If a trellis code uses four orthogonal tones with spacing $\Delta = 1/T_b$, where T_b is the bit duration, then $B = 4/T_b$. The same bandwidth results when an FH/FSK system uses two orthogonal tones, a rate-1/2 code, and binary channel symbols since $B = 2/T_s = 4/T_b$. The same bandwidth also results when a rate-1/2 binary convolutional code is used and each pair of code symbols is mapped into a 4-ary channel symbol. The performance of the 4-state, trellis-coded, 4-ary MFSK frequency-hopping system [16] indicates that it is not as strong against worst-case partial-band interference as an FH/MFSK system with a rate-1/2 convolutional code and 4-ary channel symbols or an FH/FSK system with a Reed-Solomon (32,16) code and errors-and-erasures decoding. Since the latter system is weaker than the FH/DPSK system used in Figure 14, we find that trellis-coded modulation is relatively weak against partial-band interference. The advantage of trellis-coded modulation in a frequency-hopping system is its relatively low implementation complexity.

Turbo Codes

Turbo codes provide an alternative to errors and erasures decoding for suppressing partial-band interference. A turbo-coded frequency-hopping system that uses spectrally compact channel symbols will also resist multiple-access interference. An accurate estimate of the variance of the interference plus noise,

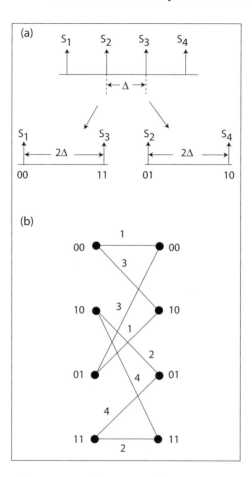

Figure 3.14: Rate-1/2, four-state trellis-coded 4-ary MFSK: (a) signal set partitioning and mapping of bits to signals, and (b) mapping of signals to state transitions.

which is modeled as zero-mean, white Gaussian noise, is always needed in the iterative turbo decoding algorithm (Chapter 1). When the channel dynamics are much slower than the hop rate, all the received symbols of a dwell interval may be used in estimating the variance associated with that dwell interval.

Consider an FH/DPSK system in which each code bit can take the values +1 or −1. The dwell interval is too short for conventional phase synchronization to be practical. The architecture of interactive turbo decoding and channel estimation is illustrated in Figure 3.15. As explained in Chapter 1, the *log-likelihood ratio* (LLR) of a bit u_k conditioned on a received sequence $\mathbf{y}_i$ of demodulator outputs applied to decoder i is defined as the natural logarithm

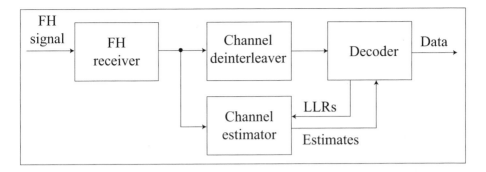

Figure 3.15: Receiver and decoder architecture for frequency-hopping system with turbo code.

of the ratio of the *a posteriori* probabilities:

$$\Lambda_{ki} = \ln\left[\frac{P(u_k = +1|\mathbf{y}_i)}{P(u_k = -1|\mathbf{y}_i)}\right], \quad i = 1, 2 \tag{3-80}$$

Successive estimates of the LLRs of the code bits are computed by each component decoder during the iterative decoding of the turbo code. The usual turbo decoding is extended to include the iterative updating of the LLRs of both the information and parity bits. After each iteration by a component decoder, its LLRs are updated and the extrinsic information is transferred to the other component decoder. The fact that

$$P(u_k = -1|\mathbf{y}_i) = 1 - P(u_k = +1|\mathbf{y}_i) \tag{3-81}$$

and (3-80) imply that the *a posteriori* probabilities are

$$p_{1ki} = P(u_k = +1|\mathbf{y}_i) = \frac{1}{1 + \exp(-\Lambda_{ki})} \tag{3-82}$$

$$p_{0ki} = P(u_k = -1|\mathbf{y}_i) = \frac{\exp(-\Lambda_{ki})}{1 + \exp(-\Lambda_{ki})} \tag{3-83}$$

These equations indicate that the channel estimator can convert a LLR transferred after a component decoder iteration into the probabilities p_{1ki} and p_{0ki}. Using these probabilities for all the bits in a dwell interval, estimates of the independent random carrier phase, the fading attenuation, and the noise variance for each dwell interval can be integrated into the iterative decoding of a turbo code if these parameters are constants over the dwell interval [17].

After the dehopping, the received signal for symbol k of a dwell interval is

$$r_k(t) = \text{Re}\left[\sqrt{2\mathcal{E}_s}u_k\psi(t)\alpha e^{j(2\pi f_0 t + \theta)}\right] + n_k(t), \quad 0 \le t \le T_s \tag{3-84}$$

where $\mathcal{E}_s$ is the symbol energy when $\alpha = 1$, T_s is the symbol duration, f_0 is the intermediate frequency, $u_k = +1$ when binary symbol k is a 1 and $u_k = -1$ when binary symbol k is a 0, $\psi(t)$ is the unit-energy symbol waveform, α is the fading attenuation, and $n_k(t)$ is independent, zero-mean, white Gaussian noise with two-sided noise power spectral density $N_0/2$. The phase shift θ is introduced by the transmission channel and is assumed to be constant during the dwell interval.   A derivation similar to that of (1-56) indicates that the conditional probability density of demodulator output y_k given the values of u_k, N_0, and $C = \sqrt{\mathcal{E}_s/2}\alpha e^{j\theta}$ is

$$f\left(y_k \mid u_k, N_0, C\right) = \frac{1}{\pi N_0/2} \exp\left(-\frac{|y_k - u_k C|^2}{N_0/2}\right), \quad k = 1, 2, \ldots, N_h \quad (3\text{-}85)$$

where N_h is the number of demodulator outputs during the dwell interval. After forming the log-likelihood function for the set of demodulator outputs during a dwell interval, the maximum-likelihood estimates of N_0 and C are found by calculating those values that maximize the log-likelihood function. Straightforward calculations indicate that the maximum-likelihood estimates are

$$\widehat{C} = \frac{1}{N_h} \sum_{k=1}^{N_h} y_k u_k \tag{3-86}$$

$$\widehat{N_0} = \frac{2}{N_h} \sum_{k=1}^{N_h} \left|y_k - u_k \widehat{C}\right|^2 \tag{3-87}$$

Since the $\{u_k\}$ are unknown, estimates are obtained by calculating approximate expected values of these expressions. If p_{0k} is the most recently computed value of p_{0k1} or p_{0k2}, then suitable estimates are

$$\widehat{C} = \frac{1}{a} \sum_{k=1}^{N_h} (1 - 2p_{0k}) y_k \tag{3-88}$$

$$\widehat{N_0} = \frac{1}{b} \sum_{k=1}^{N_h} \left[p_{0k} \left|y_k + \widehat{C}\right|^2 + (1 - p_{0k}) \left|y_k - \widehat{C}\right|^2 \right] + c \tag{3-89}$$

where $a = \sum_{k=1}^{N_h} (1 - 2p_{0k})^2$, b, and c are factors adjusted to make the estimates unbiased.   As the decoders provide progressively improved estimates of the $\{p_{0k}\}$, the estimates $\widehat{C}$ and $\widehat{N_0}$ also improve.   Substitution of these estimates into (3-85) and the evaluation of (1-135) yield

$$L(y_{sk}|u_k) = \frac{8\mathrm{Re}\left(\widehat{C}^* y_{sk}\right)}{\widehat{N_0}} \tag{3-90}$$

which represents the information about u_k provided by y_{sk}. If known symbols are inserted into the dwell interval, then we set $p_{0k} = 1$ if $u_k = +1$ and $p_{0k} = 0$

if $u_k = -1$. If the fading attenuation has the known value α, then (3-88) is still a suitable estimate if a is adjusted to ensure that the magnitude of $\widehat{C}$ is equal to $\sqrt{\mathcal{E}_s/2}\alpha$. If phase synchronization is available but the dynamics of the transmission channel are faster than the hop rate, then $\widehat{N_0}$ must be separately estimated for each symbol and, hence, (3-90) should be replaced by (1-145), as shown in Chapter 1.

A simulation of a turbo-coded FH/DPSK system [17] that uses the preceding estimates indicates that its performance is more than 2 dB better than that shown in Figure 3.10. The rate-1/3 turbo code uses two 4-state systematic recursive convolutional encoders, each with octal generator (5,7), a 200-bit turbo interleaver, ideal channel interleaving, 5 decoder iterations, $N_h = 10$, and $\mathcal{E}_b/N_0$ = 20 dB. For a sufficiently large dwell interval, the resulting performance is almost as good as theoretically possible with perfect side information about the carrier phase and the fading attenuation. Known symbols may be inserted into the transmitted code symbols to facilitate the estimation, but the energy per information bit is reduced. Increasing N_h improves the estimates because they may be based on more observations and more known symbols can be accommodated. However, since the reduction in the number of independent hops per information block of fixed size decreases the diversity, and hence the independence of errors, there is a limit on N_h beyond which a performance degradation occurs.

Although turbo codes are generally used with binary channel symbols, their error-control capabilities are strong enough to compensate for the relatively low channel-symbol energy. However, if the system latency and computational complexity of turbo codes is unacceptable, then there is a trade-off in the choice of the modulation and code.

Turbo product codes (Chapter 1) are an attractive option because of their reduced complexity compared with other turbo codes. The outer encoder fills the block interleaver row-by-row with the outer codewords. Since the interleaver columns are read by the inner encoder to provide the channel symbols, there is an inherent interleaving of the inner code. Since the outer code is not inherently interleaved, the channel interleaver of Figure 1-14 is an essential part of the transmitter. The channel interleaver precludes the possibility that sufficiently corrupted outer codewords due to dwell intervals hit by interference can undermine the iterative process in the turbo decoder, which is illustrated in Figure 1-20. Side information about whether or not a hit has occurred is obtained by hard-decision decoding of the inner codewords. The metric for determining a hit occurrence is the Hamming distance between the binary sequence resulting from the hard decisions and the codewords obtained by bounded-distance decoding. When full interleaving and side information are used, the turbo product code is competitive in performance with other turbo codes except for a slight inferiority against partial-band interference occupying a small fraction of the hopping band [18].

3.4   Frequency Synthesizers

A *frequency synthesizer* converts a standard reference frequency into a different desired frequency. In a frequency-hopping system, the frequencies of the hopset must be synthesized. In practical applications, the frequencies of the hopset have the form

$$f_{hi} = af_1 + b_i f_r , \quad i = 1, 2, ..., M \tag{3-91}$$

where a and the $\{b_i\}$ are rational numbers, f_r is the reference frequency, and f_1 is a frequency in the spectral band of the hopset. The *reference signal*, which is a tone at the reference frequency, is usually the output of a frequency divider or multiplier fed by a stable frequency source, such as an atomic or crystal oscillator. The use of a single reference signal, which even generates f_1, ensures that any output frequency of the synthesizer has the same stability and accuracy as the reference. The three fundamental types of frequency synthesizers are the direct, digital, and indirect synthesizers. Most practical synthesizers are hybrids of these fundamental types [19], [20], [21], [22].

Direct Frequency Synthesizer

A *direct frequency synthesizer* uses frequency multipliers and dividers, mixers, bandpass filters, and electronic switches to produce signals with the desired frequencies. Direct frequency synthesizers provide both very fine resolution and high frequencies, but often require a very large amount of hardware and do not provide a phase-continuous output after frequency changes. Although a direct synthesizer can be realized with programmable dividers and multipliers, the standard approach is to use the *double-mix-divide* (DMD) system illustrated in Figure 3.16. The reference signal at frequency f_r is mixed with a tone at the fixed frequency f_a. The bandpass filter selects the sum frequency $f_r + f_a$ produced by the mixer. Another mixing and filtering operation with a tone at $f_b + f_1$ produces the frequency $f_r + f_a + f_b + f_1$. If the fixed frequencies f_a and f_b are chosen so that

$$f_a + f_b = 9f_r \tag{3-92}$$

then the divider produces the output frequency $f_r + f_1/10$. In principle, a single mixer and bandpass filter could produce this output frequency, but two mixers and bandpass filters simplify the filters. Each bandpass filter must select the sum frequency while suppressing the difference frequency and the mixer input frequencies, which may enter the filter because of mixer leakage. If the sum frequency is too close to one of these other frequencies, the bandpass filter becomes prohibitively complex and expensive.

The DMD system of Figure 3.16 can be used as a module in a direct frequency synthesizer that can achieve arbitrary frequency resolution by cascading enough DMD modules. A synthesizer that provides two-digit resolution is shown in Figure 3.17. When the synthesizer is used in a frequency-hopping system, the control bits are produced by the code generator. Each decade switch passes a single tone to a DMD module. The ten tones that are available to the decade switches may be produced by applying the reference frequency to

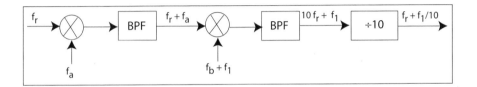

Figure 3.16: Double-mix-divide module, where BPF = bandpass filter.

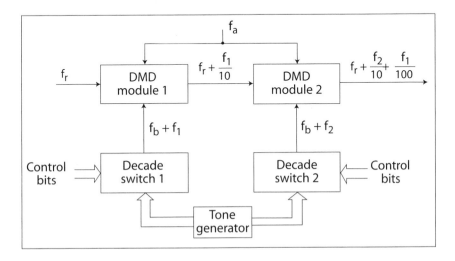

Figure 3.17: Direct frequency synthesizer with two-digit resolution.

appropriate frequency multipliers and dividers in the tone generator. Equation (3-92) ensures that the output frequency of the second bandpass filter in DMD module 2 is $10f_r + f_2 + f_1/10$. Thus, the final synthesizer output frequency is $f_r + f_2/10 + f_1/100$.

Example 1. It is desired to produce a 1.79 MHz tone. Let $f_r = 1$ MHz and $f_b = 5$ MHz. The ten tones provided to the decade switches are 5, 6, 7, ..., 14 MHz so that f_1 and f_2 can range from 0 to 9 MHz. Equation (3-92) yields $f_a = 4$ MHz. If $f_1 = 7$ MHz and $f_2 = 9$ MHz, then the output frequency is 1.79 MHz. The frequencies f_a and f_b are such that the designs of the bandpass filters inside the modules are reasonably simple. □

Digital Frequency Synthesizer

A *digital frequency synthesizer* converts the stored sample values of a sine wave into an analog sine wave with a specified frequency. The periodic and symmetric character of a sine wave implies that only values for the first quadrant need to be stored. The basic elements of a digital frequency synthesizer are shown in Figure 3.18. A set of bits, which are produced by the code generator in a frequency-

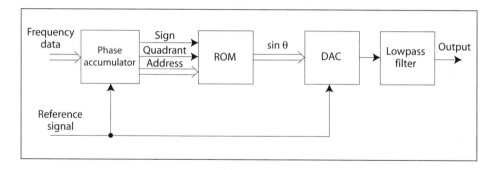

Figure 3.18: Digital frequency synthesizer.

hopping system, determine the synthesized frequency by specifying a phase increment δ. The accumulator converts the phase increment into successive samples of the phase by adding the increment to the content of an internal register after every cycle of the reference signal. A phase sample $\theta = i\delta$, $i = 1,2,\ldots,N$, defines an address in the read-only memory (ROM) at which the value $\sin\theta$ is stored. This value is applied to a digital-to-analog converter (DAC), which performs a sample-and-hold operation at a sampling rate equal to the reference frequency f_r. The converter output is applied to an anti-aliasing lowpass filter with a cutoff frequency less than f_r. The output of the lowpass filter is the desired analog signal.

The numerical capacity of the accumulator determines the maximum number of ROM addresses that the phase accumulator can specify. One sample value of $\sin\theta$ is read out after every cycle of the reference signal. If N sample values are read out during each period of $\sin\theta$, then the frequency of the analog signal produced is

$$\Delta = \frac{f_r}{N} \tag{3-93}$$

where f_r is the frequency of the reference signal. The output frequency f_{0k} is produced when every kth stored sample value is read out at the reference rate. Thus, if the phase accumulator increments by k after every cycle of the reference signal, then

$$f_{0k} = k\Delta \tag{3-94}$$

which implies that Δ is the frequency resolution and the minimum frequency that can be generated by the synthesizer.

The maximum frequency f_m that can be generated is produced by using only a few samples of $\sin\theta$ per period. From the Nyquist sampling theorem, it is known that $f_m < f_r/2$ is required to avoid aliasing. Practical DAC and lowpass filter requirements further limit f_m to approximately $0.4\ f_r$ or less. Thus, $q \geq 2.5$ samples of $\sin\theta$ per period are used in synthesizing f_m, and

$$f_m = \frac{f_r}{q} \tag{3-95}$$

The lowpass filter may be implemented with a linear phase across a flat passband extending to approximately f_m. The frequencies f_r and f_m are limited by the speed of the digital-to-analog converter.

Let ν denote the number of bits in the accumulator register. The numerical capacity of the accumulator is $2^\nu \geq N$. Suppose that f_{min} and f_{max} are specified minimum and maximum frequencies that must be produced by a synthesizer. Equations (3-93) and (3-95) imply that $f_{min} \geq f_r/N$ and $f_{max} \leq f_r/q$. Therefore, $qf_{max}/f_{min} \leq N \leq 2^\nu$ and the required number of accumulator bits is

$$\nu = \lfloor \log_2(qf_{max}/f_{min}) \rfloor + 1 \qquad (3\text{-}96)$$

where $\lfloor x \rfloor$ denotes the largest integer in x.

The ROM stores 2^n or fewer distinct n-bit words. Each word represents one possible value of $\sin \theta$ in the first quadrant or, equivalently, one possible magnitude of $\sin \theta$. The input to the ROM comprises $n + 2$ parallel bits. The two most significant bits are the *sign bit* and the *quadrant bit*. The sign bit specifies the polarity of $\sin \theta$. The quadrant bit specifies whether $\sin \theta$ is in the first or second quadrants or in the third or fourth quadrants. The n least significant bits of the input determine the address in which the magnitude of $\sin \theta$ is stored. The address specified by the n least significant bits is appropriately modified by the quadrant bit when θ is in the second or fourth quadrants. The sign bit becomes one of the $n + 1$ ROM output bits. The phase accumulator uses $\nu \geq n+2$ bits. Since $n + 2$ bits are needed by the ROM, the $\nu - n - 2$ least significant bits in the accumulator are not applied to the ROM. The memory requirements of a ROM and the number of its input bits can be reduced by using trigonometric identities and hardware multipliers.

Since n ROM output bits specify the magnitude of $\sin \theta$, the quantization error produces the worst-case noise power

$$E_q = (2^{-n})^2 = -6n \ dB \qquad (3\text{-}97)$$

in the digital-to-analog converter output. The magnitude of E_q is called the *spectral purity* of the synthesizer.

Example 2. A digital synthesizer is to be designed to cover 1 kHz to 1 MHz with a spectral purity greater than 45 dB. According to (3-97), the use of 8-bit words in the ROM is adequate for the required spectral purity. The ROM contains $2^8 = 256$ or fewer distinct words and requires $n + 2 = 10$ input bits. If $2.5 \leq q \leq 4$, then since $f_{max}/f_{min} = 10^3$, (3-96) yields $\nu = 12$. Thus, a 12-bit phase accumulator is needed. Since $2^{12} = 4096$, we may choose N = 4000. If the frequency resolution and smallest frequency is to be $\Delta = 1$ kHz, then (3-93) indicates that $f_r = 4$ MHz is required. When the frequency Δ is desired, the phase increments are so small that $2^{\nu-n-2} = 4$ increments occur before a new address is specified and a new value of $\sin \theta$ is produced. Thus, the 4 least-significant bits in the accumulator are not used in the addressing of the ROM. $\square$

The direct digital synthesizer can be easily modified to produce a modulated output when high-speed digital data is available. For amplitude modulation, the ROM output is applied to a multiplier. Phase modulation may be implemented by adding the appropriate bits to the phase accumulator output. Frequency modulation entails a modification of the accumulator input bits. For a quaternary modulation, separate sine and cosine ROMs may be used.

A digital frequency synthesizer can produce nearly instantaneous, phase-continuous frequency changes and a very fine frequency resolution despite its relatively small size, weight, and power requirements. A disadvantage is the limited maximum frequency, which restricts the bandwidth of the covered frequencies following a frequency translation of the synthesizer output. For this reason, digital frequency synthesizers are often used as components in hybrid synthesizers. Another disadvantage is the stringent requirement for the lowpass filter to suppress frequency spurs generated during changes in the synthesized frequency.

Indirect Frequency Synthesizers

An *indirect frequency synthesizer* uses voltage-controlled oscillators and feedback loops. Indirect synthesizers usually require less hardware than comparable direct ones, but require more time to switch from one frequency to another. Like digital synthesizers, indirect synthesizers inherently produce phase-continuous outputs after frequency changes. The principal components of a single-loop indirect synthesizer, which is similar in operation to a phase-locked loop, are depicted in Figure 3.19. The control bits, which determine the value of the modulus or divisor N, are supplied by a code generator. The input signal at frequency f_1 may be provided by another synthesizer. Since the feedback loop forces the frequency of the divider output, $(f_0 - f_1)/N$, to closely approximate the reference frequency f_r, the output of the voltage-controlled oscillator (VCO) is a sine wave with frequency

$$f_0 = Nf_r + f_1 \qquad (3\text{-}98)$$

where N is a positive integer. *Phase detectors* in frequency-hopping synthesizers are usually digital devices that measure zero-crossing times rather than the phase differences measured when mixers are used. Digital phase detectors have an extended linear range, are less sensitive to input-level variations, and simplify the interface with a digital divider.

Since the output frequencies change in increments of f_r, the frequency resolution of the single-loop synthesizer is f_r. For stable operation and the suppression of sidebands that are offset from f_0 by f_r, it is desirable that the loop bandwidth be on the order of 0.1 f_r. The *switching time* t_s for changing frequencies, which is inversely proportional to the loop bandwidth, is roughly approximated by

$$t_s = \frac{25}{f_r} \qquad (3\text{-}99)$$

This equation indicates that a low resolution and a low switching time may not be achievable by a single loop. The switching time t_s is less than or equal

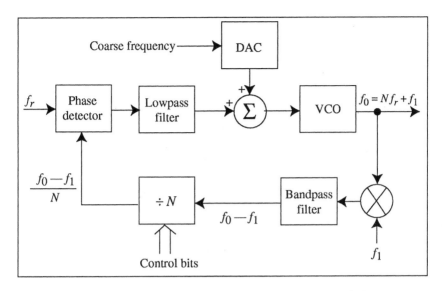

Figure 3.19: Indirect frequency synthesizer with single loop.

to T_{sw} defined previously for frequency-hopping pulses, which may have additional guard time inserted. To decrease the switching time while maintaining the frequency resolution of a single loop, a coarse steering signal can be stored in a ROM, converted into analog form by a digital-to-analog converter (DAC), and applied to the VCO (as shown in Figure 3.19) immediately after a frequency change. The steering signal reduces the frequency step that must be acquired by the loop when a hop occurs. An alternative approach is to place a fixed divider with modulus M after the loop so that the output frequency is $f_0 = N f_r / M + f_1 / M$. By this means, f_r can be increased without sacrificing resolution provided that the VCO output frequency, which equals $M f_0$, is not too large for the divider in the feedback loop. To limit the transmission of spurious frequencies, it may be desirable to inhibit the transmitter output during frequency transitions.

The switching time can be dramatically reduced by using two synthesizers that alternately produce the output frequency. One synthesizer produces the output frequency while the second one is being tuned to the next frequency following a command from the code generator. If the hop duration exceeds the switching time of each synthesizer, then the second synthesizer begins producing the next frequency before a control switch routes its output to a modulator or a dehopping mixer.

A *divider* is a binary counter that produces a square-wave output. The divider counts down by one unit every time its input crosses zero. If the modulus or divisor is the positive integer N, then after N zero crossings, the divider output crosses zero and changes state. The divider then resumes counting down from N. Programmable dividers have limited operating speeds that impair their ability to accommodate a high-frequency VCO output. A problem is avoided by the down-conversion of the VCO output by the mixer shown in Figure 3.19, but spurious components are introduced. Since fixed dividers can operate at much higher speeds than programmable dividers, one might consider placing a fixed divider before the programmable divider in the feedback loop. However, if the fixed divider has a modulus N_1, then the loop resolution becomes $N_1 f_r$ so this solution is usually unsatisfactory.

A *dual-modulus divider*, which is depicted in Figure 3.20, allows synthesizer operation at high frequencies while maintainingthe frequency resolution equal

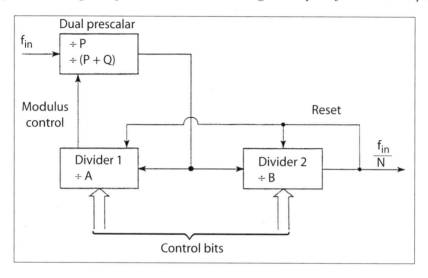

Figure 3.20: Dual-modulus divider.

to f_r. The *dual prescalar* consists of two fixed dividers with divisors equal to the positive integers P and $P + Q$. The two programmable dividers count down from the integers A and B, where $B > A$ and A is nonnegative. These programmable dividers are only required to accommodate a frequency f_{in}/P. The dual prescalar initially divides by the modulus $P + Q$. This modulus changes whenever a programmable divider reaches zero. After $(P + Q)A$ input transitions, divider 1 reaches zero, and the modulus control causes the dual prescalar to divide by P. Divider 2 has counted down to $B - A$. After $P(B - A)$ more input transitions, divider 2 reaches zero and causes an output transition. The two programmable dividers are then reset, and the dual prescalar reverts to division by $P + Q$. Thus, each output transition corresponds to $A(P + Q) + P(B - A) = AQ + PB$ input transitions, which implies that the dual-modulus divider has a modulus

$$N = AQ + PB , \quad B > A \tag{3-100}$$

and produces the output frequency f_{in}/N.

If $Q = 1$ and $P = 10$, then the dual-modulus divider is called a *10/11 divider*, and

$$N = 10B + A , \quad B > A \tag{3-101}$$

which can be increased in unit steps by changing A in unit steps. Since $B > A$ is required, a suitable range for A and minimum value of B are

$$0 \leq A \leq 9 , \quad B_{\min} = 10 \tag{3-102}$$

The relations (3-98), (3-101), and (3-102) indicate that the range of a synthesized hopset is from $f_1 + 100f_r$ to $f_1 + (10B_{\max} + 9)f_r$. Therefore, a spectral band between f_{min} and f_{max} is covered by the hopset if

$$f_1 + 100f_r \leq f_{min} \tag{3-103}$$

and

$$f_1 + (10B_{\max} + 9)f_r \geq f_{max} \tag{3-104}$$

Example 3. The *Bluetooth* communication system is used to establish wireless communications among portable electronic devices. The system has a hopset of 79 carrier frequencies, its hop rate is 1600 hops per second, its hop band is between 2400 and 2483.5 MHz, and the bandwidth of each frequency channel is 1 MHz. Consider a system in which the 79 carrier frequencies are spaced 1 MHz apart from 2402 MHz to 2480 MHz. A 10/11 divider with $f_r = 1$ MHz provides the desired increment, which is equal to the frequency resolution. Equation (3-99) indicates that $t_s = 25$ μs, which indicates that 25 potential data symbols will have to be omitted during each hop interval. Inequality (3-103) indicates that $f_1 = 2300$ MHz is a suitable choice. Then (3-104) is satisfied by $B_{max} = 18$. Therefore, dividers A and B require 4 and 5 control bits, respectively, to specify their potential values. If the control bits are stored in a ROM, then each ROM location contains 9 bits. The number of ROM addresses is at least 79, the number of frequencies in the hopset. Thus, a ROM input address requires 7 bits. $\square$

Multiple loops

A *multiple-loop frequency synthesizer* uses two or more single-loop synthesizers to obtain both fine frequency resolution and fast switching. A three-loop frequency synthesizer is shown in Figure 3.21. Loops A and B have the form of Figure 3.19, but loop A does not have a mixer and filter in its feedback. Loop C has the mixer and filter, but lacks the divider. The reference frequency f_r is chosen to ensure that the desired switching time is realized. If $A > M$, then loop C does not appreciably degrade the switching time. The divisor M is chosen so that f_r/M is equal to the desired resolution. Loop A and the divider generate increments of f_r/M while loop B generates increments of f_r. Loop C combines the outputs of loops A and B to produce the output frequency

$$f_0 = Bf_r + A\frac{f_r}{M} + f_1 \tag{3-105}$$

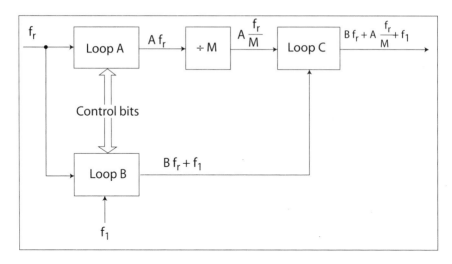

Figure 3.21: Indirect frequency synthesizer with three loops.

where B, A, and M are positive integers because they are produced by dividers. Loop C is preferable to a mixer and bandpass filter because the filter would have to suppress a closely spaced, unwanted component when Af_r/M and Bf_r were far apart. To ensure that each output frequency is produced by unique values of A and B, it is required that $A_{max} = A_{min} + M - 1$. To prevent degradation in the switching time, it is required that $A_{min} > M$. Both requirements are met by choosing

$$A_{min} = M + 1, \qquad A_{max} = 2M \tag{3-106}$$

According to (3-105), a range of frequencies from f_{min} to f_{max} is covered if

$$B_{min}f_r + A_{min}\frac{f_r}{M} + f_1 \le f_{min} \tag{3-107}$$

and

$$B_{max}f_r + A_{max}\frac{f_r}{M} + f_1 \ge f_{max} \tag{3-108}$$

Example 4. Consider the Bluetooth system of Example 3 but with the more stringent requirement that $t_s = 2.5$ μs, which only sacrifices 3 potential data symbols per hop interval. The single-loop synthesizer of Example 3 cannot provide this short switching time. The required switching time is provided by a three-loop synthesizer with $f_r = 10$ MHz. The resolution of 1 MHz is achieved by taking $M = 10$. Equations (3-106) indicate that $A_{min} = 11$ and $A_{max} = 20$. Inequalities (3-107) and (3-108) are satisfied if $f_1 = 2300$ MHz, $B_{min} = 9$, and $B_{max} = 16$. The maximum frequencies that must be accommodated by the dividers in loops A and B are $A_{max}f_r = 200$ MHz and $B_{max}f_r = 160$ MHz, respectively. Dividers A and B require 5 and 4 control bits, respectively. $\square$

Fractional-N Synthesizer

A *fractional-N synthesizer* uses a single loop and extensive auxiliary hardware to produce an output frequency given by (3-105) with $0 \leq A \leq M - 1$. Although the switching time is inversely proportional to f_r , the resolution is f_r/M, which can be made arbitrarily small in principle. The synthesis method alters the loop feedback by dividing the output frequency by B most of the time but dividing $B + 1$ every M/A output cycles so that the effective divisor is $N = B + A/M$. The main disadvantage of the fractional-N synthesizer relative to the other synthesizers of comparable performance is its production of relatively high-level spurious signals that frequency-modulate its output signal.

As shown in Figure 3.22, the number A/M is added to the content of an accumulator every output cycle. Each time the content exceeds unity, a carry

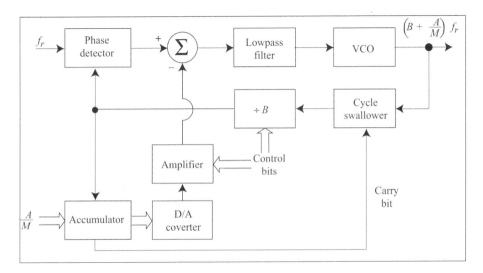

Figure 3.22: Fractional-N frequency synthesizer.

bit is generated that causes division by $B + 1$ instead of B. For example, if $f_r = 1$ MHz and it is desired to generate $f_0 = 9.15$ MHz, then $B = 9$ and $A/M = 0.15$ is added to the content of an accumulator every output cycle. The output frequency is divided by $B + 1 = 10$ every $M/A = 1/0.15 = 6.67$ output cycles on the average. The *cycle swallower* is a device that blocks one of the VCO output cycles in response to a carry bit from the accumulator. For the VCO to produce a stable output frequency, its input must be approximately a direct-current signal. However, for every reference cycle, the VCO output undergoes N cycles, and the divider output undergoes $N/B = 1 + A/BM$ cycles. Therefore, the relative phase between the two phase-detector inputs increases by $2\pi A/BM$ radians per amplifier output. Since the accumulator output increases by A/M every reference cycle, a programmable amplifier with a gain of $2\pi/B$ yields the output needed for cancellation.

Example 5. Consider a fractional-N synthesizer for the Bluetooth system of Example 4 in which $t_s = 2.5$ μs. If the output of the fractional-N synthesizer is frequency-translated by 2300 MHz, then the synthesizer itself needs to cover 102 MHz to 180 MHz. The switching time is achieved by taking $f_r = 10$ MHz. The resolution is achieved by taking $M = 10$. Equation (3-105) indicates that the required frequencies are covered by varying B from 10 to 18 and A from 0 to 9. The accumulator increases its content by $A/M = A/10$ every reference cycle. The integers B and A require 5 and 4 control bits, respectively. □

3.5   Problems

1. An n-stage feedback shift register is used as the code generator in the FH/MFSK transmitter shown in Figure 3.5(a). What is the maximum number of effective frequency channels in the hopset? What is required for message privacy?

2. Consider FH/MFSK with soft-decision decoding of repetition codes and $\mathcal{E}_b/I_{t0} \geq 3/m$. Show that L_0 is given by (3-30).

3. Consider FH/MFSK with soft-decision decoding of repetition codes and large values of $\mathcal{E}_b/I_{t0}$. Suppose that the number of repetitions is not chosen to minimize the potential impact of partial-band jamming. Show that a nonbinary modulation with m bits per symbol gives a better performance than binary modulation in the presence of worst-case partial-band jamming if $L > (m - 1)\ln(2)/\ln(m)$.

4. Draw the block diagram of a receiver for an FH/MFSK system with an independently hopped MFSK set. This system precludes sophisticated multitone jamming.

5. How many symbols per hop are required for the loss due to a phase-reference symbol to be less than 0.1 dB in an FH/DPSK system?

6. This problem illustrates the importance of a channel code to a frequency-hopping system in the presence of worst-case partial-band interference. Consider an FH/MSK system with limiter-discriminator demodulation. (a) Use (3-70) to calculate the required $\mathcal{E}_b/I_{t0}$ to obtain a bit error rate $P_b = 10^{-5}$ when no channel code is used. (b) Calculate the required $\mathcal{E}_b/I_{t0}$ for $P_b = 10^{-5}$ when a Golay (23,12) code is used. As a first step, use the first term in (1-25) to estimate the required symbol error probability. What is the coding gain?

7. Consider an FH/DPSK system with a turbo decoder. (a) Derive the maximum-likelihood estimates of (3-86) and (3-87). (b) Assume that the $\{p_{0k}\}$ are correct. Derive the value of the factor necessary for $\widehat{C}$ to be unbiased. (c) If phase synchronization is available, N_0 is the same for each symbol, and both N_0 and the attenuation α are known, show that (3-90) reduces to (1-137).

8. It is desired to cover 198 – 200 MHz in 10 Hz increments using double-mix-divide modules. (a) What is the minimum number of modules required? (b) What is the range of acceptable reference frequencies? (c) Choose a reference frequency. What are the frequencies of the required tones? (d) If an upconversion by 180 MHz follows the DMD modules, what is the range of acceptable reference frequencies? Is this system more practical?

9. It is desired to cover 100 – 100.99 MHz in 10 kHz increments with an indirect frequency synthesizer containing a single loop and a dual-modulus divider. Let $f_1 = 0$ in Figure 3.19 and $Q = 1$ in Figure 3.20. (a) What is a suitable range of values of A? (b) What are a suitable value of P and a suitable range of values of B if it is required to minimize the highest frequency applied to the programmable dividers?

10. It is desired to cover 198 – 200 MHz in 10 Hz increments with a switching time equal to 2.5 ms. An indirect frequency synthesizer with three loops in the form of Figure 3.21 is used. It is desired that $B_{max} \leq 10^4$. (a) What are suitable values of the parameters f_r , M, A_{min}, $A_{max}, B_{min},$ B_{max} , and f_1 ? (b) If the desired switching time is reduced to 250 μs and f_1 is minimized, what are the values of these parameters?

11. Specify the design parameters of a fractional-N synthesizer that covers 198 – 200 MHz in 10 Hz increments with a switching time equal to 250 μs.

3.6 References

1. P. V. Kumar, "Frequency-Hopping Code Sequence Designs Having Large Linear Span," *IEEE Trans. Inform. Theory*, vol. 34, pp. 146–151, January 1988.

2. L. Cong and S. Songgeng, "Chaotic Frequency Hopping Sequences," *IEEE Trans. Commun.*, vol. 46, pp. 1433–1437, November 1998.

3. D. J. Torrieri "Fundamental Limitations on Repeater Jamming of Frequency-Hopping Communications," *IEEE J. Select. Areas Commun.*, vol. 7, pp. 569–578, May 1989.

4. R. L. Peterson, R. E. Ziemer, and D. E. Borth, *Introduction to Spread Spectrum Communications*. Upper Saddle River, NJ: Prentice Hall, 1995.

5. M. K. Simon et al., *Spread Spectrum Communications Handbook*. Boston: McGraw-Hill, 1994.

6. D. J. Torrieri, "Frequency Hopping with Multiple Frequency-Shift Keying and Hard Decisions," *IEEE Trans. Commun.*, vol. 32, pp. 574–583, May 1984.

7. J. S. Lee, R. H. French, and L. E. Miller, "Probability of Error Analyses of a BFSK Frequency-Hopping System with Diversity under Partial-Band Jamming Interference—Part I: Performance of Square-Law Linear Combining Soft Decision Receiver," *IEEE Trans. Commun.*, vol. 32, pp. 645–653, June 1984.

8. J. S. Lee, L. E. Miller, and Y. K. Kim, "Probability of Error Analyses of a BFSK Frequency-Hopping System with Diversity under Partial-Band Jamming Interference—Part II: Performance of Square-Law Nonlinear Combining Soft Decision Receivers," *IEEE Trans. Commun.*, vol. 32, pp. 1243–1250, December 1984.

9. L. E. Miller, J. S. Lee, and A. P. Kadrichu, "Probability of Error Analyses of a BFSK Frequency-Hopping System with Diversity under Partial-Band Jamming Interference—Part III: Performance of Square-Law Self-Normalizing Soft Decision Receiver," *IEEE Trans. Commun.*, vol. 34, pp. 669–675, July 1986.

10. Y. M. Lam and P. H. Wittke, "Frequency-Hopped Spread-Spectrum Transmission with Band-Efficient Modulations and Simplified Noncoherent Sequence Estimation," *IEEE Trans. Commun.*, vol. 38, pp. 2184–2196, December 1990.

11. M. K. Simon, S. M. Hinedi, and W. C. Lindsey, *Digital Communication Techniques*. Englewood Cliffs, NJ: Prentice Hall, 1995.

12. R. F. Pawula, "Refinements to the Theory of Error Rates for Narrow-Band Digital FM," *IEEE Trans. Commun.*, vol. 36, pp. 509-513, April 1988.

13. M. B. Pursley, "The Derivation and Use of Side Information in Frequency-Hop Spread Spectrum Communications," *IEICE Trans. Commun.*, vol. E76-B, pp. 814–824, August 1993.

14. T. G. Macdonald and M. B. Pursley, "Staggered Interleaving and Iterative Errors-and-Erasures Decoding for Frequency-Hop Packet Radio," *IEEE Trans. Wireless Commun.*, vol. 2, pp. 92-98, January 2003.

15. L.-L. Yang and L. Hanzo, "Low Complexity Erasure Insertion in RS-Coded SFH Spread-Sprectum Communications With Partial-Band Interference and Nakagami-m Fading," *IEEE Trans. Commun.*, vol. 50, pp. 914-925, June 2002.

16. P. H. Wittke, Y. M. Lam, and M. J. Schefter, "The Performance of Trellis-Coded Nonorthogonal Noncoherent FSK in Noise and Jamming," *IEEE Trans. Commun.*, vol. 43, pp. 635–645, February/March/April 1995.

17. H. El Gamal and E. Geraniotis, "Iterative Chanel Estimation and Decoding for Convolutionally Coded Anti-Jam FH Signals," *IEEE Trans. Commun.*, vol. 50, pp. 321–331, February 2002.

18. Q. Zhang and T. LeNgoc, "Turbo Product Codes for FH-SS with Partial-Band Interference," *IEEE Trans. Wireless Commun.*, vol. 1, pp. 513-520, July 2002.

19. J. R. Smith, *Modern Communication Circuits, 2nd ed.* Boston: McGraw-Hill, 1998.

20. U. L. Rohde, *Microwave and Wireless Synthesizers, Theory and Design.* New York: Wiley, 1997.

21. W. F. Egan, *Frequency Synthesis by Phase Lock, 2nd ed.* New York: Wiley, 2000.

22. J. R. Alexovich and R. M. Gagliardi, "Effect of PLL Frequency Synthesizer in FSK Frequency-Hopped Communications," *IEEE Trans. Commun.*, vol. 37, pp. 268–276, March 1989.

Chapter 4

Code Synchronization

A spread-spectrum receiver must generate a spreading sequence or frequency-hopping pattern that is synchronized with the received sequence or pattern; that is, the corresponding chips or dwell intervals must precisely or nearly coincide. Any misalignment causes the signal amplitude at the demodulator output to fall in accordance with the autocorrelation or partial autocorrelation function. Although the use of precision clocks in both the transmitter and the receiver limit the timing uncertainty in the receiver, clock drifts, range uncertainty, and the Doppler shift may cause synchronization problems. *Code synchronization,* which is either sequence or pattern synchronization, might be obtained from separately transmitted pilot or timing signals. It may be aided or enabled by feedback signals from the receiver to the transmitter. However, to reduce the cost in power and overhead, most spread-spectrum receivers can acquire code synchronization from the received signal.

4.1 Acquisition of Spreading Sequences

In the first part of this chapter, we consider direct-sequence systems. To derive the maximum-likelihood estimate of the *code phase* or timing offset of the spreading sequence, several assumptions are made. Since the presence of the data modulation impedes code synchronization, the transmitter is assumed to facilitate the synchronization by transmitting the spreading sequence without any data modulation. In nearly all applications, *noncoherent code synchronization* must precede carrier synchronization because the signal energy is spread over a wide spectral band. Prior to despreading, which requires code synchronization, the signal-to-noise ratio (SNR) is unlikely to be sufficiently high for successful carrier tracking by a phase-locked loop. The received signal is

$$r(t) = s(t) + n(t) \tag{4-1}$$

where $s(t)$ is the desired signal and $n(t)$ is the additive white Gaussian noise. For a direct-sequence system with PSK modulation, the desired signal is

$$s(t) = \sqrt{2S}\, p(t - \tau) \cos\left(2\pi f_c t + 2\pi f_d t + \theta\right) \tag{4-2}$$

where S is the average power, $p(t)$ is the spreading waveform, f_c is the carrier frequency, θ is the random carrier phase, and τ and f_d are the unknown code phase and frequency offset, respectively, that must be estimated. The frequency offset may be due to a Doppler shift or to a drift or instability in the transmitter oscillator.

The coefficients in the expansion of the observed waveform in terms of orthonormal basis functions constitute the vector $\mathbf{r} = [r_1 \, r_2 \ldots r_N]$. The *likelihood function* for the unknown τ and f_d is the conditional density function of $\mathbf{r}$ given the values of τ and f_d. Since θ is a random variable, the likelihood function is

$$\Lambda(\mathbf{r}) = E_\theta \left[f \left(\mathbf{r} | \tau, f_d, \theta \right) \right] \tag{4-3}$$

where $f(\mathbf{r}|\tau, f_d, \theta)$ is the conditional density function of $\mathbf{r}$ given the values of τ, f_d, and θ, and E_θ is the expectation with respect to θ. The maximum-likelihood estimates are those values of τ and f_d that maximize $\Lambda(\mathbf{r})$.

The coefficients in the expansion of $r(t)$ in terms of the orthonormal basis functions are statistically independent. Since each coefficient is Gaussian with variance $N_0/2$,

$$f(\mathbf{r}|\tau, f_d, \theta) = \prod_{i=1}^{N} \frac{1}{\sqrt{\pi N_0}} \exp \left[-\frac{(r_i - s_i)^2}{N_0} \right] \tag{4-4}$$

where the $\{s_i\}$ are the coefficients of the signal $s(t)$ when τ, f_d, and θ are given. Substituting this equation into (4-3) and eliminating factors irrelevant to the maximum-likelihood estimation, we obtain

$$\Lambda(\mathbf{r}) = E_\theta \left\{ \exp \left[\frac{2}{N_0} \sum_{i=1}^{N} r_i s_i - \frac{1}{N_0} \sum_{i=1}^{N} s_i^2 \right] \right\} \tag{4-5}$$

Expansions in the orthonormal basis functions indicate that if $N \to \infty$, the likelihood function may be expressed in terms of the signal waveforms as

$$\Lambda[r(t)] = E_\theta \left\{ \exp \left[\frac{2}{N_0} \int_0^T r(t) s(t) dt - \frac{\mathcal{E}}{N_0} \right] \right\} \tag{4-6}$$

where $\mathcal{E}$ is the energy in the signal waveform over the observation interval of duration T. Assuming that $\mathcal{E}$ does not vary significantly over the ranges of τ and f_d considered, the factor involving $\mathcal{E}$ may be dropped from further consideration. The substitution of $s(t)$ in (4-2) into (4-6) then yields

$$\Lambda[r(t)] = E_\theta \left\{ \exp \left[\frac{2\sqrt{2S}}{N_0} \int_0^T r(t) p(t - \tau) \cos \left(2\pi f_c t + 2\pi f_d t + \theta \right) dt \right] \right\} \tag{4-7}$$

For noncoherent estimation, the received carrier phase θ is assumed to be uniformly distributed over $[0, 2\pi)$. A trigonometric expansion followed by an integration of (4-7) over θ gives (cf. (7-14))

$$\Lambda[r(t)] = I_0 \left(\frac{2\sqrt{2SR(\tau, f_d)}}{N_0} \right) \tag{4-8}$$

where $I_0(\ )$ is the modified Bessel function of the first kind and order zero given by (D-30), and

$$R\left(\tau, f_d\right) = \left[\int_0^T r(t)p(t-\tau)\cos\left(2\pi f_c t + 2\pi f_d t\right) dt\right]^2$$

$$+ \left[\int_0^T r(t)p(t-\tau)\sin\left(2\pi f_c t + 2\pi f_d t\right) dt\right]^2 \qquad (4\text{-}9)$$

Since $I_0(x)$ is a monotonically increasing function of x, (4-8) implies that $R(\tau, f_d)$ is a sufficient statistic for maximum-likelihood estimation. Ideally, the estimates are determined by considering all possible values of τ and f_d, and then choosing those values that maximize (4-9). A device that implements (4-9) is called a *noncoherent correlator*.

A practical implementation of maximum-likelihood estimation or other type of estimation is greatly facilitated by dividing synchronization into the two operations of acquisition and tracking. *Acquisition* provides coarse synchronization by limiting the choices of the estimated values to a finite number of quantized candidates. Following the acquisition, *tracking* provides and maintains fine synchronization.

One method of acquisition is to use a *parallel array* of processors, each matched to candidate quantized values of the timing and frequency offsets. The largest processor output then indicates which candidates are selected as the estimates. An alternative method of acquisition, which is much less complex, but significantly increases the time needed to make a decision, is to serially search over the candidate offsets. Since the frequency offset is usually negligible or requires only a few candidate values, the remainder of this chapter analyzes code synchronization in which only the timing offset τ is estimated. Search methods rather than parallel processing are examined. *Code acquisition* is the operation by which the phase of the receiver-generated sequence is brought to within a fraction of a chip of the phase of the received sequence. After this condition is detected and verified, the tracking system is activated. *Code tracking* is the operation by which synchronization errors are further reduced or at least maintained within certain bounds. Both the acquisition and tracking devices regulate the clock rate. Changes in the clock rate adjust the phase or timing offset of the *local sequence* generated by the receiver relative to the phase or timing offset of the received sequence.

In a benign environment, *sequential estimation* methods provide rapid acquisition [1]. Successive received chips are demodulated and then loaded into the receiver's code generator to establish its initial state. The tracking system then ensures that the code generator maintains synchronization. However, because chip demodulation is required, the usual despreading mechanism cannot be used to suppress interference during acquisition. Since an acquisition failure completely disables a communication system, an acquisition system must be capable of rejecting the anticipated level of interference. To meet this requirement, *matched-filter acquisition* and *serial-search acquisition* are the most effective techniques in general.

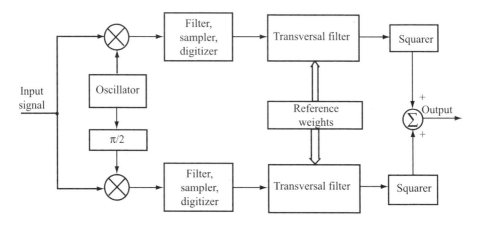

Figure 4.1: Digital matched filter.

Matched-Filter Acquisition

Matched-filter acquisition provides potentially rapid acquisition when short programmable sequences give adequate security. The matched filter in an acquisition system is matched to one period of the spreading waveform, which is usually transmitted without modulation during acquisition. The sequence length
or integration time of the matched filter is limited by frequency offsets and
chip-rate errors. The output envelope, which ideally comprises triangular autocorrelation spikes, is compared with one or more thresholds, one of which is
close to the peak value of the spikes. If the data-symbol boundaries coincide
with the beginning and end of a spreading sequence, the occurrence of a threshold crossing provides timing information used for both symbol synchronization
and acquisition. A major application of matched-filter acquisition is for *burst
communications*, which are short and infrequent communications that do not
require a long spreading sequence.

A *digital matched filter* that generates $R(\tau, 0)$ for noncoherent acquisition of
a binary spreading waveform is illustrated in Figure 4.1. The digital matched
filter offers great flexibility, but is limited in the bandwidth it can accommodate.
The received spreading waveform is decomposed into in-phase and quadrature
baseband components, each of which is applied to a separate branch. The
outputs of each digitizer are applied to a transversal filter. Tapped outputs of
each transversal filter are multiplied by stored weights and summed. The two
sums are squared and added together to produce the final matched-filter output.
A *one-bit digitizer* makes hard decisions on the received chips by observing the
polarities of the sample values. Each transversal filter is a shift register, and
the reference weights are sequence chips stored in shift-register stages. The
transversal filter contains G successive received spreading-sequence chips and a
correlator that computes the number of received and stored chips that match.
The correlator outputs are applied to the squarers.

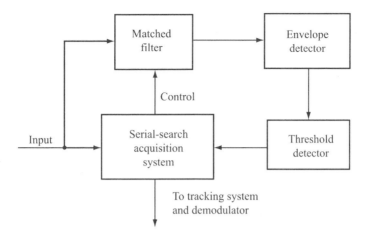

Figure 4.2: Configuration of a serial-search acquisition system enabled by a matched filter.

Matched-filter acquisition for continuous communications is useful when serial-search acquisition with a long sequence fails or takes too long. The transmission of the short sequence may be concealed by embedding it within the long sequence. The short sequence may be a subsequence of the long sequence that is presumed to be ahead of the received sequence and is stored in the programmable matched filter. Figure 4.2 depicts the configuration of a matched filter for short-sequence acquisition and a serial-search system for long-sequence acquisition. The control signal provides the short sequence that is stored or recirculated in the matched filter. The control signal activates the matched filter when it is needed and deactivates it otherwise. The short sequence is detected when the envelope of the matched-filter output crosses a threshold. The threshold-detector output starts a long-sequence generator in the serial-search system at a predetermined initial state. The long sequence is used for verifying the acquisition and for despreading the received direct-sequence signal. Several matched filters in parallel may be used to expedite the process.

4.2 Serial-Search Acquisition

Serial-search acquisition consists of a search, usually in discrete steps, among candidate code phases of a local sequence until it is determined that the local sequence is nearly synchronized with the received spreading sequence. Conceptually, the timing uncertainty covers a region that is quantized into a finite number of *cells*, which are search positions of relative code phases or timing alignments. The cells are serially tested until it is determined that a particular cell corresponds to the alignment of the two sequences to within a fraction of a chip.

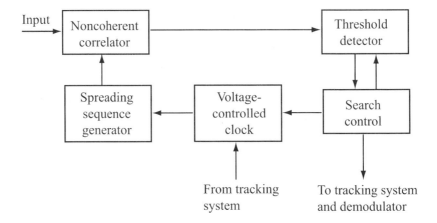

Figure 4.3: Serial-search acquisition system.

Figure 4.3 depicts the principal components of a serial-search acquisition system. The received direct-sequence signal and a local spreading sequence are applied to a noncoherent correlator that produces the statistic (4-9). If the received and local spreading sequences are not aligned, the sampled correlator output is low. Therefore, the threshold is not exceeded, the cell under test is rejected, and the phase of the local sequence is retarded or advanced, possibly by generating an extra clock pulse or by blocking one. A new cell is then tested. If the sequences are nearly aligned, the sampled correlator output is high, the threshold is exceeded, the search is stopped, and the two sequences run in parallel at some fixed phase offset. Subsequent tests verify that the correct cell has been identified. If a cell fails the verification tests, the search is resumed. If a cell passes, the two sequences are assumed to be coarsely synchronized, demodulation begins, and the tracking system is activated. The threshold-detector output continues to be monitored so that any subsequent loss of synchronization activates the serial search.

There may be several cells that potentially provide a valid acquisition. However, if none of these cells corresponds to perfect synchronization, the detected energy is reduced below its potential peak value. The *step size* is the separation between cells. If the step size is one-half of a chip, then one of the cells corresponds to an alignment within one-fourth of a chip. On the average, the misalignment of this cell is one-eighth of a chip, which may cause a negligible degradation. As the step size decreases, both the average detected energy during acquisition and the number of cells to be searched increase.

The *dwell time* is the amount of time required for testing a cell and is approximately equal to the length of the integration interval in the noncoherent correlator (Section 4.3). An acquisition system is called a *single-dwell system* if a single test determines whether a cell is accepted as the correct one. If verification testing occurs before acceptance, the system is called a *multiple-dwell system*. The dwell times either are fixed or are variable but bounded by some

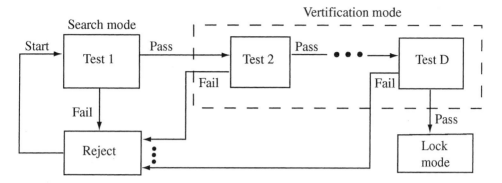

Figure 4.4: Flow graph of multiple-dwell system with consecutive-count strategy.

maximum value. The dwell time for the initial test of a cell is usually designed to be much shorter than the dwell times for verification tests. This approach expedites the acquisition by quickly eliminating the bulk of the incorrect cells. In any serial-search system, the dwell time allotted to a test is limited by the Doppler shift, which causes the received and local chip rates to differ. As a result, an initial close alignment of the two sequences may disappear by the end of the test.

A multiple-dwell system may use a *consecutive-count strategy*, in which a failed test causes a cell to be immediately rejected, or an *up-down strategy*, in which a failed test causes a repetition of a previous test. Figures 4.4 and 4.5 depict the flow graphs of the consecutive-count and up-down strategies, respectively, that require D tests to be passed before acquisition is declared. If the threshold is not exceeded during test 1, the cell fails the test, and the next cell is tested. If it is exceeded, the cell passes the test, the search is stopped, and the system enters the *verification mode*. The same cell is tested again, but the dwell time and the threshold may be changed. Once all the verification tests have been passed, the code tracking is activated, and the system enters the *lock mode*. In the lock mode, the lock detector continually verifies that code synchronization is maintained. If the lock detector decides that synchronization has been lost, *reacquisition* begins in the search mode.

The order in which the cells are tested is determined by the general search strategy. Figure 4.6(a) depicts a *uniform search* over the q cells of the *timing uncertainty*. The broken lines represent the discontinuous transitions of the search from the one part of the timing uncertainty to another. The *broken-center Z search*, illustrated in Figure 4.6(b), is appropriate when *a priori* information makes part of the timing uncertainty more likely to contain the correct cell than the rest of the region. *A priori* information may be derived from the detection of a short preamble. If the sequences are synchronized with the time of day, then the receiver's estimate of the transmitter range combined with the time of day provide the *a priori* information.

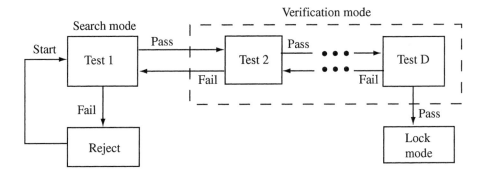

Figure 4.5: Flow graph of multiple-dwell system with up-down strategy.

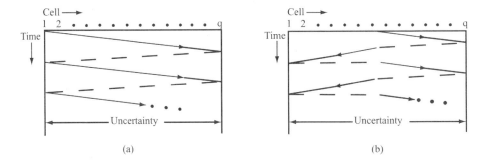

Figure 4.6: Trajectories of search positions: (a) uniform search and (b) broken-center Z search.

The *acquisition time* is the amount of time required for an acquisition system to locate the correct cell and initiate the code tracking system. To derive the statistics of the acquisition time [2], one of the q possible cells is considered the correct cell, and the other $(q-1)$ cells are incorrect. The difference in timing offsets among cells is ΔT_c, where the *step size* Δ is usually either 1 or 1/2. However, it is convenient to allow the correct cell to include two or more timing offsets or code phases. Let L denote the number of times the correct cell is tested before it is accepted and acquisition terminates. Let C denote the number of the correct cell and π_j denote the probability that $C = j$. Let $\nu(L,C)$ denote the number of incorrect cells tested during the acquisition process. The functional dependence is determined by the search strategy. Let $T_r(L,C)$ denote the total *rewinding time*, which is the time required for the search to move discontinuously within the timing uncertainty. Since an incorrect cell is always ultimately rejected, there are only three types of events that occur during a serial search. Either the nth incorrect cell is dismissed after $T_{11}(n)$ seconds, a correct cell is falsely dismissed for the mth time after $T_{12}(m)$

seconds, or a correct cell is accepted after T_{22} seconds, where the first subscript is 1 if dismissal occurs, and 2 otherwise; the second subscript is 1 if the cell is incorrect, and 2 otherwise. Each of these decision times is a random variable. If an incorrect cell is accepted, the receiver eventually recognizes the mistake and reinitiates the search at the next cell. The wasted time expended in code tracking is a random variable called the *penalty time*. These definitions imply that the acquisition time is the random variable given by

$$T_a = \sum_{n=1}^{\nu(L,C)} T_{11}(n) + \sum_{m=1}^{L-1} T_{12}(m) + T_{22} + T_r(L,C) \qquad (4\text{-}10)$$

The most important performance measures of the serial search are the mean and variance of T_a. Given $L = i$ and $C = j$, the conditional expected value of T_a is

$$E[T_a|i,j] = \nu(i,j)\bar{T}_{11} + (i-1)\bar{T}_{12} + \bar{T}_{22} + T_r(i,j) \qquad (4\text{-}11)$$

where $\bar{T}_{11}, \bar{T}_{12}$, and $\bar{T}_{22}$ are the expected values of each $T_{11}(n), T_{12}(m)$, and T_{22}, respectively. Therefore, the mean acquisition time is

$$\bar{T}_a = \bar{T}_{22} + \sum_{i=1}^{\infty} P_L(i) \sum_{j=1}^{q} \pi_j \left[\nu(i,j)\bar{T}_{11} + (i-1)\bar{T}_{12} + T_r(i,j) \right] \qquad (4\text{-}12)$$

where $P_L(i)$ is the probability that $L = i$. We assume that the test statistics are independent and identically distributed. Therefore,

$$P_L(i) = P_D \left(1 - P_D\right)^{i-1} \qquad (4\text{-}13)$$

where P_D is the probability that the correct cell is detected when it is tested during a scan of the uncertainty region. After calculating the conditional expected value of T_a^2 given that $L = i$ and $C = j$, and using the identity $\overline{x^2} = var(x) + \bar{x}^2$, we obtain

$$\overline{T_a^2} = \sum_{i=1}^{\infty} P_L(i) \sum_{j=1}^{q} \pi_j \{ [\nu(i,j)\bar{T}_{11} + (i-1)\bar{T}_{12} + \bar{T}_{22} + T_r(i,j)]^2$$
$$+ \nu(i,j)var(T_{11}) + (i-1)var(T_{12}) + var(T_{22}) \} \qquad (4\text{-}14)$$

The variance of T_a is

$$\sigma_a^2 = \overline{T_a^2} - \bar{T}_a^2 \qquad (4\text{-}15)$$

In some applications, the serial-search acquisition must be completed within a specified period of duration T_{max}. If it is not, the serial search is terminated, and special measures such as the matched-filter acquisition of a short sequence are undertaken. The probability that $T_a \le T_{max}$ can be bounded by using Chebyshev's inequality (Appendix A):

$$P[T_a \le T_{max}] \ge P[|T_a - \bar{T}_a| \le T_{max} - \bar{T}_a]$$
$$\ge 1 - \frac{\sigma_a^2}{\left(T_{max} - \bar{T}_a\right)^2} \qquad (4\text{-}16)$$

where $P[A]$ denotes the probability of the event A.

Uniform Search with Uniform Distribution

As an important application, we consider the uniform search of Figure 4.6(a) and a uniform *a priori* distribution for the location of the correct cell given by

$$\pi_j = \frac{1}{q}, \quad 1 \le j \le 1 \tag{4-17}$$

If the cells in the figure are labeled consecutively from left to right, then

$$\nu(i,j) = (i-1)(q-1) + j - 1 \tag{4-18}$$

The rewinding time is

$$T_r(i,j) = T_r(i) = (i-1)T_r \tag{4-19}$$

where T_r is the rewinding time associated with each broken line in the figure. If the timing uncertainty covers an entire sequence period, then the cells at the two edges are actually adjacent and $T_r = 0$.

To evaluate $\bar{T}_a$ and $\overline{T_a^2}$, we substitute (4-13), (4-17), (4-18), and (4-19) into (4-12) and (4-14) and use the following identities:

$$\sum_{i=0}^{\infty} r^i = \frac{1}{1-r}, \quad \sum_{i=1}^{\infty} i r^i = \frac{r}{(1-r)^2}, \quad \sum_{i=1}^{\infty} i^2 r^i = \frac{r(1+r)}{(1-r)^3}$$

$$\sum_{i=1}^{n} i = \frac{n(n+1)}{2}, \quad \sum_{i=1}^{n} i^2 = \frac{n(n+1)(2n+1)}{6} \tag{4-20}$$

where $0 \le |r| < 1$. Defining

$$\alpha = (q-1)\bar{T}_{11} + \bar{T}_{12} + T_r \tag{4-21}$$

we obtain

$$\bar{T}_a = (q-1)\left(\frac{2-P_D}{2P_D}\right)\bar{T}_{11} + \left(\frac{1-P_D}{P_D}\right)(\bar{T}_{12} + T_r) + \bar{T}_{22} \tag{4-22}$$

and

$$\overline{T_a^2} = (q-1)\left(\frac{2-P_D}{2P_D}\right) var\,(T_{11}) + \left(\frac{1-P_D}{P_D}\right) var\,(T_{12}) + var\,(T_{22})$$

$$+ \frac{(2q+1)(q+1)}{6}\bar{T}_{11}^2 + \frac{\alpha^2(1-P_D)(2-P_D)}{P_D^2} + (q+1)\alpha\left(\frac{1-P_D}{P_D}\right)$$

$$+ (q+1)\bar{T}_{11}(\bar{T}_{22} - \bar{T}_{11}) + 2\alpha\left(\frac{1-P_D}{P_D}\right)(\bar{T}_{22} - \bar{T}_{11}) + (\bar{T}_{22} - \bar{T}_{11})^2 \tag{4-23}$$

In most applications, the number of cells to be searched is large, and simpler *asymptotic forms* for the mean and variance of the acquisition time are applicable. As $q \to \infty$, (4-22) gives

$$\bar{T}_a \to q\left(\frac{2-P_D}{2P_D}\right)\bar{T}_{11}, \quad q \to \infty \tag{4-24}$$

Similarly, (4-23) and (4-15) yield

$$\sigma_a^2 \rightarrow q^2 \left(\frac{1}{P_D^2} - \frac{1}{P_D} + \frac{1}{12} \right) \bar{T}_{11}^2, \quad q \rightarrow \infty \qquad (4\text{-}25)$$

These equations must be modified in the presence of a large uncorrected Doppler shift. The fractional change in the received chip rate of the spreading sequence is equal to the fractional change in the carrier frequency due to the Doppler shift. If the chip rate changes from $1/T_c$ to $1/T_c + \delta$, then the average change in the code or sequence phase during the test of an incorrect cell is $\delta \bar{T}_{11}$. The change relative to the step size is $\delta \bar{T}_{11}/\Delta$. The number of cells that are actually tested in a sweep of the timing uncertainty becomes $q(1 + \delta \bar{T}_{11}/\Delta)^{-1}$. Since incorrect cells predominate, the substitution of the latter quantity in place of q in (4-24) and (4-25) gives approximate asymptotic expressions for $\bar{T}_a$ and σ_a^2 when the Doppler shift is significant.

Consecutive-Count Double-Dwell System

For further specialization, consider the consecutive-count double-dwell system described by Figure 4.4 with $D = 2$. Assume that the *correct cell* actually subsumes two consecutive cells with detection probabilities P_a and P_b, respectively. If the test results are assumed to be statistically independent, then

$$P_D = P_a + (1 - P_a)P_b \qquad (4\text{-}26)$$

Let τ_1, P_{F1}, P_{a1}, and P_{b1} denote the search-mode dwell time, false-alarm probability, and successive detection probabilities, respectively. Let τ_2, P_{F2}, P_{a2}, and P_{b2} denote the verification-mode dwell time, false-alarm probability, and successive detection probabilities, respectively. Let $\bar{T}_p$ denote the mean penalty time, which is incurred by the incorrect activation of the tracking mode. The flow graph indicates that since each cell must pass two tests,

$$P_a = P_{a1}P_{a2}, \quad P_b = P_{b1}P_{b2} \qquad (4\text{-}27)$$

and

$$\bar{T}_{11} = \tau_1 + P_{F1}\left(\tau_2 + P_{F2}\bar{T}_p\right) \qquad (4\text{-}28)$$

Equations (4-26) to (4-28) are sufficient for the evaluation of the asymptotic values of the mean and variance given by (4-24) and (4-25).

For a more accurate evaluation of the mean acquisition time, expressions for the conditional means $\bar{T}_{22}$ and $\bar{T}_{12}$ are needed. Expressing $\bar{T}_{22}$ as the conditional expectation of the correct-cell test duration given cell detection, enumerating the possible durations and their conditional probabilities, and then simplifying, we obtain

$$\bar{T}_{22} = \tau_1 + \tau_2 + \tau_1 \frac{(1 - P_a)\,P_b}{P_D} + \tau_2 \frac{P_{a1}\,(1 - P_{a2})\,P_b}{P_D} \qquad (4\text{-}29)$$

Similarly,

$$\bar{T}_{12} = 2\tau_1 + \tau_2 \left[\frac{P_{a1}\,(1 - P_{a2})\,(1 - P_b) + (1 - P_a)\,P_{b1}\,(1 - P_{b2})}{1 - P_D} \right] \qquad (4\text{-}30)$$

Single-Dwell and Matched-Filter Systems

Results for a single-dwell system are obtained by setting $P_{a2} = P_{b2} = P_{F2} = 1, \tau_2 = 0, P_a = P_{a1}, P_b = P_{b1}, P_{F1} = P_F$, and $\tau_1 = \tau_d$ in (4-28) to (4-30). We obtain

$$\bar{T}_{11} = \tau_d + P_F \bar{T}_p, \quad \bar{T}_{22} = \tau_d \left[1 + \frac{(1 - P_a) P_b}{P_D} \right], \quad \bar{T}_{12} = 2\tau_d \qquad (4\text{-}31)$$

Thus, (4-22) yields

$$\bar{T}_a = \frac{(q - 1)(2 - P_D)\left(\tau_d + P_F \bar{T}_p\right) + 2\tau_d\left(2 - P_a\right) + 2\left(1 - P_D\right)T_r}{2P_D} \qquad (4\text{-}32)$$

Since the single-dwell system may be regarded as a special case of the double-dwell system, the latter can provide a better performance by the appropriate setting of its additional parameters.

The approximate mean acquisition time for a matched filter can be derived in a similar manner. Suppose that many periods of a short spreading sequence with N chips per period are received, and the matched-filter output is sampled m times per chip. Then the number of cells that are tested is $q = mN$ and $T_r = 0$. Each sampled output is compared to a threshold so $\tau_d = T_c/m$ is the time duration associated with a test. For $m = 1$ or 2, it is reasonable to regard two of the cells as the correct ones. These cells are effectively tested when a signal period fills or nearly fills the matched filter. Thus, (4-26) is applicable with $P_a \approx P_b$, and (4-32) yields

$$\bar{T}_a \approx NT_c \left(\frac{2 - P_D}{2P_D} \right) (1 + mKP_F), \qquad q \gg 1 \qquad (4\text{-}33)$$

where $K = \bar{T}_p/T_c$. Ideally, the threshold is exceeded once per period, and each threshold crossing provides a timing marker.

Up-Down Double-Dwell System

For the up-down double-dwell system with two correct cells, the flow graph of Figure 4.5 with $D = 2$ indicates that

$$P_a = P_{a1} P_{a2} \sum_{i=0}^{\infty} [P_{a1}(1 - P_{a2})]^i = \frac{P_{a1} P_{a2}}{1 - P_{a1}(1 - P_{a2})} \qquad (4\text{-}34)$$

Similarly,

$$P_b = \frac{P_{b1} P_{b2}}{1 - P_{b1}(1 - P_{b2})} \qquad (4\text{-}35)$$

and P_D is given by (4-26). If an incorrect cell passes the initial test but fails the verification test, then the cell begins the testing sequence again without any memory of the previous testing. Therefore, for an up-down double-dwell system, a recursive evaluation gives

$$\bar{T}_{11} = (1 - P_{F1})\tau_1 + P_{F1}P_{F2}\left(\tau_1 + \tau_2 + \bar{T}_p\right) + P_{F1}\left(1 - P_{F2}\right)\left(\tau_1 + \tau_2 + \bar{T}_{11}\right)$$
$$= \frac{\tau_1 + P_{F1}\left(\tau_2 + P_{F2}\bar{T}_p\right)}{1 - P_{F1}\left(1 - P_{F2}\right)} \tag{4-36}$$

Substitution of (4-34) to (4-36) into (4-24) to (4-26) gives the asymptotic values of the mean and variance of the acquisition time.

From the possible durations and their conditional probabilities, we obtain

$$\bar{T}_{22} = \tau_1 + \tau_2 + P_{a1}\left(1 - P_{a2}\right)\bar{T}_{22} + \frac{(1 - P_{a1})P_{b1}P_{b2}}{P_D}\tau_1$$
$$+ \frac{(1 - P_{a1})P_{b1}\left(1 - P_{b2}\right)P_b}{P_D}\left(\tau_1 + \bar{T}_{22}'\right) \tag{4-37}$$

where $\bar{T}_{22}'$ is the expected delay for the detection of the correct cell given that the testing begins at the second correct cell. A recursive evaluation gives

$$\bar{T}_{22}' = \tau_1 + \tau_2 + P_{b1}\left(1 - P_{b2}\right)\bar{T}_{22}'$$
$$= \frac{\tau_1 + \tau_2}{1 - P_{b1}\left(1 - P_{b2}\right)} \tag{4-38}$$

Similarly, $\bar{T}_{12}$ is determined by the recursive equation

$$\bar{T}_{12} = \tau_1 + \frac{(1 - P_{a1})\left(1 - P_{a2}\right)}{1 - P_D}\tau_1 + P_{a1}\left(1 - P_{a2}\right)\left(\tau_2 + \bar{T}_{12}\right)$$
$$+ \frac{(1 - P_{a1})P_{b1}\left(1 - P_{b2}\right)\left(1 - P_b\right)}{1 - P_D}\left(\tau_1 + \tau_2 + \bar{T}_{12}'\right) \tag{4-39}$$

with

$$\bar{T}_{12}' = \frac{\tau_1 + P_{b1}\left(1 - P_{b2}\right)\tau_2}{1 - P_{b1}\left(1 - P_{b2}\right)} \tag{4-40}$$

Penalty Time

The *lock detector* that monitors the code synchronization in the lock mode performs tests to verify the lock condition. The time that elapses before the system incorrectly leaves the lock mode is called the *holding time*. It is desirable to have a large mean holding time and a small mean penalty time, but the realization of one of these goals tends to impede the realization of the other. As a simple example, suppose that each test has a fixed duration τ and that code synchronization is actually maintained. A single missed detection, which occurs with probability $1 - P_{DL}$, causes the lock detector to assume a loss of lock and to initiate a search. Assuming the statistical independence of the lock-mode tests, the mean holding time is

$$\bar{T}_h = \sum_{i=1}^{\infty} i\tau \left(1 - P_{DL}\right) P_{DL}^{i-1}$$

$$= \frac{\tau}{1 - P_{DL}} \tag{4-41}$$

This result may also be derived by recognizing that $\bar{T}_h = \tau + P_{DL}\bar{T}_h$ because once the lock mode is verified, the testing of the same cell is renewed without any memory of the previous testing. If the locally generated code phase is incorrect, the penalty time expires unless false alarms, each of which occurs with probability P_{FL}, continue to occur every τ seconds. A derivation similar to that of (4-41) yields the mean penalty time for a single-dwell lock detector:

$$\bar{T}_p = \frac{\tau}{1 - P_{FL}} \tag{4-42}$$

A trade-off between a high $\bar{T}_h$ and a low $\bar{T}_p$ exists because increasing P_{DL} tends to increase P_{FL}.

When a single test verifies the lock condition, the synchronization system is vulnerable to deep fades and pulsed interference. A preferable strategy is for the lock mode to be maintained until a number of consecutive or cumulative misses occur during a series of tests. The performance analysis is analogous to that of serial-search acquisition.

Other Search Strategies

In a Z search, no cell is tested more than once until all cells in the timing uncertainty have been tested. Both strategies of Figure 4.6 are Z searches. A characteristic of the Z search is that

$$\nu(i,j) = (i-1)(q-1) + \nu(1,j) \tag{4-43}$$

where $\nu(1,j)$ is the number of incorrect cells tested when $P_D = 1$ and, hence, $L = 1$. For simplicity, we assume that q is even. For the broken-center Z search, the search begins with cell $q/2 + 1$, and

$$\nu(1,j) = \begin{cases} j - \frac{q}{2} - 1, & j \geq \frac{q}{2} + 1 \\ q - j, & j \leq \frac{q}{2} \end{cases} \tag{4-44}$$

whereas $\nu(1,j) = j-1$ for the uniform search. If the rewinding time is negligible, then (4-12), (4-13), and (4-43) yield

$$\bar{T}_a = \frac{1 - P_D}{P_D} \left[(q-1)\bar{T}_{11} + \bar{T}_{12}\right] + \bar{T}_{22} + \bar{T}_{11}\nu(1) \tag{4-45}$$

where

$$\nu(1) = \sum_{j=1}^{q} \nu(1,j)\pi_j \tag{4-46}$$

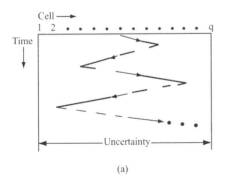
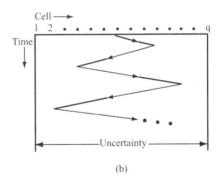

(a)   (b)

Figure 4.7: Trajectories of expanding-window search positions: (a) broken-center and (b) continuous-center search.

is the average number of incorrect cells tested when $P_D = 1$. If C has a uniform distribution, then $\nu(1)$ and, hence, $\bar{T}_a$ are the same for both strategies. If the distribution of C is symmetrical about a pronounced central peak and $P_D \approx 1$, then a uniform search gives $\nu(1) \approx q/2$. Since a broken-center Z search usually ends almost immediately or after slightly more than $q/2$ tests,

$$\nu(1) \approx 0 \left(\frac{1}{2}\right) + \frac{q}{2}\left(\frac{1}{2}\right) = \frac{q}{4} \qquad (4\text{-}47)$$

which indicates that for large values of q and P_D close to unity, the broken-center Z search reduces $\bar{T}_a$ approximately by a factor of 2 relative to its value for the uniform search.

An *expanding-window search* attempts to exploit the information in the distribution of C by continually retesting cells with high *a priori* probabilities of being the correct cell. Tests are performed on all cells within a radius R_1 from the center. If the correct cell is not found, then tests are performed on all cells within an increased radius R_2. The radius is increased successively until the boundaries of the timing uncertainty are reached. The expanding-window search then becomes a Z search. If the rewinding time is negligible and C is centrally peaked, then the broken-center search of Figure 4.7(a) is preferable to the continuous-center search of Figure 4.7(b) because the latter retests cells before testing all the cells near the center of the timing uncertainty. In an *equiexpanding search*, the radii have the form

$$R_n = \frac{nq}{2N}, \qquad n = 1, 2, \ldots, N \qquad (4\text{-}48)$$

where N is the number of sweeps before the search becomes a Z search. If the rewinding time is negligible, then it can be shown [3] that the broken-center equiexpanding-window search is optimized for $P_D \leq 0.8$ by choosing $N = 2$. For this optimized search, $\bar{T}_a$ is moderately reduced relative to its value for the broken-center Z search.

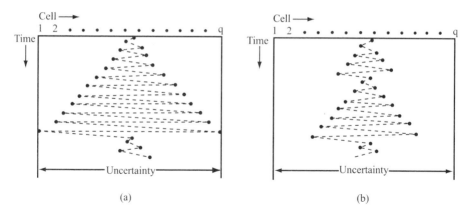

Figure 4.8: Trajectories of alternating search positions: (a) uniform search and (b) nonuniform search.

When $T_r(i,j) = 0$ and $P_D = 1$, the optimal search, which is called a *uniform alternating search*, tests the cells in order of decreasing *a priori* probability. For a symmetric, unimodal, centrally peaked distribution of C, this optimal search has the trajectory depicted in Figure 4.8(a). Once all the cells in the timing uncertainty have been tested, the search repeats the same pattern. Equations (4-43) and (4-45) are applicable. If $P_D \approx 1$ and the distribution of C has a pronounced central peak, then $\nu(1)$ is small, and a comparison with (4-47) indicates that the uniform alternating search has an advantage over the broken-center expanding-window search when $q >> 4$ and the rewinding time for any discontinuous transition is much smaller than $\bar{T}_{11}$. However, computations show that this advantage dissipates as P_D decreases [3], which occurs because all cells are tested with the same frequency without accounting for the distribution of C.

In the *nonuniform alternating search*, illustrated in Figure 4.8(b), a uniform search is performed until a radius R_1 is reached. Then a second uniform search is performed within a larger radius R_2. This process continues until the boundaries of the timing uncertainty are reached and the search becomes a uniform alternating search. Computations show that for a centrally peaked distribution of C, the nonuniform alternating search can give a significant improvement over the uniform alternating search if $P_D < 0.8$, and the radii $R_n, n = 1, 2, \ldots,$ are optimized [3]. However, if the radii are optimized for $P_D < 1$, then as $P_D \to 1$ the nonuniform search becomes inferior to the uniform search.

Density Function of the Acquisition Time

The density function of T_a, which is needed to accurately calculate $P[T_a \leq T_{max}]$ and other probabilities, may be decomposed as

$$f_a(t) = P_D \sum_{i=1}^{\infty} (1 - P_D)^{i-1} \sum_{j=1}^{q} \pi_j f_a(t|i,j) \tag{4-49}$$

where $f_a(t|i,j)$ is the conditional density of T_a given that $L = i$ and $C = j$. Let $*$ denote the convolution operation, $[f(t)]^{*n}$ denote the n-fold convolution of the density $f(t)$ with itself, $[f(t)]^{*0} = 1$, and $[f(t)]^{*1} = f(t)$. Using this notation, we obtain

$$f_a(t|i,j) = [f_{11}(t)]^{*\nu(i,j)} * [f_{12}(t)]^{*(i-1)} * [f_{22}(t)] \tag{4-50}$$

where $f_{11}(f), f_{12}(t)$, and $f_{22}(t)$ are the densities associated with T_{11}, T_{12}, and T_{22}, respectively. If one of the decision times is a constant, then the associated density is a delta function.

The exact evaluation of $f_a(t)$ is difficult [4], but an approximation usually suffices. Since the acquisition time conditioned on $L = i$ and $C = j$ is the sum of independent random variables, it is reasonable to approximate $f_a(t|i,j)$ by a truncated Gaussian density with mean

$$\mu_{ij} = \nu(i,j)\bar{T}_{11} + (i-1)\bar{T}_{12} + \bar{T}_{22} + T_r(i) \tag{4-51}$$

and variance

$$\sigma_{ij}^2 = \nu(i,j)var\,(T_{11}) + (i-1)var\,(T_{12}) + var\,(T_{22}) \tag{4-52}$$

The truncation is such that $f_a(t|i,j) \neq 0$ only if $0 \leq t \leq T_{max}$ or $0 \leq t \leq \mu_{ij} + 3\sigma_{ij}$. When P_D is large, the infinite series in (4-49) converges rapidly enough that the $f_a(t)$ can be accurately approximated by its first few terms.

Alternative Analysis

An alternative method of analyzing acquisition relies on transfer functions [5]. Each phase offset of the local code defines a *state* of the system. Of the total number of q states, $q - 1$ are states that correspond to offsets (cells) that equal or exceed a chip duration. One state is a *collective state* that corresponds to all phase offsets that are less than a chip duration and, hence, cause acquisition to be terminated and code tracking to begin. The serial-search acquisition process is represented by its *circular state diagram*, a segment of which is illustrated in Figure 4.9. The *a priori* probability distribution $\pi_j, j = 1, 2, \ldots, q$, gives the probability that the search begins in state j. The rewinding time is assumed to be negligible.

The branch labels between two states are transfer functions that contain information about the delays that may occur during the transition between the two states. Let z denote the unit-delay variable and let the power of z denote the

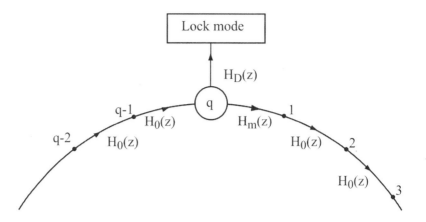

Figure 4.9: Circular state diagram for serial-search acquisition.

time delay. A single-dwell system with dwell τ, false-alarm probability P_F, and constant penalty time T_p has transfer function $H_0(z) = (1 - P_F) z^\tau + P_F z^{\tau + T_p}$ for all branches that do not originate in collective state q because the transition delay is τ with probability $1 - P_F$ and $\tau + T_p$ with probability P_F. For a multiple-dwell system, $H_0(z)$ is determined by first drawing a subsidiary state diagram representing intermediate states and transitions that may occur as the system progresses from one state to the next one in the original circular state diagram. For example, Figure 4.10 illustrates the subsidiary state diagram for a consecutive-count double-dwell system with false alarms P_{F1} and P_{F2} and delays τ_1 and τ_2 for the initial test and the verification test, respectively. Examination of all possible paths between the initial state and the next state indicates that

$$
\begin{aligned}
H_0(z) &= (1 - P_{F1}) z^{\tau_1} + P_{F1} z^{\tau_1} \left[(1 - P_{F2}) z^{\tau_2} + P_{F2} z^{\tau_2 + T_P} \right] \\
&= (1 - P_{F1}) z^{\tau_1} + P_{F1} (1 - P_{F2}) z^{\tau_1 + \tau_2} + P_{F1} P_{F2} z^{\tau_1 + \tau_2 + T_p}
\end{aligned}
\tag{4-53}
$$

Let $H_D(z)$ denote the transfer function between the collective state q and the lock mode. Let $H_M(z)$ denote the transfer function between state q and state 1, which represents the failure to recognize code-phase offsets that are less than a chip duration. These transfer functions may be derived in the same manner as $H_0(z)$. For example, consider a consecutive-count, double-dwell system with a collective state that comprises two states. Figure 4.11 depicts the subsidiary state diagram representing intermediate states and transitions that may occur as the system progresses from state q (with subsidiary states a and b) to either the lock mode or state 1. Examination of all possible paths yields

$$
\begin{aligned}
H_D(z) &= P_{a1} P_{a2} z^{\tau_1 + \tau_2} + P_{a1} (1 - P_{a2}) P_{b1} P_{b2} z^{2\tau_1 + 2\tau_2} \\
&\quad + (1 - P_{a1}) P_{b1} P_{b2} z^{2\tau_1 + \tau_2}
\end{aligned}
\tag{4-54}
$$

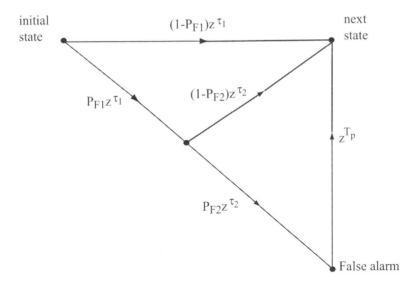

Figure 4.10: Subsidiary state diagram for determination of $H_0(z)$ for consecutive-count double-dwell system.

$$
\begin{aligned}
H_M(z) &= (1 - P_{a1})(1 - P_{b1}) z^{2\tau_1} + (1 - P_{a1}) P_{b1} (1 - P_{b2}) z^{2\tau_1 + \tau_2} \\
&\quad + P_{a1} (1 - P_{a2})(1 - P_{b1}) z^{2\tau_1 + \tau_2} \\
&\quad + P_{a1} (1 - P_{a2}) P_{b1} (1 - P_{b2}) z^{2\tau_1 + 2\tau_2}
\end{aligned}
\tag{4-55}
$$

For a single-dwell system with a collective state that comprises N states,

$$
H_D(z) = P_1 z^\tau + \sum_{j=2}^{N} P_j \left[\prod_{i=1}^{j-1} (1 - P_j) \right] z^{j\tau}
\tag{4-56}
$$

$$
H_M(z) = \left[\prod_{j=1}^{N} (1 - P_j) \right] z^{N\tau}
\tag{4-57}
$$

$$
H_0(z) = (1 - P_F) z^\tau + P_F z^{\tau + T_p}
\tag{4-58}
$$

where τ is the dwell time, P_F is the false-alarm probability, and P_j is the detection probability of state j within the collective state. To calculate the statistics of the acquisition time, we seek the *generating function* defined as the polynomial

$$
H(z) = \sum_{i=0}^{\infty} p_i (\tau_i) z^{\tau_i}
\tag{4-59}
$$

where $p_i(\tau_i)$ is the probability that the acquisition process will terminate in the lock mode after τ_i seconds. If $H(z)$ is known, then the mean acquisition time

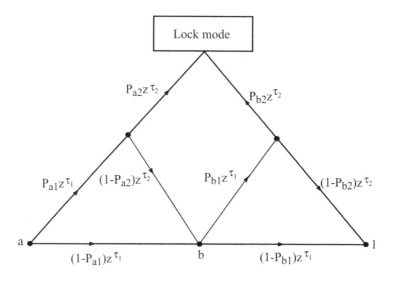

Figure 4.11: Subsidiary state diagram for calculation of $H_D(z)$ and $H_M(z)$ for consecutive-count double-dwell system with two-state collective state.

is

$$\bar{T}_a = \sum_{i=0}^{\infty} \tau_i p_i(\tau_i) = \left. \frac{dH(z)}{dz} \right|_{z=1} \tag{4-60}$$

The second derivative of $H(z)$ gives

$$\left. \frac{d^2 H(z)}{dz^2} \right|_{z=1} = \sum_{i=0}^{\infty} \tau_i (\tau_i - 1) p_i (\tau_i) = \overline{T_a^2} - \bar{T}_a \tag{4-61}$$

Therefore, the variance of the acquisition time is

$$\sigma_a^2 = \left\{ \frac{d^2 H(z)}{dz^2} + \frac{dH(z)}{dz} - \left[\frac{dH(z)}{dz} \right]^2 \right\} \Bigg|_{z=1} \tag{4-62}$$

To derive $H(z)$, we observe that it may be expressed as

$$H(z) = \sum_{j=1}^{q} \pi_j H_j(z) \tag{4-63}$$

where $H_j(z)$ is the transfer function from an initial state j to the lock mode. Since the circular state diagram of Figure 4.9 may be traversed an indefinite number of times during the acquisition process,

$$H_j(z) = H_0^{q-j}(z)H_D(z)\sum_{i=0}^{\infty}\left[H_M(z)H_0^{q-1}(z)\right]^i$$

$$= \frac{H_0^{q-j}(z)H_D(z)}{1 - H_M(z)H_0^{q-1}(z)} \qquad (4\text{-}64)$$

Substitution of this equation into (4-63) yields

$$H(z) = \frac{H_D(z)}{1 - H_M(z)H_0^{q-1}(z)}\sum_{j=1}^{q}\pi_j H_0^{q-j}(z) \qquad (4\text{-}65)$$

The generating function may be expressed as the polynomial in (4-59) by means of polynomial long division.

For the uniform *a priori* distribution given by (4-17),

$$H(z) = \frac{H_D(z)\left[1 - H_0^q(z)\right]}{q\left[1 - H_M(z)H_0^{q-1}(z)\right]\left[1 - H_0(z)\right]} \qquad (4\text{-}66)$$

Since the progression from one state to another is inevitable until the lock mode is reached, $H_0(1) = 1$. Since $H_D(1) + H_M(1) = 1$, (4-65) and (4-60) yield

$$\bar{T}_a = \frac{1}{H_D'(1)}\left\{H_D'(1) + H_M'(1) + (q-1)H_0'(1)\left[1 - \frac{H_D(1)}{2}\right]\right\} \qquad (4\text{-}67)$$

where the prime indicates differentiation with respect to z. As an example, consider a single-dwell system with a two-state collective state. The evaluation of (4-67) using (4-56) to (4-58) with $N = 2$ yields (4-32) with $T_r = 0$ if we set $P_1 = P_a, P_2 = P_b, T_p = \bar{T}_p, \tau = \tau_d$, and define P_D by (4-26).

4.3 Acquisition Correlator

The noncoherent correlator of Figure 4.3 provides the approximate maximization of $V(\tau) = R(\tau, 0)$ given by (4-9). It is assumed that chip synchronization is established by one of the standard methods of symbol synchronization. Consequently, the test interval can be defined with boundaries that coincide with chip boundaries, and we test code phases such that $\tau = \nu T_c$, where ν is an integer. Let MT_c denote the duration of the test interval, where M is a positive integer. If the Doppler shift is not estimated, f_d may be absorbed into f_c in (4-9). If the test interval begins with chip ν of the local spreading sequence, then (2-76) and (4-9) imply that the decision variable for one test of a specific code phase νT_c is

$$V = V_c^2 + V_s^2 \qquad (4\text{-}68)$$

where

$$V_c = \sum_{k=0}^{M-1} p_{k-\nu}x_k, \qquad V_s = \sum_{k=0}^{M-1} p_{k-\nu}y_k \qquad (4\text{-}69)$$

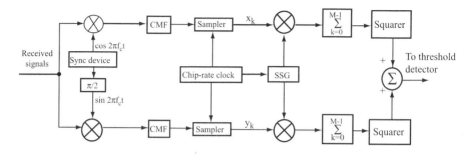

Figure 4.12: Noncoherent correlator for acquisition system. CMF = chip matched filter. SSG = spreading sequence generator.

$$x_k = \int_{kT_c}^{(k+1)T_c} r(t)\psi\left(t - kT_c\right) \cos 2\pi f_c t \; dt \tag{4-70}$$

$$y_k = \int_{kT_c}^{(k+1)T_c} r(t)\psi\left(t - kT_c\right) \sin 2\pi f_c t \; dt \tag{4-71}$$

The sequences $\{x_k\}$ and $\{y_k\}$ can be obtained by an in-phase and quadrature downconversions followed by chip-matched filters sampled at times $t = kT_c$. Thus, the acquisition correlator has the form depicted in Figure 4.12. The decision variable V is applied to a threshold detector to determine whether or not a test of a particular code phase is passed. If a quaternary data modulation is used instead of PSK, then the only modification necessary is to assign separate spreading sequence generators to the two parallel branches of the correlator.

The sequences $\{x_k\}$ and $\{y_k\}$ can be applied to multiple parallel inner products with different values of ν simultaneously. This procedure allows a parallel search of various code phases with a moderate amount of additional hardware or software. Since $p_k = \pm 1$, each inner product may be computed by either adding or subtracting each component of $\{x_k\}$ or $\{y_k\}$.

To analyze the performance of the acquisition correlator under fading conditions, we assume that the received signal is

$$r(t) = \sqrt{2S}\alpha p(t - \tau) \cos(2\pi f_c t + \theta) + n(t) \tag{4-72}$$

where α is the attenuation due to fading, S is the average power when $\alpha = 1$, $p(t)$ is the spreading waveform, f_c is the carrier frequency, θ is the random carrier phase, τ is the delay due to the unknown code phase, and $n(t)$ is the interference plus noise modeled as additive white Gaussian noise. The data modulation $d(t)$ is omitted because either it is not transmitted during acquisition or the test duration MT_c is much smaller than a symbol duration $T_s = GT_c$. In the latter case, the probability that a symbol transition occurs during a test is negligible, and the squaring operations eliminate the symbol value from V. Let $\hat{\tau} = \nu T_c$ denote the delay associated with the code phase of the local spreading sequence. The difference between $\hat{\tau}$ and τ may be expressed in the form $\hat{\tau} - \tau = NT_c + \epsilon T_c$, where N is an integer and $0 < \epsilon < 1$. For a rectangular chip waveform, (4-69), (4-70), and (4-72), $f_c T_c \gg 1$, and the definition of chip ν yield

$$V_c = \sqrt{\frac{S}{2}} \alpha T_c \cos\theta \sum_{k=0}^{M-1} p_{k-\nu} \left[(1 - \epsilon) p_{k-\nu+N} + \epsilon p_{k-\nu+N+1} \right] + \sum_{k=0}^{M-1} p_{k-\nu} n_k \tag{4-73}$$

where

$$n_k = \int_{kT_c}^{(k+1)T_c} n(t) \psi \left(t - kT_c \right) \cos 2\pi f_c t \, dt \tag{4-74}$$

The alignment of the received and local spreading sequences is close enough for acquisition if $N = -1$ or $N = 0$. If $N \neq -1$ and $N \neq 0$, then the cell may be considered incorrect. The equation for V_s is the same as (4-74) except that $-\sin\theta$ replaces $\cos\theta$, and n_k is given by (4-74) with $\sin 2\pi f_c t$ replacing $\cos 2\pi f_c t$.

The first term of V_c in (4-73) contributes *self-interference* that may cause a false alarm. The self-interference is small if the autocorrelation of the spreading sequence is sharply peaked. In a network of similar systems, interfering sequences are substantially suppressed if the cross-correlations among sequences are small, as they are if all the sequences are Gold or Kasami sequences (Chapter 6).

In the performance analysis, the spreading sequence $\{p_k\}$ is modeled as a random binary sequence and ϵ is modeled as a random variable. Thus, given the values of α and θ, the self-interference varies with respect to its mean value and, hence, degrades acquisition even when the noise term is negligible. If the variable part the self-interference is negligible, then (4-73) can be approximated by

$$V_c = E[V_c] + N_{gc} \tag{4-75}$$

where N_{gc} is the second term in (4-73). Since $n(t)$ is zero-mean, white Gaussian noise, n_k is a zero-mean Gaussian random variable. Since $p_{k-\nu} = \pm 1$ and is independent of n_k, the product $p_{k-\nu} n_k$ is zero-mean and Gaussian. The independence of the terms in the sum then indicates that N_{gc} is a zero-mean Gaussian random variable. Similarly, we obtain the approximation

$$V_s = E[V_s] + N_{gs} \tag{4-76}$$

where N_{gs} is a zero-mean, Gaussian random variable. Straightforward calculations using $f_c T_c \gg 1$ indicate that N_{gc} and N_{gs} are statistically independent with the same variance:

$$\sigma^2 = var\left(N_{gc}\right) = var\left(N_{gs}\right) = \frac{N_0 M T_c}{4} \tag{4-77}$$

To determine the condition under which the self-interference is approximated by its mean value, we calculate $var(V_c)$. Given α, θ, and ϵ, (4-73) yields

$$var\left(V_c\right) = \frac{S \alpha^2 M T_c^2 \cos^2\theta}{2} \left(2\epsilon^2 - 2\epsilon + 1\right) + \frac{N_0 M T_c}{4} \tag{4-78}$$

where $g(\epsilon) = \epsilon^2$ if $N = 0$, $g(\epsilon) = (1 - \epsilon)^2$ if $N = -1$, and $g(\epsilon) = 2\epsilon^2 - 2\epsilon + 1$ if $N \neq 0, -1$. The first term is much smaller than the second term if $\mathcal{E}_c/N_0 << 1$, where $\mathcal{E}_c = S\alpha^2 T_c$ is the energy per chip. This condition is satisfied with high probability in most practical systems, especially if N_0 incorporates the power spectral densities due to multiple-access interference and multipath signals. Accordingly, we proceed with the analysis using the approximations (4-75) and (4-76). Without these approximations, alternative approximations and assumptions are necessary or the analysis becomes much more complicated [6]. A common approximation is that $\epsilon \approx 0$.

If $\hat{\tau} - \tau \geq T_c$, then a cell is incorrect and (4-73) with $N \neq -1, 0$ implies that $E[V_c] = E[V_s] = 0$. If $\hat{\tau} - \tau < T_c$, the values of $E[V_c]$ and $E[V_s]$ depend on the step size of the serial search. The step size is the separation in chips between cells and is denoted by Δ. When $\Delta = 1$, two consecutive cells are considered correct. If $\hat{\tau} - \tau$ is increasing, then the cell corresponding to $N = -1$ occurs first and is followed by the cell corresponding to $N = 0$. If ϵ is assumed to be uniformly distributed over (0,1) and $N_1 = -1$ or 0, then (4-73) and the similar equation for V_s yield the conditional means given α and θ:

$$E\left[V_c\right] = \sqrt{\frac{S}{2}}\, \frac{\alpha M T_c \cos\theta}{2}, \quad E\left[V_s\right] = -\sqrt{\frac{S}{2}}\, \frac{\alpha M T_c \sin\theta}{2}, \quad \Delta = 1$$

$$(4\text{-}79)$$

When $\Delta = 1/2$, the two consecutive cells with the smallest values of $\hat{\tau} - \tau$ are considered the two correct cells. For all the others, we assume that $E[V_c] \approx E[V_s] \approx 0$. The first correct cell corresponds to $N = -1$ and $1/2 \leq \epsilon < 1$, whereas the second one corresponds to $N = 0$ and $0 < \epsilon \leq 1/2$. If ϵ is assumed to be uniformly distributed over the latter intervals, then for both cells, we obtain

$$E\left[V_c\right] = \sqrt{\frac{S}{2}}\, \frac{3\alpha M T_c \cos\theta}{4}, \quad E\left[V_s\right] = -\sqrt{\frac{S}{2}}\, \frac{3\alpha M T_c \sin\theta}{4}, \quad \Delta = \frac{1}{2}$$

$$(4\text{-}80)$$

Let V_1 denote the decision variable V when the correct cell is tested, and let V_0 denote V when the incorrect cell is tested. Equations (4-68), (4-75), and (4-76) and the preceding analysis indicate that V_0 is the sum of the squares of two independent, zero-mean Gaussian random variables. The results of Appendix D then indicate that V_0 has a central chi-square distribution with two degrees of freedom and probability density function

$$f_c(x) = \frac{1}{2\sigma^2} \exp\left(-\frac{x}{2\sigma^2}\right) u(x)$$

$$(4\text{-}81)$$

where $u(x) = 1$, $x \geq 0$, and $u(x) = 0$, $x < 0$, and $\sigma^2 = var(N_{gc}) = var(N_{gs})$. The false-alarm probability for a test of an incorrect cell is the probability that $V_0 > V_t$, where V_t is the threshold. The integration of (4-81) gives the false-alarm probability:

$$P_f = \exp\left(-\frac{V_t}{2\sigma^2}\right)$$

$$(4\text{-}82)$$

Similarly, given α and θ, V_1 is the sum of the squares of two independent Gaussian random variables with nonzero means. The results of Appendix D then indicate that V_1 has a noncentral chi-square distribution with two degrees of freedom and probability density function

$$f_1(x) = \frac{1}{2\sigma^2} \exp\left(-\frac{\lambda + x}{2\sigma^2}\right) I_0\left(\frac{\sqrt{\lambda x}}{\sigma^2}\right) u(x) \qquad (4\text{-}83)$$

where

$$\lambda = (E\,[V_c])^2 + (E\,[V_s])^2 = \frac{9}{32} f S \alpha^2 M^2 T_c^2 \qquad (4\text{-}84)$$

and

$$f = \begin{cases} 1, & \Delta = 1/2 \\ 4/9, & \Delta = 1 \end{cases} \qquad (4\text{-}85)$$

The detection probability for a test of a correct cell is the probability that $V_1 > V_t$. The integration of (4-83) and the substitution of (4-84) give the detection probability

$$P_d = Q_1\left(\sqrt{\xi}\alpha, \frac{\sqrt{V_t}}{\sigma}\right) \qquad (4\text{-}86)$$

where $Q_1(\ )$ is the generalized Q-function defined by (D-15),

$$\xi = \frac{9}{8} f M \frac{\mathcal{E}_c}{N_0} \qquad (4\text{-}87)$$

and $\mathcal{E}_c = ST_c$ is the signal energy per chip when fading is absent and $\alpha = 1$.
Combining (4-86) and (4-82) yields

$$P_d = Q_1\left(\sqrt{\xi}\alpha, \sqrt{-2\ln P_f}\right) \qquad (4\text{-}88)$$

Thus, if P_f is specified, P_d is given by (4-88). The threshold needed to realize a specified P_f is

$$V_t = -\frac{N_0 M T_c}{2} \ln P_f \qquad (4\text{-}89)$$

which requires an accurate estimate of N_0.

In the presence of fast Rayleigh fading, α has the Rayleigh probability density (Appendix D.4):

$$f_\alpha(x) = 2x \exp(-x^2) u(x) \qquad (4\text{-}90)$$

where $E[\alpha^2] = 1$ so that S remains the average signal power in (4-72). It is assumed that α is approximately constant during a test, but independent between one test and another. Since (4-86) is conditioned on α, the detection probability in the presence of fast fading is

$$P_d = \int_0^\infty 2x \exp\left(-x^2\right) Q_1\left(\sqrt{\xi}x, \sqrt{-2\ln P_f}\right) dx \qquad (4\text{-}91)$$

To evaluate this integral, we substitute the integral definition of $Q_1(\ )$ given by (D-15), interchange the order of integration in the resulting double integral, and then use (D-33) to evaluate one of the integrals. The remaining integration over an exponential function is elementary. The final result is

$$P_d = P_f^{2/(2+\xi)} \tag{4-92}$$

For slow Rayleigh fading with a coherence time much larger than the acquisition time, it is appropriate to use (4-86) in calculating the conditional mean acquisition time and then integrate over the Rayleigh density to obtain the mean acquisition time.

Let C denote the number of chips in the timing uncertainty. The *normalized mean acquisition time* (NMAT) is defined as $\bar{T}_a/CT_c$. The *normalized standard deviation* (NSD) is defined as σ_a/CT_c.

Example 1.   As an example of the application of the preceding results, consider a single-dwell system with a uniform search and a uniform *a priori* correct-cell location distribution. Let $\tau_d = MT_c$, where M is the number of chips per test, and $\bar{T}_p = KT_c$, where K is the number of chips in the mean penalty time. It is assumed that there are two independent correct cells with the common detection probability $P_d = P_a = P_b$. If $q \gg 1$, (4-32) and (4-26) yield the NMAT:

$$\frac{\bar{T}_a}{CT_c} = \left(\frac{2 - P_D}{2P_D}\right)\frac{q}{C}(M + KP_F) \tag{4-93}$$

where

$$P_D = 2P_d - P_d^2 \tag{4-94}$$

In a single-dwell system, $P_F = P_f$, which is given by (4-82). For step size $\Delta = 1, q/C = 1$; for $\Delta = 1/2, q/C = 2$. In the absence of fading, (4-88) relates P_d and P_f, whereas (4-92) relates them in the presence of fast Rayleigh fading.

Figure 4.13 shows the NMAT as a function of $\mathcal{E}_c/N_0$ for fast Rayleigh fading and no fading. At each value of $\mathcal{E}_c/N_0$, the values of P_f and M are selected to minimize the NMAT. The figure indicates the advantage of $\Delta = 1/2$ when $K \geq 10^5$ and the advantage of $\Delta = 1$ when $K \leq 10^4$. The large increase in the NMAT due to fast Rayleigh fading is apparent. From (4-25), it is found that each plot of the NSD is similar to that of the corresponding NMAT. $\square$

Example 2.   Consider double-dwell systems with a uniform search, a uniform *a priori* correct-cell location distribution, and two independent correct cells with $P_d = P_a = P_b, P_{a1} = P_{b1}$, and $P_{a2} = P_{b2}$. The test durations are $\tau_1 = M_1T_c$ and $\tau_2 = M_2\tau_c$. If $q \gg 1$, the NMAT is obtained from (4-24) and (4-94), where $\bar{T}_{11}$ is given by (4-28) for a consecutive-count system and (4-36) for an up-down system. By replacing P_d with P_{ai} and P_f with P_{Fi}, the probabilities P_{ai} and $P_{Fi}, i = 1$ or 2, are related through (4-88) with $\alpha = 1$ for no fading and (4-92) for fast Rayleigh fading.

Figure 4.14 shows the NMAT as a function of $\mathcal{E}_c/N_0$ for double-dwell systems in the presence of fast Rayleigh fading. The step size is $\Delta = 1/2$, which is found to be advantageous for the parameter values chosen. At each value of $\mathcal{E}_c/N_0$, the values of P_{F1}, P_{F2}, M_1, and M_2 are selected to minimize the NMAT. The figure illustrates the advantage of the up-down system in most practical applications. From (4-25), it is found that each plot of the NSD is similar to

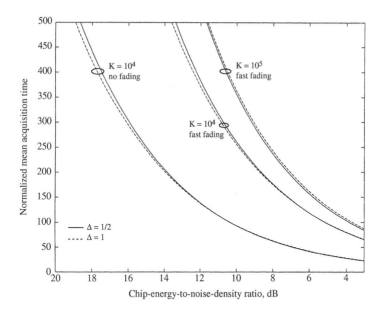

Figure 4.13: NMAT versus $\mathcal{E}_c/N_0$ for single-dwell system in presence of fast Rayleigh fading or no fading. Values of P_f and M are optimized.

that of the corresponding NMAT. A comparison of Figure 4.14 with Figure 4.13 indicates that double-dwell systems are capable of significantly lowering the NMAT relative to a single-dwell system. $\square$

The existence of two consecutive correct cells can be directly exploited in *joint two-cell detection*, which can be shown to provide a lower NMAT than the conventional cell-by-cell detection [7]. In the presence of frequency-selective fading with a large number of resolvable multipath signals, the NMAT of serial-search acquisition is usually increased because the increased self-interference is more significant than the higher number of nonconsecutive correct cells with correct phases. However, joint two-cell detection is more resistant to multiple-access interference and more robust against variations in the detection threshold, the power level of the desired signal, and the number of multipath signals. The advantages of joint two-cell detection over cell-by-cell detection are the result of the efficient combining of the energy of two adjacent correct-phase samples.

The detection threshold of (4-89) depends on an estimate of N_0, the equivalent noise-power spectral density. An accurate estimate usually requires a long observation interval. However, in mobile communication systems and in the presence of jamming, the instantaneous interference power may be rapidly varying. To cope with this environment, an adaptive threshold may be set by the instantaneous received power [8]. As a result, the mean acquisition time is lowered relative to its value for nonadaptive schemes when Rayleigh fading or

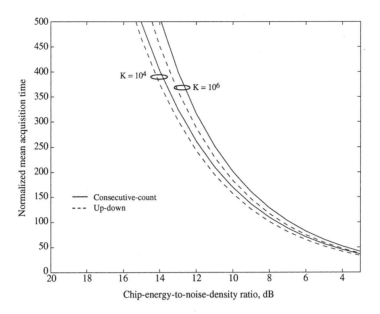

Figure 4.14: NMAT versus $\mathcal{E}_c/N_0$ for double-dwell systems in presence of fast Rayleigh fading. Step size is $\Delta = 1/2$. Values of P_{F1}, P_{F2}, M_1, and M_2 are optimized.

pulsed Gaussian noise jamming is present. When a rake receiver (Chapter 5) is used, each finger of the receiver must acquire the timing of a separate multipath signal. Whether matched filtering or a serial search is used, some mechanism is needed to ensure that each finger acquires a distinct multipath signal [9].

An alternative to acquisition tests of fixed dwell time or number of detector samples is *sequential detection*, which uses only the number necessary for a reliable decision. Thus, some sample sequences may allow a quick decision, while others may warrant using a large number of samples in the evaluation of a single phase of the spreading waveform. The *sequential probability ratio test* [2], [3] entails the recalculation of the likelihood ratio after each new detector sample is produced. This ratio is compared with both upper and lower thresholds to determine if the test is terminated and no more samples need to be extracted. If the upper threshold is exceeded, the receiver declares acquisition and the lock mode is entered. If the likelihood ratio drops below the lower threshold, the test fails, and another code phase is tested. As long as the likelihood ratio lies between the two thresholds, a decision is postponed and the ratio continues to be updated. Although the sequential detector is capable of significantly reducing the mean acquisition time relative to detectors that use a fixed number of samples, it has a number of practical limitations. Chief among them is the computational complexity of the calculation of the likelihood ratio or log-likelihood ratio.

4.4 Code Tracking

Coherent code-tracking loops operate at baseband following the coherent removal of the carrier of the received signal. An impediment to their use is
that the input SNR is usually too low for carrier synchronization prior to code
synchronization and the subsequent despreading of the received signal. Furthermore, coherent loops cannot easily accommodate the effects of data modulation.
Noncoherent loops operate directly on the received signals and are unaffected
by the data modulation.

To motivate the design of the noncoherent loop, one may adapt the statistic
(4-9). If the maximum-likelihood estimate $\hat{\tau}$ is assumed to be within the interior
of its timing uncertainty region and $R(\tau, f_d)$ is a differentiable function of τ,
then the estimate $\hat{\tau}$ that maximizes $R(\tau, f_d)$ may be found by setting

$$\left. \frac{\partial R(\tau, f_d)}{\partial \tau} \right|_{\tau=\hat{\tau}} = 0 \qquad (4\text{-}95)$$

A major problem with this approach is that $R(\tau, f_d)$ given by (4-9) is not
differentiable if the chip waveform is rectangular. This problem is circumvented
by using a difference equation as an approximation of the derivative. Thus, for
a positive δT_c, we set

$$\frac{\partial R(\tau, f_d)}{\partial \tau} \approx \frac{R(\tau + \delta T_c, f_d) - R(\tau - \delta T_c, f_d)}{2\delta T_c} \qquad (4\text{-}96)$$

This equation implies that the solution of (4-95) may be approximately obtained
by a device that finds the $\hat{\tau}$ such that

$$R(\hat{\tau} + \delta T_c, f_d) - R(\hat{\tau} - \delta T_c, f_d) = 0 \qquad (4\text{-}97)$$

To derive an alternative to this equation, we assume that no noise is present,
$f_d = 0$, and that the correct timing offset of the received signal is $\tau = 0$.
Substituting (4-2) with $\tau = 0$ into (4-9) and using trigonometry, we obtain

$$R(\hat{\tau}, 0) = \frac{S}{2} \left[\int_0^T p(t)p(t - \hat{\tau})dt \right]^2 \qquad (4\text{-}98)$$

If $p(t)$ is modeled as the spreading waveform for a random binary sequence
and the interval $[0, T]$ includes many chips, then the integral is reasonably
approximated by its expected value, which is proportional to the autocorrelation

$$R_p(\tau) = \Lambda \left(\frac{\tau}{T_c} \right) \qquad (4\text{-}99)$$

where the triangular function is defined by (2-14). Substituting this result into
(4-97), we find that the maximum-likelihood estimate is approximately obtained
by a device that finds the $\hat{\tau}$ such that

$$R_p^2(\hat{\tau} + \delta T_c) - R_p^2(\hat{\tau} - \delta T_c) = 0 \qquad (4\text{-}100)$$

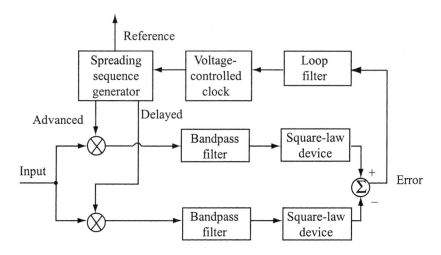

Figure 4.15: Delay-locked loop.

The noncoherent *delay-locked loop* [10], which is diagrammed in Figure 4.15, implements an approximate computation of the difference on the left-hand side of (4-100) and then continually adjusts $\hat{\tau}$ so that this difference remains near zero. The estimate is used to produce the synchronized local spreading sequence that is used for despreading the received direct-sequence signal. The code generator produces three sequences, one of which is the reference sequence used for acquisition and demodulation. The other two sequences are advanced and delayed, respectively, by δT_c relative to the reference sequence. The product δT_c is usually equal to the acquisition step size, and thus usually $\delta = 1/2$, but other values are plausible. The advanced and delayed sequences are multiplied by the received direct-sequence signal in separate branches.

For the received direct-sequence signal (4-2), the signal portion of the upper-branch mixer output is

$$s_{u1}(t) = Ad(t)p(t)p\left(t + \delta T_c - \epsilon T_c\right)\cos\left(2\pi f_c t + \theta\right) \qquad (4\text{-}101)$$

where $A = \sqrt{2S}$ and ϵT_c is the delay of the reference sequence relative to the received sequence. Although ϵ is a function of time because of the loop dynamics, the time dependence is suppressed for notational convenience. Since each bandpass filter has a bandwidth on the order of $1/T_s$, where T_s is the duration of each symbol, $d(t)$ is not significantly distorted by the filtering. Nearly all spectral components except the slowly varying expected value of $p(t)p(t + \delta T_c - \epsilon T_c)$ are blocked by the upper-branch bandpass filter. Since this expected value is the autocorrelation of the spreading sequence, the filter output is

$$s_{u2}(t) \approx Ad(t)R_p\left(\delta T_c - \epsilon T_c\right)\cos\left(2\pi f_c t + \theta\right) \qquad (4\text{-}102)$$

Any double-frequency component produced by the square-law device is ultimately suppressed by the loop filter and thus is ignored. Since $d^2(t) = 1$, the data modulation is removed, and the upper-branch output is

$$s_{u3}(t) \approx \frac{A^2}{2} R_p^2 (\delta T_c - \epsilon T_c) \qquad (4\text{-}103)$$

Similarly, the output of the lower branch is

$$s_{l3}(t) \approx \frac{A^2}{2} R_p^2 (-\delta T_c - \epsilon T_c) \qquad (4\text{-}104)$$

The difference between the outputs of the two branches is the *error signal*:

$$s_e(t) \approx \frac{A^2}{2} \left[R_p^2 (\delta T_c - \epsilon T_c) - R_p^2 (-\delta T_c - \epsilon T_c) \right] \qquad (4\text{-}105)$$

Since $R_p(\tau)$ is an even function, the error signal is proportional to the left-hand side of (4-100).

The substitution of (4-99) and (2-14) into (4-105) yields

$$s_e(t) \approx \frac{A^2}{2} S(\epsilon, \delta) \qquad (4\text{-}106)$$

where $S(\epsilon, \delta)$ is the *discriminator characteristic* or *S-curve* of the tracking loop. For $0 \leq \delta \leq 1/2$,

$$S(\epsilon, \delta) = \begin{cases} 4\epsilon(1-\delta), & 0 \leq \epsilon \leq \delta \\ 4\delta(1-\epsilon), & \delta \leq \epsilon \leq 1-\delta \\ 1 + (\epsilon - \delta)(\epsilon - \delta - 2), & 1 - \delta \leq \epsilon \leq 1 + \delta \\ 0, & 1 + \delta \leq \epsilon \end{cases} \qquad (4\text{-}107)$$

For $1/2 \leq \delta \leq 1$,

$$S(\epsilon, \delta) = \begin{cases} 4\epsilon(1-\delta), & 0 \leq \epsilon \leq 1 - \delta \\ 1 + (\epsilon - \delta)(\epsilon - \delta + 2), & 1 - \delta \leq \epsilon \leq \delta \\ 1 + (\epsilon - \delta)(\epsilon - \delta - 2), & \delta \leq \epsilon \leq 1 + \delta \\ 0, & 1 + \delta \leq \epsilon \end{cases} \qquad (4\text{-}108)$$

In both cases,

$$S(-\epsilon, \delta) = -S(\epsilon, \delta) \qquad (4\text{-}109)$$

Figure 4.16 illustrates the discriminator characteristic for $\delta = 1/2$. The filtered error signal is applied to the voltage-controlled clock. Changes in the clock frequency cause the reference sequence to converge toward alignment with the received spreading sequence. When $0 < \epsilon(t) < 1 + \delta$, the reference sequence is delayed relative to the received sequence. As shown in Figure 4.16, $S(\epsilon, \delta)$ is positive, so the clock rate is increased, and $\epsilon(t)$ decreases. The figure indicates that $s_e(t) \to 0$ as $\epsilon(t) \to 0$. Similarly, when $\epsilon(t) < 0$, we find that $s_e(t) \to 0$ as $\epsilon(t) \to 0$. Thus, the delay-locked loop tracks the received code timing once the acquisition system has finished the coarse alignment.

The discriminator characteristic of code-tracking loops differs from that of phase-locked loops in that it is nonzero only within a finite range of ϵ. Outside

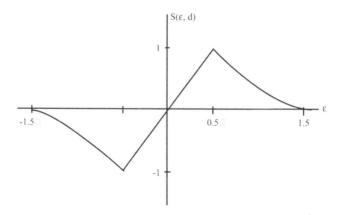

Figure 4.16: Discriminator characteristic of delay-locked loop for $\delta = 1/2$.

that range, code tracking cannot be sustained, the synchronization system loses lock, and a reacquisition search is initiated by the lock detector. Tracking resumes once the acquisition system reduces ϵ to within the range for which the discriminator characteristic is nonzero.

When short spreading sequences are used in a synchronous direct-sequence network, the reduced randomness in the multiple-access interference (Chapter 6) may cause increased tracking jitter or even an offset in the discriminator characteristic [11]. For orthogonal sequences, the interference is zero when synchronization exists, but becomes large when there is a code-phase error in the local spreading sequence. In the presence of a tracking error, the delay-locked-loop arm with the larger offset relative to the correct code phase receives relatively more noise power than the other arm. This disparity reduces the slope of the discriminator characteristic and, hence, degrades the tracking performance. Moreover, because of the nonsymmetric character of the cross-correlations among the spreading sequences, the discriminator characteristic may be biased in one direction, which will cause a tracking offset.

The noncoherent *tau-dither loop*, which is depicted in Figure 4.17, is a lower-complexity alternative to the noncoherent delay-locked loop. The dither generator produces the *dither signal* $D(t)$, a square wave that alternates between $+1$ and -1. This signal controls a switch that alternately passes an advanced or delayed version of the spreading sequence. In the absence of noise, the output of the switch can be represented by

$$s_1(t) = \left[\frac{1 + D(t)}{2}\right] p\left(t + \delta T_c - \epsilon T_c\right) + \left[\frac{1 - D(t)}{2}\right] p\left(t - \delta T_c - \epsilon T_c\right) \quad (4\text{-}110)$$

where the two factors within brackets are orthogonal functions of time and alternate between $+1$ and 0. Only one of the factors is nonzero at any instant. The received direct-sequence signal is multiplied by $s_1(t)$, filtered, and then applied to a square-law device. If the bandpass filter has a sufficiently narrow

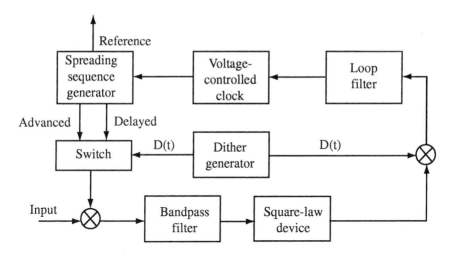

Figure 4.17: Tau-dither loop.

bandwidth, then a derivation similar to that of (4-103) indicates that the device output is

$$s_2(t) \approx \frac{A^2}{2} \left[\frac{1 + D(t)}{2} \right] R_p^2 \left(\delta T_c - \epsilon T_c \right) + \frac{A^2}{2} \left[\frac{1 - D(t)}{2} \right] R_p^2 \left(-\delta T_c - \epsilon T_c \right)$$

(4-111)

Since $D(t)[1 + D(t)] = 1 + D(t)$ and $D(t)[1 - D(t)] = -[1 - D(t)]$, the input to the loop filter is

$$s_3(t) \approx \frac{A^2}{2} \left[\frac{1 + D(t)}{2} \right] R_p^2 \left(\delta T_c - \epsilon T_c \right) - \frac{A^2}{2} \left[\frac{1 - D(t)}{2} \right] R_p^2 \left(-\delta T_c - \epsilon T_c \right)$$

(4-112)

which is a rectangular wave if the time variation of ϵ is ignored. Since the loop filter has a narrow bandwidth relative to that of $D(t)$, its output is approximately the direct-current component of $s_3(t)$, which is the average value of $s_3(t)$. Averaging the two terms of (4-112), we obtain the filter output:

$$s_4(t) \approx \frac{A^2}{4} \left[R_p^2 \left(\delta T_c - \epsilon T_c \right) - R_p^2 \left(\delta T_c - \epsilon T_c \right) \right]$$ (4-113)

The substitution of (4-99) yields the clock input:

$$s_4(t) = \frac{A^2}{4} S(\epsilon, \delta)$$ (4-114)

where the discriminator characteristic is given by (4-107) to (4-109). Thus, the tau-dither loop can track the code timing in a manner similar to that of the delay-locked loop. A detailed analysis indicates that the tau-dither loop provides less accurate code tracking [2], [3]. However, the tau-dither loop requires less hardware than the delay-locked loop and avoids the need to balance the gains and delays in the two branches of the delay-locked loop.

In the presence of frequency-selective fading, the discriminator characteristics of tracking loops are severely distorted. Much better performance is potentially available from a noncoherent tracking loop with diversity and multipath-interference cancellation [12], but a large increase in implementation complexity is required.

4.5   Frequency-Hopping Patterns

The synchronization of the reference frequency-hopping pattern produced by the receiver synthesizer with the received pattern may be facilitated by precision clocks in both the transmitter and the receiver, feedback signals from the receiver to the transmitter, or transmitted pilot signals. However, in most applications, it is necessary or desirable for the receiver to be capable of obtaining synchronization by processing the received signal. During *acquisition*, the reference pattern is synchronized with the received pattern to within a fraction of a hop duration. The *tracking* system further reduces the synchronization error, or at least maintains it within certain bounds. For communication systems that require a strong capability to reject interference, *matched-filter acquisition* and *serial-search acquisition* are the most effective techniques. The matched filter provides rapid acquisition of short frequency-hopping patterns, but requires the simultaneous synthesis of multiple frequencies. The matched filter may also be used in the configuration of Figure 4.2 to detect short patterns embedded in much longer frequency-hopping patterns. Such a detection can be used to initialize or supplement serial-search acquisition, which is more reliable and accommodates long patterns.

Matched-Filter Acquisition

Figure 4.18 shows a programmable *matched-filter acquisition system* that provides substantial protection against interference [13]. It is assumed that a single frequency channel is used during each hop interval that occurs during acquisition. One or more programmable frequency synthesizers produce tones at frequencies $f_1, f_2, \ldots, f_N$, which are offset by a constant frequency from the consecutive frequencies of the hopping pattern for code acquisition. Each tone multiplies the received frequency-hopping signal and the result is filtered so that most of the received energy is blocked, except the energy in a frequency-hopping pulse at a specific frequency. The threshold detector of branch k produces $d_k(t)$ = 1 if its threshold is exceeded, which ideally occurs only if the received signal hops to a specific frequency. Otherwise, the threshold detector produces $d_k(t) = 0$. The use of binary detector outputs prevents the system from being overwhelmed by a few strong interference signals. Input $D(t)$ of the comparator is the number of frequencies in the hopping pattern that were received in

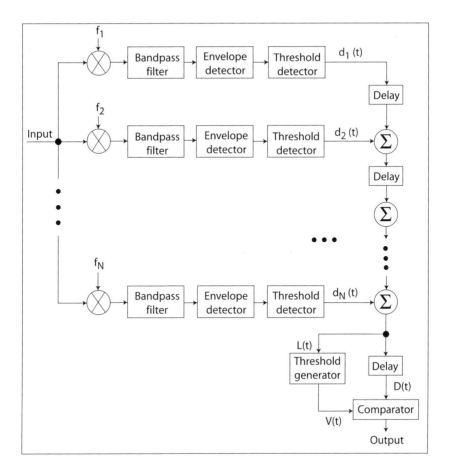

Figure 4.18: Matched-filter acquisition system with protection against interference.

succession. This discrete-valued, continuous-time function is

$$D(t) = \sum_{k=1}^{N} d_k[t - (N - k + 1)T_h] \qquad (4\text{-}115)$$

where T_h is the hop duration. These waveforms are illustrated in Figure 4.19(a) for $N = 8$. The input to the threshold generator is

$$L(t) = D(t + T_h) \qquad (4\text{-}116)$$

Acquisition is declared when $D(t) \geq V(t)$, where $V(t)$ is an adaptive threshold that is a function of $L(t)$. An effective choice is

$$V(t) = \min[L(t) + l_0, N] \qquad (4\text{-}117)$$

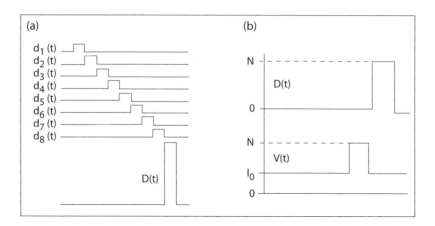

Figure 4.19: Ideal acquisition system waveforms: (a) formation of $D(t)$ when $N = 8$, and (b) comparison of $D(t)$ and $V(t)$.

where l_0 is a positive integer.

In the absence of noise and interference, $L(t) = 0$ and $V(t) = l_0$ during the hop interval in which $D(t) = N$, as illustrated in Figure 4.19(b). If j of the N frequency channels monitored by the matched filter receive strong, continuous interference, then $L(t) = j$ and $V(t) = j + l_0$ during this hop interval if $j \leq N - l_0$, and $D(t) \geq V(t)$. During other intervals, $j + l_0 \leq V(t) \leq N$, but $D(t) = j$. Therefore, $V(t) > D(t)$, and the matched filter does not declare acquisition. False alarms are prevented because $L(t)$ provides an estimate of the number of frequency channels with continuous interference.

When acquisition tone k is received, the signal in branch k of the matched filter is

$$r_k(t) = \sqrt{2S} \cos 2\pi f_0 t + \sqrt{2I} \cos(2\pi f_0 t + \phi) + n(t) \qquad (4\text{-}118)$$

where f_0 is the intermediate frequency, the first term is the desired signal with average power S, the second term represents tone interference with average power I, $n(t)$ is zero-mean, stationary Gaussian noise and interference, and ϕ is the phase shift of the tone interference relative to the desired signal. The power in $n(t)$ is

$$N_1 = N_t + N_i \qquad (4\text{-}119)$$

where N_t is power of the thermal noise and N_i is the power of the statistically independent noise interference.

Bandpass filters are used instead of filters matched to the acquisition tones because the appropriate sampling times are unknown. The passbands of the bandpass filters in the branches are assumed to be spectrally disjoint so that tone interference entering one branch has negligible effect on the other branches, and the filter outputs are statistically independent of each other. To prove the statistical independence of the noise, let $R_n(\tau)$ and $S_n(f)$ denote the autocorrelation and power spectral density, respectively, of the stationary Gaussian noise

$n(t)$ in the received signal. Let $h_1(t)$ and $h_2(t)$ denote the impulse responses and $H_1(f)$ and $H_2(f)$ the transfer functions of two bandpass filters. Since the same Gaussian noise process enters both filters, their outputs are jointly Gaussian. The cross-covariance of the jointly Gaussian, zero-mean filter outputs is

$$
\begin{aligned}
C &= E\left[\int h_1(\tau_1)n(t-\tau_1)d\tau_1 \int h_2(\tau_2)n(t-\tau_2)d\tau_2\right] \\
&= \int\int h_1(\tau_1)h_2(\tau_2)R_n(\tau_2-\tau_1)d\tau_1\,d\tau_2 \\
&= \int\int\int h_1(\tau_1)h_2(\tau_2)S(f)\exp[j2\pi f(\tau_2-\tau_1)]df\,d\tau_1\,d\tau_2 \\
&= \int S(f)H_1(f)H_2^*(f)df \quad\quad\quad\quad (4\text{-}120)
\end{aligned}
$$

where all the integrals extend over $(-\infty, \infty)$. Thus, $C = 0$, if $H_1(f)$ and $H_2(f)$ are spectrally disjoint. If the noise is white and, hence, $S(f)$ is a constant, then $C = 0$ if $H_1(f)$ and $H_2(f)$ are orthogonal. When $C = 0$ for all pairs of bandpass filters, the threshold-detector outputs in the N branches are statistically independent.

Suppose that noise interference is present in a branch, but that tone interference is absent so that $I = 0$. The stationary Gaussian noise has the representation (Appendix C.2)

$$
n(t) = n_c(t)\cos 2\pi f_0 t - n_s(t)\sin 2\pi f_0 t \quad\quad\quad (4\text{-}121)
$$

where $n_c(t)$ and $n_s(t)$ are zero-mean Gaussian processes with noise powers equal to N_1. In practice, the matched filter of Figure 4.18 would operate in continuous time so that acquisition might be declared at any moment. However, for analytical simplicity, the detection and false-alarm probabilities are calculated under the assumption that there is one sample taken per hop dwell time. From (4-119) with $I = 0$ and (4-121), it follows that

$$
r_k(t) = \sqrt{Z_1^2(t) + Z_2^2(t)}\cos[2\pi f_0 t + \psi(t)] \quad\quad\quad (4\text{-}122)
$$

where

$$
Z_1(t) = \sqrt{2S} + n_c(t)\,, \quad Z_2(t) = n_s(t)\,, \quad \psi(t) = \tan^{-1}\left[\frac{n_s(t)}{n_c(t)}\right] \quad (4\text{-}123)
$$

Since $n_c(t)$ and $n_s(t)$ are statistically independent (Appendix C.2), the joint probability density function of Z_1 and Z_2 at any specific time is

$$
g_1(z_1, z_2) = \frac{1}{2\pi N_1}\exp\left[-\frac{(z_1 - \sqrt{2S})^2 + z_2^2}{2N_1}\right] \quad\quad\quad (4\text{-}124)
$$

Let R and Θ be implicitly defined by $Z_1 = R\cos\Theta$ and $Z_2 = R\sin\Theta$. The joint density of R and Θ is

$$
g_2(r, \theta) = \frac{r}{2\pi N_1}\exp\left(-\frac{r^2 - 2r\sqrt{2S}\cos\theta + 2S}{2N_1}\right)\,, \quad r \geq 0\,,\ |\theta| \leq \pi \quad (4\text{-}125)
$$

The probability density function of the envelope $R = \sqrt{Z_1^2(t) + Z_2^2(t)}$ is obtained by integration over θ. The application of (1-59) gives

$$f_1(r) = \frac{r}{N_1} \exp\left(-\frac{r^2 - 2S}{2N_1} \right) I_0\left(r\frac{\sqrt{2S}}{N_1} \right) u(r) \qquad (4\text{-}126)$$

where $I_0(\ )$ is the modified Bessel function of the first kind and order zero, and $u(r) = 1$ if $r \geq 0$ and $u(r) = 0$ if $r < 0$.

The detection probability for the threshold detector in the branch is the probability that the envelope-detector output R exceeds the threshold η :

$$P_{11} = \int_\eta^\infty f_1(r)dr \qquad (4\text{-}127)$$

The *Marcum Q-function* is defined as

$$Q(\alpha, \beta) = \int_\beta^\infty x \exp\left(-\frac{x^2 + \alpha^2}{2} \right) I_0(\alpha x) dx \qquad (4\text{-}128)$$

Applying this definition,

$$P_{11} = Q\left(\sqrt{\frac{2S}{N_1}}, \frac{\eta}{\sqrt{N_1}} \right) \qquad (4\text{-}129)$$

In the absence of noise interference, the detection probability is

$$P_{10} = Q\left(\sqrt{\frac{2S}{N_t}}, \frac{\eta}{\sqrt{N_t}} \right) \qquad (4\text{-}130)$$

If the acquisition tone is absent, but the noise interference is present, the false-alarm probability is

$$P_{01} = \exp\left(-\frac{\eta^2}{2N_1} \right) \qquad (4\text{-}131)$$

In the absence of both the acquisition tone and the noise interference, the false-alarm probability is

$$P_{00} = \exp\left(-\frac{\eta^2}{2N_t} \right) \qquad (4\text{-}132)$$

In (4-129) to (4-132), the first subscript is 1 when the acquisition tone is present and 0 otherwise, whereas the second subscript is 1 when interference is present and 0 otherwise.

Suppose that tone interference is present in a branch. We make the pessimistic assumption that this tone has a frequency exactly equal to that of the acquisition tone, as indicted in (4-118). A trigonometric expansion of the interference term and a derivation similar to that of (4-129) indicates that given the value of ϕ, the conditional detection probability is

$$P_{11}(\phi) = Q\left(\sqrt{\frac{2(S + I + \sqrt{SI}\cos\phi)}{N_1}}, \frac{\eta}{\sqrt{N_1}} \right) \qquad (4\text{-}133)$$

If ϕ is modeled as a random variable uniformly distributed over $[0, 2\pi)$, then the detection probability is

$$P_{11} = \frac{1}{\pi} \int_0^\pi P_{11}(\phi)d\phi \qquad (4\text{-}134)$$

where the fact that $\cos\phi$ takes all its possible values over $[0, \pi]$ has been used to shorten the integration interval. If the acquisition tone is absent, but the tone interference is present, the false-alarm probability is

$$P_{01} = Q\left(\sqrt{\frac{2I}{N_1}}, \frac{\eta}{\sqrt{N_1}}\right) \qquad (4\text{-}135)$$

It is convenient to define the function

$$\beta(i, N, m, P_a, P_b) = \sum_{j=0}^i \binom{m}{j}\binom{N-m}{i-j} P_a^j(1 - P_a)^{m-j} P_b^{i-j}(1 - P_b)^{N-m-i+j}$$

$$(4\text{-}136)$$

where $\binom{b}{a} = 0$ if $a > b$. Given that m of the N matched-filter branches receive interference of equal power, let the index j represent the number of interfered channels with detector outputs above η. If $0 \le j \le i$, there are $\binom{m}{j}$ ways to choose j channels out of m and $\binom{N-m}{i-j}$ ways to choose $i - j$ channels with detector outputs above η from among the $N-m$ channels that are not interfered. Therefore, the conditional probability that $D(t) = i$ given that m channels receive interference is

$$P(D = i|m) = \beta(i, N, m, P_{h1}, P_{h0}), \quad h = 0, 1 \qquad (4\text{-}137)$$

where $h = 1$ if the acquisition tones are present and $h = 0$ if they are not. Similarly, given that m of N acquisition channels receive interference, the conditional probability that $L(t) = l$ is

$$P(L = l|m) = \beta(l, N, m, P_{h1}, P_{h0}), \quad h = 0, 1 \qquad (4\text{-}138)$$

If there are J interference signals randomly distributed among a hopset of M frequency channels, then the probability that m out of N matched-filter branches have interference is

$$P_m = \frac{\binom{N}{m}\binom{M-N}{J-m}}{\binom{M}{J}} \qquad (4\text{-}139)$$

The probability that acquisition is declared at a particular sampling time is

$$P_A = \sum_{m=0}^{\min(N,J)} P_m \sum_{l=0}^N P(L = l|m) \sum_{k=V(l)}^N P(D = k|m) \qquad (4\text{-}140)$$

When the acquisition tones are received in succession, the probability of detection is determined from (4-137) to (4-140). The result is

$$P_D = \sum_{m=0}^{\min(N,J)} \frac{\binom{N}{m}\binom{M-N}{J-m}}{\binom{M}{J}} \sum_{l=0}^{N} \beta(l, N, m, P_{01}, P_{00}) \sum_{k=V(l)}^{N} \beta(k, N, m, P_{11}, P_{10})$$

(4-141)

For simplicity in evaluating the probability of a false alarm, we ignore the sampling time preceding the peak value of $D(t)$ in Figure 4.19 because this probability is negligible at that time. Since the acquisition tones are absent, the probability of a false alarm is

$$P_F = \sum_{m=0}^{\min(N,J)} \frac{\binom{N}{m}\binom{M-N}{J-m}}{\binom{M}{J}} \sum_{l=0}^{N} \beta(l, N, m, P_{01}, P_{00}) \sum_{k=V(l)}^{N} \beta(k, N, m, P_{01}, P_{00})$$

(4-142)

If there is no interference so that $J = 0$, then (4-141) and (4-142) reduce to

$$P_D = \sum_{l=0}^{N} \binom{N}{l} P_{00}^l (1 - P_{00})^{N-1} \sum_{k=V(l)}^{N} \binom{N}{k} P_{10}^k (1 - P_{10})^{N-k} \qquad (4\text{-}143)$$

$$P_F = \sum_{l=0}^{N} \binom{N}{l} P_{00}^l (1 - P_{00})^{N-1} \sum_{k=V(l)}^{N} \binom{N}{k} P_{00}^k (1 - P_{00})^{N-k} \qquad (4\text{-}144)$$

The channel threshold η is selected to maintain a required P_F when there is no interference and the values of l_0, N, and N_t are given. The value of l_0 is then selected to maximize P_D given the values of N and S/N_t. The best choice is generally $l_0 = \lfloor N/2 \rfloor$. For example, suppose that $N = 8$, $P_F = 10^{-7}$, and the SNR is $S/N_t = 10$ dB when an acquisition tone is received. A numerical evaluation of (4-144) then yields $\eta/\sqrt{N_t} = 3.1856$ and $l_0 = 4$ as the parameter values that maintain $P_F = 10^{-7}$ while maximizing P_D in the absence of interference. The threshold pair $\eta/\sqrt{N_t} = 3.1896$, $l_0 = 4$ is the choice when a fixed comparator threshold $V(t) = l_0$ is used instead of the adaptive threshold of (4-117). If $D(t)$ and $L(t)$ are sampled once every hop dwell interval, then the false-alarm rate is P_F/T_h.

As an example, suppose that noise jamming with total power N_{it} is uniformly distributed over J matched-filter frequency channels so that

$$N_i = \frac{N_{it}}{J} \qquad (4\text{-}145)$$

is the power in each of these channels. Interference tones are absent and $N = 8$, $M = 128$, and $S/N_t = 10$ dB. To ensure that $P_F = 10^{-7}$ in the absence of jamming, we assume that $l_0 = 4$ and $\eta/\sqrt{N_t} = 3.1856$ when an adaptive comparator threshold is used, and that $l_0 = 4$ and $\eta/\sqrt{N_t} = 3.1896$ when a fixed comparator threshold is used. Since P_D is relatively insensitive to J, its effect is assessed by examining P_F. Figure 4.20 depicts P_F as a function of N_{it}/S, the jamming-to-signal ratio. The figure indicates that an adaptive threshold is much more

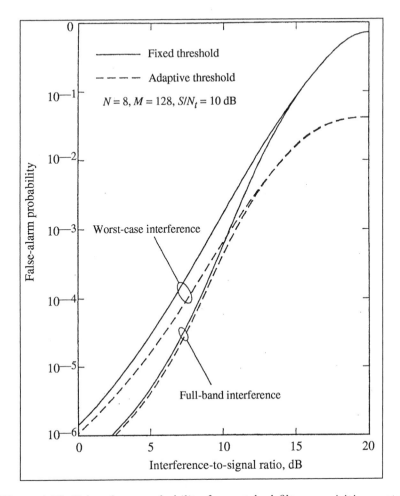

Figure 4.20: False-alarm probability for matched-filter acquisition system.

resistant to partial-band jamming than a fixed threshold when N_{it}/S is large. When $N_{it}/S < 10$ dB, the worst-case partial-band jamming causes a considerably higher P_F than full-band jamming. It is found that multitone jamming tends to produce fewer false alarms than noise jamming. Various other performance and design issues and the impact of frequency-hopping interference are addressed in [13].

Serial-Search Acquisition

As illustrated by Figure 4.21, a *serial-search acquisition system* for frequency-hopping signals determines acquisition by attempting to downconvert the received frequency-hopping pattern to a fixed intermediate frequency, and then comparing the output of an energy detector (Chapter 7) to a threshold. A

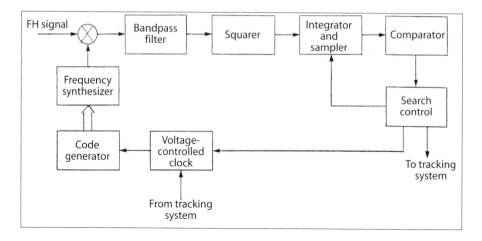

Figure 4.21: Serial-search acquisition system.

trial alignment of the frequency-hopping pattern synthesized by the receiver with the received pattern is called a *cell*. If a cell passes certain tests, acquisition is declared and the tracking system is activated. If not, the cell is rejected. A new candidate cell is produced when the reference pattern synthesized by the receiver is either advanced or delayed relative to the received pattern.

A number of search techniques are illustrated in Figure 4.22, which depicts successive frequencies in the received pattern and six possible receiver-generated patterns. Each search technique is implemented as part of a uniform or Z-search of the timing uncertainty. The small arrows indicate test times at which cells are rejected, and the large arrows indicate typical times at which acquisition is declared or subsequent verification testing begins. The step size, which is the separation in hop durations between cells, is denoted by Δ. Techniques (a) and (b) entail inhibiting the code-generator clock after each unsuccessful test. Technique (c) is the same as technique (b) but extends the test duration to 3 hops. Technique (d) advances the reference pattern by skipping frequencies in the pattern after each unsuccessful test. The inhibiting or advancing of techniques (a) to (d) or an alternation of them continues until acquisition is declared. The *small misalignment technique* (e) is effective when there is a high probability that the reference and received patterns are within r hops of each other, which usually is true immediately after the tracking system loses lock. The code generator temporarily forces the reference signal to remain at a frequency for $2r + 1$ hop intervals extending both before and after the interval in which the frequency would ordinarily be synthesized. If the misalignment is less than r hops, then acquisition occurs within $2r + 1$ hop durations. In the figure, $r = 1$, the initial misalignment is one-half hop duration, and it is assumed that the first time the reference and received frequencies coincide, detection fails, but the second time results in acquisition. Technique (f) entails waiting at a fixed synchronization frequency until this frequency is received. This

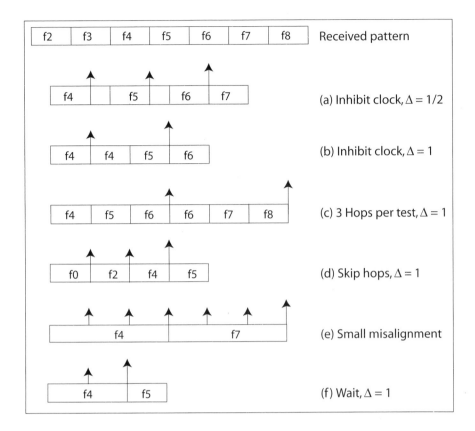

Figure 4.22: Search techniques for acquisition.

technique results in a rapid search if the reference frequency can be selected
so that it is soon reached by the received pattern. The reference frequency
is determined from an estimate of the timing uncertainty, the key bits, and
the TOD bits (Chapter 3), but must be periodically shifted by at least the
coherence bandwidth so that neither fading nor interference in any particular
frequency channel prevents acquisition.

When the period of the frequency-hopping pattern is large, special measures may be required to reduce the timing uncertainty. A reduced hopset
with a short pattern period may be used temporarily to reduce the timing uncertainty and, hence, the acquisition time. A feedback signal from the receiver
may be used to adjust the timing of the transmitted pattern. In a network,
a separate communication channel or cueing frequency may provide the TOD
to subscribers. After detection of the TOD, a receiver might use the small
misalignment technique for acquisition.

The *search control system* determines the integration intervals, the thresholds, and the logic of the tests to be conducted before acquisition is declared
and the tracking system is activated. The details of the search control strat-

egy determine the statistics of the acquisition time. The control strategy is usually a *multiple-dwell strategy* that uses an initial test to quickly eliminate improbable cells. Subsequent tests are used for verification testing of cells that pass the initial test. The multiple-dwell strategy may be a *consecutive-count strategy*, in which a failed test causes a cell to be immediately rejected, or an *up-down strategy*, in which a failed test causes a repetition of a previous test. The up-down strategy is preferable when the interference or noise level is high [14].

Since acquisition for frequency-hopping signals is analogous to acquisition for direct-sequence signals, the statistical description of acquisition given in Section 4.2 is applicable if the chips are interpreted as hops. Only the specific equations of the detection and false-alarm probabilities are sometimes different. For example, consider a single-dwell system with a uniform search, a uniform *a priori* correct-cell location distribution, two independent correct cells with the common detection probability P_d, and $q \gg 1$. In analogy with (4-93), the NMAT is

$$NMAT = \frac{\bar{T}_a}{C_h T_h} = \left(\frac{2 - P_D}{2 P_D} \right) \frac{q_h}{C_h} (M_h + K_h P_F) \qquad (4\text{-}146)$$

where M_h is the number of hops per test, K_h is the number of hops in the mean penalty time, C_h is the number of hops in the timing uncertainty, q_h is the number of cells, and

$$P_D = 2 P_d - P_d^2 \qquad (4\text{-}147)$$

For step size $\Delta = 1$, $q_h/C_h = 1$; for $\Delta = 1/2$, $q_h/C_h = 2$.

If the detector integration is over several hop intervals, strong interference or deep fading over a single hop interval can cause a false alarm with high probability. This problem is mitigated by making a hard decision after integrating over each hop interval. After N decisions, a test for acquisition is passed or failed if the comparator threshold has been exceeded l_0 or more times out of N. Let P_{Dp} and P_{Da} denote the probabilities that the comparator threshold is exceeded at the end of a hop interval when the correct cell is tested and interference is present and absent, respectively. Let P_D denote the probability that an acquisition test is passed when the correct cell is tested. If the N acquisition tones in a test are distinct, then a derivation similar to the one for matched filters yields

$$P_D = \sum_{m=0}^{\min(N,J)} \frac{\binom{N}{m}\binom{M-N}{J-m}}{\binom{M}{J}} \sum_{l=l_0}^{N} \beta(l, N, m, P_{Dp}, P_{Da}) \qquad (4\text{-}148)$$

where $l_0 \geq 0$. Similarly, the probability that an acquisition test is passed when an incorrect cell is tested and no acquisition tones are present is

$$P_F = \sum_{m=0}^{\min(N,J)} \frac{\binom{N}{m}\binom{M-N}{J-m}}{\binom{M}{J}} \sum_{l=l_0}^{N} \beta(l, N, m, P_{Fp}, P_{Fa}) \qquad (4\text{-}149)$$

where P_{Fp} and P_{Fa} are the probabilities that the threshold is exceeded when an incorrect cell is tested and interference is present and absent, respectively.

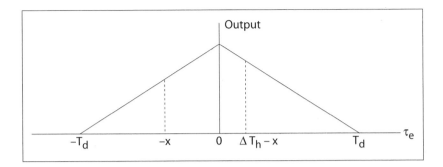

Figure 4.23: Amplitude of integrator output as function of relative pattern delay.

A suitable choice for l_0 is $\lfloor N/2 \rfloor$. Since the serial-search system of Figure 4.21 has an embedded radiometer, the performance analysis of the radiometer given in Chapter 7 can be used to obtain expressions for P_{Dp} and P_{Da}, P_{Fp}, and P_{Fa}.

Although a large step size limits the number of incorrect cells that must be tested before the correct cell is tested, it causes a loss in the average signal energy in the integrator output of Figure 4.21 when a correct cell is tested. This issue and the role of the hop dwell time are illustrated by Figure 4.23, which depicts the idealized output for a single pulse of the received and reference signals in the absence of noise. Let τ_e denote the delay of the reference pattern relative to the received pattern. Suppose that one tested cell has $\tau_e = -x$, where $0 \le x \le \Delta T_h$, and the next tested cell has $\tau_e = \Delta T_h - x$ following a cell rejection. The largest amplitude of the integrator output occurs when $|\tau_e| = y$, where

$$y = \min(x, \Delta T_h - x), \quad 0 \le x < \Delta T_h \tag{4-150}$$

Assuming that x is uniformly distributed over $(0, \Delta T_h)$, y is uniformly distributed over $(0, \Delta T_h/2)$. Therefore,

$$E[y] = \frac{\Delta T_h}{4} \tag{4-151}$$

$$E[y^2] = \frac{\Delta^2 T_h^2}{12} \tag{4-152}$$

The *correct cell* is considered to be the one for which $|\tau_e| = y$. If the output function approximates the triangular shape depicted in the figure, its amplitude when $|\tau_e| = y$ is

$$A = A_{\max}\left(1 - \frac{y}{T_d}\right) \tag{4-153}$$

Therefore, the average signal energy in the integrator output is

$$E\left[\left(1 - \frac{y}{T_d}\right)^2\right] = 1 - \frac{\Delta T_h}{2T_d} + \frac{\Delta^2 T_h^2}{12T_d^2} \tag{4-154}$$

which indicates the loss due to the misalignment of patterns when the correct cell is tested. For example, if $T_d = 0.9T_h$, then (4-154) indicates that the average loss is 1.26 dB when $\Delta = 1/2$; if $\Delta = 1$, then the loss is 2.62 dB.

The serial-search acquisition of frequency-hopping signals is faster than the acquisition of direct-sequence signals because the hop duration is much greater than a spreading-sequence chip duration for practical systems. Given the same timing uncertainty, fewer cells have to be searched to acquire frequency-hopping signals because each step covers a larger portion of the region.

Tracking System

The acquisition system ensures that the receiver-synthesized frequency-hopping pattern is aligned in time with the received pattern to within a fraction of a hop duration. The tracking system must provide a fine synchronization by reducing the residual misalignment after acquisition. Although the delay-locked and tau-dither loops used for the tracking of direct-sequence signals can be adapted to frequency-hopping signals [17], the predominant form of tracking in frequency-hopping systems is provided by the *early-late-gate tracking loop* [15]. This loop is shown in Figure 4.24 along with the ideal associated waveforms for a typical example in which there is a single carrier frequency during a hop dwell interval. If the data modulation is MFSK, then the outputs of parallel branches, each with a bandpass filter and envelope detector can be combined and applied to the early-late gate. In the absence of noise, the envelope detector produces a positive output only when the received frequency-hopping signal $r(t)$, and the receiver-generated frequency-hopping replica $r_1(t)$ are offset by the intermediate frequency f_i. The *gating signal* $g(t)$ is a square-wave clock signal with transitions from -1 to $+1$ that control the frequency transitions of $r_1(t)$. The early-late gate functions as a signal multiplier. Its output $u(t)$ is the product of the gating signal and the envelope-detector output $v(t)$. The error signal is the time integral of $u(t)$ and is a function of τ_e, the delay of $r_1(t)$ relative to $r(t)$. The error signal can be expressed as the discriminator characteristic $e(\delta)$, which is a function of $\delta = \tau_e/T_h$, the normalized delay error. For the typical waveforms shown, δ is positive, and hence so is $e(\delta)$. Therefore, the voltage-controlled clock (VCC) will increase the transition rate of the gating signal, which will bring $r_1(t)$ into better time-alignment with $r(t)$.

If the tracking system loses lock and the small-misalignment test fails, then the wait technique of Figure 4.22 can be used to expedite the reacquisition. After dehopping the received signal to baseband, demodulating, and producing oversampled information bits, the receiver establishes bit synchronization by searching for a special sequence of marker bits that match a stored reference sequence, as is often done for frame synchronization [16]. After this matching occurs, information is extracted from subsequent bits. The information could specify the time of occurrence and the spectral location of the next synchronization frequency at which the receiver waits.

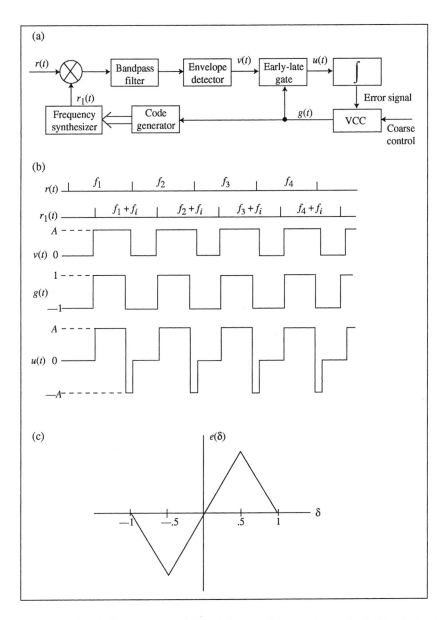

Figure 4.24: Early-late gate tracking: (a) loop, (b) signals, and (c) discriminator characteristic.

4.6   Problems

1. Use orthonormal basis functions to prove (4-4) and the statistical independence of the $\{r_i\}$.

2. Prove that for a random variable Y and a random variable X with density $f(x)$, the relation $\int [var(Y/x) f(x) dx] = var(Y)$ is not true in general. If it were, then σ_a^2 given by (4-14) and (4-15) could be simplified. Give a sufficient condition under which this relation is valid.

3. Consider a uniform search with a uniform a priori distribution for the location of the correct cell. (a) What is the average number of sweeps through the timing uncertainty during acquisition? (b) For a large number of cells, calculate an upper bound on $P(T_a > c\overline{T_a})$ as a function of P_D for $c > 1$. (c) For a large number of cells to be searched, show that the standard deviation of the acquisition time satisfies $\frac{\overline{T_a}}{\sqrt{3}} \leq \sigma_a \leq \overline{T_a}$.

4. (a) Derive (4-29) and (4-30) by first expressing T_{22} and T_{12} as conditional expectations and then enumerating the possible values of T_{22} and T_{12} and their conditional probabilities. (b) Use a similar procedure to derive (4-37) to (4-40).

5. Derive P_D in (4-33) assuming the presence of zero-mean, white Gaussian noise with two-sided power spectral density N_0. Use (4-26) and assume that $P_a = P_b$. To determine P_a, begin by writing an expression for the matched-filter output when a target signal with energy $\mathcal{E}$ completely fills the filter.

6. Consider a lock detector that uses a double-dwell consecutive-count system with equal test durations. (a) Use a recursive relation to show that $\overline{T_h} = \tau[(1 - P_D)^{-1} + (1 - P_D)^{-2}]$. (b) Use a recursive relation to show that $\overline{T_p} = \tau[(1 - P_F)^{-1} + (1 - P_F)^{-2}]$.

7. Starting with (4-73), derive $var(V_c)$ given by (4-78) for the acquisition correlator.

8. Derive (4-92) from (4-91) using the method outlined in the text.

9. Consider Example 2 of Section 4.3 leading to Figure 4.14. Assume fast fading and that $\Delta = 1/2$, $E_c/N_0 = -10dB$, $K = 10,000$, $P_{F1} = 0.03$, $P_{F2} = 0.001$, and $M_2 = 10M_1$. Plot the NMAT versus M_1 for the consecutive-count and up-down systems to determine graphically what values of M_1 minimize the NMAT.

10. Derive (4-107) and (4-108).

11. Compare the NMAT for a frequency-hopping system given by (4-146) with the NMAT for a direct-sequence system given by (4-93) when the penalty times and test durations for both systems are equal. Under the latter condition, it is reasonable to assume that P_D and P_F are roughly

equal for both systems. With these assumptions, what is the ratio of the direct-sequence NMAT to the frequency-hopping NMAT?

12. Reduce (4-148) to a single summation and simplify for the following cases. a) $l_0 = N$, and b) $J = 0$, $l_0 \geq 0$.

13. Derive P_D and P_F for serial-search acquisition of frequency-hopping signals when a single acquisition tone is used.

4.7   References

1. R. B. Ward and K. P. Y. Yiu, "Acquisition of Pseudonoise Signals by Recursion-Aided Sequential Estimation," *IEEE Trans. Commun.*, vol. 25, pp. 784–794, August 1977.

2. H. Meyr and G. Polzer, "Performance Analysis for General PN Spread-spectrum Acquisition Techniques," *IEEE Trans. Commun.*, vol. 31, pp. 1317–1319, December 1983.

3. V. M. Jovanovic, "Analysis of Strategies for Serial-Search Spread-Spectrum Code Acquisition—Direct Approach," *IEEE Trans. Commun.*, vol. 36, pp. 1208–1220, November 1988.

4. S.-M. Pan, D. E. Dodds, and S. Kumar, "Acquisition Time Distribution for Spread-Spectrum Receivers, *IEEE J. Select. Areas Commun.*, vol. 8, pp. 800-808, June 1990.

5. A. Polydoros and C. L. Weber, "A Unified Approach to Serial-Search Spread Spectrum Code Acquisition," *IEEE Trans. Commun.*, vol. 32, pp. 542–560, May 1984.

6. V. M. Jovanovic and E. S. Sousa, "Analysis of Noncoherent Correlation in DS/BPSK Spread Spectrum Acquisition," *IEEE Trans. Commun.*, vol. 43, pp. 565–573, Feb./March/April 1995.

7. L.-L. Yang and L. Hanzo, "Serial Acquisition of DS-CDMA Signals in Multipath Fading Mobile Channels," *IEEE Trans. Veh. Technol.*, vol. 50, pp. 617–628, March 2001.

8. K. Choi, K. Cheun, and T. Jung, "Adaptive PN Code Acquisition Using Instantaneous Power-Scaled Detection Threshold under Rayleigh Fading and Pulsed Gaussian Noise Jamming," *IEEE Trans. Commun.*, pp. 1232–1235, August 2001.

9. S. Glisic and M. D. Katz, "Modeling of the Code Acquisition Process for Rake Receivers in CDMA Wireless Networks with Multipath and Transmitter Diversity," *IEEE J. Select. Areas Commun.*, vol. 19, pp. 21–32, January 2001.

10. A. Polydoros and C. L. Weber, "Analysis and Optimization of Correlative Code-Tracking Loops in Spread-Spectrum Systems," *IEEE Trans. Commun.*, vol. 33, pp. 30–43, January 1985.

11. W. R. Braun, "PN Acquisition and Tracking Performance in DS/CDMA Systems with Symbol-Length Spreading Sequence," *IEEE Trans. Commun.*, vol. 45, pp. 1595–1601, December 1997.

12. W.-H. Sheen and C.-H. Tai, "A Noncoherent Tracking Loop with Diversity and Multipath Interference Cancellation for Direct-Sequence Spread-Spectrum Systems," *IEEE Trans. Commun.*, vol. 46, pp. 1516–1524, November 1998.

13. L. E. Miller, J. S. Lee, R. H. French, and D. J. Torrieri, "Analysis of an Anti-jam FH Acquisition Scheme," *IEEE Trans. Commun.*, vol. 40, pp. 160-170, January 1992.

14. C. A. Putman, S. S. Rappaport, and D. L. Schilling, "Comparison of Strategies for Serial Acquisition of Frequency-Hopped Spread-Spectrum Signals," *IEE Proc.*, vol. 133, pt. F, pp. 129–137, April 1986.

15. C. A. Putman, S. S. Rappaport, and D. L. Schilling, "Tracking of Frequency-Hopped Spread-Spectrum Signals in Adverse Environments," *IEEE Trans. Commun.*, vol. 31, pp. 955–963, August 1983.

16. D. Torrieri, *Principles of Secure Communication Systems*, 2nd ed. Boston: Artech House, 1992.

Chapter 5

Fading of Wireless Communications

5.1 Path Loss, Shadowing, and Fading

Free-space propagation losses of electromagnetic waves vary inversely with the square of the distance between a transmitter and a receiver. Analysis indicates that if a signal traverses a direct path and combines in the receiver with a multipath component that is perfectly reflected from a plane, then the composite received signal has a power loss proportional to the inverse of the fourth power of the distance. Thus, it is natural to seek a power-law variation for the average received power in a specified geographic area as a function of distance. For terrestrial wireless communications, measurements averaged over many different positions of a transmitter and a receiver in a specified geographic area confirm that the average received power, which is called the *area-mean power*, does tend to vary inversely as a power of the transmitter-receiver distance r. It is found that the area-mean power is approximately given by

$$p_a = p_0 \left(\frac{r}{R_0} \right)^{-\beta} \tag{5-1}$$

where p_0 is the average received power when the distance is $r = R_0$, and β is the *attenuation power law*. The parameters p_0 and β are functions of the carrier frequency, antenna heights, terrain characteristics, vegetation, and various characteristics of the propagation medium. Typically, the parameters vary with distance, but are constant within a range of distances. A typical value of the attenuation power law for urban areas and microwave frequencies is $\beta = 4$. The power law increases with the carrier frequency.

For a specific propagation path, the received *local-mean power* departs from the area-mean power due to *shadowing*, which is the effect of diffractions and propagation conditions that are path-dependent. Each diffraction due to obstructing terrain and each reflection from an obstacle causes the signal power to be multiplied by an attenuation factor. Thus, the received signal power is often the product of many factors, and hence the logarithm of the signal power is the sum of many factors. If each factor is modeled as a uniformly bounded, independent random variable that varies from path to path, then the central-limit theorem implies that the logarithm of the received signal power has an approximately normal distribution if the number of attenuation factors is large enough. Extensive empirical data confirms that the received local-mean power after transmission over a randomly selected propagation path with a fixed distance is approximately lognormally distributed. Thus, the shadowing model specifies that the local-mean power has the form

$$p_l = p_a 10^{\xi/10} \tag{5-2}$$

where the *shadowing factor* ξ is a zero-mean random variable with a normal distribution. The standard deviation of ξ is denoted by σ_s, which is expressed in decibels. From (5-1) and (5-2), it follows that the probability distribution function of the normalized local-mean power, p_l/p_0, is

$$F(x) = 1 - Q\left\{ \frac{\eta}{\sigma_s} \ln\left[x\left(\frac{r}{R_0} \right)^{\beta} \right] \right\} \tag{5-3}$$

where ln[ ] denotes the natural logarithm, and $\eta = (10\log_{10} e)$. The standard deviation σ_s increases with carrier frequency and terrain irregularity and often exceeds 10 dB for terrestrial communications. The value of the shadowing factor for a propagation path is usually strongly correlated with its value for a nearby propagation path. For mobile communications, the typical time interval during which the shadowing factor is nearly constant is a second or more.

Fading, which is endemic in mobile, long-distance, high-frequency, and other communication channels, causes power fluctuations about the local-mean power. Fading occurs at much faster rate than shadowing. During an observation interval in which the shadowing factor is nearly constant, the received signal power may be expressed as the product

$$p_r = p_a 10^{\xi/10} \alpha^2(t) \tag{5-4}$$

where the factor $\alpha^2(t)$ is due to the fading. Since ξ is fixed, the local-mean power is

$$p_l = E[p_r] = p_a 10^{\xi/10} E[\alpha^2(t)] \tag{5-5}$$

A signal experiences *fading* when the interaction of multipath components and time- or frequency-varying channel conditions cause significant fluctuations in its amplitude at a receiver. *Multipath components* of a signal are generated by inhomogeneities in the propagation medium or reflections from obstacles. These components travel along different paths before being recombined at the receiver. Because of the different time-varying delays and attenuations encountered by the multipath components, the recombined signal is a distorted version of the original transmitted signal. Fading may be classified as time-selective, frequency-selective, or both. *Time-selective fading* is fading caused by the movement of the transmitter or receiver or by changes in the propagation medium. *Frequency-selective fading* is fading caused by the different delays of the multipath components, which may affect certain frequencies more than others. The following concise development of fading theory[1], [2], [3] emphasizes basic physical mechanisms.

A bandpass transmitted signal can be expressed as

$$s_t(t) = \text{Re}[s(t)\exp(j2\pi f_c t)] \qquad (5\text{-}6)$$

where $s(t)$ denotes its complex envelope, f_c denotes its carrier frequency, and Re[ ] denotes the real part. Transmission over a time-varying multipath channel of $N(t)$ paths produces a received bandpass signal that consists of the sum of $N(t)$ waveforms. The ith waveform is the transmitted signal delayed by time $\tau_i(t)$, attenuated by a factor $a_i(t)$ that depends on the path loss and shadowing, and shifted in frequency by the amount $f_{di}(t)$ due to the Doppler effect. Assuming that $f_{di}(t)$ is constant during the path delays, the received signal may be expressed as

$$s_r(t) = \text{Re}[s_1(t)\exp(j2\pi f_c t)] \qquad (5\text{-}7)$$

where the received complex envelope is

$$s_1(t) = \sum_{i=1}^{N(t)} a_i(t)\exp[j\phi_i(t)]s[t - \tau_i(t)] \qquad (5\text{-}8)$$

and its phase is

$$\phi_i(t) = -2\pi f_c \tau_i(t) + 2\pi f_{di}(t)\left[t - \tau_i(t)\right] \qquad (5\text{-}9)$$

The Doppler shift arises because of the relative motion between the transmitter and the receiver. In Figure 5.1(a), the receiver is moving at speed $v(t)$ and the angle between the velocity vector and the propagation direction of an electromagnetic wave is $\psi_i(t)$. For this geometry, the received frequency is increased by the Doppler shift

$$f_{di}(t) = f_c \frac{v(t)}{c}\cos\psi_i(t) \qquad (5\text{-}10)$$

where c is the speed of an electromagnetic wave. In Figure 5.1(b), the transmitter is moving at speed $v(t)$ and there is a reflecting surface that changes the arrival angle of the electromagnetic wave at the receiver. If $\psi_i(t)$ represents the angle between the velocity vector and the initial direction of the electromagnetic wave, then (5-10) again gives the Doppler shift.

5.2 Time-Selective Fading

To analyze time-selective fading, it is assumed that $N(t) = N$ for the time interval of interest and that the differences in the time delays along the various

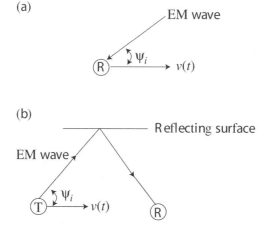

Figure 5.1: Examples of the Doppler effect: (a) receiver motion and (b) transmitter motion and reflecting surface.

paths are small compared with the inverse of the signal bandwidth. Therefore, the received multipath components overlap in time and are called *unresolvable multipath components*. If the time origin is chosen to coincide with the average arrival time of the multipath components at a receiver, then the received complex envelope of (5-8) may be expressed as

$$s_1(t) \approx s(t)r(t) \tag{5-11}$$

where the *equivalent lowpass* or *equivalent baseband channel response* is

$$r(t) = \sum_{i=1}^{N} a_i(t) \exp\left[j\phi_i(t)\right] \tag{5-12}$$

The fluctuations in this factor cause signal fading at the receiver and increase the bandwidth of the received signal. If the transmitted signal is an unmodulated tone, then $s(t) = 1$ and (5-12) represents the complex envelope of the received signal.

The channel response can be decomposed as

$$r(t) = r_c(t) + jr_s(t) \tag{5-13}$$

where $j = \sqrt{-1}$ and

$$r_c(t) = \sum_{i=1}^{N} a_i(t) \cos[\phi_i(t)] \, , \qquad r_s(t) = \sum_{i=1}^{N} a_i(t) \sin[\phi_i(t)] \tag{5-14}$$

If the range of the delay values exceeds $1/f_c$, then the sensitivity of $\phi_i(t)$ to small variations in the delay $\tau_i(t)$ makes it plausible to model the phases $\phi_i(t), i$

1, 2, ... , N, as random variables that are independent of each other and the $\{a_i(t)\}$ and are uniformly distributed over $[0, 2\pi)$ at a specific time t. Therefore,

$$E[r_c(t)] = E[r_s(t)] = 0 \tag{5-15}$$

If the amplitude factors $a_i(t), i = 1, 2, \ldots , N$, are either identically distributed or uniformly bounded independent random variables at time t, then according to the central-limit theorem, the probability distributions of both $r_c(t)$ and $r_s(t)$ approach Gaussian distributions as N increases. Thus, if N is sufficiently large, then $r(t)$ at a specific time is well modeled as a *complex Gaussian random variable*. Since the phases are independent and uniformly distributed, it follows that

$$E[r_c(t)r_s(t)] = 0 \tag{5-16}$$
$$E[r_c^2(t)] = E[r_s^2(t)] = \sigma_r^2(t) \tag{5-17}$$

where we define

$$\sigma_r^2(t) = \frac{1}{2} \sum_{i=1}^{N} E[a_i^2(t)] \tag{5-18}$$

This equation indicates that $\sigma_r^2(t)$ is equal to the sum of the local-mean powers of the multipath components. Equations (5-15) to (5-17) imply that $r_c(t)$ and $r_s(t)$ are independent, identically distributed, zero-mean Gaussian random variables.

Let $\alpha(t) = |r(t)|$ denote the envelope, and $\theta(t) = \tan^{-1}[r_s(t)/r_c(t)]$ the phase of $r(t)$ at a specific time t. Then

$$r(t) = \alpha(t)e^{j\theta(t)} \tag{5-19}$$

As shown in Appendix D.4, since $r_c(t)$ and $r_s(t)$ are Gaussian and $\alpha^2(t) = r_c^2(t) + r_s^2(t)$, $\theta(t)$ has a uniform distribution over $[0, 2\pi)$, and $\alpha(t)$ has the *Rayleigh* probability density function:

$$f_\alpha(r) = \frac{r}{\sigma_r^2} \exp\left(-\frac{r^2}{2\sigma_r^2}\right) u(r) \tag{5-20}$$

where the time-dependence has been suppressed for convenience, and $u(r) = 1$, $r \geq 0$, and $u(r) = 0$, $r < 0$. From (5-20) or directly from (5-13) and (5-17), it follows that

$$E[\alpha^2(t)] = 2\sigma_r^2(t) \tag{5-21}$$

The substitution of (5-19) and (5-11) into (5-7) gives

$$\begin{aligned} s_r(t) &= \text{Re}[\alpha(t)s(t)\exp(j2\pi f_c t + j\theta(t))] \\ &= \alpha(t)\, A(t)\cos[2\pi f_c t + \phi(t) + \theta(t)] \end{aligned} \tag{5-22}$$

where $A(t)$ is the amplitude and $\phi(t)$ the phase of $s(t)$. Equations (5-21) and (5-22) indicate that the instantaneous local-mean power is $p_l = \sigma_r^2(t)A^2(t)$.

When a line-of-sight exists between a transmitter and a receiver, one of the received multipath components may be much stronger than the others. This strong component is called the *specular component* and the other unresolvable components are called *diffuse* or *scattered components*. As a result, the multiplicative channel response of (5-12) becomes

$$r(t) = a_0(t) \exp[j\phi_0(t)] + \sum_{i=1}^{N} a_i(t) \exp[j\phi_i(t)] \qquad (5\text{-}23)$$

where the summation term is due to the diffuse components, and the first term is due to the specular component. If N is sufficiently large, then at time t the summation term is well-approximated by a zero-mean, complex Gaussian random variable. Thus, $r(t)$ at a specific time is a complex Gaussian random variable with a nonzero mean equal to the deterministic first term, and (5-13) implies that

$$E[r_c(t)] = a_0(t) \cos[\phi_0(t)] , \qquad\qquad E[r_s(t)] = a_0(t) \sin[\phi_0(t)] \qquad (5\text{-}24)$$

As shown in Appendix D.3, since $r_c(t)$ and $r_s(t)$ are Gaussian and $\alpha^2(t) = r_c^2(t) + r_s^2(t)$, the envelope $\alpha(t) = |r(t)|$ has the *Rice* probability density function:

$$f_\alpha(r) = \frac{r}{\sigma_r^2} \exp\left\{ -\frac{r^2 + a_0^2}{2\sigma_r^2} \right\} I_0\left(\frac{a_0 r}{\sigma_r^2} \right) u(r) \qquad (5\text{-}25)$$

where $I_0(\ )$ is the modified Bessel function of the first kind and order zero, and the time-dependence is suppressed for convenience. From (5-25) or directly from (5-18) and (5-23), it follows that the average envelope power is

$$\Omega = E[\alpha^2(t)] = a_0^2(t) + 2\sigma_r^2(t) \qquad (5\text{-}26)$$

The type of fading modeled by (5-23) and (5-25) is called *Ricean fading*. At a specific time, the *Rice factor* is defined as

$$\kappa = \frac{a_0^2}{2\sigma_r^2} \qquad (5\text{-}27)$$

which is the ratio of the specular power to the diffuse power. In terms of κ and Ω, the Rice density is

$$f_\alpha(r) = \frac{2(\kappa+1)}{\Omega} r \exp\left\{ -\kappa - \frac{(\kappa+1)r^2}{\Omega} \right\} I_0\left(\sqrt{\frac{\kappa(\kappa+1)}{\Omega}} 2r \right) u(r) \qquad (5\text{-}28)$$

When $\kappa = 0$, Ricean fading is the same as Rayleigh fading. When $\kappa = \infty$, there is no fading.

A more flexible fading model is created by introducing a new parameter m; the *Nakagami-m* probability density function for the envelope $\alpha(t)$ is

$$f_\alpha(r) = \frac{2}{\Gamma(m)} \left(\frac{m}{\Omega} \right)^m r^{2m-1} \exp\left(-\frac{m}{\Omega} r^2 \right) u(r) , \quad m \geq \frac{1}{2} \qquad (5\text{-}29)$$

where the gamma function $\Gamma(\ )$ is defined by (D-12). When $m = 1$, the Nakagami density becomes the Rayleigh density, and when $m \to \infty$, there is no fading. When $m = 1/2$, the Nakagami density becomes the one-sided Gaussian density. A measure of the severity of the fading is $var(\alpha^2)/(E[\alpha^2])^2$. Equating this ratio for the Rice and Nakagami densities, it is found that when $m \geq 1$, the Nakagami density closely approximates a Rice density with

$$\kappa = \frac{\sqrt{m^2 - m}}{m - \sqrt{m^2 - m}}, \quad m \geq 1 \tag{5-30}$$

Since the Nakagami-m model essentially incorporates the Rayleigh and Rice models as special cases and provides for many other possibilities, it is not surprising that this model often fits well with empirical data. Integrating over (5-29), changing the integration variable, and using (D-12), we obtain

$$E[\alpha^n] = \frac{\Gamma(m + \frac{n}{2})}{\Gamma(m)} \left(\frac{\Omega}{m}\right)^{n/2} \tag{5-31}$$

Consider a time interval small enough that $N(t) = N$, $v(t) = v$, and $\psi_i(t) = \psi_i$ are approximately constants and $a_i(t) = a_i$ and $\tau_i(t) = \tau_i$ are random variables. Then (5-9) and (5-10) yield

$$\phi_i(t + \tau) - \phi_i(t) = 2\pi f_d \tau \cos \psi_i \tag{5-32}$$

where $f_d = f_0 v/c$ is the maximum Doppler shift and τ is a time shift. The *autocorrelation of a wide-sense-stationary complex process* $r(t)$ is defined as

$$A_r(\tau) = \frac{1}{2} E[r^*(t)r(t + \tau)] \tag{5-33}$$

where the asterisk denotes the complex conjugate. The variation of the autocorrelation of the equivalent baseband channel response defined by (5-12) provides a measure of the changing channel characteristics. To interpret the meaning of (5-33), we substitute (5-13) and decompose the autocorrelation as

$$\text{Re}\{A_r(\tau)\} = \frac{1}{2}\{E[r_c(t)r_c(t + \tau)] + E[r_s(t)r_s(t + \tau)]\} \tag{5-34}$$

$$\text{Im}\{A_r(\tau)\} = \frac{1}{2}\{E[r_c(t)r_s(t + \tau)] - E[r_s(t)r_c(t + \tau)]\} \tag{5-35}$$

Thus, the real part of this autocorrelation is the average of the autocorrelations of the real and imaginary parts of $r(t)$; the imaginary part is proportional to the difference between two cross-correlations of the real and imaginary parts of $r(t)$. Substituting (5-12) into (5-33), using the independence and uniform distribution of each ϕ_i and the independence of a_i and ϕ_i, and then substituting (5-32), we obtain

$$A_r(\tau) = \frac{1}{2} \sum_{i=1}^{N} E[a_i^2] \exp(j2\pi f_d \tau \cos \psi_i) \tag{5-36}$$

If all the received multipath components have approximately the same power and the receive antenna is omnidirectional, then (5-18) implies that $E[a_i^2] \approx$

$2\sigma_r^2/N$, $i = 1, 2, \ldots, N$, and (5-36) becomes

$$A_r(\tau) = \frac{\sigma_r^2}{N} \sum_{i=1}^{N} \exp(j2\pi f_d \tau \cos\psi_i) \tag{5-37}$$

A communication system such as a mobile that receives a signal from an elevated base station may be surrounded by many scattering objects. An *isotropic scattering* model assumes that multipath components of comparable power are reflected from many different scattering objects and hence arrive from many different directions. For two-dimensional isotropic scattering, N is large, and the $\{\psi_i\}$ lie in a plane and have values that are uniformly distributed over [0, 2π). Therefore, the summation in (5-37) can be approximated by an integral; that is,

$$A_r(\tau) \approx \frac{\sigma_r^2}{2\pi} \int_0^{2\pi} \exp(j2\pi f_d \tau \cos\psi)d\psi \tag{5-38}$$

This integral has the same form as the integral representation of $J_0(\ )$, the Bessel function of the first kind and order zero. Thus, the *autocorrelation of the channel response for two-dimensional isotropic scattering* is

$$A_r(\tau) = \sigma_r^2 J_0(2\pi f_d \tau) \tag{5-39}$$

The normalized autocorrelation $A_r(\tau)/A_r(0)$, which is a real-valued function of $f_d\tau$, is plotted in Figure 5.2. It is observed that its magnitude is less than 0.3 when $f_d\tau > 1$. This observation leads to the definition of the *coherence time* or *correlation time* of the channel as

$$T_c = \frac{1}{f_d} \tag{5-40}$$

where f_d is the maximum Doppler shift or *Doppler spread*. The coherence time is a measure of the time separation between signal samples sufficient for the samples to be largely decorrelated. If the coherence time is much longer than the duration of a channel symbol, then the fading is relatively constant over a symbol and is called *slow fading*. Conversely, if the coherence time is on the order of the duration of a channel symbol or less, then the fading is called *fast fading*.

The power spectral density of a complex process is defined as the Fourier transform of its autocorrelation. From (5-39) and tabulated Fourier transforms, we obtain the *Doppler power spectrum for two-dimensional isotropic scattering*:

$$S_r(f) = \begin{cases} \frac{\sigma_r^2}{\pi\sqrt{f_d^2 - f^2}}, & |f| < f_d \\ 0, & otherwise \end{cases} \tag{5-41}$$

The normalized Doppler spectrum $S_r(f)/S_r(0)$, which is plotted in Figure 5.3 versus f/f_d, is bandlimited by the Doppler spread f_d and tends to infinity as f approaches $\pm f_d$. The Doppler spectrum is the superposition of contributions

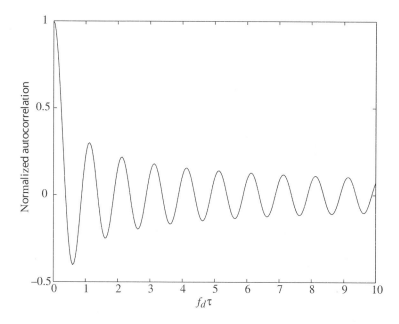

Figure 5.2: Autocorrelation of r(t) for isotropic scattering.

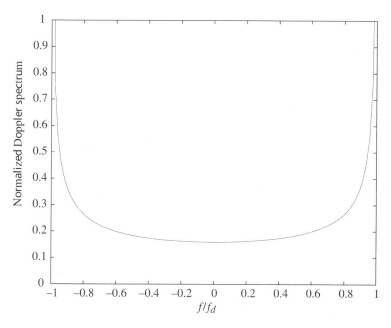

Figure 5.3: Doppler spectrum for isotropic scattering.

from multipath components, each of which experiences a different Doppler shift
upper bounded by f_d.

The received signal power spectrum may be calculated from (5-7), (5-11),
and (5-41). For an unmodulated carrier, $s(t) = 1$ and the received signal power
spectrum is

$$S_{\text{rec}}(f) = \frac{1}{2}S_r(f - f_c) + \frac{1}{2}S_r(f + f_c) \tag{5-42}$$

In general, when the scattering is not isotropic, the imaginary part of the autocorrelation $A_r(\tau)$ is nonzero, and the amplitude of the real part decreases much
more slowly and less smoothly with increasing τ than (5-39). Both the real
and imaginary parts often exhibit minor peaks for time shifts exceeding $1/f_d$.
Thus, the coherence time provides only a rough characterization of the channel
behavior.

Fading Rate and Fade Duration

The *fading rate* is the rate at which the envelope of a received fading signal
crosses below a specified level. Consider a time interval over which the fading
parameters are constant. For a level $r \geq 0$, isotropic scattering, and Ricean
fading, it can be shown that the fading rate is [1]

$$f_r = \sqrt{2\pi(\kappa + 1)}\, f_d\rho\exp[-\kappa - (\kappa + 1)\rho^2]I_0\left(2\rho\sqrt{\kappa(\kappa + 1)}\,\right) \tag{5-43}$$

where κ is the Rice factor and

$$\rho = \frac{r}{\sigma_r\sqrt{2(\kappa + 1)}} \tag{5-44}$$

For Rayleigh fading, $\kappa = 0$ and (5-43) becomes

$$f_r = \frac{\sqrt{\pi}f_d r}{\sigma_r}\exp(-r^2/2\sigma_r^2) \tag{5-45}$$

Equations (5-43) and (5-45) indicate that the fading rate is proportional to the
Doppler spread f_d. Thus, slow fading occurs when the Doppler spread is small,
whereas fast fading occurs when the Doppler spread is large.

Let T_f denote the average envelope *fade duration*, which is the amount
of time the envelope remains below the specified level r. The product $f_r T_f$
is the fraction of the time between fades during which a fade occurs. If the
time-varying envelope is assumed to be a stationary ergodic process, then this
fraction is equal to $F_\alpha(r)$, the probability that the envelope is below or equal
to the level r. Thus,

$$T_f = \frac{F_\alpha(r)}{f_r} \tag{5-46}$$

If the envelope has the Rice distribution, then integrating (5-28) and using
(5-43), (5-44), and (5-46), we obtain

$$T_f = \frac{1 - Q_1\left(\sqrt{2\kappa}, \sqrt{2(\kappa + 1)}\rho\right)}{\sqrt{2\pi(\kappa + 1)}f_d\,\rho\exp[-\kappa - (\kappa + 1)\rho^2]I_0(2\rho\sqrt{\kappa(\kappa + 1)})} \tag{5-47}$$

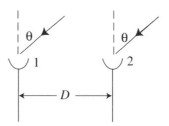

Figure 5.4: Two antennas receiving plane wave that results in a signal copy at each antenna.

where $Q_1(\ )$ is defined by (D-15). For Rayleigh fading, (5-45), (5-46), and the integration of (5-20) yields

$$T_f = \frac{\exp(r^2/2\sigma_r^2) - 1}{\sqrt{\pi} f_d r/\sigma_r} \tag{5-48}$$

For both Ricean and Rayleigh fading, the fade duration is inversely proportional to f_d.

Spatial Diversity and Fading

To obtain spatial diversity in a fading environment, the antennas in an array must be separated enough that there is little correlation between signal replicas or copies at the antennas. To determine what separation is needed, consider the reception of a signal at two antennas separated by a distance D, as illustrated in Figure 5.4. If the signal arrives as an electromagnetic plane wave, then the signal copy at antenna 1 relative to antenna 2 is delayed by $D \sin \theta / c$, where θ is the arrival angle of the plane wave relative to a line perpendicular to the line joining the two antennas. Let $\phi_{ki}(t)$ denote the phase of the complex envelope of multipath component i at antenna k. Consider a time interval small enough that $\phi_{ki}(t) = \phi_{ki}$, $N(t) = N$, $a_i(t) = a_i$, and each multipath component arrives from a fixed angle. Thus, if multipath component i of a narrowband signal arrives as a plane wave at angle ψ_i, then the phase ϕ_{2i} of the complex envelope of the component copy at antenna 2 is related to the phase ϕ_{1i} at antenna 1 by

$$\phi_{2i} = \phi_{1i} + 2\pi \frac{D}{\lambda} \sin \psi_i \tag{5-49}$$

where $\lambda = c/f_c$ is the wavelength of the signal. If the multipath component propagates over a distance much larger than the separation between the two antennas, then it is reasonable to assume that the attenuation a_i is identical at the two antennas. If the range of the delay values exceeds $1/f_c$, then the sensitivity of the phases to small delay variations makes it plausible that the phases ϕ_{ki}, $i = 1, 2, \ldots, N$, $k = 1, 2$ are well modeled as independent random variables that are uniformly distributed over $[0, 2\pi)$. From (5-12), the complex envelope r_k of the signal copy at antenna k when the signal is a tone is

$$r_k = \sum_{i=1}^{N} a_i \exp(j\phi_{ki}), \quad k = 1, 2 \tag{5-50}$$

The cross-correlation between r_1 and r_2 is defined as

$$C_{12}(D) = \frac{1}{2} E[r_1^* r_2] \tag{5-51}$$

Substituting (5-50) into (5-51), using the independence of each a_i and ϕ_{ki}, the independence of ϕ_{ki} and ϕ_{kl}, $i \neq l$, and the uniform distribution of each ϕ_{ki}, and then substituting (5-49), we obtain

$$C_{12}(D) = \frac{1}{2} \sum_{i=1}^{N} E[a_i^2] \exp(j2\pi D \sin\psi_i/\lambda) \tag{5-52}$$

This equation for the cross-correlation as a function of spatial separation clearly resembles (5-36) for the autocorrelation as a function of time delay. If all the multipath components have approximately the same power so that $E[a_i^2] \approx 2\sigma_r^2/N$, $i = 1, 2, \ldots, N$, then

$$C_{12}(D) = \frac{\sigma_r^2}{N} \sum_{i=1}^{N} \exp(j2\pi D \sin\psi_i/\lambda) \tag{5-53}$$

Applying the two-dimensional isotropic scattering model, a derivation similar to that of (5-39) gives the real-valued cross-correlation

$$C_{12}(D) = \sigma_r^2 J_0(2\pi D/\lambda) \tag{5-54}$$

This model indicates that an antenna separation of $D \geq \lambda/2$ ensures that the normalized cross-correlation $C_{12}(D)/C_{12}(0)$ is less than 0.3. A plot of the normalized cross-correlation is obtained from Figure 5.2 if the abscissa is interpreted as D/λ. When the scattering is not isotropic or the number of scattering objects producing multipath components is small, then the real and imaginary parts of the cross-correlation decrease much more slowly with D/λ. For example, Figure 5.5 shows the real and imaginary parts of the normalized cross-correlation when the $\{\psi_i\}$ are a nearly continuous band of angles between $7\pi/32$ and $9\pi/32$ radians so that (5-53) can be approximated by an integral. Figure 5.6 depicts the real and imaginary parts of the normalized cross-correlation when $N = 9$ and the $\{\psi_i\}$ are uniformly spaced throughout the first two quadrants: $\psi_i = (i - 1)\pi/8$, $i = 1, 2, \ldots, 9$. In the example of Figure 5.5, an antenna separation of at least 5λ is necessary to ensure approximate decorrelation of the signal copies and obtain spatial diversity. In the example of Figure 5.6, not even a separation of 10λ is adequate to ensure approximate decorrelation.

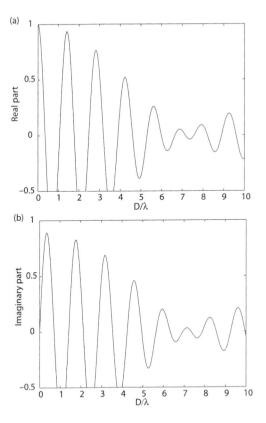

Figure 5.5: Normalized cross-correlation for multipath components arriving between $7\pi/32$ and $9\pi/32$ radians: (a) real part and (b) imaginary part.

5.3   Frequency-Selective Fading

Frequency-selective fading occurs because multipath components combine destructively at some frequencies, but constructively at others. The different path delays cause *dispersion* of a received pulse in time and cause intersymbol interference between successive symbols. When a multipath channel introduces neither time variations nor Doppler shifts, (5-8) and (5-9) indicate that the received complex envelope is

$$s_1(t) = \sum_{i=1}^{L_s} a_i \exp(-j2\pi f_c \tau_i) s(t - \tau_i) \tag{5-55}$$

The number of multipath components L_s includes only those components with power that is a significant fraction, perhaps 0.05 or more, of the power of the dominant component. The multipath *delay spread* T_d is defined as the maximum delay of a significant multipath component relative to the minimum

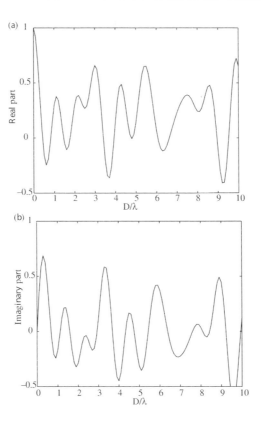

Figure 5.6: Normalized cross-correlation for N = 9 multipath components arriving from uniformly spaced angles in the first two quadrants: (a) real part, and (b) imaginary part.

delay of a component; that is,

$$T_d = \max_i \tau_i - \min_i \tau_i \ , \quad i = 1, 2, \ldots, L_s \qquad (5\text{-}56)$$

If the duration of a received symbol T_s is much larger than T_d, then the multipath components are usually unresolvable, $s(t - \tau_i) \approx s(t - \tau_1)$, $i = 1, 2, \ldots,$ L_s, and hence $s_1(t)$ is proportional to $s(t - \tau_1)$. Since all frequency components of the received signal fade nearly simultaneously, this type of fading is called *frequency-nonselective* or *flat fading* and occurs if $T_s >> T_d$. In contrast, a signal is said to experience *frequency-selective fading* if $T_s < T_d$ because then the time variation or fading of the spectral components of $s(t)$ may be different. The large delay spread may cause intersymbol interference, which is accommodated by equalization in the receiver. However, if the time delays are sufficiently different among the multipath components that they are *resolvable* at the demodulator or matched-filter output, then the independently fading components provide diversity that can be exploited by a rake receiver(Section 5.5).

It is conceptually useful to define the *coherence bandwidth* as

$$B_c = \frac{1}{T_d} \tag{5-57}$$

Let B_m denote the bandwidth of $s(t)$. In general, $B_m \geq 1/T_s$ for practical modulations, so flat fading occurs if $B_m << B_c$. Frequency-selective fading requires $B_m > B_c$.

To illustrate frequency-selective fading, consider the reception of a tone at frequency f_0 with two multipath components so that $s(t) = 1$ and $L_s = 2$ in (5-55). It then follows that the complex envelope has magnitude

$$|s_1(t)| = a_1 \left[1 + \left(\frac{a_2}{a_1} \right)^2 + 2 \left(\frac{a_2}{a_1} \right) \cos 2\pi f_0 T_d \right]^{1/2} \tag{5-58}$$

where $T_d = \tau_1 - \tau_2$. If another tone at frequency f_1 is received, then this equation is valid with f_1 substituted for f_0. Thus, the two complex envelopes can differ considerably as f_0 and f_1 both range over a spectral band with bandwidth equal to the coherence bandwidth. If $a_1 = a_2$, then the difference between the two complex-envelope magnitudes varies from 0 to $2a_1$.

Channel Impulse Response

A generalized impulse response may be used to characterize the impact of the transmission channel on the signal. The *equivalent complex-valued baseband impulse response* of the channel $h(t, \tau)$ is the response at time t due to an impulse applied τ seconds earlier. The complex envelope $s_1(t)$ of the received signal is the result of the convolution of the complex envelope $s(t)$ of the transmitted signal with the baseband impulse response:

$$s_1(t) = \int_{-\infty}^{\infty} h(t, \tau) s(t - \tau) d\tau \tag{5-59}$$

In accordance with (5-8), the impulse response is usually modeled as a complex-valued stochastic process:

$$h(t, \tau) = \sum_{i=1}^{N(t)} a_i(t) \exp[j\phi_i(t)] \delta[\tau - \tau_i(t)] \tag{5-60}$$

For most practical applications, the *wide-sense stationary, uncorrelated scattering* model is reasonably accurate. The impulse response is *wide-sense stationary* if the correlation between its value at t_1 and its value at t_2 depends only on $t_1 - t_2$. Thus, the autocorrelation of the impulse response is

$$R_h(t_1, t_2, \tau_1, \tau_2) = R_h(t_1 - t_2, \tau_1, \tau_2) = \frac{1}{2} E[h^*(t_1, \tau_1) h(t_2, \tau_2)] \tag{5-61}$$

Uncorrelated scattering implies that the gains and phase shifts associated with two different delays are uncorrelated. Extending this notion, the wide-sense stationary, uncorrelated scattering model assumes that the autocorrelation has the form

$$R_h(t_1 - t_2, \tau_1, \tau_2) = R_h(t_1 - t_2, \tau_1)\delta(\tau_1 - \tau_2) \qquad (5\text{-}62)$$

where $\delta(\ )$ is the Dirac delta function. This equation implies that $R_h(t_1 - t_2, \tau_1)$ is the result of integrating the autocorrelation over τ_2. The *multipath intensity profile* or *delay power spectrum* $S_h(\tau) = R_h(0, \tau)$ can be interpreted as the channel output power due to an impulse applied τ seconds earlier. The range of the delay τ over which the multipath intensity profile has a significant magnitude is a measure of the multipath delay spread. The multipath intensity profile has *diffuse* components if it is a piecewise continuous function and has *specular* components if it includes delta functions at specific values of the delay.

A received signal from one source can often be decomposed into the sum of signals reflected from several clusters of scatterers. Each cluster is the sum of a number of multipath components with nearly the same delay. In this model, the impulse response can be expressed as

$$h(t, \tau) = \sum_{i=1}^{L_c(t)} r_i(t)\,\delta(\tau - \tau_i(t)) \qquad (5\text{-}63)$$

where $L_c(t)$ is the number of clusters and $\tau_i(t)$ is the distinct delay associated with the ith cluster. Each complex process $r_i(t)$ has has the form of (5-23) and an envelope with a Rayleigh, Rice, or Nakagami probability density function.

The Fourier transform of the impulse response gives the *time-varying channel frequency response*:

$$H(t, f) = \int_{-\infty}^{\infty} h(t, \tau)\exp(-j2\pi f\tau)d\tau \qquad (5\text{-}64)$$

The autocorrelation of the frequency response for a wide-sense stationary channel is

$$R_H(t_1, t_2, f_1, f_2) = R_H(t_1 - t_2, f_1, f_2) = \frac{1}{2}E[H^*(t_1, f_1)H(t_2, f_2)] \qquad (5\text{-}65)$$

For the *wide-sense stationary, uncorrelated scattering* model, the substitution of (5-64), (5-61), and (5-62) into (5-65) yields

$$R_H(t_1-t_2, f_1, f_2) = R_H(t_1-t_2, f_1-f_2) = \int_{-\infty}^{\infty} R_h(t_1-t_2, \tau)\exp[-j2\pi(f_1-f_2)\tau]d\tau \qquad (5\text{-}66)$$

which is a function only of the difference $f_1 - f_2$. If $t_1 = t_2$, then the autocorrelation of the frequency response is

$$R_H(0, f_1 - f_2) = \int_{-\infty}^{\infty} S_h(\tau)\exp[-j2\pi(f_1 - f_2)\tau]d\tau \qquad (5\text{-}67)$$

which is the Fourier transform of the delay power spectrum. From a fundamental characteristic of the Fourier transform, it follows that the coherence bandwidth of the channel, which is a measure of the range of frequency shift over which the autocorrelation has a significant value, is given by the reciprocal of the multipath delay spread. Thus (5-57) is confirmed for this channel model.

The Doppler shift is the main limitation on the channel coherence time or range of values of the difference $t_d = t_1 - t_2$ for which $R_h(t_d, 0)$ is significant. Thus, the *Doppler power spectrum* is defined as

$$S_D(f) = \int_{-\infty}^{\infty} R_h(t_d, 0) \exp(-j2\pi f t_d) dt_d \qquad (5\text{-}68)$$

The spectral extent of the Doppler power spectrum is on the order of the maximum Doppler shift. Thus, (5-40) is confirmed for this channel model.

5.4 Diversity for Fading Channels

Diversity combiners for fading channels are designed to combine independently fading copies of the same signal in different branches. The combining is done in such a way that the combiner output has a power level that varies much more slowly than that of a single copy. Although useless in improving communications over the additive-white-Gaussian-noise (AWGN) channel, diversity improves communications over fading channels because the diversity gain is large enough to overcome any noncoherent combining loss. Diversity may be provided by signal redundancy that arises in a number of different ways. *Time diversity* is provided by channel coding or by signal copies that differ in time delay. *Frequency diversity* may be available when signal copies using different carrier frequencies experience independent or weakly correlated fading. If each signal copy is extracted from the output of a separate antenna in an antenna array, then the diversity is called *spatial diversity*. *Polarization diversity* may be obtained by using two cross-polarized antennas at the same site. Although this configuration provides compactness, it is not as potentially effective as spatial diversity because the received horizontal component of an electric field is usually much weaker than the vertical component.

The three most common types of diversity combining are selective, maximal-ratio, and equal-gain combining. The last two methods use linear combining with variable weights for each signal copy. Since they usually must eventually adjust their weights, maximal-ratio and equal-gain combiners can be viewed as types of adaptive arrays. They differ from other adaptive antenna arrays in that they are not designed to cancel interference signals.

Optimal Array

Consider a receiver array of L diversity branches, each of which processes a different signal copy. Each branch input is translated to baseband, and then either the baseband signal is applied to a matched filter and sampled or the sampled complex envelope is extracted (Appendix C.3). Alternatively, each branch input is translated to an intermediate frequency, and the sampled analytic signal is extracted. The subsequent analysis is valid for any of these types of branch processing. It is simplest to assume that the branch outputs are sampled complex envelopes. The branch outputs provide the inputs to a linear combiner. Let $\mathbf{x}(l)$ denote the discrete-time vector of the L complex-valued combiner inputs, where the index denotes the sample number. This vector can be decomposed as

$$\mathbf{x}(l) = \mathbf{s}(l) + \mathbf{n}(l) \qquad (5\text{-}69)$$

where $\mathbf{s}(l)$ and $\mathbf{n}(l)$ are the discrete-time vectors of the desired signal and the interference plus thermal noise, respectively. Let $\mathbf{W}$ denote the weight vector of a linear combiner applied to the input vector. The combiner output is

$$y(l) = \mathbf{W}^T \mathbf{x}(l) = y_s + y_n \qquad (5\text{-}70)$$

where T denotes the transpose of a matrix or vector,

$$y_s(l) = \mathbf{W}^T \mathbf{s}(l) \qquad (5\text{-}71)$$

is the output component due to the desired signal, and

$$y_n(l) = \mathbf{W}^T \mathbf{n}(l) \qquad (5\text{-}72)$$

is the output component due to the interference plus noise. The components of both $\mathbf{s}(l)$ and $\mathbf{n}(l)$ are modeled as discrete-time jointly wide-sense-stationary processes.

The correlation matrix of the desired signal is defined as

$$\mathbf{R}_{ss} = E\left[\mathbf{s}^*(l)\mathbf{s}^T(l)\right] \qquad (5\text{-}73)$$

and the correlation matrix of the interference plus noise is defined as

$$\mathbf{R}_{nn} = E\left[\mathbf{n}^*(l)\mathbf{n}^T(l)\right] \qquad (5\text{-}74)$$

The desired-signal power at the output is

$$p_{so} = \frac{1}{2}E\left[|y_s(l)|^2\right] = \frac{1}{2}\mathbf{W}^H \mathbf{R}_{ss}\mathbf{W} \qquad (5\text{-}75)$$

where the superscript H denotes the conjugate transpose. The interference plus noise power at the output is

$$p_n = \frac{1}{2}E\left[|y_n(l)|^2\right] = \frac{1}{2}\mathbf{W}^H \mathbf{R}_{nn}\mathbf{W} \qquad (5\text{-}76)$$

The signal-to-interference-plus-noise ratio (SINR) at the combiner output is

$$\rho_0 = \frac{p_{so}}{p_n} = \frac{\mathbf{W}^H \mathbf{R}_{ss}\mathbf{W}}{\mathbf{W}^H \mathbf{R}_{nn}\mathbf{W}} \qquad (5\text{-}77)$$

The definitions of $\mathbf{R}_{ss}$ and $\mathbf{R}_{nn}$ ensure that these matrices are Hermitian and nonnegative definite. Consequently, these matrices have complete sets of orthonormal eigenvectors, and their eigenvalues are real-valued and nonnegative. The noise power is assumed to be positive. Therefore, $\mathbf{R}_{nn}$ is positive definite and has positive eigenvalues. Since $\mathbf{R}_{nn}$ can be diagonalized, it can be expressed as [4].

$$\mathbf{R}_{nn} = \sum_{i=1}^{L} \lambda_i \mathbf{e}_i \mathbf{e}_i^H \tag{5-78}$$

where λ_i is an eigenvalue and $\mathbf{e}_i$ is the associated eigenvector.

To derive the weight vector that maximizes the SINR with no restriction on $\mathbf{R}_{ss}$, we define the Hermitian matrix

$$\mathbf{A} = \sum_{i=1}^{L} \sqrt{\lambda_i} \mathbf{e}_i \mathbf{e}_i^H \tag{5-79}$$

where the positive square root is used. Direct calculations verify that

$$\mathbf{R}_{nn} = \mathbf{A}^2 \tag{5-80}$$

and the inverse of $\mathbf{A}$ is

$$\mathbf{A}^{-1} = \sum_{i=1}^{L} \frac{1}{\sqrt{\lambda_i}} \mathbf{e}_i \mathbf{e}_i^H \tag{5-81}$$

The matrix $\mathbf{A}$ specifies an invertible transformation of $\mathbf{W}$ into the vector

$$\mathbf{V} = \mathbf{A}\mathbf{W} \tag{5-82}$$

We define the Hermitian matrix

$$\mathbf{C} = \mathbf{A}^{-1}\mathbf{R}_{ss}\mathbf{A}^{-1} \tag{5-83}$$

Then (5-77), (5-80), (5-82), and (5-83) indicate that the SINR can be expressed as

$$\rho_0 = \frac{\mathbf{V}^H \mathbf{C} \mathbf{V}}{\|\mathbf{V}\|^2} \tag{5-84}$$

where $\|\ \|$ denotes the Euclidean norm of a vector and $\|\mathbf{V}\|^2 = \mathbf{V}^H \mathbf{V}$. Equation (5-84) is a Rayleigh quotient [4], which is maximized by $\mathbf{V} = \eta \mathbf{u}$, where $\mathbf{u}$ is the eigenvector of $\mathbf{C}$ associated with its largest eigenvalue $l_{\max}$, and η is an arbitrary constant. Thus, the maximum value of ρ_0 is

$$\rho_{\max} = l_{\max} \tag{5-85}$$

From (5-82) with $\mathbf{V} = \eta \mathbf{u}$, it follows that the *optimal weight vector that maximizes the SINR* is

$$\mathbf{W}_0 = \eta \mathbf{A}^{-1} \mathbf{u} \tag{5-86}$$

The purpose of an adaptive-array algorithm is to adjust the weight vector to converge to the optimal value, which is given by (5-86) when the maximization of the SINR is the performance criterion.

When the discrete-time dependence of $\mathbf{s}(l)$ is the same for all its components, (5-86) can be made more explicit. Let $s(l)$ denote the discrete-time sampled complex envelope of the desired signal in a fixed reference branch. It is assumed henceforth that the desired signal is sufficiently narrowband that the desired-signal copies in all the branches are nearly aligned in time, and the desired-signal input vector may be represented as

$$\mathbf{s}(l) = s(l)\mathbf{S}_0 \tag{5-87}$$

where the *steering vector* is

$$\mathbf{S}_0 = [\alpha_1 \exp(j\,\theta_1) \quad \alpha_2 \exp(j\,\theta_2) \quad \ldots \quad \alpha_L \exp(j\,\theta_L)]^T \tag{5-88}$$

For independent Rayleigh fading in each branch, each phases θ_i is modeled as a random variable with a uniform distribution over $[0, 2\pi)$, and each attenuation α_i has a Rayleigh distribution function, as explained in Section 1.3.

Example 1.   Equation (5-88) can serve as a model for a narrowband desired signal that arrives at an antenna array as a plane wave and does not experience fading. Let T_i, $i = 1, 2, \ldots, L$, denote the arrival-time delay of the desired signal at the output of antenna i relative to a fixed reference point in space. Equations (5-87) and (5-88) are valid with $\theta_i = -2\pi f_c T_i, i = 1, 2, \ldots, L$, where f_c is the carrier frequency of the desired signal. The α_i, $i = 1, 2, \ldots, L$, depend on the relative antenna patterns and propagation losses. If they are all equal, then the common value can be subsumed into $s(l)$. It is convenient to define the origin of a Cartesian coordinate system to coincide with the fixed reference point. Let (x_i, y_i) denote the coordinates of antenna i. If a single plane wave arrives from direction ψ relative to the normal to the array, then

$$\theta_i = \frac{2\pi}{c} f_c(x_i \sin\psi + y_i \cos\psi), \quad i = 1, 2, \ldots, L \tag{5-89}$$

where c is the speed of an electromagnetic wave. $\square$

The substitution of (5-87) into (5-73) yields

$$\mathbf{R}_{ss} = 2p_s \mathbf{S}_0^* \mathbf{S}_0^T \tag{5-90}$$

where

$$p_s = \frac{1}{2} E[|s(l)|^2] \tag{5-91}$$

After substituting (5-90) into (5-83), it is observed that $\mathbf{C}$ may be factored:

$$\mathbf{C} = 2p_s \mathbf{A}^{-1} \mathbf{S}_0^* \mathbf{S}_0^T \mathbf{A}^{-1} = \mathbf{F}\mathbf{F}^H \tag{5-92}$$

where

$$\mathbf{F} = \sqrt{2p_s} \mathbf{A}^{-1} \mathbf{S}_0^* \tag{5-93}$$

This factorization explicitly shows that $\mathbf{C}$ is a rank-one matrix. Therefore, an eigenvector of $\mathbf{C}$ associated with the only nonzero eigenvalue is

$$\mathbf{u} = \mathbf{F} = \sqrt{2p_s} \mathbf{A}^{-1} \mathbf{S}_0^* \tag{5-94}$$

and the nonzero eigenvalue is

$$l_{\max} = \|\mathbf{F}\|^2 \tag{5-95}$$

Substituting (5-94) into (5-86), using (5-80), and then merging $\sqrt{2p_s}$ into the arbitrary constant, we obtain the *Wiener-Hopf equation* for the optimal weight vector :

$$\mathbf{W}_0 = \eta \mathbf{R}_{nn}^{-1} \mathbf{S}_0^* \tag{5-96}$$

where η is an arbitrary constant. The maximum value of the SINR, obtained from (5-85), (5-95), (5-93), and (5-80), is

$$\rho_{\max} = 2p_s \mathbf{S}_0^T \mathbf{R}_{nn}^{-1} \mathbf{S}_0^* \ . \tag{5-97}$$

Maximal-Ratio Combining

Suppose that the interference plus noise in a branch is zero-mean and uncorrelated with the interference plus noise in any of the other branches in the array. Then the correlation matrix $\mathbf{R}_{nn}$ is diagonal. The ith diagonal element has the value

$$2\sigma_i^2 = E[|n_i|^2] \tag{5-98}$$

Since $\mathbf{R}_{nn}^{-1}$ is diagonal with diagonal elements $1/2\sigma_i^2$, the Wiener-Hopf equation implies that the optimal weight vector that maximizes the SINR is

$$\mathbf{W}_m = \eta \left[\frac{S_{01}^*}{\sigma_1^2} \ \frac{S_{02}^*}{\sigma_2^2} \ \cdots \ \frac{S_{0N}^*}{\sigma_N^2} \right]^T \tag{5-99}$$

and (5-97) and (5-88) yield

$$\rho_{\max} = \sum_{i=1}^{L} \frac{p_s}{\sigma_i^2} \alpha_i^2 \tag{5-100}$$

where each term is the SINR at a branch output. Linear combining that uses $\mathbf{W}_m$ is called *maximal-ratio combining* (MRC). It is optimal only if the interference-plus-noise signals in all the diversity branches are uncorrelated. As discussed subsequently, the maximal-ratio combiner can also be derived as the maximum-likelihood estimator associated with a multivariate Gaussian density function. The critical assumption in the derivation is that the noise process in each array branch is both Gaussian and independent of the noise processes in the other branches.

In most applications, the interference-plus-noise power in each array branch is approximately equal, and it is assumed that $\sigma_i^2 = \sigma^2$, $i = 1, 2, \ldots, L$. If this common value is merged with the constant in (5-96) or (5-99), then the MRC weight vector is

$$\mathbf{W}_m = \eta \mathbf{S}_0^* \tag{5-101}$$

and the corresponding maximum SINR is

$$\rho_{\max} = \frac{p_s}{\sigma^2} \sum_{i=1}^{L} \alpha_i^2 \tag{5-102}$$

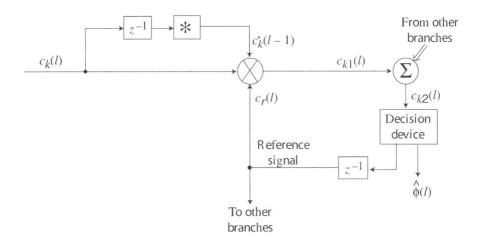

Figure 5.7: Branch k of a maximal-ratio combiner with a phase stripper.

Since the weight vector is not a function of the interference parameters, the combiner attempts no interference cancellation. The interference signals are ignored while the combiner does coherent combining of the desired signal. Equations (5-71), (5-101), (5-87), and (5-88) yield the desired part of the combiner output:

$$y_s(l) = \mathbf{W}_m^T \mathbf{s}(l) = \eta s(l) \sum_{i=1}^{L} \alpha_i^2 \qquad (5\text{-}103)$$

Since $y_s(l)$ is proportional to $s(l)$, the MRC equalizes the phases of the signal copies in the array branches, a process called *cophasing*. If cophasing can be done rapidly enough to be practical, then so can coherent demodulation.

If each α_i, $i = 1, 2, \ldots, L$, is modeled as a random variable with an identical probability distribution function, then (5-102) implies that

$$E[\rho_{\max}] = L \frac{P_s}{\sigma^2} E[\alpha_1^2] \qquad (5\text{-}104)$$

which indicates a gain in the mean SINR that is proportional to L. There are several ways to implement cophasing [5]. Unlike most other cophasing systems, the *phase stripper* does not require a pilot signal. Figure 5.7 depicts branch k of a digital version of a maximal-ratio combiner with a phase stripper. It is assumed that the interference-plus-noise power in each branch is equal so that only cophasing and amplitude multiplication are required for the MRC. In the absence of noise, the angle-modulated input signal is assumed to have the form

$$c_k(l) = \alpha_k s(l) \exp[j\theta_k] = \alpha_k \exp[j\phi(l) + j\theta_k] \qquad (5\text{-}105)$$

where α_k is the amplitude, $\phi(l)$ is the angle modulation carried by all the signal copies in the diversity branches, and θ_k is the undesired phase shift in branch k, which is assumed to be constant for at least two consecutive samples. The signal $c_k^*(l-1)$ is produced by a delay and complex conjugation. During steady-state operation following an initialization process, the reference signal is assumed to have the form

$$c_r(l) = \exp[j\phi(l-1) + j\psi] \tag{5-106}$$

where ψ is a phase angle. The three signals $c_r(l)$, $c_k(l)$, and $c_k^*(l-1)$ are multiplied together to produce

$$c_{k1}(l) = \alpha_k^2 \exp[j\phi(l) + j\psi] \tag{5-107}$$

which as been stripped of the undesired phase shift θ_k. This signal is combined with similar signals from the other diversity branches that use the same reference signal. The input to the decision device is

$$c_{k2}(l) = \sum_{k=1}^{L} \alpha_k^2 \exp[j\phi(l) + j\psi] = e^{j\psi} s(l) \sum_{k=1}^{L} \alpha_k^2 \tag{5-108}$$

which indicates that MRC has been obtained by phase equalization, as in (5-103). After extracting the phase $\phi(l) + \psi$, the decision device produces the demodulated sequence $\hat{\phi}(l)$, which is an estimate of $\phi(l)$, by some type of phase-recovery loop [6]. The device also produces the complex exponential $\exp[j\hat{\phi}(l) + j\psi]$. After a delay, the complex exponential provides the reference signal of (5-106).

Bit Error Probabilities for Coherent Binary Modulations

Suppose that the desired-signal modulation is binary PSK and consider the reception of a single binary symbol or bit. Each bit is equally likely to be a 0 or a 1 and is represented by $+\psi(t)$ or $-\psi(t)$, respectively. Each received signal copy in a diversity branch experiences independent Rayleigh fading that is constant during the signal interval. The received signal in branch i is

$$r_i(t) = \mathrm{Re}[\alpha_i e^{j\theta_i} x\psi(t) e^{j2\pi f_c t}] + n_i(t), \quad 0 \le t \le T, \quad i = 1, 2, \ldots, L \tag{5-109}$$

where $x = +1$ or -1 depending on the transmitted bit, each α_i is an amplitude, each θ_i is a phase shift, f_c is the carrier frequency, T is the bit duration, and $n_i(t)$ is the noise. It is assumed that either the interference is absent or, more generally, that the received interference plus noise in each diversity branch can be modeled as independent, zero-mean, white Gaussian noise with the same two-sided power spectral density $N_0/2$.

Although MRC maximizes the SINR after linear combining, the theory of maximum-likelihood detection is needed to determine an optimal decision variable that can be compared to a threshold. The initial branch processing before sampling could entail extraction of the complex envelope, passband matched-filtering followed by a downconversion to baseband, or, equivalently, a downconversion followed by baseband matched-filtering [6]. Since it is slightly simpler, we assume the latter in this analysis. The same results are obtained if one assumes the extraction of the complex envelope and uses the equations of Appendix C.4.

Using $2\mathrm{Re}(x) = x + x^*$ and discarding a negligible integral, it is found that after the downconversion to baseband, the matched filter in each diversity branch, which is matched to $\psi(t)$, produces the samples

$$y_i = \int_0^T 2r_i(t)e^{-j2\pi f_c t}\psi^*(t)dt$$

$$= 2\mathcal{E}\alpha_i e^{j\theta_i}x + \int_0^T 2n_i(t)e^{-j2\pi f_c t}\psi^*(t)dt, \quad i = 1, 2, \ldots, L \qquad (5\text{-}110)$$

where a factor of "2" has been inserted for analytical convenience, and the desired-signal energy per bit in the absence of fading and diversity combining is

$$\mathcal{E} = \frac{1}{2}\int_0^T |\psi(t)|^2 dt \qquad (5\text{-}111)$$

These samples provide sufficient statistics that contain all the relevant information in the received signal copies in the L diversity branches.

It is assumed that $\psi(t)$ has a spectrum confined to $|f| < f_c$. The zero-mean, real-valued, white Gaussian noise process $n_i(t)$ has autocorrelation

$$E[n_i(t)n_i(t+\tau)] = \frac{N_0}{2}\delta(\tau) \qquad (5\text{-}112)$$

where $\delta(\tau)$ is the Dirac delta function. Let N_i denote the complex-valued noise term in (5-110). Using the spectral limitations of $\psi(t)$, (5-111), and (5-112), we find that $E[N_i^2] = 0$, which indicates that the noise term is circularly symmetric (cf. Appendix C.4). Therefore, it has independent real and imaginary components with the same variance. Since $E[|N_i|^2] = 4\mathcal{E}N_0$, this variance is $2\mathcal{E}N_0$. Given x, α_i, and θ_i, the branch likelihood function or conditional probability density function of y_i is

$$f(y_i|x, \alpha_i, \theta_i) = \frac{1}{4\pi\mathcal{E}N_0}\exp\left[-\frac{|y_i - 2\mathcal{E}\alpha_i e^{j\theta_i}x|^2}{4\mathcal{E}N_0}\right], \quad i = 1, 2, \ldots, L \quad (5\text{-}113)$$

Since the branch samples are statistically independent, the log-likelihood function for the vector $\mathbf{y} = (y_1\, y_2 \ldots y_L)$ given $\boldsymbol{\alpha} = (\alpha_1\, \alpha_2\, \ldots\, \alpha_L)$ and $\boldsymbol{\theta} = (\theta_1\, \theta_2\, \ldots\, \theta_L)$ is

$$\ln[f(\mathbf{y}|x, \boldsymbol{\alpha}, \boldsymbol{\theta})] = \sum_{i=1}^L \ln[f(y_i|x, \alpha_i, \theta_i)] \qquad (5\text{-}114)$$

The receiver decides in favor of a 0 or a 1 depending on whether $x = +1$ or $x = -1$ gives the larger value of the log-likelihood function. Substituting (5-113) into (5-114) and eliminating irrelevant terms and factors that do not depend on the value of x, we find that the maximum-likelihood detector can base its decision on the single variable

$$U = \sum_{i=1}^L \mathrm{Re}(\alpha_i e^{-j\theta_i}y_i) \qquad (5\text{-}115)$$

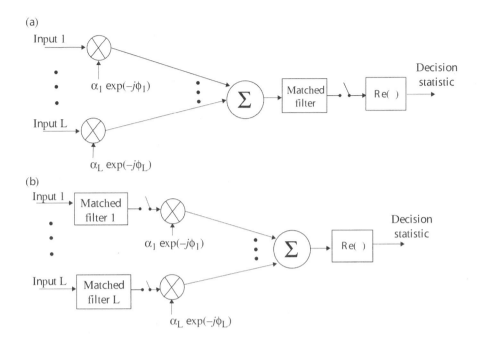

Figure 5.8: Maximal-ratio combiner for PSK with (a) predetection combining and (b) postdetection combining. Coherent equal-gain combiner for PSK omits the factors $\{\alpha_i\}$.

which is compared with a threshold equal to zero to determine the bit state. If we let $\mathbf{s}(l) = [y_1 \ y_2 \ \cdots \ y_L]^T$ and use (5-101), we find that the decision variable may be expressed as $U = Re[\mathbf{W}_m^T \mathbf{s}(l)]$. Since taking the real part of $\mathbf{W}_m^T \mathbf{s}(l)$ serves only to eliminate orthogonal noise, the decision variable U is produced by maximal-ratio combining.

Since (5-115) is computed in either case, the implementation of the maximum-likelihood detector may use either maximal-ratio *predetection combining* before the demodulation, as illustrated in Figure 5.8(a), or *postdetection combining* following the demodulation, as illustrated in Figure 5.8(b). Since the optimal coherent matched-filter or correlation demodulator performs a linear operation on the $\{y_i\}$, both predetection and postdetection combining provide the same decision variable, and hence the same performance.

If the transmitted bit is represented by x, then the substitution of 5-110 into 5-115 yields

$$U = 2\mathcal{E} \sum_{i=1}^{L} \alpha_i^2 + \sum_{i=1}^{L} \alpha_i N_i \qquad (5\text{-}116)$$

where N_i is the zero-mean Gaussian random variable

$$N_i = \mathrm{Re}\left[e^{-j\theta_i} \int_0^T 2n_i(t)e^{-j2\pi f_c t}\psi^*(t)dt \right] \tag{5-117}$$

If the $\{\alpha_i\}$ and $\{\theta_i\}$ are given, then the decision variable has a Gaussian distribution with mean

$$E(U) = 2\mathcal{E}\sum_{i=1}^{L}\alpha_i^2 \tag{5-118}$$

Since the $\{n_i(t)\}$ and, hence, the $\{N_i\}$ are independent, the variance of U is

$$\sigma_u^2 = \sum_{i=1}^{L}\alpha_i^2\,\mathrm{var}(N_i) \tag{5-119}$$

The variance of N_i can be evaluated from (5-111), (5-112), and (5-117). It then follows from (5-119) that

$$\sigma_u^2 = 2\mathcal{E}N_0\sum_{i=1}^{L}\alpha_i^2 \tag{5-120}$$

Because of the symmetry, the bit error probability is equal to the conditional bit error probability given that $x = +1$, corresponding to a transmitted 0. A decision error is made if $U < 0$. Since the decision variable has a Gaussian conditional distribution and neither $E(U)$ nor σ_u^2 depends on the $\{\theta_i\}$, a standard evaluation indicates that the conditional bit error probability given the $\{\alpha_i\}$ is

$$P_{b|\alpha}(\gamma_b) = Q(\sqrt{2\gamma_b}) \tag{5-121}$$

where the signal-to-noise ratio (SNR) for the bit is

$$\gamma_b = \frac{\mathcal{E}}{N_0}\sum_{i=1}^{L}\alpha_i^2 \tag{5-122}$$

The bit error probability is determined by averaging $P_{b|\alpha}(\gamma_b)$ over the distribution of γ_b, which depends on the $\{\alpha_i\}$ and embodies the statistics of the fading channel.

Suppose that independent Rayleigh fading occurs so that each of the $\{\alpha_i\}$ is independent with the identical Rayleigh distribution and $E[\alpha_i^2] = E[\alpha_1^2]$. As shown in Appendix D.4, α_i^2 is exponentially distributed. Therefore, γ_b is the sum of L independent, identically and exponentially distributed random variables. From (D-49), it follows that the probability density function of γ_b is

$$f_\gamma(x) = \frac{1}{(L-1)!\bar{\gamma}^L}x^{L-1}\exp\left(-\frac{x}{\bar{\gamma}}\right)u(x) \tag{5-123}$$

where the average SNR per branch is

$$\bar{\gamma} = \frac{\mathcal{E}}{N_0}E[\alpha_1^2] \tag{5-124}$$

The bit error probability is determined by averaging (5-121) over the density given by (5-123). Thus,

$$P_b(L) = \int_0^\infty Q(\sqrt{2x}) \frac{1}{(L-1)! \bar{\gamma}^L} x^{L-1} \exp\left(-\frac{x}{\bar{\gamma}}\right) dx \qquad (5\text{-}125)$$

Direct calculations verify that since L is an integer,

$$\frac{d}{dx} Q\left(\sqrt{2x}\right) = -\frac{1}{2\sqrt{\pi}} \frac{\exp(-x)}{\sqrt{x}} \qquad (5\text{-}126)$$

$$\frac{d}{dx}\left[e^{-x/\bar{\gamma}} \sum_{i=0}^{L-1} \frac{(x/\bar{\gamma})^i}{i!} \right] = -\frac{1}{(L-1)! \bar{\gamma}^L} x^{L-1} \exp\left(-\frac{x}{\bar{\gamma}}\right) \qquad (5\text{-}127)$$

Applying integration by parts to (5-125), using (5-126), (5-127), and $Q(0) = 1/2$, we obtain

$$P_b(L) = \frac{1}{2} - \sum_{i=0}^{L-1} \frac{1}{i! \bar{\gamma}^i 2\sqrt{\pi}} \int_0^\infty \exp\left[-x\left(1 + \bar{\gamma}^{-1}\right)\right] x^{i-1/2} dx \qquad (5\text{-}128)$$

This integral can be evaluated in terms of the gamma function, which is defined in (D-12). A change of variable in (5-128) yields

$$P_b(L) = \frac{1}{2} - \frac{1}{2}\sqrt{\frac{\bar{\gamma}}{1+\bar{\gamma}}} \sum_{i=0}^{L-1} \frac{\Gamma(i+1/2)}{\sqrt{\pi} i! (1+\bar{\gamma})^i} \qquad (5\text{-}129)$$

Since $\Gamma(1/2) = \sqrt{\pi}$, the bit error probability for no diversity or a single branch is

$$p = P_b(1) = \frac{1}{2}\left(1 - \sqrt{\frac{\bar{\gamma}}{1+\bar{\gamma}}}\right) \qquad \text{(PSK, QPSK)} \qquad (5\text{-}130)$$

Since $\Gamma(x) = (x-1)\Gamma(x-1)$, it follows that

$$\Gamma(k+1/2) = \frac{\sqrt{\pi}\,\Gamma(2k)}{2^{2k-1}\Gamma(k)} = \frac{\sqrt{\pi}\,k!}{2^{2k-1}}\binom{2k-1}{k}, \quad k \geq 1 \qquad (5\text{-}131)$$

Solving (5-130) to determine $\bar{\gamma}$ as a function of p and then using this result and (5-131) in (5-129) gives

$$P_b(L) = p - (1-2p) \sum_{i=1}^{L-1} \binom{2i-1}{i} [p(1-p)]^i \qquad (5\text{-}132)$$

This expression explicitly shows the change in the bit error probability as the number of diversity branches increases. Equations (5-130) and (5-132) are valid for QPSK because the latter can be transmitted as two independent binary PSK waveforms in phase quadrature.

An alternative expression for $P_b(L)$, which may be obtained by a far more complicated calculation entailing the use of the properties of the Gauss hypergeometric function, is [3], [7].

$$P_b(L) = p^L \sum_{i=0}^{L-1} \binom{L+i-1}{i} (1-p)^i \tag{5-133}$$

By using mathematical induction, this equation can be derived from (5-132) without invoking the hypergeometric function.

From a known identity for the sum of binomial coefficients [8], it follows that

$$\sum_{i=0}^{L-1} \binom{L+i-1}{i} = \binom{2L-1}{L} \tag{5-134}$$

Since $1-p \le 1$, (5-133) and (5-134) imply that

$$P_b(L) \le \binom{2L-1}{L} p^L \tag{5-135}$$

This upper bound becomes tighter as $p \to 0$. If $\bar{\gamma} >> 1$ so that $p << 1$, (5-130) implies that $p \approx 1/4\bar{\gamma}$ and (5-135) indicates that the bit error probability decreases inversely with $\bar{\gamma}^L$, thereby demonstrating the large performance improvement provided by diversity.

The advantage of MRC is critically dependent on the assumption of uncorrelated fading in each diversity branch. If there is complete correlation so that the $\{\alpha_i\}$ are all equal and the fading occurs simultaneously in all the diversity branches, then $\gamma_b = L\mathcal{E}\alpha_1^2/N_0$. Therefore, γ_b has a chi-square distribution with 2 degrees of freedom and probability density function

$$f_\gamma^c(x) = \frac{1}{L\bar{\gamma}} \exp\left(-\frac{x}{L\bar{\gamma}}\right) u(x) \tag{5-136}$$

where $\bar{\gamma}$ is defined by (5-124) and the superscript c denotes correlated fading. A derivation similar to that of (5-129) yields

$$P_b^c(L) = \frac{1}{2}\left(1 - \sqrt{\frac{L\bar{\gamma}}{1+L\bar{\gamma}}}\right) \quad \text{(PSK, QPSK)} \tag{5-137}$$

When $L\bar{\gamma} >> 1$,

$$P_b^c(L) \approx \frac{1}{4L\bar{\gamma}} \approx \frac{p}{L}, \quad p << 1 \quad \text{(PSK, QPSK)} \tag{5-138}$$

where p is given by (5-130). A comparison of (5-138) with (5-135) shows the large disparity in performance between a system with completely correlated fading and one with uncorrelated fading.

Graphs of the bit error probability for a single branch with no fading, L branches with independent fading and MRC, and L branches with completely

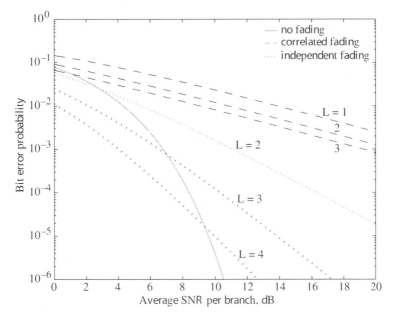

Figure 5.9: Bit error probability of PSK for no fading, completely correlated fading, and independent fading.

correlated fading and MRC are shown in Figure 5.9. Equations (5-121), (5-130), (5-132), and (5-137) are used in generating the graphs. The independent variable is the average SNR per branch for a bit, which is equal to $\bar{\gamma}$ for MRC and is equal to $\gamma_b = \mathcal{E}/N_0$ for the single branch with no fading. The average SNR per bit for MRC is $L\bar{\gamma}$. The figure demonstrates the advantage of diversity combining and independent fading.

For MFSK, one of q equal-energy orthogonal signals $s_1(t)$, $s_2(t)$, ..., $s_q(t)$, each representing $\log_2 q$ bits, is transmitted. The maximum-likelihood detector generates q decision variables corresponding to the q possible nonbinary symbols. The decoder decides in favor of the symbol associated with the largest of the decision variables. Matched filters for the q orthogonal signals are needed in every diversity branch. Because of the orthogonality, each filter matched to $s_k(t)$ has a zero response to $s_l(t)$, $l \neq k$, at the sampling time. When symbol l represented by $s_l(t)$ is received in the presence of white Gaussian noise, matched-filter k of branch i produces the sample

$$
\begin{aligned}
y_{ki} &= \int_0^T 2r_i(t)e^{-j2\pi f_c t}s_k^*(t)\, dt \\
&= 2\mathcal{E}\alpha_i e^{j\theta_i}\delta_{kl} + \int_0^T 2n_i(t)e^{-j2\pi f_c t}s_k^*(t)\, dt\ ,
\end{aligned}
\tag{5-139}
$$

where $\delta_{kl} = 1$ if $k = l$ and $\delta_{kl} = 0$ if $k \neq l$ and

$$\mathcal{E} = \frac{1}{2} \int_0^T |s_k(t)|^2 \, dt \,, \quad k = 1, 2, \ldots, q \tag{5-140}$$

It is assumed that each $s_k(t)$ has a spectrum confined to $|f| < f_c$. Using these spectral limitations and (5-112), we find that the noise term in (5-139) is circularly symmetric. Therefore, its real and imaginary components are independent and have the same variance. From the noise term, this variance is found to be $2\mathcal{E}N_0$. The conditional probability density function of y_{ki} given the values of l, α_i, and θ_i is

$$f(y_{ki}|l, \alpha_i, \theta_i) = \frac{1}{4\pi\mathcal{E}N_0} \exp\left[-\frac{|y_{ki} - 2\mathcal{E}\alpha_i e^{j\theta_i}\delta_{kl}|^2}{4\mathcal{E}N_0} \right] \tag{5-141}$$

For coherent MFSK, the $\{\alpha_i\}$ and the $\{\theta_i\}$ are assumed to be known. Since the noise in each branch is assumed to be independent, the likelihood function is the product of qL densities given by (5-141) for $k = 1, 2, \ldots, q$ and $i = 1, 2, \ldots, L$. Forming the log-likelihood function, observing that $\sum_k \delta_{kl}^2 = 1$, and eliminating irrelevant terms and factors that are independent of l, we find that the maximization of the log-likelihood function is equivalent to selecting the largest of q decision variables, one for each of $s_1(t), s_2(t), \ldots, s_q(t)$. They are

$$U_l = \sum_{i=1}^L \text{Re}\left(\alpha_i e^{-j\theta_i} y_{li} \right) \,, \quad l = 1, 2, \ldots, q \tag{5-142}$$

Consider coherent binary frequency-shift keying (FSK). Because of the symmetry of the model, $P_b(L)$ can be calculated by assuming that $s_1(t)$ was transmitted. With this assumption, the two decision variables become

$$U_1 = 2\mathcal{E} \sum_{i=1}^L \alpha_i^2 + \sum_{i=1}^L \alpha_i N_{1i} \tag{5-143}$$

$$U_2 = \sum_{i=1}^L \alpha_i N_{2i} \tag{5-144}$$

where N_{1i} and N_{2i} are independent, real-valued, Gaussian noise variables given by

$$N_{ki} = \text{Re}\left[e^{-j\theta_i} \int_0^T 2n_i(t) e^{-j2\pi f_c t} s_{ki}^*(t) \, dt \right] \,, \quad k = 1, 2 \tag{5-145}$$

A derivation similar to the one for coherent PSK indicates that (5-132) and (5-133) are again valid for coherent FSK provided that

$$p = \frac{1}{2}\left(1 - \sqrt{\frac{\bar{\gamma}}{2 + \bar{\gamma}}} \right) \qquad \text{(coherent FSK)} \tag{5-146}$$

which can also be obtained by observing the presence of two independent noise variables and, hence, substituting $\bar{\gamma}/2$ in place of $\bar{\gamma}$ in (5-130). Thus, in a fading environment, PSK retains its usual 3 dB advantage over coherent FSK.

The preceding analysis for independent Rayleigh fading can be extended to independent Nakagami fading if the parameter m is a positive integer. From (5-29) and elementary probability, it follows that the probability density function of each random variable $\gamma_i = \mathcal{E}\alpha_i^2/N_0$ is

$$f_{\gamma i}(x) = \frac{m^m}{(m-1)!\,\bar{\gamma}^m}\, x^{m-1} \exp\left(-\frac{mx}{\bar{\gamma}}\right) u(x), \quad m = 1, 2, \ldots \quad (5\text{-}147)$$

where $\bar{\gamma}$ is defined by (5-124). As indicated in Appendix D.2, the characteristic function of γ_i is

$$C_{\gamma i}(j\nu) = \frac{1}{(1 - j\frac{\bar{\gamma}}{m}\nu)^m} \quad (5\text{-}148)$$

If γ_b in (5-122) is the sum of L independent, identically-distributed random variables, then it has the characteristic function

$$C_{\gamma}(j\nu) = \frac{1}{(1 - j\frac{\bar{\gamma}}{m}\nu)^{mL}} \quad (5\text{-}149)$$

The inverse of this function yields the probability density function

$$f_{\gamma}(x) = \frac{1}{(mL-1)!\,(\bar{\gamma}/m)^{mL}}\, x^{mL-1} \exp\left(-\frac{mx}{\bar{\gamma}}\right) u(x), \quad m = 1, 2, \ldots \quad (5\text{-}150)$$

The form of this expression is the same as that in (5-123) except that L and $\bar{\gamma}$ are replaced by mL and $\bar{\gamma}/m$, respectively. Consequently, the derivation following (5-123) is valid once the replacements are made, and

$$P_b(L) = p - (1-2p) \sum_{i=1}^{mL-1} \binom{2i-1}{i} [p(1-p)]^i \quad (5\text{-}151)$$

where

$$p = \frac{1}{2}\left(1 - \sqrt{\frac{\bar{\gamma}}{m+\bar{\gamma}}}\right) \quad \text{(PSK, QPSK)} \quad (5\text{-}152)$$

$$p = \frac{1}{2}\left(1 - \sqrt{\frac{\bar{\gamma}}{2m+\bar{\gamma}}}\right) \quad \text{(coherent FSK)} \quad (5\text{-}153)$$

These results can be approximately related to Ricean fading by using (5-30). Figure 5.10 displays the bit error probability for Nakagami fading with $m = 4$, PSK, and $L = 1, 2, 3,$ and 4 diversity branches.

Equal-Gain Combining

Coherent equal-gain combining (EGC) performs cophasing, but does not correct for unequal values of α_i/σ_i^2, $i = 1, 2, \ldots, L$, where $\alpha_i = |S_{0i}|$. Thus, when a

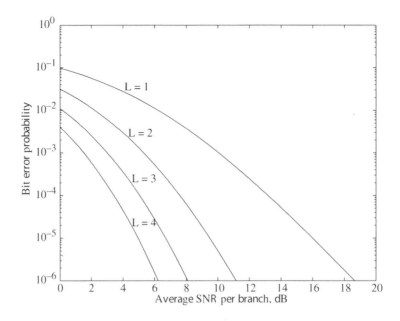

Figure 5.10: Bit error probability of PSK for Nakagami fading with $m = 4$.

narrowband desired signal experiences fading, instead of (5-99) and (5-88), the EGC weight vector is

$$\mathbf{W}_e = \eta[\exp(-j\theta_1) \ \exp(-j\theta_2) \ \ldots \ \exp(-j\theta_L)]^T \qquad (5\text{-}154)$$

where θ_i is the phase shift of the desired signal in branch i. When MRC is optimal and the values of the $\{\alpha_i/\sigma_i^2\}$ are unequal, EGC is suboptimal, but requires much less information about the channel. If the interference plus noise in each array branch is zero-mean and uncorrelated with the other branches and $E[|n_i|^2] = 2\sigma^2, i = 1, 2, \ldots, L$, then $\mathbf{R}_{nn}$ is diagonal, and (5-77), (5-88), and (5-90) with $\mathbf{W} = \mathbf{W}_e$ give the output SINR

$$\rho_0 = \frac{p_s}{L\sigma^2} \left(\sum_{i=1}^{L} \alpha_i \right)^2 \qquad (5\text{-}155)$$

It can be verified by applying the Schwarz inequality for inner products that this SINR is less than or equal to $\rho_{\max}$ given by (5-102). Figure 5.8 displays EGC with predetection and postdetection combining if the factors $\{\alpha_i\}$ are omitted.

In a Rayleigh-fading environment, each α_i, $i = 1, 2, \ldots, L$, has a Rayleigh probability distribution function. If the desired signal in each array branch is uncorrelated with the other branches and has identical average power, then using (D-36), we obtain

$$E[\alpha_i^2] = E[\alpha_1^2], \ \ E[\alpha_i] = \left\{ \frac{\pi}{4} E[\alpha_1^2] \right\}^{1/2}, \ i = 1, 2, \ldots, L \qquad (5\text{-}156)$$

$$E[\alpha_i \alpha_k] = E[\alpha_i]\, E[\alpha_k] = \frac{\pi}{4} E[\alpha_1^2]\,,\ i \neq k \tag{5-157}$$

These equations and (5-155) give

$$E[\rho_0] = \left[1 + (L-1)\frac{\pi}{4}\right]\frac{p_s}{\sigma^2}\, E[\alpha_1^2] \tag{5-158}$$

which exceeds $\pi/4$ times $E[\rho_{\max}]$ given by (5-104) for MRC. Thus, the loss associated with using EGC instead of MRC is on the order of 1 dB.

Example 2. In some environments, MRC is identical to EGC but distinctly suboptimal. Consider narrowband desired and interference signals that do not experience fading and arrive as plane waves. The array antennas are sufficiently close that the steering vector $\mathbf{S}_0$ of the desired signal and the steering vector $\mathbf{J}_0$ of the interference signal can be represented by

$$\mathbf{S}_0 = \left[e^{-j2\pi f_0\tau_1}\ e^{-j2\pi f_0\tau_2}\ \cdots\ e^{-j2\pi f_0\tau_L}\right]^T \tag{5-159}$$

$$\mathbf{J}_0 = \left[e^{-j2\pi f_0\delta_1}\ e^{-j2\pi f_0\delta_2}\ \cdots\ e^{-j2\pi f_0\delta_L}\right]^T \tag{5-160}$$

The correlation matrix for the interference plus noise is

$$\mathbf{R}_{nn} = 2p_n\, \mathbf{I} + 2p_i\, \mathbf{J}_0^*\, \mathbf{J}_0^T \tag{5-161}$$

where p_n and p_i are the noise and interference powers, respectively, in each array branch. This equation shows explicitly that the interference in one branch is correlated with the interference in the other branches. A direct matrix multiplication using $\|\mathbf{J}_0\|^2 = L$ verifies that

$$\mathbf{R}_{nn}^{-1} = \frac{1}{2p_n}\left(\mathbf{I} - \frac{g\,\mathbf{J}_0^*\,\mathbf{J}_0^T}{Lg+1}\right) \tag{5-162}$$

where $g = p_i/p_n$ is the interference-to-noise ratio in each array branch. After merging $1/2p_n$ with the constant in (5-96), it is found that the optimal weight vector is

$$\mathbf{W}_0 = \eta\left(\mathbf{S}_0^* - \frac{\xi L g}{Lg+1}\mathbf{J}_0^*\right) \tag{5-163}$$

where ξ is the normalized inner product

$$\xi = \frac{1}{L}\mathbf{J}_0^T\mathbf{S}_0^* \tag{5-164}$$

The corresponding maximum SINR, which is calculated by substituting (5-159), (5-162), and (5-164) into (5-97), is

$$\rho_{\max} = L\gamma_s\left(1 - \frac{|\xi|^2 L g}{Lg+1}\right) \tag{5-165}$$

where $\gamma_s = p_s/p_n$ is the SNR in each branch. Equations (5-159), (5-160), and (5-164) indicate that $0 \leq |\xi| \leq 1$ and $|\xi| = 1$ if $L = 1$. Equation (5-165) indicates

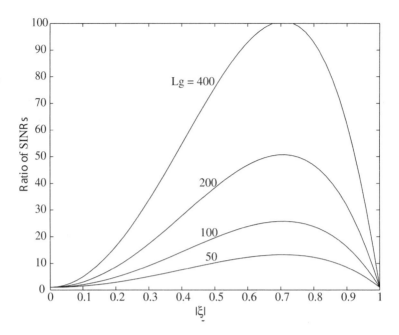

Figure 5.11: Ratio of the maximum SINR to the maximal-ratio-combiner SINR.

that $\rho_{\max}$ decreases as $|\xi|$ increases if $L \geq 2$ and is nearly directly proportional to L if $g \gg 1$.

Since the values of the $\{\alpha_i/\sigma_i^2\}$ are all equal, both MRC and EGC use the weight vector of (5-154) with $\theta_i = -2\pi f_c \tau_i, i = 1, 2, \ldots, L$, which gives $\mathbf{W} = \eta \mathbf{S}_0^*$. Substituting (5-90), (5-159)–(5-161), and (5-164) into (5-77) gives the SINR for MRC and EGC:

$$\rho_0 = \frac{L\,\gamma_s}{1 + |\xi|^2 L\,g} \tag{5-166}$$

Both $\rho_{\max}$ and ρ_0 equal $L\gamma_s$, the peak value, when $\xi = 0$. They both equal $L\gamma_s/(Lg+1)$ when $|\xi| = 1$, which occurs when both the desired and interference signals arrive from the same direction or $L = 1$. Using calculus, it is determined that the maximum value of $\rho_{\max}/\rho_0$, which occurs when $|\xi| = 1/\sqrt{2}$, is

$$\left(\frac{\rho_{\max}}{\rho_0}\right)_{\max} = \frac{(Lg/2+1)^2}{Lg+1} \;,\quad L \geq 2 \tag{5-167}$$

This ratio approaches $Lg/4$ for large values of Lg. Thus, an adaptive array based on the maximization of the SINR has the potential to significantly outperform MRC or EGC if $Lg \gg 1$ under the conditions of the nonfading environment assumed. Figure 5.11 displays $\rho_{\max}/\rho_0$ as a function of $|\xi|$ for various values of Lg. □

When accurate phase estimation is unavailable so that neither cophasing nor coherent demodulation is possible, then postdetection combining following

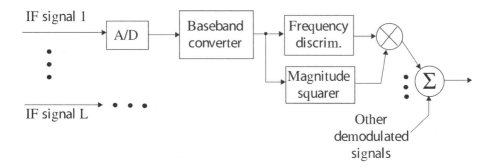

Figure 5.12: Postdetection combining with frequency discriminator.

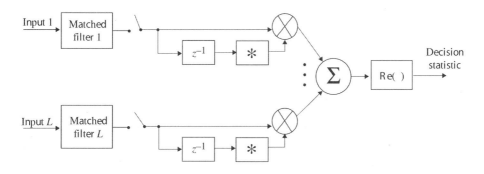

Figure 5.13: Equal-gain combiner for DPSK with postdetection combining.

noncoherent demodulation can provide a significant performance improvement over a system with no diversity. For FSK or minimum-shift keying, postdetection combining with a frequency discriminator is illustrated in Figure 5.12. Each intermediate frequency (IF) signal is sampled, converted to a discrete-time complex baseband signal, and then demodulated by a digital frequency discriminator [9]. The square of the magnitude or possibly the magnitude of the discrete-time complex baseband signal is used to weight the output of each branch. If the noise power in each branch is approximately the same and much smaller then the desired-signal power, then this weighting is a good approximation of the weighting used in MRC, but it is suboptimal since cophasing is absent.

An alternative is postdetection EGC. However, when the desired-signal power is very low in a branch, then that branch contributes only noise to the EGC output. This problem is eliminated if each branch has a threshold device that blocks the output of that branch if the desired-signal power falls below the threshold.

A block diagram of a DPSK receiver with postdetection EGC is depicted in Figure 5.13. For equally likely binary symbols, the error probability is the same regardless of whether two consecutive symbols are the same or different. Assuming that they are the same and that the fading is constant over two symbols, the EGC decision statistic is

$$U = \text{Re} \left[\sum_{i=1}^{L} \left(2\mathcal{E}\alpha_i e^{j\theta_i} + N_{1i} \right) \left(2\mathcal{E}\alpha_i e^{-j\theta_i} + N_{2i}^* \right) \right] \tag{5-168}$$

where N_{1i} and N_{2i} are independent, complex-valued, Gaussian noise variables arising from two consecutive symbol intervals. A derivation [3] indicates that if the $\{\alpha_i\}$ are independent but have identical Rayleigh distributions, then $P_b(L)$ is given by (5-132), (5-133), and (5-135) with the single-branch bit error probability

$$p = \frac{1}{2(1+\bar{\gamma})} \qquad \text{(DPSK)} \tag{5-169}$$

where $\bar{\gamma}$ is given by (5-124). Equation (5-169) can be directly derived by observing that the conditional bit error probability for DPSK with no diversity is $\frac{1}{2}\exp(-\gamma_b)$ and then integrating the equation over the density (5-123) with $L = 1$. A comparison of (5-169) with (5-146) indicates that DPSK with EGC and coherent FSK with MRC give nearly the same performance in a Rayleigh-fading environment if $\bar{\gamma} \gg 1$.

To derive a noncoherent MFSK receiver from the maximum-likelihood criterion, we assume that the $\{\alpha_i\}$ and the $\{\theta_i\}$ in (5-139) are random variables. We expand the argument of the exponential function in (5-141), assume that θ_i is uniformly distributed over $[0, 2\pi)$, and integrate over the density of θ_i. The integral may be evaluated by expressing y_{ki} in polar form, using (D-30), and observing that the integral is over one period of a periodic integrand. Thus, we obtain the conditional density function

$$f(y_{ki}|l, \alpha_i) = \frac{1}{4\pi\mathcal{E}N_0} \exp\left[-\frac{|y_{ki}|^2 + 4\mathcal{E}^2\alpha_i^2\delta_{kl}}{4\mathcal{E}N_0} \right] I_0 \left(\frac{\alpha_i|y_{ki}|\delta_{kl}}{N_0} \right) \tag{5-170}$$

Assuming that α_i has the Rayleigh probability density function given by (5-20) with $2\sigma_r^2 = E[\alpha_i^2]$, the density $f(y_{ki}|l)$ may be evaluated by using the identity (D-33). The likelihood function is the product of qL densities for $k = 1, 2, \ldots, q$, and $i = 1, 2, \ldots, L$. Forming the log-likelihood function and eliminating irrelevant terms and factors that are independent of l, we find that the maximization of the log-likelihood function is equivalent to selecting the largest of the q decision variables

$$U_l = \sum_{i=1}^{L} |y_{li}|^2 \left(\frac{\bar{\gamma}_i}{1+\bar{\gamma}_i} \right), \quad l = 1, 2, \ldots, q \tag{5-171}$$

where

$$\bar{\gamma}_i = \frac{\mathcal{E}}{N_0} E[\alpha_i^2], \quad i = 1, 2, \ldots, L \tag{5-172}$$

If it is assumed that all the $\{\bar{\gamma}_i\}$ are equal, then we obtain the *Rayleigh metric*:

$$U_l = \sum_{i=1}^{L} |y_{li}|^2, \quad l = 1, 2, \ldots, q \tag{5-173}$$

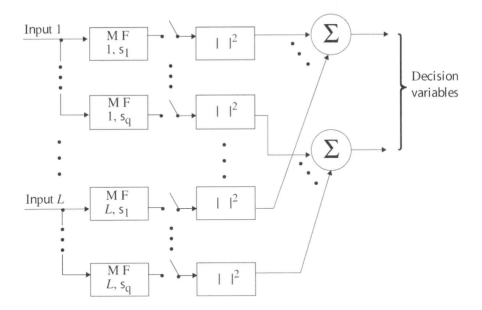

Figure 5.14: Equal-gain combiner for noncoherent MFSK with postdetection combining.

This metric implies a noncoherent MFSK receiver with postdetection square-law EGC, which is illustrated in Figure 5.14. Each branch contains filters matched to the equal-energy orthogonal signals $s_1(t)$, $s_2(t)$, ..., $s_q(t)$. If the $\{\bar{\gamma}_i\}$ are unequal, then the Rayleigh metric is inferior to the maximum-likelihood metric. However, when $\bar{\gamma}_i$ is large, the corresponding terms in the two metrics are nearly equal; when $\bar{\gamma}_i$ is small, the corresponding terms in the two metrics tend to be insignificant. Thus, there is little penalty in using the Rayleigh metric, as is confirmed by numerical evaluations [11].

Consider noncoherent binary FSK. Because of the symmetry of the signals, $P_b(L)$ can be calculated by assuming that $s_1(t)$ was transmitted. Given that $s_1(t)$ was transmitted, the two decision variables at the combiner output are

$$
U_1 = \sum_{i=1}^{L} |2\mathcal{E}\alpha_i e^{j\theta_i} + N_{1i}|^2
$$

$$
= \sum_{i=1}^{L} \left(2\mathcal{E}\alpha_i \cos\theta_i + N_{1i}^R\right)^2 + \sum_{i=1}^{L} \left(2\mathcal{E}\alpha_i \sin\theta_i + N_{1i}^I\right)^2 \qquad (5\text{-}174)
$$

$$
U_2 = \sum_{i=1}^{L} |N_{2i}|^2 = \sum_{i=1}^{L} \left(N_{2i}^R\right)^2 + \sum_{i=1}^{L} \left(N_{2i}^I\right)^2 \qquad (5\text{-}175)
$$

where N_{1i} and N_{2i} are the independent, complex-valued, zero-mean, Gaussian noise variables defined by

$$N_{ki} = \int_0^T 2n_i(t)e^{-j2\pi f_c t} s_k^*(t)dt \ , \quad k = 1, 2, \quad i = 1, 2, \ldots, L \qquad (5\text{-}176)$$

and N_{ki}^R and N_{ki}^I are the real and imaginary parts of N_{ki}, respectively.

Since each $n_i(t)$ in (5-176) is a zero-mean, white Gaussian noise process with the same two-sided power spectral density $N_0/2$, (5-112), (5-176), and the spectral limitations of each $s_k(t)$ imply that $E[N_{ki}^2] = 0$; that is, N_{ki} is circularly symmetric. By calculating $E[|N_{ki}|^2]$, we obtain

$$E[(N_{ki}^R)^2] = E[(N_{ki}^I)^2] = 2\mathcal{E}N_0 \ , \quad k = 1, 2, \quad i = 1, 2, \ldots, L \qquad (5\text{-}177)$$

The circular symmetry implies that N_{ki}^R and N_{ki}^I are uncorrelated, zero-mean, jointly Gaussian random variables and, hence, are independent of each other. Similarly, it can be verified by using the independence of $n_i(t)$ and $n_l(t)$, $i \neq l$, and the orthogonality of $s_1(t)$ and $s_2(t)$ that all $4L$ random variables in the sets $\{N_{ki}^R\}$ and $\{N_{ki}^I\}$ are statistically independent of each other. When independent, identically distributed, Rayleigh fading occurs in each branch, $\alpha_i \cos\theta_i$ and $\alpha_i \sin\theta_i$ are zero-mean, independent, Gaussian random variables with the same variance equal to $E[\alpha_i^2]/2 = E[\alpha_1^2]/2$, $i = 1, 2, \ldots, L$, as shown in Section D.4. Therefore, both U_1 and U_2 have central chi-square distributions with $2L$ degrees of freedom. From (D-18), the density function of U_k is

$$f_k(x) = \frac{1}{(2\sigma_k^2)^L(L-1)!} x^{L-1} \exp\left(-\frac{x}{2\sigma_k^2}\right) u(x), \quad k = 1, 2 \qquad (5\text{-}178)$$

where (5-177) and (5-124) give

$$\sigma_2^2 = E[(N_{2i}^R)^2] = 2\mathcal{E}N_0 \qquad (5\text{-}179)$$
$$\sigma_1^2 = E[(2\mathcal{E}\alpha_1 \cos\theta_i + N_{1i}^R)^2] = 2\mathcal{E}N_0(1 + \bar{\gamma}) \qquad (5\text{-}180)$$

Since an erroneous decision is made if $U_2 > U_1$,

$$P_b(L) = \int_0^\infty \frac{x^{L-1}\exp\left(-\frac{x}{2\sigma_1^2}\right)}{(2\sigma_1^2)^L(L-1)!} \left[\int_x^\infty \frac{y^{L-1}\exp\left(-\frac{y}{2\sigma_2^2}\right)}{(2\sigma_2^2)^L(L-1)!} dy\right] dx \qquad (5\text{-}181)$$

Using (5-127) inside the brackets and integrating, we obtain

$$P_b(L) = \int_0^\infty \exp\left(-\frac{x}{2\sigma_1^2}\right) \sum_{i=0}^{L-1} \frac{(x/2\sigma_2^2)^i}{i!} \frac{x^{L-1}\exp\left(-\frac{x}{2\sigma_2^2}\right)}{(2\sigma_1^2)^L(L-1)!} dx \qquad (5\text{-}182)$$

Changing variables, applying (D-12), and simplifying gives (5-133), where the bit error probability for $L = 1$ is

$$p = \frac{1}{2 + \bar{\gamma}} \qquad \text{(noncoherent FSK)} \qquad (5\text{-}183)$$

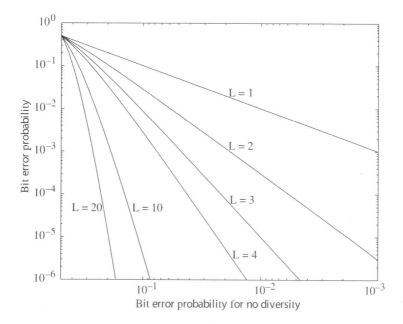

Figure 5.15: Bit error probability for MRC with PSK and coherent FSK and for EGC with DPSK and noncoherent FSK.

and $\bar{\gamma}$ is given by (5-124). Thus, $P_b(L)$ is once again given by (5-132). Equations (5-183) and (5-169) indicate that 3 dB more power is needed for noncoherent FSK to provide the same performance as DPSK. As discussed subsequently in Chapter 6, the performance of DPSK is approximately equaled by using minimum-shift keying and the configuration shown in Figure 5.12.

Equation (5-132) is valid for MRC and PSK or coherent FSK and also for EGC and DPSK or noncoherent FSK. Once the bit error probability in the absence of diversity combining, p, is determined, the bit error probability for diversity combining in the presence of independent Rayleigh fading, $P_b(L)$, can be calculated from (5-132). A plot of $P_b(L)$ versus p for different values of L is displayed in Figure 5.15. This figure illustrates the diminishing returns obtained as L increases. A plot of $P_b(L)$ versus $\bar{\gamma}$, the SNR per branch for one bit, is displayed in Figure 5.16 for MRC with PSK and EGC with DPSK and noncoherent FSK. The plot for MRC with coherent FSK is nearly the same as that for EGC with DPSK. Since (5-135) is valid for all these modulations, we find that $P_b(L)$ is asymptotically proportional to $\bar{\gamma}^{-L}$ with only the proportionality constant differing among the modulation types.

For noncoherent q-ary orthogonal signals such as MFSK with $L \geq 2$, it can be shown that the symbol error probability $P_s(L)$ decreases slightly as q increases [3], [7]. The price for this modest improvement is an increase is transmission bandwidth.

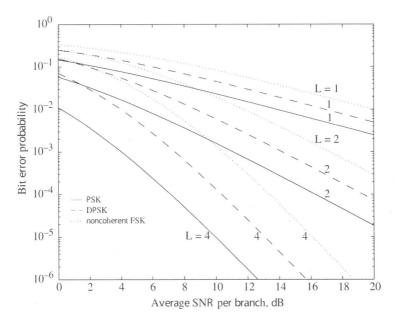

Figure 5.16: Bit error probability for MRC with PSK and for EGC with DPSK and noncoherent FSK.

Selection Diversity

A *selection-diversity system* or *predetection selection-combining system* selects the diversity branch that has the largest SNR and forwards the signal in this branch for further processing. In a fading environment, selection diversity is sensible only if the selection rate is much faster than the fading rate. If the noise and interference levels in all the branches are nearly the same, then the total signal-plus-noise power in each branch rather than the SNR can be measured to enable the selection process, thereby allowing a major simplification. Selection diversity does not provide a performance as good as maximal-ratio combining or equal-gain combining when the interference plus noise in each branch is uncorrelated with that in the other branches. However, selection diversity requires only a single demodulator, and when noises or interference signals are correlated, then selection diversity may become more competitive.

If the noise in each diversity branch is zero-mean and $E[|n_i|^2] = 2\sigma_i^2$, then the SNR in branch i is $\rho_i = p_s \alpha_i^2 / \sigma_i^2$. If each of the $\{\alpha_i\}$ has a Rayleigh distribution and $\sigma_i^2 = \sigma^2$, $i = 1, 2, \ldots, L$, then the SNR in each branch has the same expected value

$$\bar{\rho} = \frac{p_s}{\sigma^2} \, E[\alpha_1^2] \tag{5-184}$$

The results of Appendix D.4 for the square of a Rayleigh-distributed random variable indicate that each SNR has the exponential probability density function

$$f_\rho(x) = \frac{1}{\bar{\rho}} \exp\left(-\frac{x}{\bar{\rho}}\right) u(x) \tag{5-185}$$

The corresponding probability distribution function is

$$F_\rho(x) = \left[1 - \exp\left(-\frac{x}{\bar{\rho}}\right)\right] u(x) \tag{5-186}$$

The branch with the largest SNR is selected. The probability that the SNR of the selected branch is less than or equal to x is equal to the probability that all the branch SNR's are simultaneously less than or equal to x. Therefore, the probability distribution function of the SNR of the selected branch is

$$F_{\rho 0}(x) = \left[1 - \exp\left(-\frac{x}{\bar{\rho}}\right)\right]^L u(x) \tag{5-187}$$

The corresponding probability density function is

$$f_{\rho 0}(x) = \frac{L}{\bar{\rho}} \exp\left(-\frac{x}{\bar{\rho}}\right) \left[1 - \exp\left(-\frac{x}{\bar{\rho}}\right)\right]^{L-1} u(x) \tag{5-188}$$

The average SNR obtained by selection diversity is calculated by integrating the SNR over the density given by (5-188). The result is

$$
\begin{aligned}
E[\rho_0] &= \int_0^\infty \frac{L}{\bar{\rho}} x \exp\left(-\frac{x}{\bar{\rho}}\right) \left[1 - \exp\left(-\frac{x}{\bar{\rho}}\right)\right]^{L-1} dx \\
&= \bar{\rho} L \int_0^\infty x e^{-x} \left(\sum_{i=0}^{L-1} \binom{L-1}{i} (-1)^i e^{-xi}\right) dx \\
&= \bar{\rho} \sum_{i=1}^{L} \binom{L}{i} \frac{(-1)^{i+1}}{i}
\end{aligned}
\tag{5-189}
$$

The second equality results from a change of variable and the substitution of the binomial expansion. The third equality results from a term-by-term integration using (D-12) and an algebraic simplification. Substituting (5-184) and using a known series identity [8], we obtain

$$E[\rho_0] = \frac{p_s}{\sigma^2} E[\alpha_1^2] \sum_{i=1}^{L} \frac{1}{i} \tag{5-190}$$

Thus, the average SNR for selection diversity with $L \geq 2$ is less than that for MRC and the EGC, as indicated by (5-104) and (5-158), respectively. Approximating the summation in (5-190) by an integral, it is observed that the ratio of the average SNR for MRC to that for selection diversity is approximately $L/\ln L$ for $L \geq 2$.

Suppose that the modulation is PSK and optimal coherent demodulation follows the selection process. From (5-113), it follows that the conditional bit error probability is again given by the right-hand side of (5-121) with

$$\gamma_b = \frac{\mathcal{E}}{N_0} \max_i \left(\alpha_i^2 \right) \tag{5-191}$$

If the $\{\alpha_i\}$ have identical Rayleigh distribution functions, then a derivation similar to the one leading to (5-188) indicates that the density function of γ_b is given by (5-188) with $\bar{\gamma}$ in place of $\bar{\rho}$, where $\bar{\gamma}$ is defined by (5-124). Therefore, using the binomial expansion, the bit error probability is

$$
\begin{aligned}
P_b(L) &= \int_0^\infty Q(\sqrt{2x}) \frac{L}{\bar{\gamma}} \exp\left(-\frac{x}{\bar{\gamma}}\right) \left[1 - \exp\left(-\frac{x}{\gamma}\right)\right]^{L-1} dx \\
&= \sum_{i=0}^{L-1} \binom{L-1}{i} (-1)^i \frac{L}{\bar{\gamma}} \int_0^\infty Q(\sqrt{2x}) \exp\left[-x\left(\frac{1+i}{\bar{\gamma}}\right)\right] dx
\end{aligned}
\tag{5-192}
$$

The last integral can be evaluated in the same manner as the one in (5-125). After regrouping factors, the result is

$$P_b(L) = \frac{1}{2} \sum_{i=0}^{L-1} \binom{L}{i+1} (-1)^i \left(1 - \sqrt{\frac{\bar{\gamma}}{i+1+\bar{\gamma}}}\right) \qquad \text{(PSK, QPSK)} \tag{5-193}$$

This equation is valid for QPSK since it can be implemented as two parallel binary PSK waveforms.

For coherent FSK, the conditional bit error probability is $P_b(\gamma_b) = Q(\sqrt{\gamma_b})$. Therefore, it is found that

$$P_b(L) = \frac{1}{2} \sum_{i=0}^{L-1} \binom{L}{i+1} (-1)^i \left(1 - \sqrt{\frac{\bar{\gamma}}{2i+2+\bar{\gamma}}}\right) \qquad \text{(coherent FSK)} \tag{5-194}$$

Again, 3 dB more power is needed to provide to the same performance as PSK.

When DPSK is the data modulation, the conditional bit error probability is $\exp(-\gamma_b)/2$. Thus, selection diversity provides the bit error probability

$$P_b(L) = \int_0^\infty \frac{L}{2\bar{\gamma}} \exp\left[\left(-x\frac{1+\bar{\gamma}}{\bar{\gamma}}\right)\right] \left[1 - \exp\left(-\frac{x}{\bar{\gamma}}\right)\right]^{L-1} dx \tag{5-195}$$

The *beta function* is defined as

$$B(x,y) = \int_0^1 t^{x-1}(1-t)^{y-1} dt \ , \quad x > 0, \quad y > 0 \tag{5-196}$$

If y is a positive integer n, then the substitution of the binomial expansion of $(1-t)^{n-1}$ and the evaluation of the resulting integral yields

$$B(x,n) = \sum_{i=0}^{n-1} \binom{n-1}{i} \frac{(-1)^i}{i+x} \ , \quad n \geq 1, \quad x > 0 \tag{5-197}$$

Using $t = \exp(-x/\bar{\gamma})$ to change the integration variable in (5-195) and then using (5-196) gives

$$P_b(L) = \frac{L}{2}B(\bar{\gamma}+1, L) \qquad \text{(DPSK)} \qquad (5\text{-}198)$$

For noncoherent MFSK, the conditional symbol error probability given the $\{\alpha_i\}$ is obtained from (1-84):

$$P_{s|\alpha}(\gamma_b) = \sum_{i=1}^{q-1} \frac{(-1)^{i+1}}{i+1}\binom{q-1}{i}\exp\left(-\frac{i\gamma_b}{i+1}\right) \qquad (5\text{-}199)$$

Therefore, a derivation similar to that of (5-198) yields the symbol error probability

$$P_s(L) = L\sum_{i=1}^{q-1}\frac{(-1)^{i+1}}{i+1}\binom{q-1}{i}B\left(1+\frac{i\bar{\gamma}}{i+1}, L\right) \qquad \text{(noncoherent MFSK)}$$

$$(5\text{-}200)$$

For binary FSK, the bit error probability is

$$P_b(L) = \frac{L}{2}B\left(\frac{\bar{\gamma}}{2}+1, L\right) \qquad \text{(noncoherent FSK)} \qquad (5\text{-}201)$$

which exhibits the usual 3 dB disadvantage compared with DPSK.

Asymptotic forms of (5-198) and (5-201) may be obtained by substituting

$$B(a, b) = \frac{\Gamma(a)\Gamma(b)}{\Gamma(a+b)} \qquad (5\text{-}202)$$

To prove this identity, let $y = z^2$ in the integrand of the gamma function defined in (D-12). Express the product $\Gamma(a)\Gamma(b)$ as a double integral, change to polar coordinates, integrate over the radius to obtain a result proportional to $\Gamma(a+b)$, and then change the variable in the remaining integral to obtain $B(a, b)\Gamma(a+b)$.

For DPSK, the substitution of (5-202) and (5-169) into (5-198) and the use of $\Gamma(\bar{\gamma}+L+1) = (\bar{\gamma}+L)(\bar{\gamma}+L-1)\ldots(\bar{\gamma}+1)\Gamma(\bar{\gamma}+1) \geq (\bar{\gamma}+1)^L\Gamma(\bar{\gamma}+1)$ give

$$P_b(L) \leq 2^{L-1}L!\,p^L \qquad (5\text{-}203)$$

For noncoherent FSK, a similar derivation using (5-183) and (5-201) yields the same upper bound, which is tight when $\bar{\gamma} \gg L$. The upper bound on $P_b(L)$ for DPSK and noncoherent FSK with EGC is given by (5-135). Comparing the latter with (5-203) indicates the disadvantage of selection diversity relative to EGC when $\bar{\gamma} \gg L$ and $L \geq 2$.

Figure 5.17 shows $P_b(L)$ as a function of the average SNR per branch, assuming selection diversity with PSK, DPSK, and noncoherent FSK. A comparison of Figures 5.17 and 5.16 indicates the reduced gain provided by selection diversity relative to MRC and EGC.

A fundamental limitation of selection diversity is made evident by the plane-wave example in which the signal and interference steering vectors are given by

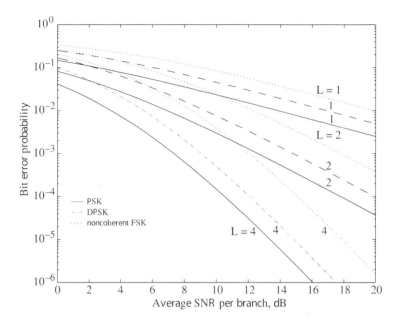

Figure 5.17: Bit error probability for selection diversity with PSK, DPSK, and noncoherent FSK.

(5-159) and (5-160). In this example, the SNR's are equal in all the diversity branches. Consequently, selection diversity can give no better performance than no diversity combining or the use of a single branch. In contrast, (5-166) indicates that EGC can improve the SINR significantly.

Other types of selection diversity besides predetection selection combining are sometimes of interest. *Postdetection selection combining* entails the selection of the diversity branch with the largest signal plus noise power after detection. It outperforms predetection selection combining in general but requires as many matched filters as diversity branches. Thus, its complexity is not much less than that required for EGC. *Switch-and-stay combining(SSC)* or *switched combining* entails processing the output of a particular diversity branch as long as its quality measure remains above a fixed threshold. When it does not, the receiver selects another branch output and continues processing this output until the quality measure drops below the threshold. In *predetection SSC*, the quality measure is the instantaneous SNR of the connected branch. Since only one SNR is measured, predetection SSC is less complex than selection combining but suffers a performance loss. In *postdetection SSC*, the quality measure is the same output quantity used for data detection. The optimal threshold depends on the average SNR per branch. Postdetection SSC provides a lower bit error rate than predetection SSC, and the improvement increases with both the average SNR and less severe fading [11].

5.5 Rake Receiver

In a fading environment, the principal means for a direct-sequence system to obtain the benefits of diversity combining is by using a rake receiver. A *rake receiver* provides *path diversity* by coherently combining resolvable multipath components that are often present during frequency-selective fading. This receiver is the standard type for direct-sequence systems used in mobile communication networks.

Consider a multipath channel with frequency-selective fading slow enough that its time variations are negligible over a signaling interval. To harness the energy in all the multipath components, a receiver should decide which signal was transmitted among M candidates, $s_1(t)$, $s_2(t)$, ..., $s_M(t)$, only after processing all the received multipath components of the signal. Thus, the receiver selects among the M baseband signals or complex envelopes

$$v_k(t) = \sum_{i=1}^{L} c_i s_k(t - \tau_i), \quad k = 1, 2, \ldots, M , \quad 0 \le t \le T + T_d \qquad (5\text{-}204)$$

where T is the duration of the transmitted signal, T_d is the multipath delay spread, L is the number of multipath components, τ_i is the delay of component i, and the channel parameter c_i is a complex number representing the attenuation and phase shift of component i. An idealized sketch of the output of a baseband matched filter that receives three multipath components of the signal to which it is matched is shown in Figure 5.18. If a signal has bandwidth W,

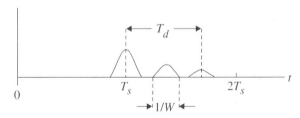

Figure 5.18: Response of matched filter to input with three resolvable multipath components.

then the duration of the matched-filter response to this signal is on the order of $1/W$. Multipath components that produce distinguishable matched-filter output pulses are said to be *resolvable*. Thus, three multipath components are resolvable if their relative delays are greater than $1/W$, as depicted in the figure. A necessary condition for at least two resolvable multipath components is that duration $1/W$ is less than the delay spread T_d. From (5-57) it follows that $W > B_c$ is required, which implies that frequency-selective fading and resolvable multipath components are associated with wideband signals. There are at most $\lfloor T_d W \rfloor + 1$ resolvable components, where $\lfloor x \rfloor$ denotes the largest integer in x. As observed in the figure, intersymbol interference at the sampling times is not significant if $T_d + 1/W$ is less than the symbol duration T_s.

For the following analysis, it is assumed that the M possible signals are orthogonal to each other and that the data symbols are independent of each other so that the maximum-likelihood receiver makes symbol-by-symbol decisions [3], [6]. This receiver uses a separate baseband matched filter or correlator for each possible desired signal including its multipath components. Thus, if $s_k(t)$ is the kth symbol waveform, $k = 1, 2, \ldots, M$, then the kth matched filter is matched to the signal $v_k(t)$ in (5-204) with $T = T_s$, the symbol duration. Each matched-filter output sampled at $t = T_s + T_d$ provides a decision variable. A derivation similar to that of (5-142) indicates that the kth decision variable is

$$U_k = \text{Re}\left[\sum_{i=1}^{L} c_i^* \int_0^{T_s+T_d} r(\tau)s_k^*(\tau - \tau_i)d\tau\right] \qquad (5\text{-}205)$$

where $r(t)$ is the received signal, including the noise, after downconversion to baseband. A receiver implementation based on this equation would require a separate transversal filter or delay line and a matched filter for each possible waveform $s_k(t)$. An alternative form that requires only a single transversal filter and M matched filters is derived by changing variables in (5-205) and using the fact that $s_k(t)$ is zero outside the interval $[0, T_s)$. The result is

$$U_k = \text{Re}\left[\sum_{i=1}^{L} c_i^* \int_0^{T_s} r(\tau + \tau_i)s_k^*(\tau)d\tau\right] \qquad (5\text{-}206)$$

For frequency-selective fading and resolvable multipath components, a simplifying assumption is that each delay is an integer multiple of $1/W$. Accordingly, L is increased to equal the maximum number of resolvable components, and we set $\tau_i = (i-1)/W$, $i = 1, 2, \ldots, L$, and $(L-1)/W \approx \tau_m$, where τ_m is the maximum delay. As a result, some of the $\{c_i\}$ may be equal to zero. The decision variables become

$$U_k = \text{Re}\left[\sum_{i=1}^{L} c_i^* \int_0^{T_s} r(\tau + (i-1)/W)s_k^*(\tau)d\tau\right], \quad k = 1, 2, \ldots, M \qquad (5\text{-}207)$$

A receiver based on these decision variables, which is called a *rake receiver*, is diagrammed in Figure 5.19. Since $r(t)$ is designated as the output of the final tap, the sampling occurs at $t = T_s$. Each tap output contains at most one multipath component of $r(t)$.

The rake receiver requires that the channel parameters $\{c_i\}$ be known or estimated. An estimation might be done by applying each tap output to M parallel matched filters after a one-symbol delay. The previous symbol decision is used to select one matched-filter output for each tap output. The L matched-filter outputs are lowpass-filtered to provide estimates of the channel parameters. The estimates must be updated at a rate exceeding the fade rate of (5-43) or (5-45).

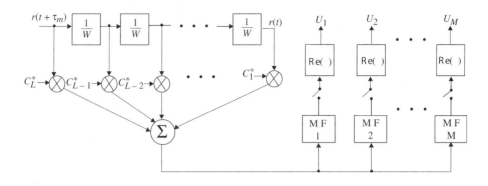

Figure 5.19: Rake receiver for M orthogonal pulses. MF denotes a matched filter.

An alternative configuration to that of Figure 5.19 uses a separate transversal filter for each decision variable and has the corresponding matched filter in the front, as shown in Figure 5.20(a). The matched-filter or correlator output is applied to L_s parallel *fingers*, the outputs of which are recombined and sampled to produce the decision variable. The number of fingers L_s, where $L_s \leq L$, is equal to the number the resolvable components that have significant power. The matched filter produces a number of output pulses in response to the multipath components, as illustrated in Figure 5.18. Each finger delays and weights one of these pulses by the appropriate amount so that all the finger output pulses are aligned in time and can be constructively combined after weighting, as shown in Figure 5.20(b). Digital devices can be used because the sampling immediately follows the matched filtering.

The delay of each significant multipath component may be estimated by using envelope detectors and threshold devices. Let t_e denote the time required to estimate the relative delay of a multipath component, and let v denote the relative radial velocity of a receiver relative to a transmitter. Then vt_e/c is the change in delay that occurs during the estimation procedure, where c is the speed of an electromagnetic wave. This change must be much less than the duration of a multipath output pulse shown in Figure 5.18 if the delay estimate is to be useful. Thus, with v interpreted as the maximum speed of a mobile in a mobile communications network,

$$t_e << \frac{c}{vW} \tag{5-208}$$

is required of the multipath-delay estimation.

Suppose that $s_k(t)$ is a direct-sequence signal with chip duration $T_c = 1/W$. If the processing gain T_s/T_c is large, the spreading sequence has a small autocorrelation when the relative delay is T_c or more, and

$$\int_0^{T_s} s_k(t + (i-1)/W)s_k(t)dt << \int_0^{T_s} |s_k(t)|^2 dt \ , \ i \geq 2 \tag{5-209}$$

(a)

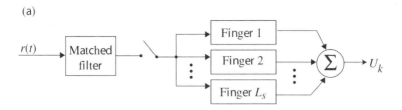

(b)

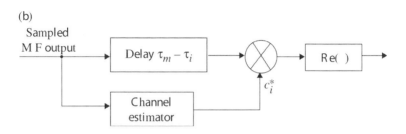

Figure 5.20: Rake receiver: (a) basic configuration for generating a decision variable and (b) a single finger.

When the data modulation is binary antipodal or PSK, only a single symbol waveform $s_1(t)$ and its associated decision variable U_1 are needed. After downconversion to baseband, the received signal is

$$r(t) = \left\{ \text{Re}\left[v_1(t)e^{j2\pi f_c t} \right] + n(t) \right\} 2e^{-j2\pi f_c t} \qquad (5\text{-}210)$$

where $v_1(t)$ is given by (5-204) and $n(t)$ is zero-mean white Gaussian noise. Let $\alpha_i = |c_i|$ and set $k = 1$ and $\tau_i = (i-1)/W$, $i = 1, 2, \ldots, L$ in (5-204). Substituting (5-210) and (5-204) into (5-207) with $k = 1$ and then using (5-209), we again obtain (5-116). Thus, the rake receiver produces MRC, and the conditional bit error probability given the $\{\alpha_i\}$ is provided by (5-121). However, for a rake receiver, each of the $\{\alpha_i\}$ is associated with a different multipath component, and hence each $E[\alpha_i]$ has a different value in general. Therefore, the derivation of $P_b(L)$ must be modified.

Equation (5-122) may be expressed as

$$\gamma_b = \sum_{i=1}^{L} \gamma_i \,, \quad \gamma_i = \frac{\mathcal{E}}{N_0}\alpha_i^2 \qquad (5\text{-}211)$$

If each α_i has a Rayleigh distribution then each γ_i has the exponential probability density function (Appendix D.4)

$$f_{\gamma_i}(x) = \frac{1}{\overline{\gamma}_i} \exp\left(-\frac{x}{\overline{\gamma}_i} \right) u(x), \quad i = 1, 2, \ldots, L \qquad (5\text{-}212)$$

where the average SNR for a bit in branch i is

$$\bar{\gamma}_i = \frac{\mathcal{E}}{N_0} E[\alpha_i^2] \, , \ i = 1, 2, \ldots, L \tag{5-213}$$

If each multipath component fades independently so that each of the $\{\gamma_i\}$ is statistically independent, then γ_b is the sum of independent, exponentially distributed random variables. The results of Appendix D.5 indicate that the probability density function of γ_b is

$$f_{\gamma_b}(x) = \sum_{i=1}^{L} \frac{A_i}{\bar{\gamma}_i} \exp\left(-\frac{x}{\bar{\gamma}_i}\right) u(x) \tag{5-214}$$

where

$$A_i = \begin{cases} \displaystyle\prod_{\substack{k=1 \\ k\neq i}}^{L} \frac{\bar{\gamma}_i}{\bar{\gamma}_i - \bar{\gamma}_k} \, , & L \geq 2 \\ 1 \, , & L = 1 \end{cases} \tag{5-215}$$

The bit error probability is determined by averaging the conditional bit error probability $P_2(\gamma_b) = Q(\sqrt{2\gamma_b})$ over the density given by (5-214). A derivation similar to that leading to (5-129) yields

$$P_b(L) = \frac{1}{2} \sum_{i=1}^{L} A_i \left(1 - \sqrt{\frac{\bar{\gamma}_i}{1 + \bar{\gamma}_i}}\right) \qquad \text{(PSK, QPSK)} \tag{5-216}$$

The number of fingers in an ideal rake receiver equals the number of significant resolvable multipath components, which is constantly changing in a mobile communications receiver. Rather than attempting to implement all the required fingers that may sometimes be required, a more practical alternative is to implement a fixed number of fingers independent of the number of multipath components. *Generalized selection diversity* entails selecting the L_c strongest resolvable components among the L available ones and then applying MRC or EGC of these L_c components, thereby discarding the $L - L_c$ components with the lowest SNRs. Analysis [2] indicates that diminishing returns are obtained as L_c increases, but for a fixed value of L_c, the performance improves as L increases.

An increase in the number of resolved components L is potentially beneficial if it is caused by natural changes in the physical environment that generate additional multipath components. However, an increase in L due to an increase in the bandwidth W is not always beneficial [12]. Although new components provide additional diversity and may exhibit the more favorable Ricean fading rather than Rayleigh fading, the average power per multipath component decreases because some composite components fragment into more numerous but weaker components. Hence, the estimation of the channel parameters becomes more difficult, and the fading of some multipath components may be highly correlated rather than independent.

The estimation of the channel parameters needed in a rake receiver becomes more difficult as the fading rate increases. When the estimation errors are

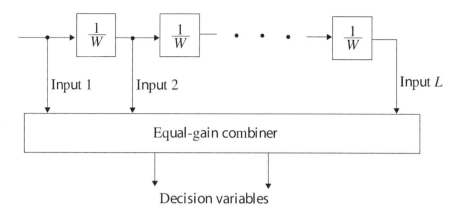

Figure 5.21: Rake receiver that uses equal-gain combiner to avoid channel-parameter estimation.

large, it may be preferable to use a rake receiver that avoids channel-parameter estimation by abandoning MRC and using noncoherent postdetection EGC. The form of this rake receiver for binary signals is depicted in Figure 5.21. Each tap output of the transversal filter provides an input to the equal-gain combiner, which may have the form of Figure 5.13 or Figure 5.14.

For two orthogonal signals that satisfy (5-209) and the rake receiver of Figures 5.21 and 5.14, the decision variables are given by (5-174) and (5-175). Since U_2 has a central chi-square distribution with $2L$ degrees of freedom, the probability density function of U_2 is given by (5-178) and (5-179). Equation (5-174) can be expressed as

$$U_1 = \sum_{i=1}^{L} \left[(2\mathcal{E}\alpha_i \cos\theta_i + N_{1i}^R)^2 + (2\mathcal{E}\alpha_i \sin\theta_i + N_{1i}^I)^2 \right] \tag{5-217}$$

Each phase θ_i is assumed to be statistically independent and uniformly distributed over $[0, 2\pi)$. Since each α_i has a Rayleigh distribution, $\alpha_i \cos\theta_i$ and $\alpha_i \sin\theta_i$ have zero-mean, independent, Gaussian distributions. Therefore, as indicated in Appendix D.4, each term of U_1 has an exponential distribution with mean

$$m_i = 4\mathcal{E}N_0(1 + \bar{\gamma}_i) \tag{5-218}$$

where $\bar{\gamma}_i$ is defined by (5-213). Since the statistical independence of the $\{\alpha_i\}$ and $\{\theta_i\}$ implies the statistical independence of the terms of U_1, the probability density function of U_1 for distinct values of the $\{\bar{\gamma}_i\}$ is given by (D-45) and (D-46) with $N = L$. Since an erroneous decision is made if $U_2 > U_1$,

$$P_b(L) = \sum_{i=1}^{L} \frac{B_i}{m_i} \int_0^\infty \exp\left(-\frac{x}{m_i}\right) \int_x^\infty \frac{y^{L-1}\exp\left(-\frac{y}{2\sigma_2^2}\right)}{(2\sigma_2^2)^L(L-1)!} dy\, dx \tag{5-219}$$

Integrating by parts to eliminate the inner integral, changing the remaining integration variable, applying (D-12), and simplifying yields the bit error probability for orthogonal signals and a rake receiver with noncoherent postdetection EGC:

$$P_b(L) = \sum_{i=1}^{L} B_i \left[1 - \left(\frac{1 + \bar{\gamma}_i}{2 + \bar{\gamma}_i} \right)^L \right] \quad \text{(orthogonal signals)} \tag{5-220}$$

where

$$B_i = \begin{cases} \displaystyle\prod_{\substack{k=1 \\ k \neq i}}^{L} \frac{1 + \bar{\gamma}_i}{\bar{\gamma}_i - \bar{\gamma}_k} \quad , & L \geq 2 \\ 1 \, , & L = 1 \end{cases} \tag{5-221}$$

An alternative derivation of (5-220) using the direct-conversion receiver modeled in Appendix C.3 is given in [13]. Equation (5-220) is more compact and considerably easier to evaluate than the classical formula [3], which is derived in a different way.

Another way to avoid channel-parameter estimation is to use DPSK and the diversity receiver of Figure 5.13 in Figure 5.21. The classical analysis [3] verifies that $P_b(L)$ is given by (5-220) and (5-221) with $\bar{\gamma}_i$ replaced by $2\bar{\gamma}_i$.

For dual rake combining with orthogonal signals, (5-220) reduces to

$$P_b(2) = \frac{8 + 5\bar{\gamma}_1 + 5\bar{\gamma}_2 + 3\bar{\gamma}_1\bar{\gamma}_2}{(2 + \bar{\gamma}_1)^2(2 + \bar{\gamma}_2)^2} \tag{5-222}$$

If $\bar{\gamma}_2 = 0$, then

$$P_b(2) = \frac{2 + \frac{5}{4}\bar{\gamma}_1}{(2 + \bar{\gamma}_1)^2} \geq \frac{1}{2 + \bar{\gamma}_1} = P_b(1) \tag{5-223}$$

This result illustrates the performance degradation that results when a rake combiner uses an input that provides no desired-signal component, which may occur when EGC is used rather than MRC. In the absence of a desired-signal component, this input contributes only noise to the combiner. For large values of $\bar{\gamma}_1$, the extraneous noise causes a loss of almost 1 dB.

If an adaptive array produces a directional beam to reject interference or enhance the desired signal, it also reduces the delay spread of the multipath components of the desired signal because components arriving from angles outside the beam are greatly attenuated. As a result, the potential benefit of a rake receiver diminishes. Another procedure is to assign a separate set of adaptive weights to each significant multipath component. Consequently, the adaptive array can form separate array patterns, each of which enhances a particular multipath component while nulling other components. The set of enhanced components are then applied to the rake receiver [14].

5.6 Error-Control Codes

If the channel symbols are interleaved to a depth beyond the coherence time of the channel, then the symbols fade independently. As a result, an error-control code provides a form of time diversity for direct-sequence systems. Interleaving over many hop intervals enables an error-control code to provide a form of frequency diversity for frequency-hopping systems with frequency channels separated by more than the coherence bandwidth of the channel.

Consider an (n, k) linear block code with soft-decision decoding, where n is the number of code symbols and k is the number of information symbols. Let $\mathbf{y}$ denote the n-dimensional vector of noisy output samples $y_i, i = 1, 2, \ldots, n$, produced by a demodulator that receives a sequence of n symbols and samples them at the symbol rate. Let $\mathbf{x}_l$ denote the lth codeword vector with symbols $x_{li}, i = 1, 2, \ldots, n$. Let $f(\mathbf{y}|\mathbf{x}_l)$ denote the likelihood function, which is the conditional probability density function of $\mathbf{y}$ given that $\mathbf{x}_l$ was transmitted. Let q denote the alphabet size of the code symbols. As explained in Chapter 1, if the demodulator outputs are statistically independent, then the likelihood function is the product of n conditional probability density functions, and the log-likelihood function *or maximum-likelihood metric* for each of the q^k possible codewords is

$$\ln\left[f(\mathbf{y}|\mathbf{x}_l)\right] = \sum_{i=1}^{n} \ln\left[f(y_i|x_{li})\right], \quad l = 1, 2, \ldots, q^k \qquad (5\text{-}224)$$

where $f(y_i|x_{li})$ is the conditional probability density function of y_i given the value of x_{li}. In the subsequent analysis, it is always assumed that perfect symbol interleaving or sufficiently fast fading ensures the statistical independence of the demodulator outputs so that (5-224) is applicable.

The subsequent analysis for binary PSK is applicable to direct-sequence signals if it is assumed that the despread interference and noise are well approximated by white Gaussian noise. With this assumption, the analysis may be applied by substituting N_{0e}, the equivalent noise-power spectral density, in place of N_0 in the subsequent results.

For binary PSK over a fading channel in which the fading is constant over a symbol interval, the received signal representing symbol i of codeword l is

$$r_i(t) = \text{Re}[\alpha_i e^{j\theta_i} x_{li} \psi(t) e^{j2\pi f_c t}] + n_i(t) , \quad i = 1, 2, \ldots, n \qquad (5\text{-}225)$$

where α_i is a random variable that includes the effects of the fading, $x_{li} = +1$ when binary symbol i is a 1 and $x_{li} = -1$ when binary symbol i is a 0, and $\psi(t)$ is the symbol waveform. The noise process $n_i(t)$ is independent, zero-mean, white Gaussian noise with autocorrelation given by (5-112). When codeword l is received in the presence of white Gaussian noise, it is downconverted, and then the matched-filter or correlator, which is matched to $\psi(t)$, produces the samples

$$y_i = \int_0^{T_s} 2r_i(t) e^{-j2\pi f_c t} \psi^*(t) dt$$

$$= 2\mathcal{E}_s \alpha_i e^{j\theta_i} x_{li} + \int_0^{T_s} 2n_i(t) e^{-j2\pi f_c t} \psi^*(t) dt , \quad i = 1, 2, \ldots, n$$

$$(5\text{-}226)$$

where T_s denotes the symbol duration and the symbol energy is

$$\mathcal{E}_s = \frac{1}{2} \int_0^{T_s} |\psi(t)|^2 dt \qquad (5\text{-}227)$$

Since $\psi(t)$ is the sole basis function for the signal space, these samples provide sufficient statistics; that is, they contain all the relevant information in the received signal [3], [6].

The spectrum of $\psi(t)$ is assumed to be confined to $|f| < f_c$. Using this assumption, (5-227), and (5-112), we find that the Gaussian noise term in (5-226) is circularly symmetric and, hence, has independent real and imaginary components with the same variance, which is calculated to be $2\mathcal{E}_s N_0$. Therefore, the conditional probability density function of y_i given the values of x_{li}, α_i, and θ_i is,

$$f(y_i|x_{li}, \alpha_i, \theta_i) = \frac{1}{4\pi\mathcal{E}_s N_0} \exp\left[-\frac{|y_i - 2\mathcal{E}_s \alpha_i e^{j\theta_i} x_{li}|^2}{4\mathcal{E}_s N_0} \right], \quad i = 1, 2, \ldots, n,$$

$$l = 1, 2, \ldots, 2^k \qquad (5\text{-}228)$$

Substituting this equation into (5-224) and then eliminating irrelevant terms and factors that do not depend on the codeword l, we obtain the *maximum-likelihood metric for PSK*:

$$U(l) = \sum_{i=1}^{n} \text{Re}\left[y_i \alpha_i e^{-j\theta_i} x_{li} \right], \quad l = 1, 2, \ldots, 2^k \qquad (5\text{-}229)$$

which serve as decision variables and require knowledge of the $\{x_{li}\}$, $\{\alpha_i\}$, and $\{\theta_i\}$.

For a linear block code, the error probabilities may be calculated by assuming that the all-zero codeword denoted by $l = 1$ was transmitted. The comparison of the metrics $U(1)$ and $U(l)$, $l \neq 1$, depends only on the d terms that differ, where d is the weight of codeword l. The two-codeword error probability is equal to the probability that $U(1) < U(l)$. If each of the $\{\alpha_i\}$ is independent with the identical Rayleigh distribution and $E[\alpha_i^2] = E[\alpha_1^2]$, $i = 1$, $2, \ldots, n$, the average SNR per binary code symbol is

$$\bar{\gamma}_s = \frac{\mathcal{E}_s}{N_0} E[\alpha_1^2] = \frac{r\mathcal{E}_b}{N_0} E[\alpha_1^2] = r\bar{\gamma}_b \quad \text{(binary symbols)} \qquad (5\text{-}230)$$

where $\mathcal{E}_b$ is the information-bit energy, r is the code rate, and $\bar{\gamma}_b$ is the average SNR per bit. A derivation similar to the one leading to (5-132) indicates that the two-codeword error probability is

$$P_2(d) = P_s - (1 - 2P_s) \sum_{i=1}^{d-1} \binom{2i-1}{i} [P_s(1 - P_s)]^i \qquad (5\text{-}231)$$

where the symbol error probability is

$$P_s = \frac{1}{2}\left(1 - \sqrt{\frac{\bar{\gamma}_s}{1 + \bar{\gamma}_s}} \right) \quad \text{(PSK, QPSK)} \qquad (5\text{-}232)$$

The same equations are valid for both PSK and QPSK because the latter can be transmitted as two independent binary PSK waveforms in phase quadrature. A derivation analogous to that of (5-135) indicates that

$$P_2(d) \leq \binom{2d-1}{d} P_s^d \qquad (5\text{-}233)$$

For q-ary orthogonal symbol waveforms $s_1(t)$, $s_2(t)$, ..., $s_q(t)$, q matched filters are needed. The observation vector is $\mathbf{y} = [\mathbf{y}_1 \ \mathbf{y}_2 \cdots \mathbf{y}_q]$, where each $\mathbf{y}_k$ is an n-dimensional row vector of output samples y_{ki}, $i = 1, 2, \ldots, n$, from matched-filter k, which is matched to $s_k(t)$. Suppose that symbol i of codeword l uses $s_\nu(t)$. Because the symbol waveforms are orthogonal, when codeword l is received in the presence of white Gaussian noise, matched-filter k produces the samples

$$y_{ki} = 2\mathcal{E}_s \alpha_i e^{j\theta_i} \delta_{k\nu} + \int_0^{T_s} 2n_i(t) e^{-j2\pi f_c t} s_k^*(t) dt, \quad i = 1, 2, \ldots, n,$$

$$k = 1, 2, \ldots, q \qquad (5\text{-}234)$$

where $\delta_{k\nu} = 1$ if $k = \nu$ and $\delta_{k\nu} = 0$ otherwise. The symbol energy for all the waveforms is

$$\mathcal{E}_s = \frac{1}{2} \int_0^{T_s} |s_k(t)|^2 dt, \quad k = 1, 2, \ldots, q \qquad (5\text{-}235)$$

Since each symbol waveform represents $\log_2 q$ bits, the average SNR per code symbol is

$$\bar{\gamma}_s = (\log_2 q) r \bar{\gamma}_b \qquad (5\text{-}236)$$

which reduces to (5-230) when $q = 2$. If the spectra of the $\{s_k(t)\}$ are confined to $|f| < f_c$, then (5-235) and (5-112) imply that the Gaussian noise term in (5-234) is circularly symmetric and, hence, its real and imaginary components are independent and have the same variance, which is calculated to be $2\mathcal{E}_s N_0$. Therefore, the conditional probability density function of y_{ki} given the values of l, α_i, and θ_i is

$$f(y_{ki}|l, \alpha_i, \theta_i) = \frac{1}{4\pi\mathcal{E}_s N_0} \exp\left[-\frac{|y_{ki} - 2\mathcal{E}_s \alpha_i e^{j\theta_i} \delta_{k\nu}|^2}{4\mathcal{E}_s N_0} \right] \qquad (5\text{-}237)$$

The orthogonality of the $\{s_k(t)\}$, the independence of the white noise from symbol to symbol, and (5-112) imply the conditional independence of the $\{y_{ki}\}$.

For coherent MFSK, the $\{\alpha_i\}$ and the $\{\theta_i\}$ are assumed to be known, and the likelihood function is the product of qn densities given by (5-237) for $k = 1$, 2, ..., q and $i = 1, 2, \ldots, n$. Forming the log-likelihood function and eliminating irrelevant terms that are independent of l, we obtain the *maximum-likelihood metric for coherent MFSK*:

$$U(l) = \sum_{i=1}^{n} \mathrm{Re}\left[\alpha_i e^{-j\theta_i} V_{li} \right], \quad l = 1, 2, \ldots, q^k \qquad (5\text{-}238)$$

where $V_{li} = y_{\nu i}$ is the sampled output of the filter matched to $s_\nu(t)$, the signal representing symbol i of codeword l. For independent, identically distributed Rayleigh fading of each codeword symbol, a derivation similar to the one for PSK indicates that the two-codeword error probability $P_2(d)$ is again given by (5-231) provided that

$$P_s = \frac{1}{2}\left(1 - \sqrt{\frac{\bar{\gamma}_s}{2 + \bar{\gamma}_s}}\right) \quad \text{(coherent MFSK)} \qquad (5\text{-}239)$$

where $\bar{\gamma}_s$ is given by (5-236). A comparison of (5-232) and (5-239) indicates that for large values of $\bar{\gamma}_s$ and the same block code, PSK and QPSK have a 3 dB advantage over coherent binary FSK in a fading environment.

The preceding analysis can be extended to Nakagami fading if the fading parameter m is a positive integer. It is found that the preceding equations for the error probabilities remain valid except that d in (5-231) is replaced by md and P_s is given by the right-hand sides of (5-152) or (5-153) with $\bar{\gamma} = r\bar{\gamma}_b$.

When fast fading makes it impossible to obtain accurate estimates of the $\{\alpha_i\}$ and $\{\theta_i\}$, noncoherent MFSK is a suitable modulation. Expanding the argument of the exponential function in (5-237), assuming that θ_i is uniformly distributed over $[0, 2\pi)$, expressing y_{ki} in polar form, observing that the integral over θ_i is over one period of the integrand, and using the identity (D-30), we obtain the conditional probability density function of y_{ki} given l and α_i:

$$f(y_{ki}|l,\alpha_i) = \frac{1}{4\pi\mathcal{E}_s N_0}\exp\left[-\frac{|y_{ki}|^2 + 4\mathcal{E}_s^2\alpha_i^2\delta_{k\nu}}{4\mathcal{E}_s N_0}\right]I_0\left(\frac{\alpha_i|y_{ki}|\delta_{k\nu}}{N_0}\right) \qquad (5\text{-}240)$$

Assuming that each α_i is statistically independent and has the same Rayleigh probability density function given by (5-20), $f(y_{ki}|l)$ can be evaluated by using the identity (D-33). Calculating the log-likelihood function and eliminating irrelevant terms and factors, we obtain the *Rayleigh metric* for noncoherent MFSK:

$$U(l) = \sum_{i=1}^{n} R_{li}^2, \quad l = 1, 2, \ldots, q^k \qquad (5\text{-}241)$$

where $R_{li} = |y_{\nu i}|$ denotes the envelope produced by the filter matched to the transmitted signal for symbol i of codeword l. Assuming that the all-zero codeword was transmitted, a derivation similar to the one preceding (5-183) again verifies (5-231) with

$$P_s = \frac{1}{2 + \bar{\gamma}_s} \quad \text{(noncoherent MFSK)} \qquad (5\text{-}242)$$

where $\bar{\gamma}_s$ is given by (5-236). A comparison of (5-232) and (5-242) indicates that for large values of $\bar{\gamma}_b$ and the same block code, PSK and QPSK have an approximate 6 dB advantage over noncoherent binary FSK in a fading environment. Thus, the fading accentuates the advantage that exists for the AWGN channel.

As indicated in (1-49), an upper bound on the information-symbol error probability for soft-decision decoding is given by

$$P_{is} \leq \sum_{d=d_m}^{n} \frac{d}{n} A_d P_2(d) \tag{5-243}$$

and the information-bit error probability P_b is given by (1-27). A comparison of (5-132) with (5-231) and the first term on the right-hand side of (5-243) indicates that a binary block code with maximum-likelihood decoding provides an equivalent diversity equal to d_m if $P_b = P_{is}$ is low enough that the first term in (5-243) dominates. For hard-decision decoding, the symbol error probability P_s is given by (5-232) for coherent PSK, (5-239) for coherent MFSK, (5-242) for noncoherent MFSK, or (5-169) for DPSK. For loosely packed codes, P_{is} is approximated by (1-26) whereas it is approximated by (1-25) for tightly packed codes.

Figure 5.22 illustrates $P_b = P_{is}$ for an extended Golay (24,12) code with $L = 1$ and P_b for MRC with $L = 1, 4, 5$, and 6 diversity branches. A Rayleigh fading channel and binary PSK are assumed. The extended Golay (24,12) code is tightly packed with 12 information bits, $r = 1/2$, $d_m = 8$, and $t = 3$. The values of A_d in (5-243) are listed in Table 1.3. The MRC graphs assume that a single bit is transmitted. The SNR per code symbol $\bar{\gamma}_s = \bar{\gamma}/2$, where $\bar{\gamma}$ is the average SNR per bit and branch. The figure indicates the benefits of coding particularly when the desired P_b is low. At $P_b = 10^{-3}$, the (24,12) code with hard decisions provides on 11 dB advantage over uncoded PSK; with soft decisions, the advantage becomes 16 dB. The advantage of soft-decision decoding relative to hard-decision decoding increases to more than 10 dB at $P_b = 10^{-7}$, a vast gain over the approximately 2 dB advantage of soft-decision decoding for the AWGN channel. At $P_b = 10^{-9}$, the Golay (24,12) code with soft decisions outperforms MRC with $L = 5$ and is nearing the performance of MRC with $L = 6$. However, since $A_{d_m} = A_8 = 759$, the equivalent diversity will not reach $L = 8$ even for very low P_b. For noncoherent binary FSK, all the graphs in the figure are shifted approximately 6 dB to the right when $P_b \leq 10^{-3}$.

Since the soft-decision decoding of long block codes is usually impractical, convolutional codes are more likely to give a good performance over a fading channel. The metrics are basically the same as they are for block codes with the same modulation, but they are evaluated over path segments that diverge from the correct path through the trellis and then merge with it subsequently. The linearity of binary convolutional codes ensures that all-zero path can be assumed to be the correct one when calculating the decoding error probability. Let d denote the Hamming distance of an incorrect path from the correct all-zero path. If perfect symbol interleaving is used, then the probability of error in the pairwise comparison of two paths with an unmerged segment is $P_2(d)$, which is given by (5-231). As shown in Chapter 1, the probability of an information-bit error in soft-decision decoding is upper bounded by

$$P_b \leq \frac{1}{k} \sum_{d=d_f}^{\infty} B(d) P_2(d) \tag{5-244}$$

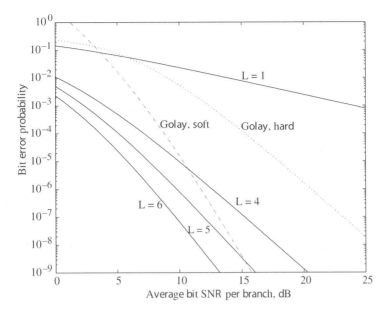

Figure 5.22: Information-bit error probability for extended Golay (24,12) code with soft and hard decisions, coherent PSK modulation, and Rayleigh fading, and for maximal-ratio combining with L = 1, 4, 5, and 6.

where $B(d)$ is the number of information-bit errors over all paths with unmerged segments at Hamming distance d, k is the number of information bits per trellis branch, and d_f is the minimum free distance, which is the minimum Hamming distance between any two convolutional codewords. This upper bound approaches $B_{d_f} P_2(d_f)/k$ as $P_b \to 0$ so the equivalent diversity is d_f if P_b and $B(d_f)/k$ are small.

In general, d_f increases with the constraint length of the convolutional code. However, if each encoder output bit is repeated n_r times, then the minimum distance of the convolutional code increases to $n_r d_f$ without a change in the constraint length, but at the cost of a bandwidth expansion by the factor n_r. From (5-244), we infer that for the code with repeated bits,

$$P_b \le \frac{1}{k} \sum_{d=d_f}^{\infty} B(d) P_2(n_r d) \tag{5-245}$$

where $B(d)$ refers to the original code.

Figure 5.23 illustrates P_b as a function of $\bar{\gamma}_b$ for the Rayleigh-fading channel and binary convolutional codes with different values of the constraint length K, the code rate r, and the number of repetitions n_r. Relations (5-245) and (5-231) with $k = 1$ are used, and the $\{B(d)\}$ are taken from the listings for seven terms in Tables 1.4 and 1.5. The figure indicates that an increase in the constraint length provides a much greater performance improvement for the

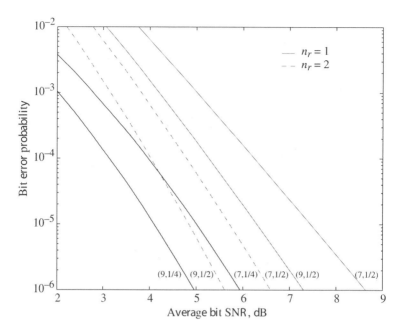

Figure 5.23: Information-bit error probability for Rayleigh fading, coherent PSK, and binary convolutional codes with various values of (K, r) and n_r.

Rayleigh-fading channel than the increase does for the AWGN channel [16]. For a fixed constraint length, the rate-1/4 codes give a better performance than the rate-1/2 codes with $n_r = 2$, which require the same bandwidth but are less complex to implement. The latter two codes require twice the bandwidth of the rate-1/2 code with no repetitions.

The issues are similar for trellis-coded modulation (Chapter 1), which provides a coding gain without a bandwidth expansion. However, if parallel state transitions occur in the trellis, then $d_f = 1$, which implies that the code provides no diversity protection against fading. Thus, for fading communications, a conventional trellis code with distinct transitions from each state to all other states must be selected. Since Rayleigh fading causes large amplitude variations, multiphase PSK is usually a better choice than multilevel quadrature amplitude modulation (QAM) for the symbol modulation. However, the optimum trellis decoder uses coherent detection and requires an estimate of the channel attenuation.

Whether a block, convolutional, or trellis code is used, the results of this section indicate that the minimum Hamming distance rather than the minimum Euclidean distance is the critical parameter in designing an effective code for the Rayleigh fading channel.

Turbo codes or serially concatenated codes with iterative decoding based on the *maximum a posteriori* criterion can provide excellent performance. However, the system must be able to accommodate considerable decoding delay and

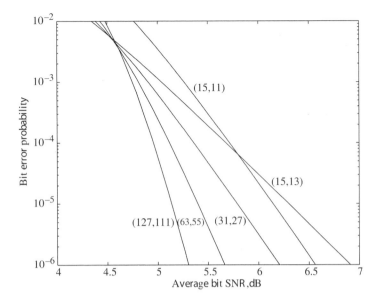

Figure 5.24: Information-bit error probability for Rayleigh fading, coherent PSK, soft decisions, and concatenated codes comprising an inner binary convolutional code with K = 7 and $r_1 = 1/2$, and various Reed-Solomon (n, k) outer codes.

computational complexity. Even without iterative decoding, a serially concatenated code with an outer Reed-Solomon code and an inner binary convolutional code (Chapter 1) can be effective against Rayleigh fading. In the worst case, each output bit error of the inner decoder causes a separate symbol error at the input to the Reed-Solomon decoder. Therefore, an upper bound on P_b is given by (1-128) and (1-127). For coherent PSK modulation with soft-decision decoding, $P_2(d)$ is given by (5-231) and (5-232), and $\bar{\gamma}_s$ is given by (5-230). The concatenated code has a code rate $r = r_1 r_0$, where r_1 is the inner-code rate and r_0 is the outer-code rate.

Figure 5.24 depicts examples of the upper bound on P_b as a function $\bar{\gamma}_b$ for Rayleigh fading, coherent PSK, soft decisions, an inner binary convolutional code with $K = 7$, $r_1 = 1/2$, and $k = 1$, and various Reed-Solomon (n, k) outer codes. The required bandwidth is B_u/r, where B_u is the uncoded PSK bandwidth. Thus, the codes of the figure require a bandwidth less than $3B_u$.

Diversity and Spread Spectrum

Some form of diversity is crucial in compensating for the effects of fading. Spread-spectrum systems exploit the different types of diversity that are available. A direct-sequence receiver exploits time diversity through the small number of branches or demodulators in its rake receiver. These demodulators must be synchronized to the path delays of the multipath components. The effectiveness of the rake receiver depends on the concentration of strong diffuse and specular components in the vicinity of resolvable path delays, which becomes more likely as the chip rate increases. A large Doppler spread is beneficial to a direct-sequence system because it decreases the channel coherence time. If the coherence time is less than the interleaving depth, the performance of the error-control decoder is enhanced. Equation (5-10) indicates that the Doppler spread increases with the carrier frequency and the speed of the receiver relative to the transmitter.

Frequency-hopping systems rely on frequency diversity. Interleaving of the code symbols over many dwell intervals provides a large level of diversity to slow frequency-hopping systems operating over a frequency-selective fading channel. These systems are usually insensitive to variations in the Doppler spread of the channel because any additional diversity due to improved time-selective fading is insignificant. The relative performance of comparable frequency-hopping and direct-sequence systems ultimately depends primarily on the size of the Doppler spread and the degree to which most of the power in the received signal is concentrated in a small number of resolvable multipath components [15]. A large Doppler spread and strong specular multipath components tend to favor direct-sequence systems.

5.7   Problems

1. Give an alternative derivation of (5-41). First, observe that the total received Doppler power $S_r(f) \mid f \mid$ in the spectral band $[f, f + df]$ corresponds to arrival angles determined by $f_d \cos\theta = f$. If the received power arrives uniformly spread over all angles $\mid \theta \mid \leq \pi$, then $S(f) \mid df \mid = P(\theta) \mid d\theta \mid + P(\text{-}\theta) \mid d\theta \mid$, where $P(\theta)$ is the power density arriving from angle θ.

2. Assume that a propagation channel has the impulse response given by (5-63) with $L_c(t) = N$ and $\tau_i(t) = \tau_i$. Each $r_i(t)$ is a wide-sense stationary process and uncorrelated scattering occurs. Derive expressions for $R_h(t_1 - t_2, \tau_1)$ and $S_h(\tau)$. Show that the multipath intensity profile indicates specular components.

3. Following the guidance in the text, derive the maximum-likelihood decision variable (5-115) and its variance (5-120) for PSK over the Rayleigh fading channel.

4. Consider the maximum-likelihood detection of coherent MFSK for the Rayleigh fading channel. The independent noise in each diversity branch has power spectral density N_{0i}, $i = 1, 2, \ldots, L$. Find the q decision variables and show that they are given by (5-142) when the $\{N_{0i}\}$ are all equal.

5. Consider the maximum-likelihood detection of noncoherent MFSK for the Rayleigh fading channel. The independent noise in each diversity branch

has power spectral density N_{0i}, $i = 1, 2, \ldots, L$. Find the q decision variables and show that they are given by (5-171) when the $\{N_{0i}\}$ are all equal.

6. Suppose that diversity L is achieved by first dividing the symbol energy into L equal parts so that the SNR per branch in (5-124) is reduced by the factor L. For the four modulations of Figure 5.15 and $p \to 0$, by what factor does $P_b(L)$ increase relative to its value when the energy is not divided?

7. For noncoherent q-ary orthogonal signals such as those with MFSK, use the union bound to derive an upper bound on the symbol error probability as a function of q and the diversity L.

8. For dual rake combining, PSK, MRC, and Rayleigh fading, find $P_b(2)$ as both γ_1 and $\gamma_2 \to \infty$. Find $P_b(2)$ for dual rake combining, noncoherent orthogonal signals, EGC, and Rayleigh fading as both γ_1 and $\gamma_2 \to \infty$. What advantage does PSK have?

9. Three multipath components arrive at a direct-sequence receiver moving at 30 m/s relative to the transmitter. The second and third multipath components travel over paths 200 m and 250 m longer than the first component. If the chip rate is equal to the bandwidth of the received signal, what is the minimum chip rate required to resolve all components? How much time can the receiver allocate to the estimation of the component delays?

10. Consider a system that uses PSK or MFSK, an (n, k) block code, and the maximum-likelihood metric in the presence of Rayleigh fading. Show by successive applications of various bounds that the word error probability for soft-decision decoding satisfies

$$P_w < q^k \binom{2d_m - 1}{d_m} P_s^{d_m}$$

where q is the alphabet size and d_m is the minimum distance between codewords.

5.8   References

1. G. Stuber, *Principles of Mobile Communication, 2nd ed.* Boston: Kluwer Academic, 2001.

2. M. K. Simon and M.-S. Alouini, *Digital Communication over Fading Channels.* New York: Wiley, 2000.

3. J. G. Proakis, *Digital Communications, 4th ed.* New York: McGraw-Hill, 2001.

4. S. J. Leon, *Linear Algebra with Applications, 6th ed.* Upper Saddle River, NJ: Prentice-Hall, 2002.

5. J. D. Parsons and J. G. Gardiner, *Mobile Communication Systems.* New York: Halsted Press, 1989.

6. D. R. Barry, E. A. Lee, and D. G. Messerschmitt, *Digital Communication, 3rd ed.* Boston: Kluwer Academic, 2004.

7. J. S. Lee and L. E. Miller, *CDMA Systems Engineering Handbook.* Boston: Artech House, 1998.

8. I. S. Gradsteyn and I. M. Ryzhik, *Tables of Integrals, Series and Products, 6th ed.* San Diego: Academic Press, 2000.

9. M. E. Frerking, *Digital Signal Processing in Communication Systems.* New York: Van Nostrand Reinhold, 1994.

10. M. K. Simon and M.-S. Alouini, "Average Bit-Error Probability Performance for Optimum Diversity Combining of Noncoherent FSK over Rayleigh Fading Channels," *IEEE Trans. Commun.*, vol. 51, pp. 566-569, April 2003.

11. M.-S. Alouini and M.K.Simon, "Postdetection Switched Combining–A Simple Diversity Scheme with Improved BER Performance," *IEEE Trans. Commun.*, vol. 51, pp. 1591-1602, September 2003.

12. K. Higuchi et al., "Experimental Evaluation of Combined Effect of Coherent Rake Combining and SIR-Based Fast Transmit Power Control for Reverse Link of DS-CDMA Mobile Radio," *IEEE J. Select. Areas Commun.*, vol. 18, pp. 1526–1535, August 2000.

13. D. Torrieri, "Simple Formula for Error Probability of Rake Demodulator for Noncoherent Binary Orthogonal Signals and Rayleigh Fading," *IEEE Trans. Commun.*, vol. 50, pp. 1734-1735, November 2002.

14. S. Tanaka et al., "Experiments on Coherent Adaptive Antenna Array Diversity for Wideband DS-CDMA Mobile Radio," *IEEE J. Select. Areas Commun.*, vol. 18, pp. 1495–1504, August 2000.

15. J. H. Gass, D. L. Noneaker, and M. B. Pursley, "A Comparison of Slow-Frequency-Hop and Direct-Sequence Spread-Spectrum Packet Communications Over Doubly Selective Fading Channels," *IEEE Trans. Commun.*, vol. 50, pp. 1236–1239, August 2002.

Chapter 6

Code-Division Multiple Access

Multiple access is the ability of many users to communicate with each other while sharing a common transmission medium. Wireless multiple-access communications are facilitated if the transmitted signals are orthogonal or separable in some sense. Signals may be separated in time (*time-division multiple access* or TDMA), frequency (*frequency-division multiple access* or FDMA), or code (*code-division multiple access* or CDMA). CDMA is realized by using spread-spectrum modulation while transmitting signals from multiple users in the same frequency band at the same time. All signals use the entire allocated spectrum, but the spreading sequences or frequency-hoppong patterns differ. Information theory indicates that in an isolated cell, CDMA systems achieve the same spectral efficiency as TDMA or FDMA systems only if optimal multiuser detection is used. However, even with single-user detection, CDMA is advantageous for cellular networks because it eliminates the need for frequency and timeslot coordination among cells and allows carrier-frequency reuse in adjacent cells. Frequency planning is vastly simplified. A major CDMA advantage exists in networks accommodating voice communications. A voice-activity detector activates the transmitter only when the user is talking. Since typically fewer than 40% of the users are talking at any given time, the number of telephone users can be increased while maintaining a specified average interference power. Another major CDMA advantage is the ease with which it can be combined with multibeamed antenna arrays that are either adaptive or have fixed patterns covering cell sectors. There is no practical means of reassigning time slots in TDMA systems or frequencies in FDMA systems to increase capacity by exploiting intermittent voice signals or multibeamed arrays, and reassignments to accommodate variable data rates are almost always impractical. These general advantages and its resistance to jamming, interception, and multipath interference make CDMA the choice for most mobile communication networks. The two principal types of spread-spectrum CDMA are *direct-sequence* CDMA (DS/CDMA) and *frequency-hopping* CDMA (FH/CDMA).

6.1   Spreading Sequences for DS/CDMA

Consider a DS/CDMA network with K users in which every receiver has the form of Figure 2.14. The multiple-access interference that enters a receiver synchronized to a desired signal is modeled as

$$i(t) = \sum_{i=1}^{K-1} \sqrt{2I_i} d_i(t - \tau) q_i(t - \tau_i) \cos(2\pi f_c t + \phi_i) \tag{6-1}$$

where $K - 1$ is the number of interfering direct-sequence signals, and I_i is the average power, $d_i(t)$ is the code-symbol modulation, $q_i(t)$ is the spreading waveform, τ_i is the relative delay, and ϕ_i is the phase shift of interference signal i including the effect of carrier time delay. The spreading waveform of the desired signal is

$$p(t) = \sum_{i=-\infty}^{\infty} p_i \psi(t - iT_c) \tag{6-2}$$

where $p_i \in \{-1, 1\}$. Each spreading waveform of an interference signal has the form

$$q_i(t) = \sum_{j=-\infty}^{\infty} q_j^{(i)} \psi(t - jT_c) \tag{6-3}$$

where $q_j^{(i)} \in \{-1, 1\}$. The chip waveforms are assumed to be identical throughout the network and have unit energy:

$$\frac{1}{T_c} \int_0^{T_c} \psi^2(t) dt = 1 \tag{6-4}$$

In a DS/CDMA network, the spreading sequences are often called *signature sequences*. As shown in Chapter 2, the interference component of the demodulator output due to a received symbol is

$$V_1 = \sum_{\nu=0}^{G-1} p_\nu J_\nu \tag{6-5}$$

where

$$J_\nu = \int_{\nu T_c}^{(\nu+1)/T_c} i(t) \psi(t - \nu T_c) \cos 2\pi f_c t \, dt \tag{6-6}$$

Substituting (6-1) into (6-6) and (6-5) and then using (6-2), we obtain

$$V_1 = \sum_{i=1}^{K-1} \sqrt{\frac{I_i}{2}} \cos \phi_i \int_0^{T_s} d_i(t - \tau_i) q_i(t - \tau_i) p(t) dt \tag{6-7}$$

where a double-frequency term is neglected.

Orthogonal Sequences

Suppose that the communication signals are synchronous so that all data symbols have duration T_s, symbol and chip transitions are aligned at the receiver input, and short spreading sequences with period $N = G$ extend over each data symbol. Then $\tau_i = 0, i = 1, 2, \ldots, K - 1$, and $d_i(t) = d_i$ is constant over the integration interval $[0, T_s]$. The *cross-correlation* between $q_i(t)$ and $p(t)$ is defined as

$$C_{pi}(\tau) = \frac{1}{T_s} \int_0^{T_s} p(t) q_i(t - \tau) dt \qquad (6\text{-}8)$$

Thus, for synchronous communications, (6-7) may be expressed as

$$V_1 = \sum_{i=1}^{K-1} \lambda_i C_{pi}(0) \qquad (6\text{-}9)$$

where

$$\lambda_i = \sqrt{\frac{I_i}{2}} d_i T_s \cos \phi_i \qquad (6\text{-}10)$$

Substituting (6-3) and (6-2) into (6-8) and then using (6-4) and $G = T_s/T_c$, we obtain

$$C_{pi}(0) = \frac{1}{G} \sum_{j=1}^{G} p_j q_j^{(i)} \qquad (6\text{-}11)$$

where the right-hand side is the *periodic cross-correlation* between the sequences $\{q_j^{(i)}\}$ and $\{p_j\}$. Let $\mathbf{a}$ and $\mathbf{b}_i$ denote the binary sequences with components $a_j, b_j^{(i)} \in GF(2)$, respectively, that map into the binary antipodal sequences with components $p_j = (-1)^{a_j+1}$ and $q_j^{(i)} = (-1)^{b_j^{(i)}+1}$. Then a derivation similar to that in (2-34) gives

$$C_{pi}(0) = \frac{A_i - D_i}{G} \qquad (6\text{-}12)$$

where A_i denotes the number of agreements in the corresponding bits of $\mathbf{a}$ and $\mathbf{b}_i$, and D_i denotes the number of disagreements. The sequences are *orthogonal* if $C_{pi}(0) = 0$. If the spreading sequence $\mathbf{a}$ is orthogonal to all the spreading sequences $\mathbf{b}_i, i = 1, 2, \ldots, K$, then $V_1 = 0$ and the multiple-access interference $i(t)$ is suppressed at the receiver. A large number of multiple-access interference signals can be suppressed in a network if each such signal has its chip transitions aligned and the spreading sequences are mutually orthogonal.

Two binary sequences, each of length two, are orthogonal if each sequence is described by one of the rows of the 2×2 matrix

$$\mathbf{H}_1 = \begin{bmatrix} 0 & 0 \\ 0 & 1 \end{bmatrix} \qquad (6\text{-}13)$$

because $A = D = 1$. A set of 2^n sequences, each of length 2^n, is obtained by using the rows of the matrix

$$\mathbf{H}_n = \begin{bmatrix} \mathbf{H}_{n-1} & \mathbf{H}_{n-1} \\ \mathbf{H}_{n-1} & \bar{\mathbf{H}}_{n-1} \end{bmatrix}, \quad n = 2, 3, \ldots \qquad (6\text{-}14)$$

where $\bar{\mathbf{H}}_{n-1}$ is the *complement* of $\mathbf{H}_{n-1}$, obtained by replacing each 1 and 0 by 0 and 1, respectively, and $\mathbf{H}_1$ is defined by (6-13). Any pair of rows in $\mathbf{H}_n$ differ in exactly 2^{n-1} columns, thereby ensuring orthogonality of the corresponding sequences. The $2^n \times 2^n$ matrix $\mathbf{H}_n$, which is called a *Hadamard* matrix, can be used to generate 2^n orthogonal spreading sequences for synchronous direct-sequence communications. The orthogonal spreading sequences generated from a Hadamard matrix are called *Walsh sequences*.

In CDMA networks for multimedia applications, the data rates for various services and users often differ. If the transmitted signal bandwidth is the same for all signals, then so is the chip rate. For synchronous communications, it is desirable to use spreading sequences that are orthogonal to each other despite differences in the processing gains, which are often called *spreading factors* in CDMA networks. Starting with a set of Walsh sequences, a tree-structured set of orthogonal Walsh sequences called the *orthogonal variable-spreading-factor codes* can be generated recursively for this purpose. Let $\mathbf{C}_N(n)$ denote the row vector representing the *n*th sequence with spreading factor N, where $n = 1, 2, ..., N$, and $N = 2^k$ for some positive integer k. The set of N sequences with N chips is derived by concatenating sequences from the set of N/2 sequences with N/2 chips:

$$\mathbf{C}_N(1) = [\mathbf{C}_{N/2}(1) \ \mathbf{C}_{N/2}(1)]$$
$$\mathbf{C}_N(2) = [\mathbf{C}_{N/2}(1) \ \overline{\mathbf{C}}_{N/2}(1)]$$

$$\vdots \qquad\qquad\qquad (6\text{-}15)$$

$$\mathbf{C}_N(N-1) = [\mathbf{C}_{N/2}(N/2) \ \mathbf{C}_{N/2}(N/2)]$$
$$\mathbf{C}_N(N) = [\mathbf{C}_{N/2}(N/2) \ \overline{\mathbf{C}}_{N/2}(N/2)]$$

For example, $\mathbf{C}_{16}(4)$ is produced by concatenating $\mathbf{C}_8(2)$ and $\overline{\mathbf{C}}_8(2)$, thereby doubling the number of chips per data symbol to 16. A sequence used in the recursive generation of a longer sequence is called a *mother code* of the longer sequence. Equation (6-15) indicates that the sequences with N chips are orthogonal to each other, and each $\mathbf{C}_N(n)$ is orthogonal to concatenations of all sequences $\mathbf{C}_{N/2}(n'), \mathbf{C}_{N/4}(n''), \ldots$ and their complements except for its mother codes. For example, $\mathbf{C}_{16}(4)$ is not orthogonal to $\mathbf{C}_8(2)$ or $\mathbf{C}_4(1)$. Synchronous signals with a judicious selection of orthogonal variable-spreading-factor codes enable the receiver to completely suppress multiple-access interference.

As an alternative to the Walsh sequences, consider the set of $2^m - 1$ maximal sequences generated by a primitive polynomial of degree m and the $2^m - 1$ different initial states of the shift register. Equation (2-34) implies that by appending a 0 at the end of each period of each sequence, we obtain a set of $2^m - 1$ orthogonal sequences of period 2^m. Without the appending of symbols, a set of nearly orthogonal sequences for a synchronous network may be generated from different time displacements of a single maximal sequence because its autocorrelation, which is given by (2-35), determines the cross-correlations among the sequences of the set. The low values of the autocorrelation for nonzero delay causes the rejection of multipath signals. In contrast, the Walsh sequences do not have such favorable autocorrelation functions.

Sequences with Small Cross-Correlations

The symbol transitions of *asynchronous* multiple-access signals at a receiver are not simultaneous, usually because of changing path-length differences among the various communication links. Since the spreading sequences are shifted relative to each other, sets of periodic sequences with small cross-correlations for any relative shifts are desirable to limit the effect of multiple-access interference. Maximal sequences, which have the longest periods of sequences generated by a linear feedback shift register of fixed length, are often inadequate. Let $\mathbf{a} = (\ldots, a_0, a_1, \ldots)$ and $\mathbf{b} = (\ldots, b_0, b_1, \ldots)$ denote binary sequences with components in $GF(2)$. The sequences $\mathbf{a}$ and $\mathbf{b}$ are mapped into antipodal sequences $\mathbf{p}$ and $\mathbf{q}$, respectively, with components in $\{-1, +1\}$ by means of the transformation

$$p_i = (-1)^{a_i+1} \quad , \quad q_i = (-1)^{b_i+1} \tag{6-16}$$

The *periodic cross-correlation* of periodic binary sequences $\mathbf{a}$ and $\mathbf{b}$ with the same period N is defined as the periodic cross-correlation of the antipodal sequences $\mathbf{p}$ and $\mathbf{q}$, which is defined as

$$\theta_{pq}(j) = \frac{1}{N} \sum_{i=0}^{N-1} p_i q_{i+j}, \quad j = 1, 2, \ldots, N - 1 \tag{6-17}$$

A calculation similar to that in (2-34) yields the periodic cross-correlation

$$\theta_{pq}(j) = \frac{A_j - D_j}{N} \tag{6-18}$$

where A_j denotes the number of agreements in the corresponding components of $\mathbf{a}$ and the shift sequence $\mathbf{b}(j) = (\ldots, b_j, b_{j+1}, \ldots, b_{j+N-1}, \ldots)$, and D_j denotes the number of disagreements.

In the presence of asynchronous multiple-access interference for which $\tau_i \neq 0$, the interference component of the correlator output is given by (6-7). If we assume that the data modulation is absent so that we may set $d_i(t) = 1$ in (6-7), then it is observed that interference signal i produces a term in V_1 that is proportional to $C_{pi}(\tau_i)$ given by (6-8). Let $\tau_i = N_i T_c + \epsilon_i$, where N_i is a nonnegative integer and $0 \leq \epsilon_i < T_c$. A derivation similar to the one leading to (2-40) gives

$$C_{pi}(N_i T_c + \epsilon_i) = \left(1 - \frac{\epsilon_i}{T_c}\right) \theta_{pi}(N_i) + \frac{\epsilon_i}{T_c} \theta_{pi}(N_i + 1) \tag{6-19}$$

where $\theta_{pi}(N_i)$ is the periodic cross-correlation of the sequence $\mathbf{p}$ and $\mathbf{q}_i$ and is given by (6-18). Thus, ensuring that the periodic cross-correlations are always small is a critical *necessary* condition for the success of asynchronous multiple-access communications. Although the data modulation may be absent during acquisition, it will be present during data transmission, and $d_i(t)$ may change polarity during an integration interval. Thus, the effect of asynchronous multiple-access interference will exceed that predicted from (6-19).

For a set S of M periodic antipodal sequences of length N, let $\theta_{\max}$ denote the peak magnitude of the cross-correlations or autocorrelations:

$$\theta_{\max} = \max\{|\theta_{pq}(k)| : 0 \le k \le N - 1; \ \mathbf{p}, \mathbf{q} \in S; \ p \ne q \text{ or } k \ne 0\} \quad (6\text{-}20)$$

Theorem. *A set S of M periodic antipodal sequences of length N has*

$$\theta_{\max} \ge \sqrt{\frac{M - 1}{MN - 1}} \quad (6\text{-}21)$$

Proof: Consider an extended set S_e of MN sequences $\mathbf{p}^{(i)}, i = 1, 2, \ldots, MN$, that comprises the N distinct shifted sequences derived from each of the sequences in S. The cross-correlation of sequences $\mathbf{p}^{(i)}$ and $\mathbf{p}^{(j)}$ in S_e is

$$\psi_{ij} = \frac{1}{N} \sum_{n=1}^{N} p_n^{(i)} p_n^{(j)} \quad (6\text{-}22)$$

and

$$\theta_{\max} = \max\left\{|\psi_{ij}|, \mathbf{p}^{(i)} \epsilon \, S_e, \ \mathbf{p}^{(j)} \, \epsilon \, S_e, i \ne j\right\} \quad (6\text{-}23)$$

Define the double summation

$$Z = \sum_{i=1}^{MN} \sum_{j=1}^{MN} \psi_{ij}^2 \quad (6\text{-}24)$$

Separating the MN terms for which $\psi_{ii} = 1$ and then bounding the remaining $MN(MN - 1)$ terms yields

$$Z \le MN + MN(MN - 1)\theta_{\max}^2 \quad (6\text{-}25)$$

Substituting (6-22) into (6-24), interchanging summations, and omitting the terms for which $m \ne n$, we obtain

$$Z = \frac{1}{N^2} \sum_{n=1}^{N} \sum_{m=1}^{N} \sum_{i=1}^{MN} p_n^{(i)} p_m^{(i)} \sum_{j=1}^{MN} p_n^{(j)} p_m^{(j)} = \frac{1}{N^2} \sum_{n=1}^{N} \sum_{m=1}^{N} \left(\sum_{i=1}^{MN} p_n^{(i)} p_m^{(i)}\right)^2$$

$$\ge \frac{1}{N^2} \sum_{n=1}^{N} \left[\sum_{n=1}^{MN} \left(p_n^{(i)}\right)^2\right]^2 = M^2 N \quad (6\text{-}26)$$

Combining this inequality with (6-25) gives (6-21). $\square$

The lower bound in (6-21) is known as the *Welch bound*. It approaches $1/\sqrt{N}$ for large values of M and N. Only small subsets of maximal sequences can be found with $\theta_{\max}$ close to this lower bound. The same is true for Walsh sequences.

Large sets of sequences with $\theta_{\max}$ approaching the Welch bound can be obtained by combining maximal sequences with sampled versions of these sequences. If q is a positive integer, the new binary sequence $\mathbf{b}$ formed by taking every qth bit of binary sequence $\mathbf{a}$ is known as a *decimation* of $\mathbf{a}$ by q, and the components of the two sequences are related by $b_i = a_{qi}$. Let $gcd(x, y)$ denote the greatest common divisor of x and y. If the original sequence $\mathbf{a}$ has a period N and the new sequence $\mathbf{b}$ is not identically zero, then $\mathbf{b}$ has period $N/gcd(N, q)$. If $gcd(N, q) = 1$, then the decimation is called a *proper decimation*. Following a proper decimation, the bits of $\mathbf{b}$ do not repeat themselves until every bit of $\mathbf{a}$ has been sampled. Therefore, $\mathbf{b}$ and $\mathbf{a}$ have the same period N, and it can be shown that if $\mathbf{a}$ is maximal, then $\mathbf{b}$ is a maximal sequence [1]. A *preferred pair* of maximal sequences with period $2^m - 1$ are a pair with a periodic cross-correlation that takes only the three values $-t(m)/N, -1/N$, and $[t(m) - 2]/N$, where

$$t(m) = 2^{\lfloor (m+2)/2 \rfloor} + 1 \qquad (6\text{-}27)$$

and $\lfloor x \rfloor$ denotes the integer part of the real number x. The *Gold sequences* are a large set of sequences with period $N = 2^m - 1$ that may be generated by the modulo-2 addition of preferred pairs when m is odd or $m = 2$ modulo-4 [1]. One sequence of the preferred pair is a decimation by q of the other sequence. The positive integer q is either $q = 2^k + 1$ or $q = 2^{2k} - 2^k + 1$, where k is a positive integer such that $gcd(m, k) = 1$ when m is odd and $gcd(m, k) = 2$ when $m = 2$ modulo-4.

Since the cross-correlation between any two Gold sequences in a set can take only three values, the peak magnitude of the periodic cross-correlation between any two Gold sequences of period $N = 2^m - 1$ is

$$\theta_{\max} = \frac{t(m)}{2^m - 1} \qquad (6\text{-}28)$$

For large values of m, $\theta_{\max}$ for Gold sequences exceeds the Welch bound by a factor of $\sqrt{2}$ for m odd and a factor of 2 for m even.

One form of a Gold sequence generator is shown in Figure 6.1. If each maximal sequence generator has m stages, different Gold sequences in a set are generated by selecting the initial state of one maximal sequence generator and then shifting the initial state of the other generator. Since any shift from 0 to $2^m - 2$ results in a different Gold sequence, $2^m - 1$ different Gold sequences can be produced by the system of Figure 6.1. Gold sequences identical to maximal sequences are produced by setting the state of one of the maximal sequence generators to zero. Altogether, there are $2^m + 1$ different Gold sequences, each with a period of $2^m - 1$, in the set.

An example of a set of Gold sequences is the set generated by the preferred pair specified by the primitive characteristic polynomials

$$f_1(x) = 1 + x^3 + x^7, \quad f_2(x) = 1 + x + x^2 + x^3 + x^7 \qquad (6\text{-}29)$$

Since $m = 7$, there are 129 Gold sequences of period 127 in this set, and (6-28) gives $\theta_{\max} = 0.134$. Equation (2-66) indicates that there are only 18 maximal

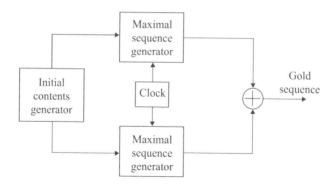

Figure 6.1: Gold sequence generator.

sequences with $m = 7$. For this set of 18 sequences, calculations [1] indicate that $\theta_{\max} = 0.323$. If $\theta_{\max} = 0.134$ is desired for a set of maximal sequences with $m = 7$, then one finds that the set has only 6 sequences. This result illustrates the much greater utility of Gold sequences in CDMA networks with many subscribers.

Consider a Gold sequence generated by using the characteristic functions $f_1(x)$ and $f_2(x)$ of degree m. The generating function for the Gold sequence is

$$
\begin{aligned}
G(x) &= \frac{\phi_1(x)}{f_1(x)} + \frac{\phi_2(x)}{f_2(x)} \\
&= \frac{\phi_1(x)f_2(x) + \phi_2(x)f_1(x)}{f_1(x)f_2(x)}
\end{aligned}
\tag{6-30}
$$

where $\phi_1(x)$ and $\phi_2(x)$ have the form specified by the numerator of (2-60). Since the degrees of both $\phi_1(x)$ and $\phi_2(x)$ are less than m, the degree of the numerator of $G(x)$ must be less than $2m$. Since the product $f_1(x)f_2(x)$ has the form of a characteristic function of degree $2m$ given by (2-56), this product defines the feedback coefficients of a single linear feedback shift register with $2m$ stages that can generate the Gold sequences. The initial state of the register for any particular sequence can be determined by equating each coefficient in the numerator of (6-30) with the corresponding coefficient in (2-60) and then solving $2m$ linear equations.

A *small set of Kasami sequences* comprises $2^{m/2}$ sequences with period $2^m - 1$ if m is even [1]. To generate a set, a maximal sequence **a** with period $N = 2^m - 1$ is decimated by $q = 2^{m/2} + 1$ to form a binary sequence **b** with period $N/gcd(N, q) = 2^{m/2} - 1$. The modulo-2 addition of **a** and any cyclic shift of **b** from 0 to $2^{m/2} - 2$ provides a Kasami sequence. By including sequence **a**, we obtain a set of $2^{m/2}$ Kasami sequences with period $2^m - 1$. The periodic cross-correlation between any two Kasami sequences in a set can only take the values $-s(m)/N, -1/N$, or $[s(m) - 2]/N$, where

$$s(m) = 2^{m/2} + 1 \tag{6-31}$$

The peak magnitude of the periodic cross-correlation between any two Kasami sequences is

$$\theta_{\max} = \frac{s(m)}{N} = \frac{2^{m/2} + 1}{2^m - 1} = \frac{1}{2^{m/2} - 1} \tag{6-32}$$

For $m \geq 2$ and $M = 2^{m/2}$, the use of $NM - 1 > NM - N$ in the Welch bound gives $\theta_{\max} > 1/\sqrt{N}$. Since $N = 2^m - 1$,

$$N\theta_{\max} > \sqrt{2^m - 1} > 2^{m/2} - 1 \tag{6-33}$$

Since N is an odd integer, $A_j - D_j$ in (6-18) must be an odd integer. Therefore, the definition of $\theta_{\max}$ and (6-18) indicate that $N\theta_{\max}$ must be an odd integer. Inequality (6-33) then implies that for $M = 2^{m/2}, N = 2^m - 1$, and even values of m,

$$N\theta_{\max} \geq 2^{m/2} + 1 \tag{6-34}$$

A comparison of this result with (6-32) indicates that the Kasami sequences are optimal in the sense that $\theta_{\max}$ has the minimum value for any set of sequences of the same size and period.

As an example, let $m = 10$. There are 60 maximal sequences, 1025 Gold sequences, and 32 Kasami sequences with period 1023. The peak cross-correlations are 0.37, 0.06, and 0.03, respectively.

A *large set of Kasami sequences* comprises $2^{m/2}(2^m + 1)$ sequences if $m = 2$ modulo-4 and $2^{m/2}(2^m + 1) - 1$ sequences if $m = 0$ modulo-4 [1]. The sequences have period $2^m - 1$. To generate a set, a maximal sequence **a** with period $N = 2^m - 1$ is decimated by $q = 2^{m/2} + 1$ to form a binary sequence **b** and then decimated by $q = 2^{(m+2)/2} + 1$ to form another binary sequence **c**. The modulo-2 addition of **a**, a cyclic shift of **b**, and a cyclic shift of **c** provides a Kasami sequence with period N. The periodic cross-correlations between any two Kasami sequences in a set can only take the values $-1/N, -t(m)/N, [t(m) - 2]/N, -s(m)/N$, or $[s(m) - 2]/N$. A large set of Kasami sequences includes both a small set of Kasami sequences and a set of Gold sequences as subsets. Since $t(m) \geq s(m)$, the value of $\theta_{\max}$ for a large set is the same as that for Gold sequences (6-28). This value is suboptimal, but the large size of these sets makes them an attractive option for asynchronous CDMA networks.

Symbol Error Probability

Let $\mathbf{d}_i = (d_{-1}^{(i)}, d_0^{(i)})$ denote the vector of the two symbols of asynchronous multiple-access interference signal i that are received during the detection of a symbol of the desired signal. A straightforward evaluation of (6-7) gives

$$V_1 = \sum_{i=1}^{K-1} \sqrt{\frac{I_i}{2}} \cos\phi_i \left[d_{-1}^{(i)} R_{pi}(\tau_i) + d_0^{(i)} \hat{R}_{pi}(\tau_i) \right] \tag{6-35}$$

where the *continuous-time partial cross-correlation functions* are

$$R_{pi}(\tau) = \int_0^\tau p(t)q_i(t-\tau)dt \tag{6-36}$$

$$\hat{R}_{pi}(\tau) = \int_\tau^T p(t)q_i(t-\tau)dt \tag{6-37}$$

For rectangular chip waveforms and spreading sequences of period N, straightforward calculations yield

$$R_{pi}(\tau) = A_{pi}(l-N)T_c + [A_{pi}(l+1-N) - A_{pi}(l-N)](\tau-lT_c) \tag{6-38}$$

$$\hat{R}_{pi}(\tau) = A_{pi}(l)T_c + [A_{pi}(l+1) - A_{pi}(l)](\tau-lT_c) \tag{6-39}$$

where $l = \lfloor \tau/T_c \rfloor$ and the *aperiodic cross-correlation function* is defined by

$$A_{pi}(l) = \begin{cases} \sum_{j=0}^{N-1-l} p_j q_{j+l}^{(i)}, & 0 \leq l \leq N-1 \\ \sum_{j=0}^{N-1+l} p_j q_j^{(i)}, & 1-N \leq l < 0 \end{cases} \tag{6-40}$$

and $A_{pi}(l) = 0$ for $|l| \geq N$. These equations indicate that the aperiodic cross-correlations are more important than the related periodic cross-correlations defined by (6-17) in determining the interference level and, hence, the symbol error probability. Without careful selection of the sequences, the aperiodic cross-correlations may be much larger than the periodic cross-correlation. If all the spreading sequences are short with $N = G$, and the power levels of all received signals are equal, then the symbol error probability can be approximated and bounded [2], [3], but the process is complicated. An alternative approach is to model the spreading sequences as random binary sequences, as is done for long sequences.

In a network with multiple-access interference, code acquisition depends on both the periodic and aperiodic cross-correlations. In the absence of data modulations, V_c in (4-73) has additional terms, each of which is proportional to the periodic cross-correlation between the desired signal and an interference signal. When data modulations are present, some or all of these terms entail aperiodic cross-correlations.

Complex-Valued Quaternary Sequences

Quaternary direct-sequence system may use pairs of short binary sequences, such as Gold or Kasami sequences, to exploit the favorable periodic autocorrelation and cross-correlation functions. However, Gold sequences do not attain the Welch bound, and Kasami sequences that do are limited in number. To support many users and to facilitate the unambiguous synchronization to particular signals in a CDMA network, one might consider complex-valued quaternary sequences that are not derived from pairs of standard binary sequences but have better periodic correlation functions.

For q-ary PSK modulation, sequence symbols are powers of the complex qth root of unity, which is

$$\Omega = \exp\left(j\frac{2\pi}{q}\right) \qquad (6\text{-}41)$$

where $j = \sqrt{-1}$. The complex spreading or signature sequence $\mathbf{p}$ of period N has symbols given by

$$p_i = \Omega^{a_i} e^{j\phi_r}, \qquad a_i \in Z_q = \{0, 1, 2, \ldots, q-1\}, \qquad i = 1, 2, \ldots, N \qquad (6\text{-}42)$$

where ϕ_r is an arbitrary phase chosen for convenience. If p_i is specified by the exponent a_i and q_i is specified by the exponent b_i, then the periodic cross-correlation between sequences $\mathbf{p}$ and $\mathbf{q}$ is defined as

$$\theta_{pq}(k) = \frac{1}{N}\sum_{i=0}^{N-1} p_{i+k} q_i^* = \frac{1}{N}\sum_{i=0}^{N-1} \Omega^{a_{i+k}-b_i} \qquad (6\text{-}43)$$

The maximum magnitude θ_{max} defined by (6-20) must satisfy the Welch bound of (6-21). For a positive integer m, a family $\mathcal{A}$ of $M = N+2$ quaternary or Z_4 sequences, each of period $N = 2^m - 1$, with θ_{max} that asymptotically approaches the Welch bound has been identified [4]. In contrast, a small set of binary Kasami sequences has only $\sqrt{N+1}$ sequences

The sequences in a family $\mathcal{A}$ are determined by the *characteristic polynomial*, which is defined as

$$f(x) = 1 + \sum_{i=0}^{m} c_i x^i \qquad (6\text{-}44)$$

where coefficients $c_i \in Z_4$ and $c_m = 1$. The output sequence satisfies the linear recurrence relation of (2-20). For example, the characteristic polynomial $f(x) = 1+2x+3x^2+x^3$ has $m = 3$ and generates a family with period $N = 7$. A feedback shift register that implements the sequence of the family is depicted in Figure 6.2(a), where all operations are modulo-4. The generation of a particular sequence is illustrated in Figure 6.2(b). Different sequences may be generated by loading the shift register with any nonzero initial contents and then cycling the shift register through its full period $N = 2^m - 1$. Since the shift register has $4^m - 1$ nonzero states, there are $M = (4^m-1)/(2^m-1) = 2^m+1$ *cyclically distinct* members of the family. Each family member may be generated by loading the shift register with any nonzero triple that is not a state occurring during the generation of another family member.

By setting $\phi_r = \pi/4$ in (6-42), a complex-valued data symbol in the family $\mathcal{A}$ may be represented by $d = d_1+jd_2$, where d_1 and d_2 are antipodal symbols with values $\pm 1/\sqrt{2}$. If a complex-valued chip of the spreading sequence is $p = p_1+jp_2$, then the complex multiplication of the data and spreading sequences produces a complex-valued sequence with each chip of the form $y = y_1 + jy_2 = dp$. The implementation of this product is shown in Figure 6.3, in which real-valued

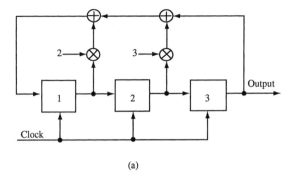

(a)

| | Contents | | |
Shift	Stage 1	Stage 2	Stage 3
Initial	0	0	1
1	1	0	0
2	2	1	0
3	3	2	1
4	1	3	2
5	1	1	3
6	0	1	1
7	0	0	1

(b)

Figure 6.2: (a) Feedback shift register for a quaternary sequence and (b) contents after successive shifts.

inputs d_1, d_2, p_1 and p_2 produce the two real-valued outputs y_1 and y_2. The equation $y = dp$ gives a compact complex-variable representation of the real variable equations:

$$y_1 = d_1 p_1 - d_2 p_2, \quad y_2 = d_2 p_1 + d_1 p_2 \qquad (6\text{-}45)$$

Each chip y_1 modulates the in-phase carrier, and each chip y_2 modulates the quadrature carrier. The transmitted signal may be represented as

$$s(t) = \mathrm{Re}\left\{ Ad(t)p(t)e^{j2\pi f_c t} \right\} \qquad (6\text{-}46)$$

where $\mathrm{Re}\{x\}$ denotes the real part of x, A is the amplitude, and $d(t)$ and $p(t)$ are waveforms modulated by the data and spreading sequences.

A representation of the receiver in terms of complex variables is illustrated in Figure 6.4. If $f_c T_c \gg 1$, two cross-correlation terms are negligible, and the actual implementation can be done by the architectures of Figures 6.17

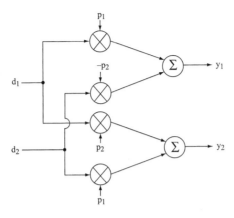

Figure 6.3: Product of quaternary data and spreading sequences.

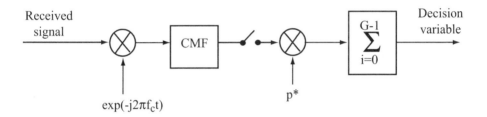

Figure 6.4: Receiver for direct-sequence system with complex quaternary spreading sequences. CMF is chip-matched filter.

and 6.19 except that the final multiplications in the two branches are replaced by a complex multiplication. Thus, y is extracted by separate in-phase and quadrature demodulation. Since the complex quaternary symbols have unity magnitude, the despreading entails the complex multiplication of y by p^* to produce $d|p|^2 = d$ along with the residual interference and noise. As illustrated in Figure 6.4, the summation of G multiplications produces the decision variable, where G is the number of chips per bit.

Although some complex-valued quaternary sequences have more favorable periodic autocorrelations and cross-correlations than pairs of standard binary sequences, they do not provide significantly smaller error probabilities in multiple-access systems [5]. The reason is that system performance is determined by the complex aperiodic functions. However, complex sequences have the potential to provide better acquisition performance than the Gold or Kasami sequences because of their superior periodic autocorrelations.

Complex-valued quaternary sequences ensure balanced power in the in-phase and quadrature branches of the transmitter, which limits the peak-to-average power fluctuations. Let $d(t) = d_1(t) + jd_2(t)$ represent a complex-valued data signal. Suppose that different bit rates or quality-of-service requirements make it desirable for $d_1(t)$ and $d_2(t)$ to have unequal amplitudes. Multiplication by a complex-valued spreading waveform $p(t) = p_1(t) + jp_2(t)$ produces $y(t) = y_1(t) + jy_2(t) = p(t)d(t)$. If the symbols of $d_1(t)$ and $d_2(t)$ are zero-mean, antipodal, and independent, and $p_1^2(t) = p_2^2(t) = p_1^2$, a constant, then $E[y_1^2(t)] = E[y_2^2(t)] = p_1^2(d_1^2 + d_2^2)$. This result indicates that the power in the in-phase and quadrature components after the spreading are equal despite any disparity between d_1^2 and d_2^2.

6.2   Systems with Random Spreading Sequences

If all the spreading sequences in a network of asynchronous CDMA systems have a common period equal to the data-symbol duration, then by the proper selection of the sequences and their relative phases, one can obtain a system performance better than that theoretically attainable with random sequences. However, the number of suitable sequences is too small for many applications, and long sequences that extend over many data symbols provide more system security. Furthermore, long sequences ensure that successive data symbols are covered by different sequences, thereby limiting the time duration of an unfavorable cross-correlation due to multiple-access interference. Even if short sequences are used, the random-sequence model gives fairly accurate performance predictions.

Direct-Sequence Systems with PSK

Consider the direct-sequence receiver of Figure 2.14 when the modulation is PSK and multiple-access interference is present. If the spreading sequence of the desired signal is modeled as a random binary sequence and the chip waveform confined to $[0, T_c)$, then the input V to the decision device is given by (2-84) and has mean value

$$E[V] = d_0 \sqrt{\frac{S}{2}}\, T_s \tag{6-47}$$

The interference component is given by (6-5), (6-6), and (6-1). Since the data modulation $d_i(t)$ in an interference signal is modeled as a random binary sequence, it can be subsumed into $q_i(t)$ given by (6-3) with no loss of generality. Since $q_i(t)$ is determined by an independent, random spreading sequence, only time delays modulo-T_c are significant and, thus, we can assume that $0 \le \tau_i < T_c$ in (6-1) without loss of generality.

Since $\psi(t)$ is confined to $[0, T_c]$ and $f_c T_c >> 1$, the substitution of (6-1) and

(6-3) into (6-6) yields

$$J_\nu = \sum_{i=1}^{K-1} \sqrt{\frac{I_i}{2}} \cos \phi_i \left\{ q_{\nu-1}^{(i)} \int_{\nu T_c}^{\nu T_c + \tau_i} \psi(t - \nu T_c) \psi[t - (\nu - 1)T_c - \tau_i] dt \right.$$

$$\left. + q_\nu^{(i)} \int_{\nu T_c + \tau_i}^{(\nu+1)T_c} \psi(t - \nu T_c) \psi(t - \nu T_c - \tau_i) dt \right\} \qquad (6\text{-}48)$$

The *partial autocorrelation* for the chip waveform is defined as

$$R_\psi(s) = \int_0^s \psi(t) \psi(t + T_c - s) dt, \quad 0 \le s < T_c \qquad (6\text{-}49)$$

Substitution into (6-48) and appropriate changes of variables in the integrals yield

$$J_\nu = \sum_{i=1}^{K-1} \sqrt{\frac{I_i}{2}} \cos \phi_i \left[q_{\nu-1}^{(i)} R_\psi(\tau_i) + q_\nu^{(i)} R_\psi(T_c - \tau_i) \right] \qquad (6\text{-}50)$$

For rectangular chips in the spreading waveform,

$$\psi(t) = \begin{cases} 1 & , \quad 0 \le t \le T_c \\ 0 & , \quad \text{otherwise.} \end{cases} \qquad (6\text{-}51)$$

Consequently,

$$R_\psi(s) = s, \quad \textit{rectangular chip} \qquad (6\text{-}52)$$

For sinusoidal chips in the spreading waveform,

$$\psi(t) = \begin{cases} \sqrt{2} \sin \left(\frac{\pi}{T_c} t \right) & , \quad 0 \le t \le T_c \\ 0 & , \quad \text{otherwise.} \end{cases} \qquad (6\text{-}53)$$

Substituting this equation into (6-49), using a trigonometric identity, and performing the integrations, we obtain

$$R_\psi(s) = \frac{T_c}{\pi} \sin \left(\frac{\pi}{T_c} s \right) - s \cos \left(\frac{\pi}{T_c} s \right), \quad \textit{sinusoidal chip} \qquad (6\text{-}54)$$

Since both J_ν and $J_{\nu+1}$ contain the same random variable $q_\nu^{(i)}$, it does not appear at first that the terms in (6-50) are statistically independent even when $\phi = (\phi_1, \phi_2, \ldots, \phi_{K-1})$ and $\tau = (\tau_1, \tau_2, \ldots, \tau_{K-1})$ are given. The following lemma [6] resolves this issue.

Lemma. Suppose that $\{\alpha_i\}$ and $\{\beta_i\}$ are statistically independent, random binary sequences. Let x and y denote arbitrary constants. Then $\alpha_i \beta_j x$ and $\alpha_i \beta_k y$ are statistically independent random variables when $j \ne k$.

Proof: Let $P(\alpha_i \beta_j x = a, \alpha_i \beta_k y = b)$ denote the joint probability that $\alpha_i \beta_j x = a$ and $\alpha_i \beta_k y = b$ where $|a| = |x|$ and $|b| = |y|$. From the theorem of total probability, it follows that

$$P(\alpha_i \beta_j x = a, \alpha_i \beta_k y = b)$$
$$= P(\alpha_i \beta_j x = a, \alpha_i \beta_k y = b, \alpha_i = 1) + P(\alpha_i \beta_j x = a, \alpha_i \beta_k y = b, \alpha_i = -1)$$
$$= P(\beta_j x = a, \beta_k y = b, \alpha_i = 1) + P(\beta_j x = -a, \beta_k y = -b, \alpha_i = -1)$$

From the independence of $\{\alpha_i\}$ and $\{\beta_j\}$ and the fact that they are random binary sequences, we obtain a simplification for $j \neq k$, $x \neq 0$, and $y \neq 0$:

$$P(\alpha_i \beta_j x = a, \alpha_i \beta_k y = b)$$

$$= P(\beta_j x = a)P(\beta_k y = b)P(\alpha_i = 1) + P(\beta_j x = -a)P(\beta_k y = -b)P(\alpha_i = -1)$$

$$= \frac{1}{2} P\left(\beta_j = \frac{a}{x}\right) P\left(\beta_k = \frac{b}{y}\right) + \frac{1}{2} P\left(\beta_j = -\frac{a}{x}\right) P\left(\beta_k = -\frac{b}{y}\right)$$

Since β_j equals $+1$ or -1 with equal probability, $P(\beta_j = a/x) = P(\beta_j = -a/x)$ and thus

$$P(\alpha_i \beta_j x = a, \alpha_i \beta_k y = b) = P\left(\beta_j = \frac{a}{x}\right) P\left(\beta_k = \frac{b}{y}\right)$$

$$= P(\beta_j x = a)P(\beta_k y = b)$$

A similar calculation gives

$$P(\alpha_i \beta_j x = a)P(\alpha_i \beta_k y = b) = P(\beta_j x = a)P(\beta_k y = b)$$

Therefore,

$$P(\alpha_i \beta_j x = a, \alpha_i \beta_k y = b) = P(\alpha_i \beta_j x = a)P(\alpha_i \beta_k y = b)$$

which satisfies the definition of statistical independence of $\alpha_i \beta_j x$ and $\alpha_i \beta_k y$. The same relation is trivial to establish for $x = 0$ or $y = 0$. $\square$

The lemma indicates that when ϕ and τ are given, the terms in (6-5) are statistically independent. Since $p_\nu^2 = 1$, the conditional variance is

$$\mathrm{var}(V_1) = \sum_{\nu=0}^{G-1} \mathrm{var}(J_\nu) \qquad (6\text{-}55)$$

The independence of the K spreading sequences, the independence of successive terms in each random binary sequence, and (6-50) imply that the conditional variance of J_ν is independent of ν and, therefore,

$$\mathrm{var}(V_1) = \sum_{i=1}^{K-1} \frac{1}{2} G I_i \cos^2 \phi_i [R_\psi^2(\tau_i) + R_\psi^2(T_c - \tau_i)] \qquad (6\text{-}56)$$

Since the terms of V_1 in (6-5) are independent, zero-mean random variables that are uniformly bounded and $\mathrm{var}(V_1) \to \infty$ as $G \to \infty$, the central limit theorem implies that $V_1/\sqrt{\mathrm{var}(V_1)}$ converges in distribution to a Gaussian random variable with mean 0 and variance 1. Thus, when ϕ and τ are given, the conditional distribution of V_1 is approximately Gaussian when G is large. Since the noise component V_2 in (2-84) has a Gaussian distribution and is independent of V_1, $V = V_1 + V_2$ has an approximate Gaussian distribution with mean given by (6-47), and $\mathrm{var}(V) = \mathrm{var}(V_1) + \mathrm{var}(V_2)$.

A straightforward derivation using the Gaussian distribution of the decision statistic V indicates that the conditional symbol error probability given ϕ and τ is

$$P_s(\phi, \tau) = Q\left(\sqrt{\frac{2\mathcal{E}_s}{N_{0e}(\phi, \tau)}}\right) \tag{6-57}$$

where $\mathcal{E}_s = ST_s$ is the *energy per symbol* in $d(t)$, and the *equivalent-noise power spectral density* is defined as

$$N_{0e}(\phi, \tau) = N_0 + \sum_{i=1}^{K-1} 2\frac{I_i}{T_c} \cos^2 \phi_i [R_\psi^2(\tau_i) + R_\psi^2(T_c - \tau_i)] \tag{6-58}$$

For a rectangular chip waveform, this equation simplifies to

$$N_{0e}(\phi, \tau) = N_0 + \sum_{i=1}^{K-1} 2I_i T_c \cos^2 \phi_i \left(1 - 2\frac{\tau_i}{T_c} + 2\frac{\tau_i^2}{T_c^2}\right) \tag{6-59}$$

Numerical evaluations [6] give strong evidence that the error in (6-57) due to the Gaussian approximation is negligible if $G \geq 50$. For an asynchronous network, it is assumed that the time delays are independent and uniformly distributed over $[0, T_c)$ and that the phase angles $\theta_i, i = 1, 2, \ldots, K - 1$, are uniformly distributed over $[0, 2\pi)$. Therefore, the symbol error probability is

$$P_s = \left(\frac{2}{\pi T_c}\right)^{K-1} \int_0^{\pi/2} \cdots \int_0^{\pi/2} \int_0^{T_c} \cdots \int_0^{T_c} P_s(\phi, \tau) d\phi \, d\tau \tag{6-60}$$

where the fact that $\cos^2 \phi_i$ takes all its possible values over $[0, \pi/2]$ has been used to shorten the integration intervals. The absence of sequence parameters ensures that the amount of computation required for (6-60) is much less than the amount required to compute P_s when the spreading sequence is short. Nevertheless, the computational requirements are large enough that it is highly desirable to find an accurate approximation that entails less computation. The conditional symbol error probability given ϕ is defined as

$$P_s(\phi) = \left(\frac{1}{T_c}\right)^{K-1} \int_0^{T_c} \cdots \int_0^{T_c} P_s(\phi, \tau) d\tau \tag{6-61}$$

A closed-form approximation to $P_s(\phi)$ greatly simplifies the computation of P_s, which reduces to

$$P_s = \left(\frac{2}{\pi}\right)^{K-1} \int_0^{\pi/2} \cdots \int_0^{\pi/2} P_s(\phi) d\phi \tag{6-62}$$

To approximate $P_s(\phi)$, we first obtain upper and lower bounds on it.

For either rectangular or sinusoidal chip waveforms, elementary calculus establishes that

$$R_\psi^2(\tau_i) + R_\psi^2(T_c - \tau_i) \leq T_c^2 \tag{6-63}$$

Using this upper bound successively in (6-58), (6-57), and (6-61), and performing the trivial integrations that result, we obtain

$$P_s(\phi) \le Q\left(\sqrt{\frac{2\mathcal{E}_s}{N_{0u}(\phi)}}\right) \tag{6-64}$$

where

$$N_{0u}(\phi) = N_0 + \sum_{i=1}^{K-1} 2I_i T_c \cos^2 \phi_i \tag{6-65}$$

To apply Jensen's inequality (2-144), the successive integrals in (6-60) are interpreted as the evaluation of expected values. Consider the random variable

$$X = R_\psi^2(\tau_i) + R_\psi^2(T_c - \tau_i) \tag{6-66}$$

Since τ_i is uniformly distributed over $[0, T_c)$, straightforward calculations using (6-52) and (6-54) give

$$E[X] = \frac{1}{T_c}\int_0^{T_c}[R_\psi^2(\tau_i) + R_\psi^2(T_c - \tau_i)]d\tau_i = hT_c^2 \tag{6-67}$$

where

$$h = \begin{cases} \dfrac{2}{3}, & \text{rectangular chip} \\ \dfrac{1}{3} + \dfrac{5}{2\pi^2}, & \text{sinusoidal chip} \end{cases} \tag{6-68}$$

The function (6-57) has the form given by (2-145). Equations (6-58), (6-63), and $\cos^2 \phi_i \le 1$ yield a sufficient condition for convexity:

$$\mathcal{E}_s \ge \frac{3}{2}\left[N_0 + \sum_{i=1}^{K-1} 2I_i T_c\right] \tag{6-69}$$

Application of Jensen's inequality successively to each component of τ in (6-61) yields

$$P_s(\phi) \ge Q\left(\sqrt{\frac{2\mathcal{E}_s}{N_{0l}(\phi)}}\right) \tag{6-70}$$

where

$$N_{0l}(\phi) = N_0 + \sum_{i=1}^{K-1} 2hI_i T_c \cos^2 \phi_i \tag{6-71}$$

If N_0 is negligible, then (6-71) and (6-65) give $N_{0l}/N_{0u} = h$. Thus, a good approximation is provided by

$$P_s(\phi) \approx Q\left(\sqrt{\frac{2\mathcal{E}_s}{N_{0a}(\phi)}}\right) \tag{6-72}$$

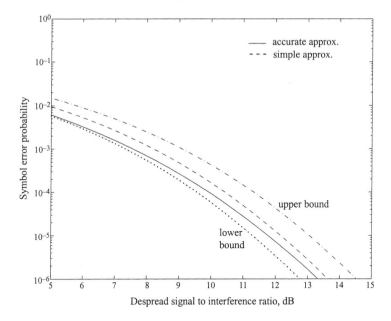

Figure 6.5: Symbol error probability of direct-sequence system with PSK in presence of single multiple-access interference signal and $\mathcal{E}_s/N_0 = 15$ dB.

where

$$N_{0a}(\phi) = N_0 + \sum_{i=1}^{K-1} 2\sqrt{h} I_i T_c \cos^2 \phi_i \qquad (6\text{-}73)$$

If N_0 is negligible, then $N_{0u}/N_{0a} = N_{0a}/N_{0l} = 1/\sqrt{h}$. Therefore, in terms of the value of $\mathcal{E}_s$ needed to ensure a given $P_s(\phi)$, the error in using approximation (6-72) instead of (6-61) is bounded by $10 \log_{10}(1/\sqrt{h})$ in decibels, which equals 0.88 dB for rectangular chip waveforms and 1.16 dB for sinusoidal chip waveforms. In practice, the error is expected to be only a few tenths of a decibel because $N_0 \neq 0$ and P_s coincides with neither the upper nor the lower bound.

As an example, suppose that rectangular chip waveforms are used, $\mathcal{E}_s/N_0 = 15$ dB, and $K = 2$. Figure 6.5 illustrates four different evaluations of P_s as a function of $G\mathcal{E}_s/IT_s = GS/I$, the *despread signal-to-interference ratio*, which is the signal-to-interference ratio after taking into account the beneficial results from the despreading in the receiver. The accurate approximation is computed from (6-57) and (6-60), the upper bound from (6-64) and (6-62), the lower bound from (6-70) and (6-62), and the simple approximation from (6-72) and (6-62). The figure shows that the accurate approximation moves from the lower bound toward the simple approximation as the symbol error probability decreases. For $P_s = 10^{-5}$, the simple approximation is less than 0.3 dB in error relative to the accurate approximation.

Figure 6.6 compares the symbol error probabilities for $K = 2$ to $K = 4$,

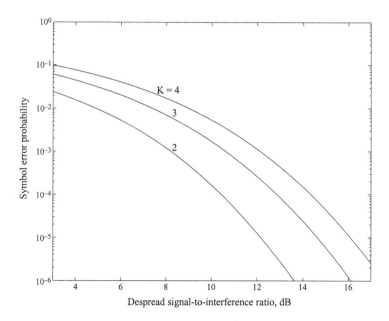

Figure 6.6: Symbol error probability of direct-sequence system with PSK in presence of $K - 1$ equal-power multiple-access interference signals and $\mathcal{E}_s/N_0 = 15$ dB.

rectangular chip waveforms and $\mathcal{E}_s/N_0 = 15$ dB. The simple approximation is used for P_s, and the abscissa shows GS/I where I is the interference power of each equal-power interfering signal. The figure shows that P_s increases with K, but the shift in P_s is mitigated somewhat because the interference signals tend to partially cancel each other.

The preceding bounding methods can be extended to the bounds on $P_s(\boldsymbol{\phi})$ by observing that $\cos^2 \phi_i \leq 1$ and setting $X = \cos^2 \phi_i$ during the successive applications of Jensen's inequality, which is applicable if (6-69) is satisfied. After evaluating (6-65), we obtain

$$Q\left(\sqrt{\frac{2\mathcal{E}_s}{N_0 + hI_tT_c}}\right) \leq P_s \leq Q\left(\sqrt{\frac{2\mathcal{E}_s}{N_0 + 2I_tT_c}}\right) \qquad (6\text{-}74)$$

where

$$I_t = \sum_{i=1}^{K-1} I_i \qquad (6\text{-}75)$$

A simple approximation is provided by

$$P_s \approx Q\left(\sqrt{\frac{2\mathcal{E}_s}{N_0 + \sqrt{2h}\, I_tT_c}}\right) \qquad (6\text{-}76)$$

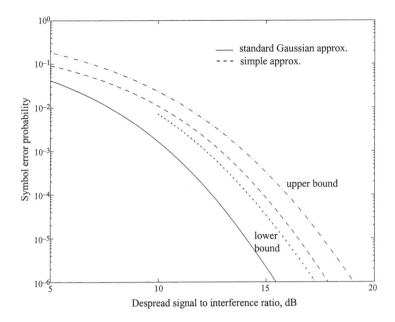

Figure 6.7: Symbol error probability of direct-sequence system with PSK in presence of 3 equal-power multiple-access interference signals and $\mathcal{E}_s/N_0 = 15$ dB.

If P_s is specified, then the error in the required $\mathcal{E}_s/I_t$ caused by using (6-76) instead of (6-60) is bounded by $10\log_{10}\sqrt{2/h}$ in decibels. Thus, the error is bounded by 2.39 dB for rectangular chip waveforms and 2.66 dB for sinusoidal ones.

The lower bound in (6-74) gives the same result as that often called the *standard Gaussian approximation*, in which V_1 in (6-5) is assumed to be approximately Gaussian, each ϕ_i in (6-50) is assumed to be uniformly distributed over $[0, 2\pi)$, and each τ_i is assumed to be uniformly distributed over $[0, T_c)$. This approximation, gives an optimistic result for P_s that can be as much as 4.77 dB in error for rectangular chip waveforms according to (6-74). The substantial improvement in accuracy provided by (6-72) or (6-57) is due to the application of the Gaussian approximation only after conditioning V_1 on given values of ϕ and τ. The accurate approximation given by (6-57) is a version of what is often called the *improved Gaussian approximation*.

Figure 6.7 illustrates the symbol error probability for 3 interferers, each with power I, rectangular chip waveforms, and $\mathcal{E}_s/N_0 = 15$ dB as a function of GS/I. The graphs show the standard Gaussian approximation of (6-74), the simple approximation of (6-76), and the upper and lower bounds given by (6-64), (6-70), and (6-62). The large error in the standard Gaussian approximation is evident. The simple approximation is reasonably accurate if $10^{-6} \leq P_s \leq 10^{-2}$.

For *synchronous* networks, (6-57) and (6-58) can be simplified because the

$\{\tau_i\}$ are all zero. For either rectangular or sinusoidal chip waveforms, we obtain

$$P_s(\phi) = Q\left(\sqrt{\frac{2\mathcal{E}_s}{N_{0e}(\phi)}}\right) \qquad (6\text{-}77)$$

where

$$N_{0e}(\phi) = N_0 + \sum_{i=1}^{K-1} 2I_i T_c \cos^2 \phi_i \qquad (6\text{-}78)$$

A comparison with (6-64) and (6-65) indicates that P_s for a synchronous network equals or exceeds P_s for a similar asynchronous network when random spreading sequences are used. This phenomenon is due to the increased bandwidth of a despread asynchronous interference signal, which allows increased filtering in the receiver.

The accurate approximation of (6-57) follows from the standard central limit theorem, which is justified by the lemma. This lemma depends on the restriction of the chip waveform to the interval $[0, T_c]$. If the chip waveform extends beyond this interval but is time-limited, as is necessary for implementation with digital hardware, then an extension of the central limit theorem for m-dependent sequences can be used to derive an improved Gaussian approximation [7]. Alternatives to the analysis in this section and the next one abound in the literature, but they are not as amenable to comparisons among systems.

Quadriphase Direct-Sequence Systems

Consider a network of quadriphase direct-sequence systems, each of which uses dual QPSK and random spreading sequences. Each direct-sequence signal is given by (2-123) with $t_0 = 0$. The multiple-access interference is

$$i(t) = \sum_{i=1}^{K-1} [\sqrt{I}\, q_{1i}(t-\tau_i)\cos(2\pi f_c t+\phi_i) + \sqrt{I}\, q_{2i}(t-\tau_i)\sin(2\pi f_c t+\phi_i)] \quad (6\text{-}79)$$

where $q_{1i}(t)$ and $q_{2i}(t)$ both have the form of (6-3) and incorporate the data modulation. The decision variables are given by (2-124) and (2-126) with $G = T_s/T_c$. A straightforward calculation using (6-6) indicates that

$$\begin{aligned} J_\nu = \sum_{i=1}^{K-1} \sqrt{I_i}\{\cos\phi_i [q_{\nu-1}^{(1i)} R_\psi(\tau_i) + q_\nu^{(1i)} R_\psi(T_c - \tau_i)] \\ - \sin\phi_i [q_{\nu-1}^{(2i)} R_\psi(\tau_i) + q_\nu^{(2i)} R_\psi(T_c - \tau_i)]\} \end{aligned} \qquad (6\text{-}80)$$

The statistical independence of the two sequences, the preceding lemma, and analogous results for J_ν' in (2-127) yield the variances of the interference terms of the decision variables:

$$\mathrm{var}(V_1) = \mathrm{var}(U_1) = \sum_{i=1}^{K-1} \frac{1}{2}\frac{T_s}{T_c} I_i [R_\psi^2(\tau_i) + R_\psi^2(T_c - \tau_i)] \qquad (6\text{-}81)$$

The noise variances and the means are given by (2-130) and (2-129). Since all variances and means are independent of ϕ, the Gaussian approximation yields a $P_s(\phi, \tau)$ that is independent of ϕ:

$$P_s = \left(\frac{1}{T_c}\right)^{K-1} \int_0^{T_c} \cdots \int_0^{T_c} Q\left(\sqrt{\frac{2\mathcal{E}_s}{N_{0e}(\tau)}}\right) d\tau \qquad (6\text{-}82)$$

where

$$N_{0e}(\tau) = N_0 + \sum_{i=1}^{K-1} \frac{I_i}{T_c}[R_\psi^2(\tau_i) + R_\psi^2(T_c - \tau_i)] \qquad (6\text{-}83)$$

Since a similar analysis for direct-sequence systems with balanced QPSK yields (6-83) again, *both quadriphase systems perform equally well against multiple-access interference.*

Application of the previous bounding and approximation methods to (10-79) yields

$$Q\left(\sqrt{\frac{2\mathcal{E}_s}{N_0 + hI_tT_c}}\right) \leq P_s \leq Q\left(\sqrt{\frac{2\mathcal{E}_s}{N_0 + I_tT_c}}\right) \qquad (6\text{-}84)$$

where the total interference power I_t is defined by (6-75). A sufficient condition for the validity of the lower bound is

$$\mathcal{E}_s \geq \frac{3}{2}(N_0 + I_tT_c) \qquad (6\text{-}85)$$

A simple approximation that limits the error in the required $\mathcal{E}_s/I_t$ for a specified P_s to $10\log_{10}(1/\sqrt{h})$ is

$$P_s \approx Q\left(\sqrt{\frac{2\mathcal{E}_s}{N_0 + \sqrt{h}I_tT_c}}\right) \qquad (6\text{-}86)$$

This approximation introduces errors bounded by 0.88 dB and 1.16 dB for rectangular and sinusoidal chip waveforms, respectively. In (6-84) and (6-86), only the total interference power is relevant, not how it is distributed among the individual interference signals.

Figure 6.8 illustrates P_s for a quadriphase direct-sequence system in the presence of 3 interferers, each with power I, rectangular chip waveforms, and $\mathcal{E}_s/N_0 = 15$ dB. The graphs represent the accurate approximation of (6-82), the simple approximation of (6-86), and the bounds of (6-84) as functions of GS/I. A comparison of Figures 6.8 and 6.7 indicates the advantage of a quadriphase system.

For *synchronous networks* with either rectangular or sinusoidal chip waveforms, we set the $\{\tau_i\}$ equal to zero in (6-82) and obtain

$$P_s = Q\left(\sqrt{\frac{2\mathcal{E}_s}{N_0 + I_tT_c}}\right) \qquad (6\text{-}87)$$

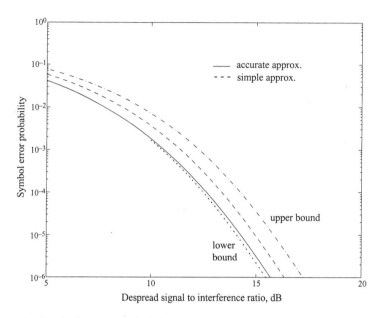

Figure 6.8: Symbol error probability of quadriphase direct-sequence system in presence of 3 equal-power multiple-access interference signals and $\mathcal{E}_s/N_0 = 15$ dB.

Since this equation coincides with the upper bound in (6-84), we conclude that asynchronous networks accommodate more multiple-access interference than similar synchronous networks using quadriphase direct-sequence signals with random spreading sequences. To compare asynchronous quadriphase direct-sequence systems with asynchronous systems using binary PSK, we find a lower bound on P_s for direct-sequence systems with PSK. Substituting (6-57) into (6-60) and applying Jensen's inequality successively to the integrations over $\phi_i, i = 1, 2, \ldots, K - 1$, we find that a lower bound on P_s is given by the right-hand side of (6-82) if (6-85) is satisfied. This result implies that asynchronous quadriphase direct-sequence systems are more resistant to multiple-access interference than asynchronous direct-sequence systems with binary PSK.

The equations for P_s allow the evaluation of the information-bit error probability P_b for error-correcting codes with hard-decision decoders. To facilitate the analysis of soft-decision decoding, two assumptions are necessary. Assume that K is large enough that the multiple-access interference after despreading is approximately Guassian rather than conditionally Gaussian. Since the equivalent noise is a zero-mean process, the equivalent-noise power spectral density N_{0e} can be obtained by averaging $N_{0e}(\phi, \tau)$ over the distributions of ϕ and τ. For asynchronous communications, (6-83) and (6-87) yield

$$N_{0e} = N_0 + h I_t T_c \tag{6-88}$$

This equation is also valid for synchronous communications if we set $h = 1$.

Thus, for a binary convolutional code with rate r, constraint length K, and minimum free distance d_f, P_b is upper-bounded by (1-112) with

$$P_2(l) = Q\left(\sqrt{\frac{2\mathcal{E}_s l}{N_{0e}}}\right) = Q\left(\sqrt{\frac{2r\mathcal{E}_b l}{N_0 + hI_t T_c}}\right) \tag{6-89}$$

The *network capacity* is the number of equal-power users in a network of identical systems that can be accommodated while achieving a specified P_b. For equal-power users, $I_t = (K-1)\mathcal{E}_s/T_s$. Let γ_1 denote the value of $\mathcal{E}_s/N_{0e}$ necessary for a specific error-control code to achieve the specified P_b. Equation (6-88) implies that the network capacity is

$$K = \left\lfloor 1 + \frac{G}{h}\left(\frac{1}{\gamma_1} - \frac{1}{\gamma_0}\right)\right\rfloor , \quad \gamma_0 \geq \gamma_1 \tag{6-90}$$

where $\lfloor x \rfloor$ is the integer part of x, $\gamma_0 = \mathcal{E}_s/N_0$, $G = T_s/T_c$ is the processing gain, and the requirement $\gamma_0 \geq \gamma_1$ is necessary to ensure that the specified P_b can be achieved for some value of K. Since $h < 1$ in general, the factor G/h reflects the increased gain due to the random distributions of interference phases and delays. If they are not random but $\phi = \tau = 0$, then $h = 1$ and the number of users accommodated is reduced. Thus, synchronous CDMA systems require orthogonal spreading sequences.

As an example, consider a network with systems that resemble those used for the synchronous downlinks of an IS-95 CDMA network. We assume the absence of fading and calculate the network capacity for power-controlled users within a single cell. The data modulation is balanced QPSK. $G = 64$, and $h = 1$. The error-control code is a rate-1/2 binary convolutional code with constraint length 9. If $P_b = 10^{-5}$ or better is desired, the performance curve of Figure 1.8 for the convolutional code indicates that $\mathcal{E}_b/N_{0e} \approx 3.5$ dB and thus $\gamma_1 \approx 0.5$ dB is required. Equation (6-90) then indicates that the network capacity is $K = 51$ if $\gamma_0 = 10$ dB and $K = 57$ if $\gamma_0 = 20$ dB.

6.3 Wideband Direct-Sequence Systems

A direct-sequence system is called *wideband* if it uses a spectral band with a bandwidth that exceeds the coherence bandwidth of a frequency-selective fading channel. The two most commonly proposed types of wideband direct-sequence systems are single-carrier and multicarrier systems. A *single-carrier system* uses a single carrier frequency to transmit signals. A *multicarrier system* partitions the available spectral band among multiple direct-sequence signals, each of which has a distinct carrier frequency. The main attractions of the multi-carrier system are its potential ability to operate over disjoint, noncontiguous spectral regions and its ability to avoid transmissions in spectral regions with strong interference or where the multicarrier signal might interfere with other signals. These features have counterparts in frequency-hopping systems.

A single-carrier system provides diversity by using a rake receiver that combines several multipath signals. A multicarrier system provides diversity by the maximal-ratio combining of the parallel correlator outputs, each of which is associated with a different carrier. Bit error probabilities are determined subsequently for ideal multicarrier and single-carrier systems with lossless diversity combining in the presence of white Gaussian noise and Rayleigh fading.

Multicarrier Direct-Sequence System

A typical multicarrier system divides a spectral band of bandwidth W into L frequency channels or *subchannels*, each of bandwidth W/L. The carrier associated with a subchannel is called a *subcarrier*. In one type of system, which is diagrammed in Figure 6.9, this bandwidth is approximately equal to the coherence bandwidth because a larger one would allow frequency-selective fading in each subchannel, while a smaller one would allow correlated fading among the subcarriers [8], [9]. It is assumed that the spacing between adjacent subcarriers is β/T_c, where $\beta \geq 1$. Equation (5-57) indicates that the coherence bandwidth is approximately $1/T_d$, where T_d is the delay spread. Thus, $T_c \leq \beta T_d$ is required to ensure that each subcarrier signal is subject to independent fading. If the bandwidth of a subcarrier signal is on the order of $1/T_c$, then $T_c \geq T_d$ is required for the subcarrier signals to experience no significant frequency selectivity. The two preceding inequalities imply that $1 \leq \beta \leq 2$ is required. If the chip waveforms are rectangular and $\beta = 1$ or $\beta = 2$, then the subcarrier frequencies are orthogonal, which can be verified by a calculation similar to that leading to (3-59). Although the orthogonality prevents *self-interference* among the subcarrier signals, its effectiveness is reduced by multipath components and Doppler shifts. One may use bandlimited subcarrier signals to minimize the self-interference without requiring orthogonality. If $\beta = 2$ and the chip waveforms are rectangular, then the spectral mainlobes of the subcarrier signals have no overlap. Furthermore, a spacing of $2/T_c$ limits the significant multiple-access interference in a subchannel to subcarrier signals from other users that have the same subcarrier frequency.

In the transmitter, the product $d(t)p(t)$ of the data modulation $d(t)$ and the spreading waveform $p(t)$ simultaneously modulates L subcarriers, each of which has its frequency in the center of one of the L spectral regions, as illustrated in Figure 6.9(a). The receiver has L parallel demodulators, one for each subcarrier, the outputs of which are suitably combined, as indicated in Figure 6.9(b). The total signal power is divided equally among the L subcarriers. The chip rate and, hence, the processing gain for each subcarrier of a multicarrier direct-sequence system is reduced by the factor L. However, if strong interference exists in a subchannel, the gain used in maximal-ratio combining is small. Alternatively, the associated subcarrier can be omitted and the saved power redistributed among the remaining subcarriers. Error-control codes and interleaving can be used to provide both time diversity and coding gain. Since the spectral regions are defined so that the fading in each of them is independent and frequency nonselective, rake combining is not possible, but the frequency diversity provided by the regions can be exploited in a diversity combiner. Whether or not the diversity gain exceeds that of a single-carrier system using the entire

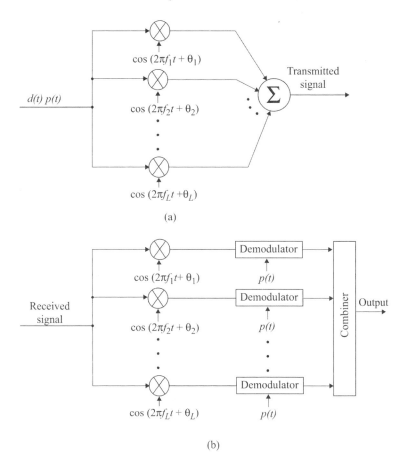

Figure 6.9: Multicarrier direct-sequence system: (a) transmitter and (b) receiver.

spectral band and rake combining depends on the multipath intensity profile of the single-carrier system.

Consider a multicarrier system that uses binary PSK to modulate each subcarrier. Each received signal copy with a different subcarrier frequency experiences independent Rayleigh fading that is constant during a symbol interval. The received signal for a symbol in branch i is

$$r_i(t) = \mathrm{Re}\left[\alpha_i e^{j\theta_i} x\psi(t) e^{j2\pi f_i t}\right] + n_i(t), \quad 0 \le t \le T_s, \quad i = 1, 2, \ldots, L \quad (6\text{-}91)$$

where $x = +1$ or -1 depending on the transmitted symbol, each α_i is a fading amplitude, each θ_i is a phase shift, f_i is the subcarrier frequency, T_s is the symbol duration, and $n_i(t)$ is the noise. Assume that the received interference plus noise in each diversity branch can be modeled as independent, zero-mean, white Gaussian noise with the same equivalent two-sided power spectral density $N_{0e}/2$.

Ideal lossless power splitting among the L subcarriers is assumed. Let $\mathcal{E} = \mathcal{E}_s/L$ denote the received symbol energy per subcarrier in the absence of fading, where $\mathcal{E}_s$ is the total received energy per symbol. Assume that the spectral division among the subcarriers prevents significant interference among them in the receiver. For coherent detection and maximal-ratio combining, the analysis of Section 5.4 is directly applicable. The conditional bit or symbol error probability given the $\{\alpha_i\}$ is

$$P_{s|\alpha}(\gamma_s) = Q\left(\sqrt{2\gamma_s}\right) \tag{6-92}$$

where

$$\gamma_s = \frac{\mathcal{E}}{N_{0e}} \sum_{i=1}^{L} \alpha_i^2 \tag{6-93}$$

The symbol error probability is determined by averaging $P_{s|\alpha}(\gamma_s)$ over the distribution of γ_s, which depends on the $\{\alpha_i\}$ and embodies the statistics of the fading channel. If each of the $\{\alpha_i\}$ is independent with the identical Rayleigh distribution and $E[\alpha_i^2] = E[\alpha^2]$, then the average signal-to-noise ratio (SNR) per branch is

$$\bar{\gamma} = \frac{\mathcal{E}}{N_{0e}}E[\alpha^2] = \frac{\mathcal{E}_s E[\alpha^2]}{L N_{0e}} \tag{6-94}$$

As shown in Section 5.4, the symbol error probability for a single subcarrier is

$$p = P_s(1) = \frac{1}{2}\left(1 - \sqrt{\frac{\bar{\gamma}}{1 + \bar{\gamma}}}\right) \qquad \text{(PSK, QPSK)} \tag{6-95}$$

The symbol error probability for L subcarriers is

$$P_s(L) = p - (1 - 2p)\sum_{i=1}^{L-1}\binom{2i-1}{i}[p(1-p)]^i \tag{6-96}$$

This expression explicitly shows the change in the symbol error probability as the number of diversity branches increases; it is valid for QPSK because the latter can be transmitted as two independent binary PSK waveforms in phase quadrature.

Figure 6.10 plots $P_s(L)$ for multicarrier systems as a function of $\mathcal{E}_s\overline{\alpha^2}/N_{0e}$, the average symbol SNR. The diminishing returns as the diversity level L increases is apparent. If the required bit error probability is 10^{-6} or more, than increasing L beyond $L = 32$ is not likely to be useful because of the hardware requirements and the losses entailed in the power division in the transmitter.

To evaluate N_{0e} for a network of K multicarrier direct-sequence systems, we assume that the mutual interference among the L subcarriers of a single signal is negligible and that K is large enough that the multiple-access interference after despreading is approximately Gaussian. It is assumed that only subcarriers at the same frequency cause significant interference in a subchannel. For QPSK

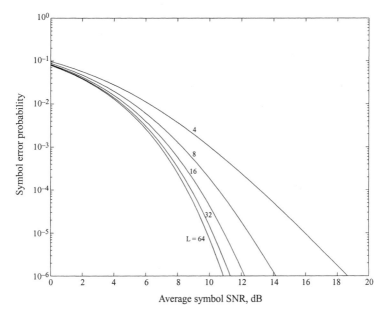

Figure 6.10: Symbol error probability for multicarrier systems with L carriers.

modulation, the power division among the subcarrier signals implies that (6-88) must be replaced by

$$N_{0e} = N_0 + h\frac{I_t}{L}T_c \qquad (6\text{-}97)$$

where h is given by (6-68) for asynchronous communications and $h = 1$ for synchronous communications, and T_c is the chip duration in each branch or subchannel of the multicarrier system. The division by L is due to the equipartition of the interference power among the L subcarriers. Let $G_0 = LG = LT_s/T_c$ denote the *overall processing gain* of the system. For equal-power users subject to the same fading statistics, $I_t = (K-1)\mathcal{E}_s\overline{\alpha^2}/T_s$ so (6-97) implies that the network capacity is

$$K = \left[1 + \frac{G_0}{h}\left(\frac{1}{\overline{\gamma}_1} - \frac{1}{\overline{\gamma}_0}\right)\right] \ , \quad \overline{\gamma}_0 \geq \overline{\gamma}_1 \qquad (6\text{-}98)$$

where $\overline{\gamma}_0 = \mathcal{E}_s\overline{\alpha^2}/N_0$ and $\overline{\gamma}_1$ is the required $\mathcal{E}_s\overline{\alpha^2}/N_{0e}$ necessary for a specific error-control code to achieve the specified P_b.

Single-Carrier Direct-Sequence System

Consider a direct-sequence signal that has a random spreading sequence and is accompanied by multipath components in addition to the direct-path signal. If the multipath components are delayed by more than one chip, then the independence of the chips ensures that the multipath interference is suppressed by at least the processing gain. However, since multipath signals carry information, they are a potential resource to be exploited rather than merely rejected. A *rake receiver* (Section 5.5) provides *path diversity* by coherently combining the resolvable multipath components present during frequency-selective fading, which occurs when the chip rate of the spreading sequence exceeds the coherence bandwidth.

Consider a multipath channel with frequency-selective fading slow enough that its time variations are negligible over a signaling interval. When the data modulation is binary PSK, only a single symbol waveform and its associated decision variable are needed. Assume the presence of zero-mean, white Gaussian noise with two-sided power spectral density $N_{0e}/2$. As indicated in Section 5.5, if $\alpha_i = |c_i|$, then for a rake receiver with perfect tap weights, the conditional bit or symbol error probability given the $\{\alpha_i\}$ is provided by (6-92). However, for a rake receiver, each of the $\{\alpha_i\}$ is associated with a different multipath component, and hence each $E[\alpha_i^2]$ has a different value in general. Since there is only a single carrier, we may set $\mathcal{E} = \mathcal{E}_s$ in (6-93), which may be expressed as

$$\gamma_s = \sum_{i=1}^{L} \gamma_i, \quad \gamma_i = \frac{\mathcal{E}_s}{N_{0e}} \alpha_i^2 \tag{6-99}$$

The average SNR for a symbol in branch i is

$$\bar{\gamma}_i = \frac{\mathcal{E}_s}{N_{0e}} E\left[\alpha_i^2\right], \quad i = 1, 2, \ldots, L \tag{6-100}$$

If each multipath component experiences independent Rayleigh fading so that each of the $\{\gamma_i\}$ is statistically independent, then the analysis of Section 5.5 gives the symbol error probability:

$$P_s(L) = \frac{1}{2} \sum_{i=1}^{L} A_i \left(1 - \sqrt{\frac{\bar{\gamma}_i}{1 + \bar{\gamma}_i}} \right) \tag{6-101}$$

where

$$A_i = \begin{cases} \prod_{\substack{k=1 \\ k \neq i}}^{L} \frac{\bar{\gamma}_i}{\bar{\gamma}_i - \bar{\gamma}_k}, & L \geq 2 \\ 1, & L = 1 \end{cases} \tag{6-102}$$

Since only white Gaussian noise is present, the processing gain of the system is irrelevant under this model.

The processing of a multipath component requires channel estimation. When a practical channel estimator is used, measurements indicate that only four or fewer components are likely to have a sufficient signal-to-interference ratio to be useful in the rake combining [10]. To assess the potential performance of the rake receiver, it is assumed that the largest multipath component has $\bar{\gamma}_1 = \bar{\gamma}$ and that $L \leq 4$ components are received and processed. The other three or fewer minor multipath components have relative average symbol SNRs specified by the *multipath intensity vector*

$$\Gamma = \left(\frac{\bar{\gamma}_2}{\bar{\gamma}_1}, \frac{\bar{\gamma}_3}{\bar{\gamma}_1}, \frac{\bar{\gamma}_4}{\bar{\gamma}_1} \right) \tag{6-103}$$

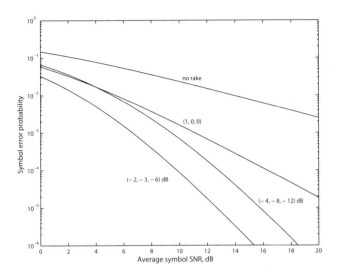

Figure 6.11: Symbol error probability for single-carrier systems and $L \leq 4$ multipath components with different multipath intensity vectors.

Figure 6.11 plots the symbol error probability as a function of $\bar{\gamma}_1 = \mathcal{E}_s \overline{a_1^2}/N_{0e}$, the average symbol SNR of the main component, for multipath intensity vectors occurring in mobile CDMA networks. Typically, three significant multipath components are available. Expressing the components in decibels, the multipath intensity vector $(-4, -8, -12)$ dB represents the minor multipath intensities typical of a rural environment. The vector $(-2, -3, -6)$ dB represents a typical urban environment. This figure and other numerical data establish two basic features of single-carrier systems with rake receivers.

1. System performance improves as the total energy in the minor multipath components increases.

2. When the total energy in the minor multipath components is fixed, the system performance improves as the number of resolved multipath components L increases and as the energy becomes uniformly distributed among these components.

For QPSK modulation and multiple-access interference, N_{0e} is given by (6-88). It follows that the system capacity is given by (6-90), where $G_0 = T_s/T_c$, $\bar{\gamma}_0 = \mathcal{E}_s \overline{\alpha_1^2}/N_0$, and $\bar{\gamma}_1$ is the required $\mathcal{E}_s \overline{\alpha_1^2}/N_{0e}$ necessary for a specific error-control code to achieve the specified P_b.

A comparison of Figures 6.10 and 6.11 indicates that a multicarrier system with diversity $L = 32$ outperforms single-carrier systems with diversity $L = 4$ if $\bar{\gamma}_1$ is sufficiently large. However, this value of $\bar{\gamma}_1$ is much larger than is required in practical systems. To make a more realistic comparison, we assume that an error-correcting code with ideal channel-symbol interleaving is used. For a loosely packed, binary block code and hard-decision decoding with a bounded-distance decoder, the information-bit error probability is (Chapter 1)

$$P_b \approx \sum_{i=t+1}^{n} \binom{n-1}{i-1} P_s^i (1 - P_s)^{n-1} \tag{6-104}$$

where n is the code length, t is the number of symbol errors that the decoder can correct, and P_s is the channel-symbol error probability. The signal energy per channel symbol is $\mathcal{E}_s = r\mathcal{E}_b$, where $r = k/n$ is the code rate, k is the number of information bits per codeword, and $\mathcal{E}_b$ is the energy per information bit. We may evaluate P_s by using the expressions for $P_s(L)$ with $\bar{\gamma}_1 = r\bar{\gamma}_b$, where $\bar{\gamma}_b = \mathcal{E}_b \overline{\alpha_1^2}/N_0$ is the average bit SNR.

As an example, we assume that a BCH (63, 36) code with $n = 63$, $k = 36$, and $t = 5$ is used. Figure 6.12 plots P_b for a multicarrier system with $L = 32$ and single-carrier systems with $\Gamma_1 = (-4, -8, -12)$ dB and $\Gamma_2 = (-2, -3, -6)$ dB, which are typical for rural and urban environments, respectively. If $P_b = 10^{-5}$ is required, then the multicarrier system is slightly advantageous in a rural environment, but rake combining provides a roughly 1.9 dB advantage in an urban environment characterized by Γ_2. For the multicarrier system, $\bar{\gamma}_b \approx 6.7$ dB and, hence, $\bar{\gamma}_1 \approx 4.27$ dB are required. Suppose that $G_0 = G = 64$ and $\bar{\gamma}_0 = 20$ dB. The chip waveform is rectangular so $h = 2/3$. Then (6-98) indicates that the network capacity is 35. For an urban single-carrier system, $\bar{\gamma}_b \approx 4.8$ dB and, hence, $\bar{\gamma}_1 \approx 2.37$ dB are required. Then (6-90) indicates that the network capacity is 55, which illustrates the potential power of ideal rake combining to overcome the detrimental effects of fading. A more powerful code, such as a concatenated or turbo code would give rake combining a performance advantage even in a rural environment.

The preceding results imply that in a benign environment, devoid of partial-band interference, a multicarrier system suffers a potential performance loss relative to the less costly single-carrier system. The underlying reason is that the rake receiver of the single-carrier system harnesses energy that would otherwise be unavailable. In contrast, the multicarrier receiver recovers energy that has been redistributed among the L carriers but is available to the single-carrier system even without rake combining. Despite its potential disadvantage in a benign urban environment, a multicarrier system will often be preferable to a single-carrier system because of its substantially superior performance against partial-band interference [8], [9].

Multicarrier DS/CDMA System

Various multicarrier direct-sequence systems that accommodate multiple-access interference have been proposed [11] for CDMA networks. The *multicarrier DS/CDMA system* is a candidate for both the uplinks and downlinks of fourth-generation cellular CDMA networks. One version of its transmitter is shown in

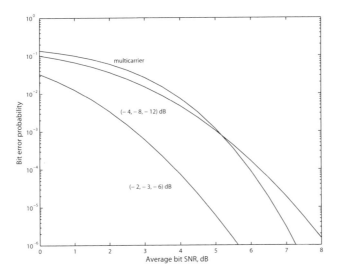

Figure 6.12: Information-bit error probability for multicarrier system with $L = 32$ and for single-carrier systems with typical rural and urban multipath intensity vectors. Error-control code is BCH (63, 36).

Figure 6.13. This system uses a serial-to-parallel converter to convert a stream of data symbols into multiple parallel substreams. Thus, the multicarrier modulation reduces the data-symbol rate and, hence, the multipath interference of the direct-sequence signal in each subchannel. The receiver is similar in form to that of Figure 6.9(b) except that the combiner is replaced by a parallel-to-serial converter. If the subcarriers are separated by $2/T_c$, then the interchannel interference and multiple-access interference from subcarrier signals are minimized. The efficient processing of orthogonal frequency-division multiplexing (OFDM) may be implemented by sampling each subchannel signal after the spreading by $p(t)$ and then applying the set of L samples in parallel to an OFDM processor [12]. The cost of this efficiency is a high peak-to-average power ratio for the transmitted signal. In contrast to the system of Figure 6.12, the multicarrier DS/CDMA system of Figure 6.13 cannot exploit frequency diversity because each subcarrier is modulated by a different data symbol. However, the processing gain of each subchannel signal is increased by the factor L, which can be exploited in the suppression of multiple-access interference. Rake combining might be possible in the subchannels if $T_c \leq 2T_d$. For synchronous communications, such as those transmitted by a base station in a cellular network, the spreading sequences of the network users may be drawn from a set of orthogonal Walsh sequences. For asynchronous communications, Gold or Kasami sequences are preferable because of their superior cross-correlation characteristics.

Another multicarrier direct-sequence system applies the spread signal $d(t)p(t)$

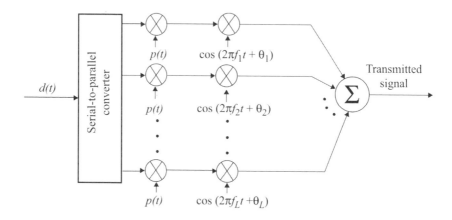

Figure 6.13: Multicarrier DS/CDMA transmitter.

to a serial-to-parallel converter, which produces G parallel data-modulated chips, where G is the number of chips per data symbol. Each of these G chips modulates a different subcarrier. Thus, the spreading occurs in the frequency domain. This system provides the same degree of diversity gain as the system of Figure 6.9, but the latter is less expensive if $L < G$ and provides nearly the same performance if $L \geq 32$.

Frequency hopping may be added to almost any communication system to strengthen it against interference or fading. Thus, the set of carriers used in a multicarrier DS/CDMA system or the subcarriers of an OFDM system may be hopped in a variety of ways[11].

6.4   Cellular Networks and Power Control

In a *cellular network*, a geographic region is partitioned into cells, as illustrated in Figure 6.14. A base station that includes a transmitter and receiver is located at the center of each cell. Ideally, the cells have equal hexagonal areas. Each *mobile* (user or subscriber) in the network transmits omnidirectionally and communicates with the base station from which it receives the largest average power. Typically, most of the mobiles in a cell communicate with the base station at the center of the cell, and only a few communicate with more distant ones. The base stations act as switching centers for the mobiles and communicate among themselves by wirelines in most applications. Cellular networks with DS/CDMA allow universal frequency reuse in that the same carrier frequency and spectral band is shared by all the cells. Distinctions among the direct-sequence signals are possible because each signal is assigned a unique spreading sequence.

Cells may be divided into *sectors* by using several directional sector antennas or arrays at the base stations. Only mobiles in the directions covered by a sector

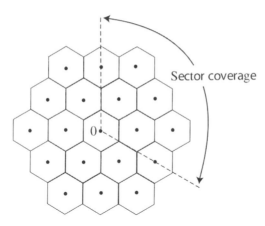

Figure 6.14: Geometry of cellular network with base station at center of each hexagon. Two concentric tiers of cells surrounding a central cell are shown.

antenna can cause multiple-access interference on the *reverse link* or *uplink* from a mobile to its associated sector antenna. Only a sector antenna serving a cell sector oriented toward a mobile can cause multiple-access interference on the *forward link* or *downlink* from the mobile's associated sector antenna to the mobile. Thus, the numbers of interfering signals on both the uplink and the downlink are reduced approximately by a factor equal to the number of sectors.

To facilitate the identification of a base station controlling communications with a mobile, each spreading sequence for a downlink is formed as the product or concatenation of two sequences often called the scrambling and channelization codes. A *scrambling code* is a sequence that identifies a particular base station when the code is acquired by mobiles associated with the base station and its cell or sector. A long sequence is preferable to minimize the possibility of a prolonged outage due to an unfavorable cross-correlation. If the set of base stations use the Global Positioning System or some other common timing source, then each scrambling code may be a known phase shift of a common long pseudonoise sequence. If a common timing source is not used, then at the cost of increased acquisition time or complexity, the scrambling codes may comprise a set of long Gold sequences that approximate random binary sequences. A *channelization code* is designed to allow each mobile receiver to extract its messages while blocking messages intended for other mobiles within the same cell or sector. Walsh or other orthogonal sequences are suitable as channelization codes for synchronous downlinks. For the uplinks, channelization codes are not strictly necessary, and the scrambling codes that identify the mobiles may be drawn from a set of long Gold sequences.

The principal difficulty of DS/CDMA is called the *near-far problem*. If all mobiles transmit at the same power level, then the received power at a base station is higher for transmitters near the receiving antenna. There is a near-far problem because transmitters that are far from the receiving antenna may be at a substantial power disadvantage, and the spread-spectrum processing gain may not be enough to allow satisfactory reception of their signals. A similar problem may also result from large differences in received power levels due to differences in the shadowing experienced by signals traversing different paths or due to independent fading.

In cellular communication networks, the near-far problem is critical only on the uplink because on the downlink, the base station transmits orthogonal signals synchronously to each mobile associated with it. For cellular networks, the usual solution to the near-far problem of uplinks is *power control*, whereby all mobiles regulate their power levels. By this means, power control potentially ensures that the power arriving at a common receiving antenna is almost the same for all transmitters. Since solving the near-far problem is essential to the viability of a DS/CDMA network, the accuracy of the power control is a crucial issue.

In networks with *peer-to-peer communications*, there is no cellular or hierarchical structure. Communications between two mobiles are either direct or are relayed by other mobiles. Since there is no feasible method of power control to prevent the near-far problem, DS/CDMA systems are not as attractive an option as FH/CDMA systems in these networks.

An *open-loop method* of power control in a cellular network causes a mobile to adjust its transmitted power to be inversely proportional to the received power of a *pilot signal* transmitted by the base station. An open-loop method is used to initiate power control, but its subsequent effectiveness requires that the propagation losses on the forward and reverse links be nearly the same. Whether they are or not depends on the duplexing method used to allow simultaneous or nearly simultaneous transmissions on both links. *Frequency-division duplexing* assigns different frequencies to an uplink and its corresponding downlink. *Time-division duplexing* assigns closely spaced but distinct time slots to the two links. When frequency-division duplexing is used, as in the IS-95 and Global System for Mobile (GSM) standards, the frequency separation is generally wide enough that the channel transfer functions of the uplink and downlink are different. This lack of *link reciprocity* implies that power measurements over the downlink do not provide reliable information for subsequent uplink transmissions. When time-division duplexing is used, the received local-mean power levels for the uplink and the downlink will usually be nearly equal when the transmitted powers are the same, but the Rayleigh fading may subvert link reciprocity. For these reasons, a *closed-loop method* of power control, which is more flexible than an open-loop method, is desirable. A closed-loop method requires the base station to transmit power-control information to each mobile based on the power level received from the mobile or the signal-to-interference ratio.

When closed-loop power control is used, each base station attempts to either directly or indirectly track the received power of a desired signal from a mobile and dynamically transmit a power-control signal [13], [14]. The effect of increasing the carrier frequency or the mobile speeds is to increase the fading rate. As the fading rate increases, the tracking ability and, hence, the power-control accuracy decline. This problem is often dismissed by invoking the putative trade-off between the power control and the bit or symbol interleaving. It is asserted that the large fade durations during slow fading enable effective power control, whereas the imperfect power control in the presence of fast fading is compensated by the increased time diversity provided by the interleaving and channel coding. However, this argument ignores both the potential severity of the near-far problem and the limits of compensation as the fading rate increases. If the power control breaks down completely, then close interfering mobiles can cause frequent error bursts of duration long enough to overwhelm the ability of the deinterleaver to disperse the errors so that the decoder can eliminate them. Thus, some degree of power control must be maintained as the vehicle speeds or the carrier frequency increases. The degree required when the interleaving is perfect is quantified subsequently.

The following performance analysis of the uplink [15] begins with the derivation of the intercell interference factor, which is the ratio of the intercell interference power to the intracell interference power. The intercell interference arrives from mobiles associated with different base stations than the one receiving a desired signal. The intracell interference arrives from mobiles that are associated with the same base station receiving a desired signal. The performance is evaluated using two different criteria: the outage and the bit error rate. The outage criterion has the advantage that it simplifies the analysis and does not require specification of the data modulation or channel coding. The bit-error-rate criterion has the advantage that the impact of the channel coding can be calculated. For both criteria, the fading is flat and no explicit diversity or rake combining is assumed. Since the interference signals arrive asynchronously, they cannot be suppressed by using orthogonal spreading sequences.

Intercell Interference of Uplink

To account for the fading and instantaneous power control in a mathematically tractable way, the shadowing and fading factors in (5-4) are approximated [16] by a lognormal random variable. Thus, at a particular time it is assumed that the *equivalent shadowing factor* η implicitly defined by

$$10^{\eta/10} = 10^{\xi/10}\alpha^2 \qquad (6\text{-}105)$$

has a probability density function that is approximately Gaussian. This equation, the statistical independence of ξ and α, and the fact that $E[\xi] = 0$ imply that

$$E[\eta] = \frac{2}{b}E[\ln \alpha] \qquad (6\text{-}106)$$

$$E[\eta^2] = E[\xi^2] + \frac{4}{b^2}E[(\ln \alpha)^2] \qquad (6\text{-}107)$$

where $b = (\ln 10)/10$. To evaluate these equations when α has the Nakagami-m density function of (5-29), we express the expectations as integrals, change the integration variables, and apply the identities [17]

$$\int_0^\infty x^{\nu-1} e^{-\mu x} \ln x \, dx = \frac{\Gamma(\nu)}{\mu^\nu} [\psi(\nu) - \ln \mu] \tag{6-108}$$

$$\int_0^\infty x^{\nu-1} e^{-\mu x} (\ln x)^2 \, dx = \frac{\Gamma(\nu)}{\mu^\nu} \{ [\psi(\nu) - \ln \mu]^2 + \zeta(2,\nu) \} \tag{6-109}$$

where $\mathrm{Re}(\mu) > 0$, $\mathrm{Re}(\nu) > 0$, $\psi(\nu)$ is the psi function given by

$$\psi(\nu) = \sum_{i=1}^{\nu-1} \frac{1}{i} - C , \quad C \cong 0.5772 \tag{6-110}$$

when ν is a positive integer, and $\zeta(2,\nu)$ is the Riemann zeta function given by

$$\zeta(2,\nu) = \sum_{i=0}^\infty \frac{1}{(\nu+i)^2}, \quad \nu \neq 0, -1, -2, \ldots \tag{6-111}$$

Let σ_η^2 denote the variance of η. Since $E[\xi^2] = \sigma_s^2$, the variance of ξ, we find that

$$E[\eta] = \frac{1}{b} [\psi(m) - \ln(m)] \tag{6-112}$$

$$\sigma_\eta^2 = \sigma_s^2 + \frac{\zeta(2,m)}{b^2} \tag{6-113}$$

The impact of the fading declines with increasing m. For Rayleigh fading, $m = 1$ and $\zeta(2, 1) = 1.65$, so $E[\eta] = -2.5$ and $\sigma_\eta^2 = \sigma_s^2 + 31.0$. For $m = 5$, which approximates Ricean fading with Rice factor $\kappa = 8.47$, $E[\eta] = -0.45$ and $\sigma_\eta^2 = \sigma_s^2 + 4.2$.

Consider a cellular network in which each base station is located at the center of a hexagonal area, as illustrated in Figure 6.14. To analyze uplink interference, it is assumed that the desired signal arrives at base station 0, while the other base stations are labeled 1, 2, ..., N_B. The directions covered by one of three sectors associated with base station 0 are indicated in the figure. Each mobile in the network transmits omnidirectionally and is associated with the base station from which it receives the largest average short-term or *instantaneous power*. This base station establishes the uplink power control of the mobile. If a mobile is associated with base station i, then (5-1), (5-4), and (6-105) indicate that the instantaneous power received by base station j is

$$D_{ij} = p_{0i} \left(\frac{r_j}{R_0} \right)^{-\beta} 10^{\eta_j/10} = p_{0i} \left(\frac{r_j}{R_0} \right)^{-\beta} \exp(b\,\eta_j) \tag{6-114}$$

where r_j is the distance to base station j, η_j is the equivalent shadowing factor, p_{0i} is the area-mean power at $r_j = R_0$, and it is assumed that the attenuation power-law β is the same throughout the network. If the power control exerted by base station i ensures that it receives unit instantaneous power from each mobile associated with it, then $D_{ii} = 1$. Consequently, $p_{0i} = (r_i/R_0)^\beta \exp(-b\eta_i)$, and

$$D_{ij} = \left(\frac{r_i}{r_j}\right)^\beta \exp[b(\eta_j - \eta_i)] \qquad (6\text{-}115)$$

Assuming a common fading model for all of the $\{\eta_i\}$, (6-106) implies that they all have the same mean value. The form of (6-115) then indicates that this common mean value is irrelevant to the statistics of D_{ij} and hence can be ignored without penalty in the subsequent statistical analysis of D_{ij}. The simplifying approximation is made that the base station with which a mobile is associated receives more instantaneous power than any other station, and hence $D_{ij} \leq 1$. This inequality is exact if the propagation losses on the uplink and downlink are the same.

The probability distribution function of the interference power D_{i0} at base station 0 given that the mobile producing the interference is associated with base station i is

$$F_i(x) = P[D_{i0} \leq x \mid D_{ij} \leq 1, 0 \leq j \leq N_B] = \frac{\phi_i(x)}{\phi_i(1)} \qquad (6\text{-}116)$$

where

$$\phi_i(x) = P[D_{i0} \leq x; D_{ij} \leq 1, 0 \leq j \leq N_B]. \qquad (6\text{-}117)$$

and $P[A]$ denotes the probability of the event A [18]. Thus, $F_i(x) = 0$ if $x < 0$, and $F_i(x) = 1$ if $x \geq 1$. Let

$$\phi_i(x \mid \eta_i, r_i, \theta_i) = P[D_{i0} \leq x; D_{ij} \leq 1, 0 \leq j \leq N_B \mid \eta_i, r_i, \theta_i] \qquad (6\text{-}118)$$

where this probability is conditioned on η_i, the equivalent shadowing factor for the controlling base station, and the polar coordinates r_i, θ_i of the mobile relative to base station i. It is assumed that each of the $\{\eta_j\}$ is statistically independent with the common variance σ_η^2. Therefore, given η_i, D_{ij} and $D_{ik}, j \neq k$, are statistically independent. Since each of the $\{\eta_j\}$ has a Gaussian probability density function, (6-115) implies that for $0 \leq x \leq 1$,

$$\phi_i(x \mid \eta_i, r_i, \theta_i) = Q_c\left(\frac{b\eta_i + \beta \ln(r_0/r_i) + \ln x}{b\sigma_\eta}\right) \prod_{\substack{j=1 \\ j \neq i}}^{N_B} Q_c\left(\frac{b\eta_i + \beta \ln(r_j/r_i)}{b\sigma_\eta}\right)$$

$$(6\text{-}119)$$

where $Q_c(x) = 1 - Q(x)$, and r_j, $0 \leq j \leq N_B$, is a function of r_i, θ_i, and the location of base station j.

The probability $\phi_i(x)$, and hence the distribution $F_i(x)$, can be determined by evaluating the expected value of (6-119) with respect to the random variables η_i, r_i, and θ_i. If a mobile is associated with base station i, then its location is assumed to be uniformly distributed within a circle of radius R_b surrounding the base station. Therefore,

$$\phi_i(x) = \int_0^{2\pi} d\theta \int_0^{R_b} dr \int_{-\infty}^{\infty} d\eta \, \frac{r \exp\left(-\frac{\eta^2}{2\sigma_\eta^2}\right)}{\sqrt{2\pi}\,\sigma_\eta \pi R_b^2} \, \phi_i(x \mid \eta, r, \theta) \qquad (6\text{-}120)$$

Table 6.1: Interference and variance factors when var$[K] = 0$.

σ_η, dB	g	g_1
3	0.460	0.137
4	0.486	0.143
$6/\sqrt{2}$	0.493	0.145
5	0.519	0.153
$8/\sqrt{2}$	0.544	0.162
6	0.558	0.167
7	0.598	0.183
$10/\sqrt{2}$	0.601	0.184
8	0.634	0.189

which determines the distribution function in (6-116).

Let I_{te} denote the total intercell interference relative to the unit desired-signal power that each base station attempts to maintain by power control. Let K denote the number of active mobiles associated with a base station or sector antenna, which may be a random variable because of voice-activity detection or the movement of mobiles among the cells. Since $E[D_{i0}]$ and var$[D_{i0}]$ are the same for all mobiles associated with base station i, a straightforward calculation yields

$$E[I_{te}] = E[K] \sum_{i=1}^{N_B} E[D_{i0}] \tag{6-121}$$

$$\text{var}[I_{te}] = E[K] \sum_{i=1}^{N_B} \text{var}[D_{i0}] + \text{var}[K] \left(\sum_{i=1}^{N_B} E[D_{i0}] \right)^2 \tag{6-122}$$

In general, $E[I_{te}]$ and var$[I_{te}]$ decrease as the attenuation power law β increases. The *intercell interference factor*, $g = E[I_{te}]/E[K]$, is the ratio of the average intercell interference power to the average intracell interference power. Table 6.1, calculated in [18], lists g versus σ_η when $N_B = 60$ cells in four concentric tiers surrounding a central cell, R_b is five times the distance from a base station to the corner of its surrounding hexagonal cell, and $\beta = 4$. The dependence of g on the specific fading model is exerted through (6-113), which relates σ_η to m and σ_s. Table 6.1 also lists the *variance factor* $g_1 = \text{var}[I_{te}]/E[K]$ assuming that var$[K] = 0$.

The results in Table 6.1 depend on the pessimistic assumption that the equivalent shadowing factors from a mobile to two different base stations are independent random variables. Suppose, instead, that each factor is the sum of a common component and an equal-power independent component that depends on the receiving base station. Then (6-115) implies that the common component cancels. As a result, in determining g from Table 6.1, the effective value of σ_η is reduced by a factor of $\sqrt{2}$ relative to what it would be without the common component.

Since $\sigma_\eta^2 = \sigma_s^2 + 31.0$ for Rayleigh fading and Table 6.1 indicates that g increases slowly with σ_η, the effect of the fading is unimportant or negligible if $\sigma_s \geq 6$ dB, which is usually satisfied in practical networks. If it is assumed, as is tacitly done by many authors, that the power control is based on a long-term-average power estimate that averages out the fading, then the preceding equations and Table 6.1 are valid with $\sigma_\eta = \sigma_s$.

Outage Analysis

For a DS/CDMA system, it is assumed that the total power I_t of the multiple-access interference after the despreading is approximately uniformly distributed over its bandwidth, which is approximately equal to $1/T_c$. For instantaneous power control, the instantaneous SINR is defined to be $E_s/(N_0+I_tT_c)$, the ratio of the received energy per symbol E_s to the equivalent power spectral density of the interference plus noise. An *outage* is said to occur if the instantaneous SINR is less than a specified threshold Z, which may be adjusted to account for any diversity or rake combining. In this section, the interference is assumed to arise from $K-1$ other active mobiles in a single cell or sector. Let $E_i = I_iT_s$, $i = 1, 2, \ldots, K-1$, denote the received energy in a symbol due to interference signal i with power I_i. These definitions imply that an outage occurs if

$$E_sZ^{-1} < N_0 + \frac{1}{G}\sum_{i=1}^{K-1} E_i \tag{6-123}$$

where $G = T_s/T_c$ is the processing gain. Let E_{s0} denote the common desired energy per symbol for all the signals associated with the base station of a single cell sector. When instantaneous power control is used, $E_s = E_{s0}\,\epsilon_0$ and $E_i = E_{s0}\,\epsilon_i$, $i = 1, 2, \ldots, K-1$, where ϵ_0 and ϵ_i are random variables that account for imperfections in the power control. Substitution into (6-123) yields the outage condition

$$G(Z^{-1}\epsilon_0 - \gamma_0^{-1}) < X \tag{6-124}$$

where $\gamma_0 = E_{s0}/N_0$ is the energy-to-noise density ratio of the desired signal when the power control is perfect, and we define

$$X = \sum_{i=1}^{K-1} \epsilon_i \tag{6-125}$$

By analogy with the lognormal spatial variation of the local-mean power, each of the $\{\epsilon_i\}$ is modeled as an independent lognormal random variable. Therefore,

$$\epsilon_i = 10^{\xi_i/10} = \exp(b\xi_i), \quad i = 0, 1, 2, \ldots, K-1 \tag{6-126}$$

where each of the $\{\xi_i\}$ is a zero-mean Gaussian random variable with common variance σ_e^2. The moments of ϵ_i can be derived by direct integration or from the moment-generating function of ξ_i. We obtain

$$E[\epsilon_i] = \exp\left(\frac{b^2\sigma_e^2}{2}\right), \quad E[\epsilon_i^2] = \exp(2b^2\sigma_e^2) \tag{6-127}$$

If K is a constant, then the mean $\bar{X}$ and the variance σ_x^2 of X in (6-125) are

$$\bar{X} = (K-1)\exp\left(\frac{b^2\sigma_e^2}{2}\right), \quad \sigma_x^2 = (K-1)[\exp(2b^2\sigma_e^2) - \exp(b^2\sigma_e^2)] \quad (6\text{-}128)$$

The random variable X is the sum of $K-1$ lognormally distributed random variables. Since the distribution of X cannot be compactly expressed in closed form when $K > 3$, two approximate methods are adopted. The first method is based on the central limit theorem, and the second method is based on the assumption that σ_e is small. Since X is the sum of $K-1$ independent, identically distributed random variables, each with a finite mean and variance, the central limit theorem implies that the probability distribution function of X is approximately Gaussian when K is sufficiently large. Consequently, given the values of K and ϵ_0, the conditional probability of outage may be calculated from (6-124). Using (6-126) and integrating over the Gaussian density function of ξ_0, we then obtain the conditional probability of outage given the value of $K \gg 1$:

$$P_{\text{out}}(K) = \int_{-\infty}^{\infty} Q\left[\frac{G(Z^{-1}e^{b\xi} - \gamma_0^{-1}) - \bar{X}}{\sigma_x}\right] \frac{\exp(-\xi^2/2\sigma_e^2)}{\sqrt{2\pi}\sigma_e} d\xi \quad (6\text{-}129)$$

As $\sigma_e \to 0$ and hence $\sigma_x \to 0$, $P_{\text{out}}(K)$ approaches a step function.

In the second approximate method, it is assumed that σ_e is sufficiently small and K is sufficiently large that $\sigma_x \ll \bar{X}$. From (6-128), it is observed that a sufficient condition for this assumption is that

$$\sqrt{K-1} \gg \exp\left(\frac{b^2\sigma_e^2}{2}\right) \quad (6\text{-}130)$$

The assumption implies that X is well approximated by the constant $\bar{X}$ given by (6-128). Since the only remaining random variable in (6-124) is $\epsilon_0 = \exp(b\xi_0)$, it follows that

$$P_{\text{out}}(K) = Q\left\{-\frac{\ln[(K-1)G^{-1}Z\exp(b^2\sigma_e^2/2) + Z\gamma_0^{-1}]}{b\sigma_e}\right\} \quad (6\text{-}131)$$

Variations in the Number of Active Mobiles

In the derivations of (6-129) and (6-131), the number of mobiles actively transmitting, K, is held constant. However, it is appropriate to model K as a random variable because of the movement of mobiles into and out of each sector and the changing of the cell or sector antenna with which a mobile communicates. Furthermore, a potentially active mobile may not be transmitting; for voice communications with voice-activity detection, energy transmission typically is necessary only roughly 40% of the time. As is shown below, a discrete random variable K with a Poisson distribution incorporates both of these effects.

To simplify the analysis, it is assumed that the average number of mobiles associated with each cell or sector antenna is the same and that the location of a mobile is uniformly distributed throughout a region. Let q denote the probability that a potentially transmitting mobile is actively transmitting. Then the probability that an active mobile is associated with a particular cell or sector antenna is $\mu q/N_r$, where N_r is the number of mobiles in the region and μ is the average number of mobiles per sector. If the N_r mobiles are independently located in the region, then the probability of $K = k$ active mobiles being associated with a sector antenna is given by the binomial distribution

$$P(N_r, k) = \binom{N_r}{k}\left(\frac{\lambda}{N_r}\right)^k\left(1 - \frac{\lambda}{N_r}\right)^{N_r-k} \tag{6-132}$$

where $\lambda = \mu q$ is assumed to be a constant. This equation can be expressed as

$$P(N_r, k) = \frac{(1 - 1/N_r)(1 - 2/N_r)\ldots(1 - (k-1)/N_r)}{k!}\lambda^k\left(1 - \frac{\lambda}{N_r}\right)^{-k}\left(1 - \frac{\lambda}{N_r}\right)^{N_r} \tag{6-133}$$

As $N_r \to \infty$, the initial fraction $\to 1/k!$, $(1-\lambda/N_r)^{-k} \to 1$, and $(1- \lambda/N_r)^{N_r} \to \exp(-\lambda)$. Therefore, $P(N_r, k)$ approaches

$$P_u(k) = \frac{\exp(-\lambda)\lambda^k}{k!}, \quad k = 0, 1, 2, \ldots \tag{6-134}$$

which is the Poisson distribution function. Since the desired mobile is assumed to be present, it is necessary to calculate the conditional probability that $K = k$ given that $K \geq 1$. From the definition of a conditional probability and (6-134), it follows that this probability is

$$P_c(k) = \frac{\exp(-\lambda)\lambda^k}{[1 - \exp(-\lambda)]k!}, \quad k = 1, 2, \ldots \tag{6-135}$$

and $P_c(0) = 0$. Using this equation, the probability of outage is

$$P_{\text{out}} = \sum_{k=1}^{\infty} \frac{\exp(-\lambda)\lambda^k}{[1 - \exp(-\lambda)]k!}P_{\text{out}}(k) \tag{6-136}$$

where $P_{\text{out}}(k)$ is given by (6-129) or (6-131).

The intercell interference from mobiles associated with other base stations introduces an additional average power equal to $g\mu q(E_{s0}/T_s)$ into a given base station, where g is the intercell interference factor. Accordingly, the impact of the intercell interference is modeled as equivalent to an average of $g\mu$ additional mobiles in a sector [19]. When intercell interference is taken into account, the equations of Section 3.7 for a single cell or sector are modified. The parameter μ is replaced by $\mu(1+g)$, and λ becomes the *equivalent number of mobiles defined as*

$$\lambda = \mu q(1 + g) \tag{6-137}$$

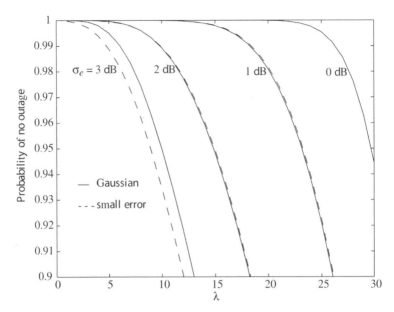

Figure 6.15: Probability of no outage for instantaneous power control, $G/Z = 40$, $\gamma_0/G = 0.5$, and $\sigma_e = 0, 1, 2, 3 \ dB$.

Figure 6.15 illustrates the probability of no outage, $1 - P_{\text{out}}$, as a function of λ for various values of σ_e. Both approximate models, which give (6-129) and (6-131), are used in (6-136) to calculate the graphs. Equations (6-129) and (6-131) indicate that the outage probability depends on the ratios G/Z and γ_0/G rather than on G, Z, and γ_0 separately. The parameter values for Figure 6.15 are $G/Z = 40$ and $\gamma_0/G = 0.5$, which could correspond to $Z = 7$ dB, $G = 23$ dB, and $\gamma_0 = 20$ dB. The closeness of the results for the two models indicates that when $\sigma_e \leq 2$ dB both models give accurate outage probabilities and the effect of power-control errors in the interference signals is unimportant. As an example of the application of the figure, suppose that the attenuation power law is $\beta = 4$, $\sigma_\eta = 8$ dB, $\sigma_e = 1.0$ dB, and $1 - P_{\text{out}} = 0.95$ is desired. Table 6.1 gives $g = 0.63$. The figure indicates that $\lambda = 23$ is needed. If $q = 0.4$ due to voice-activity detection, the average number of mobiles per sector that can be accommodated is $\mu = 35.3$. For data communications, the network capacity is lower. For example, if $q = 1$, then the average number of mobiles per sector that can be accommodated is 14.1.

Local-Mean Power Control

When the instantaneous signal power cannot be tracked because of the fast multipath fading, one might consider measuring the local-mean power, which is a long-term-average power obtained by averaging out the fading component. This measurement enables the system to implement *local-mean power control*.

Two different analyses of the effects of local-mean power control are presented.

In the first analysis, which explores the potential effectiveness of local-mean power control, all received signals experience Rayleigh fading and the local-mean power control is perfect. Therefore, the received energy levels are proportional to the squares of Rayleigh-distributed random variables and, hence, are exponentially distributed, as shown in Appendix D.4. Thus, $E_s = E_{s0}\,\epsilon_0$ and $E_i = E_{s0}\,\epsilon_i$, where each ϵ_i, $i = 0, 1, 2, \ldots, K-1$, is an independent random variable with the exponential probability density function:

$$f_s(x) = \exp(-x)u(x) \tag{6-138}$$

and E_{s0} is the desired value of the average energy per symbol after averaging over the fading. The probability distribution function of the sum of $K-1$ independent random variables, each with the exponential density of (6-138), is given by (D-50). Therefore, X in (6-125) has the distribution

$$F_X(x) = 1 - \exp(-x) \sum_{i=0}^{K-2} \frac{x^i}{i!}\ , \quad x \geq 0 \tag{6-139}$$

Conditioning on the value of ϵ_0, using (6-139) to evaluate the probability of the outage condition (6-124), and then removing the conditioning by using (6-138) yields

$$P_{\text{out}}(K) = \int_0^\infty e^{-\xi} \exp[-c(\xi)] \sum_{i=0}^{K-2} \frac{[c(\xi)]^i}{i!}\ d\xi \tag{6-140}$$

where

$$c(\xi) = GZ^{-1}\xi - G\gamma_0^{-1} \tag{6-141}$$

Replacing $[c(\xi)]^i$ by its binomial expansion, we obtain a double summation of integrals that can be evaluated using the gamma function defined by (D-12). After simplification, we obtain

$$P_{\text{out}}(K) = \sum_{i=0}^{K-2} \sum_{l=0}^{i} \frac{\exp(G\gamma_0^{-1})(GZ^{-1})^l(-G\gamma_0^{-1})^{i-l}}{(i-l)!(1+GZ^{-1})^{l+1}} \tag{6-142}$$

Interchanging the two sums and changing their limits accordingly, the inner sum is over a geometric series. Evaluating it, we obtain the final result:

$$P_{\text{out}}(K) = \exp(G\gamma_0^{-1}) \sum_{l=0}^{K-2} \frac{(-G\gamma_0^{-1})^l}{l!} \left[1 - \left(\frac{GZ^{-1}}{1+GZ^{-1}} \right)^{K-1-l} \right] \tag{6-143}$$

The probability of outage is determined by substitution into (6-136). When $\gamma_0 = \infty$, only the $l = 0$ term in (6-143) is nonzero. Substitution into (6-136) and evaluation of the sum yields

$$P_{\text{out}} = 1 - \frac{\exp\left(\frac{\lambda}{1+G^{-1}Z}\right) - 1}{\exp(\lambda) - 1}(1 + G^{-1}Z)\ , \quad \gamma_0 = \infty \tag{6-144}$$

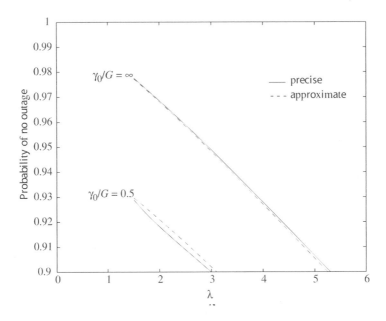

Figure 6.16: Probability of no outage for perfect local-mean power control, $G/Z = 40$, and $\gamma_0/G = 0.5, \infty$.

For perfect local-mean power control and Rayleigh fading, (6-138) gives $E[\epsilon_i] = 1$ and $\mathrm{var}(\epsilon_i) = 1$. Therefore, a sufficient condition for $\sigma_x \ll \bar{X}$ is that $\sqrt{K-1} \gg 1$. If this condition is satisfied, then X is well approximated by $\bar{X} = K - 1$, which is equivalent to ignoring the fading of the multiple-access interference signals. With this approximation, the only remaining random variable in (6-124) is exponentially distributed, and hence the conditional probability of outage given K is

$$P_{\mathrm{out}}(K) = 1 - \exp[-(K-1)G^{-1}Z - \gamma_0^{-1}Z] \qquad (6\text{-}145)$$

Substituting this equation into (6-136) and evaluating the sum, we obtain the approximation

$$P_{\mathrm{out}} = 1 - \frac{\exp(-\gamma_0^{-1}Z + G^{-1}Z)[\exp(\lambda \exp(-G^{-1}Z)) - 1]}{\exp(\lambda) - 1} \qquad (6\text{-}146)$$

Figure 6.16 illustrates the probability of no outage as a function of λ for $G/Z = 40$ and two values of γ_0/G using either the approximation(6-146) or the more precise (6-144), (6-143), and (6-136). It is observed that neglecting the fading of the interference signals and using the approximation makes little difference in the results. The effect of $\gamma_0 = E_{s0}/N_0$ is considerable. A comparison of Figures 6.15 and 6.14 indicates that when Rayleigh fading occurs, even perfect local-mean power control is not as useful as imperfect instantaneous power control unless σ_e is very large.

Since accurate power measurements require a certain amount of time, whether a power-control scheme is instantaneous, local mean, or something intermediate depends on the fading rate. To reduce the fading rate so that the power control is instantaneous and accurate, one might minimize the carrier frequency or limit the size of cells if these options are available.

The second analysis of the effects of local-mean power control uses the preceding results to develop a simple approximation to alternative performance calculations [19], [20]. This analysis has the advantages that the fading statistics do not have to be explicitly defined and the effect of imperfect local-mean power control is easily calculated. Let E_{sl} denote the local-mean energy per symbol, which is defined as the average energy per symbol after averaging over the fading. Similarly, let I_{tl} denote the total local-mean interference power in the receiver, and let E_{il} denote the local-mean received energy per symbol due to interference signal i. The local-mean SINR is defined to be $E_{sl}/(N_0 + I_{tl}T_c)$. For this analysis, a *local-mean outage* is said to occur if the local-mean SINR is less than a specified threshold Z_l, which may be adjusted to account for the fading statistics and any diversity or rake combining. When the local-mean power control is imperfect, $E_{sl} = E_{s0}\epsilon_0$ and $E_{il} = E_{s0}\epsilon_i$, $i = 1, 2, \ldots, K - 1$, where ϵ_0 and ϵ_i are lognormally distributed random variables with the common variance σ_{le}^2. A derivation similar to that leading to (6-131) indicates that if (6-130) is satisfied, then

$$P_{\text{out}}(K) = Q\left\{ -\frac{\ln[(K-1)G^{-1}Z_l \exp(b^2\sigma_{le}^2/2) + Z_l\gamma_0^{-1}]}{b\sigma_{le}} \right\} \qquad (6\text{-}147)$$

and P_{out} is calculated by using (6-136) and (6-137). The intercell interference factor g can be determined by setting $\sigma_\eta = \sigma_s$ since the fading statistics do not affect the local-mean SINR. For adequate network performance in practical applications, Z_l must be set much higher than the threshold Z in (6-131) because the local-mean SINR changes much more slowly than the instantaneous SINR.

The following example is used to compare the results of evaluating (6-136), (6-137), and (6-147) with the results obtained in a far more elaborate analysis [20]. Consider a cellular network with three sectors, $Z_l = 7$ dB, $\sigma_s = 6$ dB, and $q = 3/8$ due to the voice activity. Table 6.1 gives $g = 0.558$. A spectral band of bandwidth $W = 1.25$ MHz is occupied by the DS/CDMA signals. The symbol rate is $1/T_s = 8$ kb/s so that the processing gain is $G = 156.5$. The local-mean SNR before the despreading is -1 dB and $\gamma_0 = 20.94$ dB after the despreading. Figure 6.17 shows the local-mean outage probability versus the average number of mobiles per cell, $3\,\mu$, which is triple the average number of mobiles per sector. The results of [20] for outage probabilities of 10^{-1}, 10^{-2}, and 10^{-3} are indicated by the open circles. The proximity of these points to the graphs indicates that the simple equations (6-136), (6-137), and (6-147) closely approximate the local-mean outage probability.

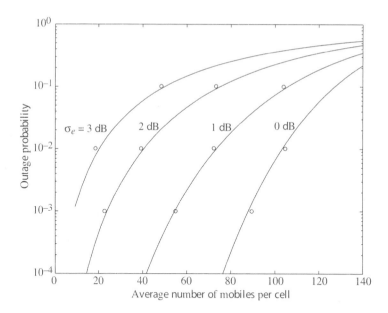

Figure 6.17: Local-mean outage probability for $Z_l = 7dB$, $q = 3/8$, $g = 0.558$, $G = 156.5$, and $\gamma_0 = 20.94$ dB with $\sigma_e = 0, 1, 2, 3$ dB. Other theoretical results are indicated by the open circles.

Bit-Error-Probability Analysis

Uplink capacity is the number of mobiles per cell that can be accommodated over the uplink at a specified information-bit error rate. Assuming a conventional correlation receiver and typical conditions for cellular communications, the subsequent results indicate that when imperfect power control causes the standard deviation of the received power from each mobile to increase beyond 2 dB, the uplink capacity rapidly collapses. When the instantaneous signal level cannot be tracked, one might consider measuring the local-mean power. Accurate local-mean power control eliminates the near-far problem and shadowing effects, but not the effects of the fading. In the subsequent analysis, it is confirmed that tracking the local-mean power is less useful than attempting to track the instantaneous signal level even if the latter results in large errors.

Consider a CDMA cell or sector with K active mobiles. The direct-sequence signals use QPSK modulation. Equation (6-86) indicates that the conditional symbol error probability given E_s and $E_t = I_t T_s$ is approximately given by

$$P_s(E_s, E_t) = Q\left(\sqrt{\frac{2E_s}{N_0 + \sqrt{h}G^{-1}E_t}}\right) \qquad (6\text{-}148)$$

It is assumed that the distribution of E_s and E_t and the values of $G = T_s/T_c$ and N_0 are such that (6-85), which is used in the derivation of (6-86), is satisfied with high probability in the subsequent analysis. We consider three models for power control: perfect instantaneous power control (perfect ipc), imperfect instantaneous power control (imperfect ipc) with lognormally distributed errors, and perfect local-mean power control (perfect lmpc).

If the power control is instantaneous and perfect, then $E_i = E_s = E_{s0}$, $i = 1, 2, \ldots, K - 1$, and $E_t = (K - 1)E_{s0}$. Equation (6-148) implies that the conditional symbol error probability given K is

$$P_s(K) = Q\left(\sqrt{\frac{2}{\gamma_0^{-1} + \sqrt{h}G^{-1}(K - 1)}}\right) \quad \text{(perfect ipc)} \qquad (6\text{-}149)$$

where $\gamma_0 = E_{s0}/N_0$ is the energy-to-noise-density ratio when the power control is perfect. If the power control is imperfect with lognormally distributed errors, then

$$E_s = E_{s0}\epsilon_0, \quad E_t = E_{s0}X \qquad (6\text{-}150)$$

and (6-125) to (6-128) are applicable. If (6-130) is satisfied, then $\bar{X} >> \sigma_x$, and X is well-approximated by $\bar{X}$. Since $\epsilon_0 = \exp(b\xi_0)$ and ξ_0 has a Gaussian density, (6-148) and an integration over this density yield

$$P_s(K) = \int_{-\infty}^{\infty} \frac{\exp(-x^2/2\sigma_e^2)}{\sqrt{2\pi}\sigma_e} Q\left(\sqrt{\frac{2\exp(bx)}{\gamma_0^{-1} + \sqrt{h}G^{-1}(K - 1)\exp(b^2\sigma_e^2/2)}}\right) dx$$

$$\text{(imperfect ipc)} \qquad (6\text{-}151)$$

Suppose that instead of the instantaneous signal power, the local-mean power averaged over the fast fading is tracked. If this tracking provides perfect power control of the local-mean power at a specific level, then a received signal still exhibits fast fading relative to this level. If the fast fading has a Rayleigh distribution but the fading level is constant over a symbol interval, then the received energy per symbol is $E_s = E_{s0}\epsilon_0$, where ϵ_0 has the exponential probability density function given by (6-138). Therefore, (6-148) implies that the conditional symbol error probability given E_t is

$$P_s(E_t) = \int_0^{\infty} \exp(-x)Q\left(\sqrt{\frac{2x}{\gamma_0^{-1} + \sqrt{h}G^{-1}E_t/E_{s0}}}\right) dx$$

$$= \frac{1}{2} - \frac{1}{2}\left(1 + \gamma_0^{-1} + \sqrt{h}G^{-1}E_t/E_{s0}\right)^{-1/2} \qquad (6\text{-}152)$$

where the integral is evaluated in the same way as (5-125). The total interference energy E_t is given by (6-150) and (6-125), where each ϵ_i is an independent, exponentially distributed random variable with mean equal to unity. Therefore, E_t/E_{s0} has a gamma probability density function given by (D-49) with $N = K - 1$, and for $K \geq 2$ the conditional symbol error probability given K is

$$P_s(K) = \frac{1}{2} - \frac{1}{2}\int_0^{\infty} \frac{x^{K-2}\exp(-x)}{(K - 2)!\left(1 + \gamma_0^{-1} + \sqrt{h}G^{-1}x\right)^{1/2}} dx \quad \text{(perfect lmpc)}$$

$$(6\text{-}153)$$

Perfect symbol interleaving is defined as interleaving that causes independent symbol errors in a codeword. Assuming that fast fading enables perfect symbol interleaving, the information-bit error probability $P_b(K)$ for hard-decision decoding can be calculated by substituting (6-149), (6-151), or (6-153) into (1-25), (1-26), and (1-27) or into (6-104) for a loosely packed binary code. If r is the code rate of a binary code and E_b is the energy per bit that is available when the channel symbols are uncoded, then $\gamma_0 = rE_b/N_0$ in (6-149), (6-151), and (6-153). As was done previously, the impact of the intercell interference is modeled by replacing K with $K(1+g)$ in the preceding equations, where g is obtained from Table 6.1. Averaging over K by using (6-135), we obtain

$$P_b = \sum_{k=1}^{\infty} \frac{\exp(-\lambda)\lambda^k}{[1 - \exp(-\lambda)]k!} P_b(k) \qquad (6\text{-}154)$$

where the equivalent number of mobiles λ is given by (6-137).

Suppose that the fading is slow enough that the interleaving is ineffective and, hence, the error in the instantaneous power control is fixed over a codeword duration. Then an approximation similar to that preceding (6-151) implies that the information-bit error probability for hard-decision decoding of a block code given K is

$$P_b(K) = \int_{-\infty}^{\infty} \frac{\exp(-x^2/2\sigma_e^2)}{\sqrt{2\pi}\sigma_e} P_b(K, P_s(x)) \, dx \qquad (6\text{-}155)$$

where $P_b(K, P_s(x))$ is given by (6-104) with P_s replaced by

$$P_s(x) = Q\left(\sqrt{\frac{2\exp(bx)}{\gamma_0^{-1} + \sqrt{h}G^{-1}(K-1)\exp(b^2\sigma_e^2/2)}}\right) \qquad (6\text{-}156)$$

Equations (6-154) to (6-156) give the information-bit error probability for slow fading.

Graphs of the information-bit error probability versus λ for instantaneous power control, $\gamma_0 = 13$ dB, $G = 128$, a rectangular chip waveform with $h = 2/3$, and various values of σ_e in decibels are illustrated in Figure 6.18. The block code is the binary BCH (63,30) code, for which $d_m = 21$ and $t = 10$. Equations (6-155) and (6-156) are used for slow fading, and (6-149), (6-151), and (6-104) are used for fast fading. When the fading is slow and the interleaving is ineffective, the coding is, as expected, less effective than when the fading is fast and the interleaving is perfect, provided that σ_e remains the same. However, σ_e increases with the fading rate, as shown subsequently. The figure indicates that when $\sigma_e > 2$ dB, there is a severe uplink capacity loss for slow fading and a substantial one for fast fading. The results for other block codes are qualitatively similar.

The use of spatial diversity or, in the presence of frequency-selective fading, a rake receiver will improve the performance of a DS/CDMA system during both slow and fast fading, but the improvement is much greater when the fading is slow. As the fading rate increases, the accuracy of the estimation of the channel parameters used in the rake or diversity combiner becomes more

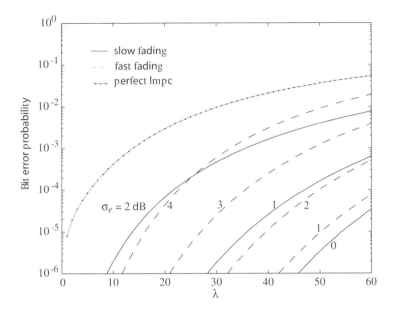

Figure 6.18: Information-bit error probability for instantaneous power control and perfect local-mean power control, $\gamma_0 = 13$ dB, G = 128, and the BCH (63,30) code with various values of σ_e in decibels.

difficult. When the channel-parameter estimation errors are too large to be accommodated, the coherent maximal-ratio combiner must be replaced by the suboptimal noncoherent equal-gain combiner, which does not require the estimation of channel parameters.

In Figure 6.18, the information-bit error probability is depicted for perfect local-mean power control with the same parameter values and coding as for instantaneous power control. It is assumed that fast fading permits perfect interleaving so that (6-153) and (6-104) are applicable. The figure confirms that tracking of the local-mean power level is an inferior strategy for obtaining a large capacity compared with tracking of the instantaneous power level unless the inaccuracy of the latter is substantial. Another problem with local-mean power control is that it requires time that may be unavailable for sporadic data.

Apart from power control, instantaneous power measurements can be used to facilitate adaptive coding or adaptive transmit diversity. Both of these techniques require timely information about the impact of the fading, and this information is inherent in the instantaneous power measurements.

Impact of Doppler Spread on Power-Control Accuracy

When the received instantaneous power of the desired signal from a mobile is tracked, there are four principal error components. They are the quantization error due to the stepping of the transmitted power level, the error introduced in the decoding of the power-control information at the mobile, the error in the power measurement at the base station, and the error caused by the processing and propagation delay. Let $\sigma_q, \sigma_d, \sigma_m$, and σ_p denote the standard deviations of these errors, respectively, expressed in decibels relative to the received power. Usually, σ_m and σ_p are much larger than σ_q and σ_d [13]. The processing and propagation delay is a source of error because the multipath propagation conditions change during the execution of the closed-loop power-control algorithm.

Assuming that the error sources are independent, the variance of the power-control error can be decomposed as

$$\sigma_e^2 = \sigma_m^2 + \sigma_p^2 + \sigma_q^2 + \sigma_d^2 \qquad (6\text{-}157)$$

If σ_e is to be less than 2 dB and σ_m is typically more than 1.5 dB [13], then even if σ_q and σ_d are small, $\sigma_p < 1.3$ dB is required. Let v denote the maximum speed of a mobile in the network, f_c the carrier frequency of its direct-sequence transmitted signal, and c the speed of an electromagnetic wave. It is assumed that this signal has a bandwidth that is only a few percent of f_c so that the effect of the bandwidth is negligible. The maximum Doppler shift or Doppler spread is

$$f_d = f_c v / c \qquad (6\text{-}158)$$

which is proportional to the fading rate. To obtain $\sigma_p < 1.3$ dB requires nearly constant values of the channel attenuation during the processing and propagation delay. Thus, this delay must be much less than the coherence time, which is approximately equal to $1/f_d$, as indicated in (5-40). Examination of attenuation graphs for representative multipath scenarios indicates that this delay must be less than α/f_d, where $\alpha \approx 0.1$ or less if $\sigma_p < 1.3$ dB is to be attained. The propagation delay for closed-loop power control is $2d/c$, where d is the distance between the mobile and the base station. Therefore, the processing delay T_p must satisfy

$$T_p < \frac{\alpha}{f_d} - \frac{2d}{c} \qquad (6\text{-}159)$$

Since T_p must be positive, this inequality and (6-158) imply that $\sigma_p < 1.3$ dB is only possible if $f_c < \alpha c^2 / 2dv$. Thus, if the carrier frequency or maximum vehicle speed is too high, then the propagation delay alone makes it impossible for the system to attain the required σ_p throughout the network. If $v = 25$ m/s, $d = 10$ km, $\alpha = 0.1$, and $f_c = 850$ MHz, then (6-159) and (6-158) give $T_p < 1.34$ ms. The IS-95 system, which must accommodate similar parameter values, uses $T_p = 1.25$ ms.

Let p_m denote the measured power level of a received signal in decibels; thus, p_m is an estimate of $10 \log p_0$, where p_0 is the average received signal power from a mobile and the logarithm is to the base 10. Let σ_{m1}^2 denote the variance of an estimate of $\ln p_0$, the natural logarithm of p_0. It follows that the variance of p_m is

$$\sigma_m^2 = (10 \log e)^2 \sigma_{m1}^2 \qquad (6\text{-}160)$$

It is assumed that power variations in a received signal at the base station are negligible during the measurement interval T_m, which is a large component of the processing delay T_p. Errors in the power measurement occur because of the presence of multiple-access interference and white Gaussian noise. A lower bound on σ_{m1}^2 can be determined by assuming that the power control is effective enough that the received powers from the mobiles in the cell or sector are approximately equal. The multiple-access interference is modeled as a Gaussian process that increases the one-sided noise power spectral density from N_0 to

$$N_t = N_0 + \frac{p_0}{B}(K-1)(1+g) \qquad (6\text{-}161)$$

where p_0 is the common signal power of each mobile at the base station and B is the bandwidth of the receiver.

The received signal from a mobile that is to be power-controlled has the form $\sqrt{p_0}s(t)$, where $s(t)$ has unity power. Thus,

$$\int_0^{T_m} s^2(t)dt = T_m \qquad (6\text{-}162)$$

The received signal can be expressed as

$$\sqrt{p_0}s(t) = \exp\left(\frac{y}{2}\right)s(t) \qquad (6\text{-}163)$$

where $y = \ln p_0$. The Cramer-Rao bound [21] provides a lower bound on the variance of any unbiased estimate or measurement of $\ln p_0$. This bound and (6-163) give

$$\sigma_{m1}^2 \geq \left\{\frac{2}{N_t}\int_0^{T_m}\left[\frac{\partial}{\partial y}e^{y/2}s(t)\right]^2 dt\right\}^{-1} \qquad (6\text{-}164)$$

Evaluating (6-164) and using (6-160) and (6-161), we obtain

$$\sigma_m^2 \geq \frac{200(\log e)^2}{p_0 T_m}\left[\frac{N_0}{p_0} + \frac{(K-1)(1+g)}{B}\right] \qquad (6\text{-}165)$$

Let $T_1 = T_p - T_m$ denote the part of the processing delay in excess of the measurement interval. Substituting (6-157) and (6-159) into (6-165), we obtain

$$\sigma_e^2 > 200(\log e)^2\left(\frac{\alpha}{f_d} - \frac{2d}{c} - T_1\right)^{-1}\left[\frac{N_0}{p_0} + \frac{(K-1)(1+g)}{B}\right] + \sigma_p^2 + \sigma_q^2 + \sigma_d^2 \qquad (6\text{-}166)$$

This lower bound indicates that σ_e^2 increases with f_d and, hence, the fading rate when the power estimation is ideal.

Inequality (6-166) indicates that an increase in the Doppler spread f_d can be offset by an increase in the bandwidth B. This observation clarifies why third-generation cellular CDMA systems such as WCDMA or cdma 2000 exhibit no more sensitivity to power-control errors than the IS-95 system despite the substantial increase in the fading rate due to the increased carrier frequency. The physical reason is that an expansion of the bandwidth of the direct-sequence signals allows enough interference suppression to more than compensate for the increased Doppler spread. Furthermore, the potential effect of power-control errors on third-generation CDMA systems is mitigated by the use of convolutional and turbo codes more powerful than the IS-95 codes.

Consider a network of CDMA systems that do not expand the bandwidth when the Doppler spread changes, but adjust T_p so that (6-159) provides a tight bound. Ideal power estimation is assumed so that the lower bound in (6-166) approximates σ_e^2. If the other parameters are unchanged as the Doppler spread changes from f_{d1} to f_{d2}, then σ_e^2 is only affected by the *Doppler factor* defined as

$$D = \frac{f_{d2}}{f_{d1}} \tag{6-167}$$

An example of the impact of the Doppler factor is illustrated in Figure 6.19, which shows the upper bounds on P_b for instantaneous power control and the BCH (63,30) code. The network experiences slow fading and a Doppler spread $f_{d1} = 100$ Hz. The Doppler factor is $D = 1$. When the Doppler factor is $D = 2$, 3, or 4, perhaps because of increased vehicular speeds, the network is assumed to experience fast fading. The parameter values are $\alpha = 0.1$, $d = 10$ km, $T_1 = 100$ μs, $B = 1/T_c = 1.25$ MHz, $N_0/p_0 = 5$ μs, $\sigma_p^2 + \sigma_q^2 + \sigma_d^2 = 0.5$ (dB)^2, $h = 2/3$, $G = 128$, and $\gamma_0 = p_0 T_s/N_0 = (p_0/N_0)(G/B) = 20 = 13$ dB. The calculations use (6-166), (6-151), (6-104), and (6-154) to (6-156). In this example, $D \geq 2.5$ causes a significant performance degradation despite the improved time diversity during the fast fading.

When fast fading causes large power-control errors, a DS/CDMA network exhibits a significant performance degradation, notwithstanding the exploitation of time diversity by interleaving and channel coding. Adopting long-term-average instead of instantaneous power control will not cure the problem. A better approach is to increase the bandwidth of the direct-sequence signals. If the bandwidth cannot be increased enough, then the Doppler spread might be reduced by minimizing the carrier frequency of the direct-sequence signals. Another strategy is to limit the size of cells so that the network must cope with the more benign Ricean fading rather than Rayleigh fading, which is more likely to cause large power-control errors.

It follows from (6-165) and $T_p = T_m + T_1$ that a specified σ_m can be attained if

$$T_p \geq \frac{200(\log e)^2}{\sigma_m^2}\left[\left(\frac{N_0}{p_0}\right) + K_1(1+g)\right] + T_1 \tag{6-168}$$

where $K_1 = (K-1)/B$ is the number of interfering active mobiles per unit bandwidth in the cell or sector. Inequalities (6-168) and (6-159) restrict the range of feasible values for T_p. Combining (6-158), (6-159), and (6-168) and assuming that K is large enough that $K_1 \approx K/B$, we conclude that to attain $\sigma_e < 2$ dB for vehicles at speed v or less, an approximate upper bound on the

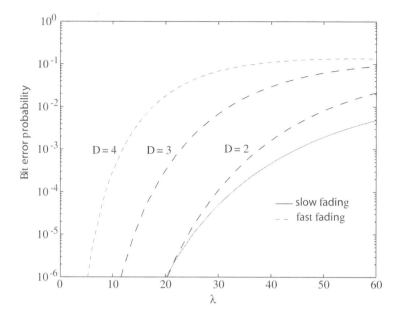

Figure 6.19: Information-bit error probability for slow fading and fast fading with different Doppler factors D. Instantaneous power control and the BCH (63,30) code are used.

uplink capacity per unit bandwidth in a cell or sector is given by

$$K_1 < (1+g)^{-1} \left[\frac{\sigma_m^2}{200(\log e)^2} \left(\frac{\alpha c}{v f_c} - \frac{2d}{c} - T_1 \right) - \left(\frac{N_0}{p_0} \right) \right] \qquad (6\text{-}169)$$

For typical parameter values, this upper bound is approximately inversely proportional to both the carrier frequency f_c and the maximum vehicle speed v.

Figure 6.20 illustrates the upper bound on the uplink capacity per megahertz as a function of frequency f_c for $\alpha = 0.1$, $\sigma_m = 1.5$ dB, $\sigma_\eta = 8$ dB, $N_0/p_0 = 5$ μs, $T_1 = 100$ μs, and representative values of d and v. Table 6.1 gives $g = 0.634$. The figure indicates the limitations on K_1 due to power control as the carrier frequency increases if σ_m and the other parameters remain fixed. If K_1 exceeds the upper bound, then the network performance will be severely degraded. The uplink capacity $K_1 B$ can be maintained by expanding the bandwidth.

Downlink Power Control and Outage

Along with all the signals transmitted to mobiles associated with it, a base station transmits a pilot signal over the downlinks. A mobile, which is usually associated with the base station from which it receives the largest pilot signal, uses the pilot to identify a base station or sector, to initiate uplink power control, to estimate the attenuation, phase shift, and delay of each significant multipath

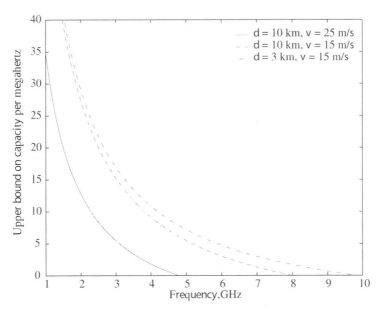

Figure 6.20: Upper bound on uplink capacity per megahertz for $\alpha = 0.1$, σ_m = 1.5 dB, g = 0.634, $N_0/p_0 = 5$ μs, and $T_1 = 100$ μs.

component, and to assess the power-allocation requirement of the mobile.

A base station synchronously combines and transmits the pilot and all the signals destined for mobiles associated with it. Consequently, all the signals fade together, and the use of orthogonal spreading sequences will prevent intracell interference and, hence, a near-far problem on a downlink, although there will be interference caused by asynchronously arriving multipath components. The orthogonal sequences can be generated from the rows of a Hadamard matrix. The orthogonality, the energy-saving sharing of the same pilot at all covered mobiles, and the coherent demodulation of all transmitted signals are major advantages of the downlinks. However, interference signals from other base stations arrive at a mobile asynchronously and fade independently, thereby significantly degrading performance.

Although there is no near-far problem on the downlinks, power control is still desirable to enhance the received power during severe fading or when a mobile is near a cell edge. However, this power enhancement increases inter-cell interference. Downlink power control entails power allocation by the base station in a manner that meets the requirements of the individual mobiles associated with it. Let C_{ij} denote the total power received by mobile i from base station j. If this mobile is associated with base station 0, then the SINR at the mobile is

$$\text{SINR} = \frac{\beta\,\phi_i\,C_{i0}\,T_s}{N_0 + \sum_{j=1}^{N_B-1} C_{ij}\,T_c} \qquad (6\text{-}170)$$

where N_B is the total number of base stations that produce significant power at mobiles in a cell or sector, β is the fraction of the base-station power that is assigned to mobiles rather than to the pilot, and ϕ_i is the fraction of the total power for mobiles in a cell or sector that is allocated to mobile i. Typically, one might set $\beta = 0.8$, which entails a 1-dB loss due to the pilot. Let R denote the SINR required by network mobiles for acceptable performance. Inverting (6-170), it is found that R is achieved by all mobiles in a cell or sector if

$$\phi_i \geq \frac{R}{\beta\, C_{i0}\, G} \left(\frac{N_0}{T_c} + \sum_{j=1}^{N_B-1} C_{ij} \right), \quad i = 1, 2, \dots, K \tag{6-171}$$

An outage occurs if the demands of all K mobiles in a cell or sector cannot be met simultaneously. Thus, no outage occurs if (6-171) is satisfied and

$$\sum_{i=1}^{K} \chi_i\, \phi_i \leq 1 \tag{6-172}$$

where χ_i is the voice-activity indicator such that $\chi_i = 1$ with probability q and $\chi_i = 0$ with probability $1 - q$. If the left-hand side of (6-172) is strictly less than unity, then the transmitted power produced by base station 0 can be safely lowered to reduce the interference in other cells or sectors. Combining (6-171) and (6-172), a necessary condition for no outage is

$$\sum_{i=1}^{K} \frac{\chi_i}{C_{i0}} \left(\frac{N_0}{T_c} + \sum_{j=1}^{N_B-1} C_{ij} \right) \leq \frac{\beta\, G}{R} \tag{6-173}$$

The assignment of mobile i to base station 0 implies the constraint that $C_{ij} \leq C_{i0}$, $j = 1, 2, \dots, N_B - 1$, except possibly during a soft handoff. A complete performance analysis with this constraint is difficult. Simulation results [19] indicate that the downlink capacity potentially exceeds the uplink capacity if the orthogonal signaling is not undermined by excessive multipath.

6.5 Multiuser Detectors

The conventional single-user direct-sequence receiver of Figure 2.14 is optimal against multiple-access interference only if the spreading sequences of all the interfering signals and the desired signal are orthogonal. Orthogonality is possible in a synchronous communication network, but in an asynchronous network, it is not possible to find sequences that remain orthogonal for all relative delays. Thus, the conventional single-user receiver, which only requires knowledge of the spreading sequence of the desired signal, is suboptimal against asynchronous multiple-access interference. The price of the suboptimality might be minor if the spreading sequences are carefully chosen and the noise is relatively high, especially if an error-control code and a sector antenna or adaptive array are used. If a potential near-far problem exists, power control may be used to limit its impact. However, power control is imperfect, entails a substantial overhead cost, and is not feasible for peer-to-peer communication networks. Even if the power control is perfect, the remaining interference causes a nonzero *error floor*, which is a minimum bit error probability that exists when the thermal noise is zero. Thus, an alternative to the conventional receiver is desirable.

A *multiuser detector* is a receiver that exploits the deterministic structure of multiple-access interference or uses joint processing of a set of multiple-access signals. An optimum multiuser detector almost completely eliminates the multiple-access interference and, hence, the near-far problem, thereby rendering power control unnecessary, but such a detector is prohibitively complex to implement, especially when long spreading sequences are used. A more practical multiuser detector alleviates but does not eliminate the power-control requirements of a cellular network on its uplinks. Even if a multiuser detector rejects *intracell interference* from mobiles within a cell, it cannot reject *intercell interference*, which arrives from mobiles associated with different base stations than the one receiving a desired signal. Since intercell interference is typically more than one-third of the total interference on an uplink, even ideal multiuser detection will increase network capacity by a factor less than three. Multipath components can be accommodated as separate interference signals or rake combining may precede the multiuser detection. Though suboptimal compared with ideal multiuser detection, multiuser interference cancellers bear a much more moderate implementation burden and still provide considerable interference suppression. However, it appears that accurate power control is still needed at least for initial synchronization and to avoid overloading the front end of the receiver. Third-generation CDMA systems use adaptive interference cancellation but retain a closed-loop power-control subsystem.

Optimum Detectors

Consider a DS/CDMA network with K users, each of which uses PSK to transmit a block of N binary symbols. A *jointly optimum* detector makes collective symbol decisions for K received signals based on the *maximum a posteriori* (MAP) criterion. The *individually optimum* detector selects the most probable set of symbols of a single desired signal from one user based on the MAP criterion, thereby providing the minimum symbol error probability. In nearly all applications, jointly optimum decisions would be preferable because of their lower complexity and because both types of decisions will agree with very high probability unless the symbol error probability is very high. Assuming equally likely symbols are transmitted, the jointly optimum MAP detector is the same as the jointly optimum maximum-likelihood detector, which is henceforth referred to as the *optimum detector*.

For synchronous communications in the presence of white Gaussian noise, the symbols are aligned in time, and the detection of each symbol of the desired signal is independent of the other symbols. Thus, the optimum detector can be determined by considering a single symbol interval $0 \leq t \leq T_s$. Let d_k denote the symbol transmitted by user k. The customary (highly idealized) assumption is that a perfect carrier synchronization enables the receiver to remove a common carrier frequency and phase. Thus, the composite baseband received signal is

$$r(t) = \sum_{k=1}^{K} A_k d_k p_k(t) + n(t), \qquad 0 \le t \le T_s \qquad (6\text{-}174)$$

where A_k is the received symbol amplitude from user k, $p_k(t)$ is the unit-energy spreading waveform of user k, $d_k = \pm 1$, and $n(t)$ is the baseband Gaussian noise. If it is assumed that each of the K signals has a common carrier frequency but a distinct phase relative to the phase of the receiver-generated synchronization signal, then each A_k in (6-174) is replaced by $A_k \cos \phi_k$, where ϕ_k is the relative phase of the signal from mobile k.

Assuming that all possible values of the symbol vector $\mathbf{d} = [d_1 \ldots d_K]^T$ are equally likely, the optimum detector is the *maximum-likelihood detector* [22], [23], which selects the value of $\mathbf{d}$ that minimizes the log-likelihood function

$$\Lambda(\mathbf{d}) = \int_0^{T_s} \left[r(t) - \sum_{k=1}^{K} A_k d_k p_k(t) \right]^2 dt \qquad (6\text{-}175)$$

subject to the constraint that $d_k = +1$ or -1. The vector of the cross correlations between $r(t)$ and the spreading sequences is denoted by $\boldsymbol{\theta} = [r_1 \ r_2 \ldots r_K]^T$, where

$$r_k = \int_0^{T_s} r(t) p_k(t) dt , \qquad k = 1, 2, \ldots, K \qquad (6\text{-}176)$$

Let $\mathbf{A}$ denote the $K \times K$ diagonal matrix with diagonal components $A_1, A_2, \ldots, A_K$. Let $\mathbf{R}$ denote the $K \times K$ *correlation matrix* with elements

$$R_{ik} = \int_0^{T_s} p_i(t) p_k(t) dt , \qquad i, k = 1, 2, \ldots, K \qquad (6\text{-}177)$$

where $R_{ii} = 1$ and $|R_{ik}| \le 1$ because the spreading waveforms are normalized to unit energy. Expanding (6-175), dropping an integral that is irrelevant to the selection of $\mathbf{d}$, and then substituting (6-176) and (6-177), we find that the maximum-likelihood detector selects the value of $\mathbf{d}$ that maximizes the *correlation metric*

$$C = 2\mathbf{d}^T \mathbf{A}\boldsymbol{\theta} - \mathbf{d}^T \mathbf{A}\mathbf{R}\mathbf{A}\mathbf{d} \qquad (6\text{-}178)$$

subject to the constraint that $d_k = +1$ or $-1, k = 1, 2, \ldots, K$.

This equation implies that the optimum detector uses a filter bank of K parallel correlators. Correlator k computes r_k given by (6-176) and can be implemented as the single-user detector of Figure 6.15. Equation (6-178) also indicates that the K spreading sequences must be known so that $\mathbf{R}$ can be calculated, and the K signal amplitudes must be estimated. Short spreading sequences are necessary or $\mathbf{R}$ must change with each symbol. The optimum detector is capable of making joint symbol decisions for all K signals or merely the symbol decisions for a single signal.

As an example, consider synchronous communications with $K = 2$ and $R_{12} = \rho$. After the elimination of terms irrelevant to the selection, (6-178) implies that the optimum detector evaluates $C = 2A_1 d_1 \theta_1 + 2A_2 d_2 \theta_2 - 2\rho A_1 A_2 d_1 d_2$ for the four pairs with $d_1 = \pm 1$ and $d_2 \pm 1$. The pair that maximizes C provides the decisions for d_1 and d_2.

For asynchronous communications over the AWGN channel, the derivation of the maximum-likelihood detector is analogous but more complicated [22], [23]. A major difference is that a desired symbol overlaps two consecutive symbols from each interference signal. The optimum detector uses a filter bank of K parallel correlators, but N symbols from each correlator must be processed to make decisions about NK binary symbols. The vector $\mathbf{d}$ is $NK \times 1$ with the first N elements representing the symbols of signal 1, the second N elements representing the symbols of signal 2, etc. The detector must estimate the transmission delays of all K multiple-access signals, and the $NK \times NK$ correlation matrix $\mathbf{R}$ has components that are partial cross correlations among the signals. In principle, the detector must compute 2^{NK} correlation metrics and then select K symbol sequences, each of length N, corresponding to the largest correlation metric. The Viterbi algorithm simplifies computations by exploiting the fact that each received symbol overlaps at most $2(K-1)$ other symbols. Nevertheless, the computational complexity increases exponentially with K.

In view of both the computational requirements and the parameters that must be estimated, it is highly unlikely that the optimum multiuser detector will have practical applications. Subsequently, alternative suboptimal multiuser detectors are considered. All of them follow carrier removal with a filter bank of correlators.

Decorrelating detector

The decorrelating detector may be derived by maximizing the correlation metric of (6-178) without any constraint on $\mathbf{d}$. For this purpose, the gradient of f with respect to the n-dimensional, real-valued vector $\mathbf{x}$ is defined as the column vector $\nabla_{\mathbf{x}} f$ with components $\partial f / \partial x_i$, $i = 1, 2, \ldots, n$. From this definition, it follows that for column vectors $\mathbf{x}$ and $\mathbf{y}$

$$\nabla_{\mathbf{x}} \left(\mathbf{x}^T \mathbf{y} \right) = \nabla x \left(\mathbf{y}^T \mathbf{x} \right) = \mathbf{y} \qquad (6\text{-}179)$$

If $\mathbf{A}$ is an $n \times n$ symmetric matrix, then expressing $\mathbf{x}^T \mathbf{A} \mathbf{x}$ in component form and using the chain rule yields

$$\nabla_{\mathbf{x}} \left(\mathbf{x}^T \mathbf{A} \mathbf{x} \right) = 2\mathbf{A}\mathbf{x} \qquad (6\text{-}180)$$

Applying (6-179) and (6-180) to the correlation metric, we find that $\nabla_{\mathbf{x}} C = \mathbf{0}$ implies that C is maximized by $\mathbf{d} = \mathbf{d}'$, where

$$\mathbf{A}\mathbf{d}' = \mathbf{R}^{-1} \boldsymbol{\theta} \qquad (6\text{-}181)$$

provided that $\mathbf{R}$ is invertible. Since each component of the vector $\mathbf{A}\mathbf{d}'$ is a positive multiple of the corresponding component of $\mathbf{d}'$, there is no need to

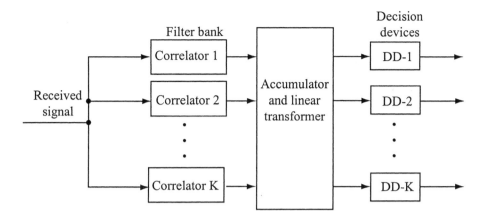

Figure 6.21: Architecture of decorrelating detector and MMSE detecter. Filter bank comprises parallel correlators.

solve for $\mathbf{d}'$. A suitable estimate of the transmitted bits is

$$\hat{\mathbf{d}} = \text{sgn}\left(\mathbf{R}^{-1}\boldsymbol{\theta}\right) \tag{6-182}$$

where each component of the vector $\text{sgn}(\mathbf{x})$ is the signum function of the corresponding component of the vector $\mathbf{x}$. The *signum function* is defined as $\text{sgn}(x) = 1, x \geq 0$, and $\text{sgn}(x) = -1, x < 0$. The *decorrelating detector*, which implements (6-182), has the form diagrammed in Figure 6.21. For asynchronous communications, each of the K correlators in the filter bank produces N bits. The accumulator constructs the NK-dimensional vector $\boldsymbol{\theta}$ and the linear transformer computes $\mathbf{R}^{-1}\boldsymbol{\theta}$. The decision devices evaluate (6-182) to produce $\hat{\mathbf{d}}$.

A second derivation of the decorrelating detector assumes that the detector has the filter bank as its first stage. If (6-174) gives the input, then the output of this stage is

$$\boldsymbol{\theta} = \mathbf{R}\mathbf{A}\mathbf{d} + \mathbf{n} \tag{6-183}$$

where $\mathbf{n}$ is the NK-dimensional noise vector. This equation indicates that the coupling among components of $\mathbf{d}$, which causes the correlation among components of $\boldsymbol{\theta}$, is due solely to the presence of the matrix $\mathbf{R}$. The effect of this matrix is removed by computing

$$\mathbf{R}^{-1}\boldsymbol{\theta} = \mathbf{A}\mathbf{d} + \mathbf{R}^{-1}\mathbf{n} \tag{6-184}$$

If (6-182) is used to determine the NK transmitted bits, the multiple-access interference is completely decorrelated from $\hat{\mathbf{d}}$.

A third derivation assumes the presence of the filter bank. If zero-mean, white Gaussian noise with two-sided power spectral density $N_0/2$ enters each correlator, then a straightforward calculation indicates that the $NK \times NK$

covariance matrix of $\boldsymbol{\theta}$ is

$$E\left[\mathbf{nn}^T\right] = \frac{N_0}{2}\mathbf{R} \tag{6-185}$$

The probability density function of $\boldsymbol{\theta}$ given $\mathbf{Ad}$ is

$$f(\boldsymbol{\theta}|\mathbf{Ad}) = (N_0\pi \det \mathbf{R})^{-K/2} \exp\left[-N_0^{-1}(\boldsymbol{\theta} - \mathbf{RAd})^T \mathbf{R}^{-1}(\boldsymbol{\theta} - \mathbf{RAd})\right] \tag{6-186}$$

The maximum-likelihood estimate of $\mathbf{Ad}$ is the estimate that maximizes (6-186) or, equivalently, minimizes the log-likelihood function

$$\Lambda(\mathbf{Ad}) = (\boldsymbol{\theta} - \mathbf{RAd})^T \mathbf{R}^{-1}(\boldsymbol{\theta} - \mathbf{RAd}) \tag{6-187}$$

Using (6-179) and (6-180), we again obtain the estimate given by (6-181), which leads to $\hat{\mathbf{d}}$ given by (6-182).

Although the decorrelating detector eliminates the multiple-access interference, it increases the noise by changing $\mathbf{n}$ to $\mathbf{R}^{-1}\mathbf{n}$. From (6-185), and the symmetric character of $\mathbf{R}$, it follows that the covariance matrix of the noise vector $\mathbf{R}^{-1}\mathbf{n}$ entering the decision devices is

$$E\left[\mathbf{R}^{-1}\mathbf{nn}^T\mathbf{R}^{-1}\right] = \frac{N_0}{2}\mathbf{R}^{-1} \tag{6-188}$$

The variance of the noise that accompanies one of the symbols of user k is $(N_0/2)(\mathbf{R}^{-1})_{kk}$. Therefore, the symbol error probability is

$$P_s(k) = Q\left(\sqrt{\frac{2\mathcal{E}_k}{N_0\left(\mathbf{R}^{-1}\right)_{kk}}}\right) \quad , \quad k = 1, 2, \ldots, K \tag{6-189}$$

where $\mathcal{E}_k = A_k^2$ is the symbol energy. The symbol error probability for single-user detection by user k in the absence of multiple access interference is

$$P_s(k) = Q\left(\sqrt{\frac{2\mathcal{E}_k}{N_0}}\right) \quad , \quad k = 1, 2, \ldots, K \tag{6-190}$$

Thus, the presence of multiple-access interference requires an increase of energy by the factor $(\mathbf{R}^{-1})_{kk}$ when the decorrelating detector is used if a specified P_s is to be maintained.

As an example, consider synchronous communications with $K = 2$ and $R_{12} = R_{21} = \rho$. The correlation matrix and its inverse are

$$\mathbf{R} = \begin{bmatrix} 1 & \rho \\ \rho & 1 \end{bmatrix}, \quad \mathbf{R}^{-1} = \frac{1}{1-\rho^2}\begin{bmatrix} 1 & -\rho \\ -\rho & 1 \end{bmatrix} \tag{6-191}$$

Equation (6-182) indicates that the symbol estimates are $\hat{d}_1 = \text{sgn}(r_1 - \rho r_2)$ and $\hat{d}_2 = \text{sgn}(r_2 - \rho r_1)$. Since $(\mathbf{R}^{-1})_{11} = (\mathbf{R}^{-1})_{22} = (1 - \rho^2)^{-1}$,

$$P_s(k) = Q\left(\sqrt{\frac{2\mathcal{E}_k(1 - \rho^2)}{N_0}}\right) \quad , \quad k = 1, 2 \tag{6-192}$$

If $\rho \leq 1/2$, the required energy increase or shift in each P_s curve is less than 1.25 dB.

To demonstrate analytically the advantage of the decorrelating detector, consider synchronous communications and a receiver with a filter bank of K conventional detectors. Each conventional detector is a single-user matched filter. If perfect carrier synchronization removes a common phase shift of all the signals and produces the baseband received signal of (6-174), then (6-176) implies, that the output of detector k is

$$r_k = d_k A_k + \sum_{\substack{i=1 \\ i \neq k}}^{K} d_i A_i R_{ik} + \int_0^{T_s} n(t) p_k(t) dt \qquad (6\text{-}193)$$

The set of K symbols is estimated by

$$\hat{d} = \text{sgn}(\theta) \qquad (6\text{-}194)$$

By symmetry, we can assume that $d_k = 1$ in the evaluation of the symbol error probability. Let $\mathbf{D}_k$ denote the $(K-1)$-dimensional vector of all the $d_i, i \neq k$. Conditioning on $\mathbf{D}_k$ and calculating that $\text{var}(r_k) = N_0/2$, we find that the conditional symbol error probability for user k is

$$P_s(k|\mathbf{D}_k) = Q\left(\sqrt{\frac{2\mathcal{E}_k}{N_0}} B_k\right) \qquad (6\text{-}195)$$

where

$$B_k = 1 + \sum_{\substack{i=1 \\ i \neq k}}^{K} d_i R_{ik} \frac{A_i}{A_k} \qquad (6\text{-}196)$$

If all symbol sets are equally likely, then the symbol error probability for user k is

$$P_s(k) = 2^{-(K-1)} \sum_{j=1}^{2^{K-1}} P_s(k|\mathbf{D}_{kj}) \qquad (6\text{-}197)$$

where $\mathbf{D}_{kj}$ is the jth choice of the vector $\mathbf{D}_k$, which can take 2^{K-1} values.

For $K = 2$ with $R_{12} = R_{21} = \rho$ and $A_2/A_1 = \sqrt{\mathcal{E}_2}/\sqrt{\mathcal{E}_1}$, (6-195) to (6-197) yield the symbol error probability for user 1:

$$
\begin{aligned}
P_s(1) &= \frac{1}{2} Q\left(\sqrt{\frac{2\mathcal{E}_1}{N_0}}\left(1 - \rho\sqrt{\frac{\mathcal{E}_2}{\mathcal{E}_1}}\right)\right) + \frac{1}{2} Q\left(\sqrt{\frac{2\mathcal{E}_1}{N_0}}\left(1 + \rho\sqrt{\frac{\mathcal{E}_2}{\mathcal{E}_1}}\right)\right) \\
&= \frac{1}{2} Q\left(\sqrt{\frac{2\mathcal{E}_1}{N_0}}\left(1 - |\rho|\sqrt{\frac{\mathcal{E}_2}{\mathcal{E}_1}}\right)\right) + \frac{1}{2} Q\left(\sqrt{\frac{2\mathcal{E}_1}{N_0}}\left(1 + |\rho|\sqrt{\frac{\mathcal{E}_2}{\mathcal{E}_1}}\right)\right)
\end{aligned}
$$

$$(6\text{-}198)$$

The symbol error probability for user 2 is given by (6-198) with the roles of $\mathcal{E}_1$ and $\mathcal{E}_2$ interchanged. The second term in (6-198) is usually negligible compared with the first one if $\rho \neq 0$. Thus, if

$$\rho^2 \mathcal{E}_2 / \mathcal{E}_1 > \left(1 - \sqrt{1 - \rho^2}\right)^2 \tag{6-199}$$

then a comparison of (6-198) with (6-192) indicates that decorrelating detector usually outperforms the conventional detector. However, if $\mathcal{E}_2$ is sufficiently small, then the conventional detector gives a lower P_s than the decorrelating detector.

In a more realistic model of the decorrelating and conventional detectors, the received signal in Figure 6.21 is passband. Correlator k uses a synchronized carrier to remove carrier k at the common frequency f_c. Since each carrier has a distinct phase ϕ_k, the elements of the correlation matrix are

$$R_{ik} = \cos\left(\phi_k - \phi_i\right) \int_0^{T_c} p_i(t) p_k(t) dt \ , \qquad i, k = 1, 2, \ldots, k \tag{6-200}$$

if $f_c T_c \gg 1$. For synchronous communications with $K = 2$, (6-192) and (6-198) with $\rho = R_{12} = R_{21}$ then represent the *conditional* symbol error probability given the value of $\phi = \phi_k - \phi_i$. Averaging over ϕ is necessary to obtain $P_s(1)$ and $P_s(2)$.

Compared with the optimum detector, the decorrelating detector offers greatly reduced, but still formidable, computational requirements. There is no need to estimate the signal amplitudes, but the transmission delays of asynchronous signals must still be estimated. The inversion of the correlation matrix $\mathbf{R}$ in real time is not possible for asynchronous signals with practical values of NK. Suboptimal partitioning and short spreading sequences are generally necessary and degrade the theoretical performance given by (6-189).

Minimum-Mean-Square-Error Detector

The *minimum-mean-square-error* (MMSE) detector is the receiver that results from a linear transformation of $\boldsymbol{\theta}$ by the $K \times K$ matrix $\mathbf{L}$ such that the metric

$$M = E\left[\|\mathbf{d} - \mathbf{L}\boldsymbol{\theta}\|^2\right] \tag{6-201}$$

is minimized. Let $\mathbf{L}_0$ denote the solution of the equation

$$E\left[(\mathbf{d} - \mathbf{L}_0\boldsymbol{\theta})\,\boldsymbol{\theta}^T\right] = \mathbf{0} \tag{6-202}$$

Let $tr(\mathbf{B})$ denote the trace of the matrix $\mathbf{B}$. Since $\|\mathbf{x}\|^2 = tr(\mathbf{x}\mathbf{x}^T)$ for a vector $\mathbf{x}$,

$$\|\mathbf{d} - \mathbf{L}\boldsymbol{\theta}\|^2 = tr\left\{[\mathbf{d} - \mathbf{L}_0\boldsymbol{\theta} + (\mathbf{L}_0 - \mathbf{L})\,\boldsymbol{\theta}][\mathbf{d} - \mathbf{L}_0\boldsymbol{\theta} + (\mathbf{L}_0 - \mathbf{L})\,\boldsymbol{\theta}]^T\right\}$$

$$= \|\mathbf{d} - \mathbf{L}_0\boldsymbol{\theta}\|^2 + \|(\mathbf{L}_0 - \mathbf{L})\,\boldsymbol{\theta}\|^2 + 2tr\left[(\mathbf{d} - \mathbf{L}_0\boldsymbol{\theta})\,\boldsymbol{\theta}^T\,(\mathbf{L}_0 - \mathbf{L})^T\right]$$

$$\tag{6-203}$$

Substitution of this equation into (6-201) and the application of (6-202) yields

$$M = E\left[\|\mathbf{d} - \mathbf{L}_0\boldsymbol{\theta}\|^2\right] + E\left[\|(\mathbf{L}_0 - \mathbf{L})\boldsymbol{\theta}\|^2\right] \geq E\left[\|\mathbf{d} - \mathbf{L}_0\boldsymbol{\theta}\|^2\right] \qquad (6\text{-}204)$$

which proves that $\mathbf{L}_0$ minimizes M. If the data symbols are independent and equally likely to be $+1$ or -1, then $E[\mathbf{dd}^\mathbf{T}] = \mathbf{I}$, where $\mathbf{I}$ is the identity matrix. Using this result, (6-183), (6-185), $E[\mathbf{n}] = \mathbf{0}$, and the independence of $\mathbf{d}$ and $\mathbf{n}$, we obtain

$$E\left[\mathbf{d}\boldsymbol{\theta}^T\right] = \mathbf{AR} , \qquad E\left[\boldsymbol{\theta}\boldsymbol{\theta}^T\right] = \mathbf{RA}^2\mathbf{R} + \frac{N_0}{2}\mathbf{R} \qquad (6\text{-}205)$$

Substitution of these equations into (6-202) yields

$$\mathbf{L}_0 = \left(\mathbf{RA} + \frac{N_0}{2}\mathbf{A}^{-1}\right)^{-1} = \mathbf{A}^{-1}\left(\mathbf{R} + \frac{N_0}{2}\mathbf{A}^{-2}\right)^{-1} \qquad (6\text{-}206)$$

provided that the inverses exist. Since $\mathbf{A}$ and, hence, $\mathbf{A}^{-1}$ are diagonal matrices with positive diagonal components if all signals are active, (6-206) may be simplified to the linear transformation matrix

$$\mathbf{L}_0 = \left(\mathbf{R} + \frac{N_0}{2}\mathbf{A}^{-2}\right)^{-1} \qquad (6\text{-}207)$$

without any change in the MMSE estimate of the transmitted symbols:

$$\hat{d} = \text{sgn}(\mathbf{L}_0\boldsymbol{\theta}) \qquad (6\text{-}208)$$

The MMSE detector has the structure of Figure 6.21.

The MMSE and decorrelating detectors have almost the same computational requirements, and they both have equalizer counterparts, but they differ in several ways. The MMSE detector is near-far resistant, but does not obliterate the multiple-access interference and, hence, does not completely eliminate the near-far problem. However, it does not accentuate the noise to the degree that the decorrelating detector does. As $N_0 \to 0$, $\mathbf{L}_0 \to \mathbf{R}^{-1}$ and the MMSE estimate approaches the decorrelating detector estimate. As N_0 increases, the MMSE estimate approaches that of the conventional detector given by (6-194), and the symbol error probability generally tends to be lower than that provided by the decorrelating detector. A disadvantage of the MMSE detector is that the signal amplitudes must be estimated so that $\mathbf{A}$ in (6-207) can be computed.

For either the MMSE or decorrelating linear detectors to be practical, it is necessary for the spreading sequences to be short. Short sequences ensure that the correlation matrix $\mathbf{R}$ is approximately constant for significant time durations if the communication channel and the amplitudes of the interference signals are slowly varying. The price of short sequences is a security loss and the occasional but sometimes persistent performance loss due to a particular set of relative signal delays. Even with short spreading sequences, adaptive versions of the MMSE detector are much more practical than the nonadaptive versions of either linear detector.

An *adaptive multiuser detector* [23] is one that does not require explicit knowledge of either the spreading sequences or the timing of the multiple-access interference signals. The receiver samples the output of a wideband filter at the chip rate. The use of short spreading sequences affords the opportunity for the adaptive detector to essentially learn the sequence cross-correlations and thereby to suppress the interference. The learning is accomplished by processing a known *training sequence* of symbols for the desired signal during a *training phase*. This operational phase is followed by a *decision-directed phase* that continues the adaptation by feeding back symbol decisions. Adaptive detectors potentially can achieve much better performance than conventional ones at least if the transmission channel is time-invariant, but coping with fast fading and interference changes requires elaborate modifications. A *blind adaptive detector* [24] is one that does not require training sequences. These detectors are desirable for applications such as system recovery but entail some performance loss and complexity increase relative to other adaptive detectors. Long sequences do not possess the cyclostationarity that makes possible many of the advanced signal processing techniques used for blind multiuser detection and adaptive channel estimation.

Interference Cancellers

An *interference canceller* is a multiuser detector that explicitly estimates the interference signals and then subtracts them from the received signal to produce the desired signal. Interference cancellers may be classified as *successive interference cancellers* in which the subtractions are done sequentially, *parallel interference cancellers* in which the subtractions are done simultaneously, or hybrids of these types. Only the basic structures and features of the successive and parallel cancellers are presented subsequently. A large number of alternative versions, some of them hybrids, adaptive, or blind, have been proposed in the literature [23]. Some type of interference canceller is by far the most practical multiuser detector for an asynchronous DS/CDMA network, especially if long spreading sequences are planned [25].

Successive Interference Canceller

Figure 6.22 is a functional block diagram of a successive interference canceller, which uses nonlinear replica generations and subtractions to produce estimates of the symbol streams $\hat{d}_1, \hat{d}_2, \ldots, \hat{d}_K$ transmitted by the K users. The input may be the sampled outputs of a chip-matched filter for PSK modulation or the complex-valued samples derived from a quadrature demodulator for quaternary modulation. Detector-generator i produces a replica of the signal transmitted by user i. Its basic structure is depicted in Figure 6.23. The correlator, which comprises a multiplier and summer as in Figure 6.15, despreads signal i. The channel estimate is a stream of complex numbers that are applied to the correlator output to remove the effects of the propagation channel. The decision device produces the estimated symbols transmitted by user i. These symbols are remodulated and modified to account for the effects of the channel. After

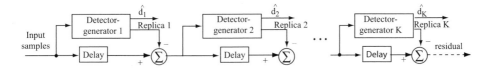

Figure 6.22: Successive interference canceller with K detector-generators to produce signal estimates for subtraction.

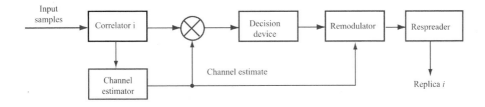

Figure 6.23: Structure of detector-generator for signal i.

a respreading, the replica of signal i is generated and sent to the corresponding subtracter. The input or output of the decision device may produce the estimated symbol stream, depending on whether the decoder uses soft or hard decisions. The channel estimator uses known pilot or training symbols to determine the effect of the channel. Hard decisions are used in the replica generation, but may not be appropriate if the channel estimate is inaccurate because a symbol error doubles the amplitude of the interference that enters the next stage of the canceller of Figure 6.22. The enhanced interference adversely affects subsequent symbol estimates and replicas.

The outputs of a set of K correlators and level detectors are applied to a device that orders the K received signals according to their power levels. This ordering determines the placement of the detector-generators in Figure 6.22. Detector-generator $i, i = 1, 2, \ldots, K$, corresponds to the ith strongest signal.

The first canceller stage eliminates the strongest signal, thereby immediately alleviating the near-far problem while exploiting the superior detectability of the strongest signal. The first difference signal is applied to the detector-generator for the second strongest received signal, etc. The amount of interference removal from a signal tends to increase from the strongest received signal to the weakest one. The delay introduced, the impact of cancellation errors, and the implementation complexity may limit the number of canceller stages to fewer than K, and a set of conventional detectors may be needed to estimate some of the symbol streams. At a low SNR, inaccurate cancellations may cause the canceller to lose its advantage over the conventional detector. The successive interference canceller of Figure 6.22 requires known spreading sequences and timing of all signals.

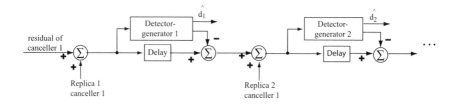

Figure 6.24: Second canceller of multistage canceller using successive interference cancellers.

A *multistage interference canceller* comprising more than one successive interference canceller potentially improves performance by repeated cancellations if the delay and complexity can be accommodated. The second canceller or stage of a multistage canceller is illustrated in Figure 6.24. The input is the residual of canceller 1, which is shown in Figure 6.22. Replica 1 of canceller 1 is added to the input and then an improved replica of signal 1 is subtracted. Subsequently, other replicas from canceller 1 are added and corresponding improved replicas are subtracted. The symbol streams are produced by the final canceller. Rake combining of multipath components may be incorporated into a multistage or single-stage canceller to improve performance in a fading environment [26].

Parallel Interference Canceller

A parallel interference canceller detects, generates, and then subtracts all multiple-access interference signals simultaneously, thereby avoiding the delay inherent in successive interference cancellers. A parallel interference canceller for two signals is diagrammed in Figure 6.25. Each detector-generator pair may be implemented as shown in Figure 6.23. Each of the final detectors includes a digital matched filter and a decision device that produce soft or hard decisions, which are applied to the decoder. Since all signals are processed in the same manner and the initial detections influence the final ones, the parallel interference canceller is not as effective in suppressing the near-far problem as the successive interference canceller unless the CDMA network uses power control. Power control also relieves the timing synchronization requirements. A better suppression of the near-far problem is provided by the *multistage parallel interference canceller*, in which each stage is similar but has an improved input that results in an improved output. Figure 6.26 shows the multistage canceller for two signals. Each stage has the form of Figure 6.25 without the final detectors.

Multiuser Detector for Frequency Hopping

Multiuser detection is more challenging for frequency-hopping systems than for direct-sequence systems, but it is possible in principle. The hopping patterns

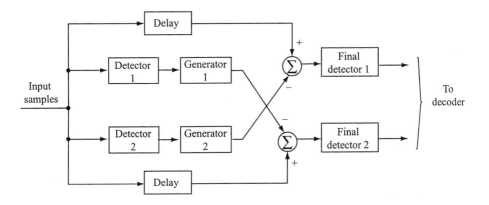

Figure 6.25: Parallel interference canceller for two signals.

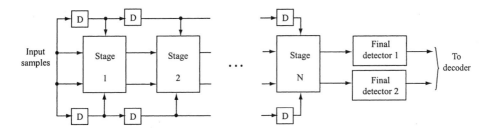

Figure 6.26: Multistage parallel interference canceller for two signals. D=delay.

of all users to be detected must be known.   In an asynchronous frequency-hopping network, the hop transition times serve as identifying parameters for the network users and must be known by a receiver, which uses them to determine when collisions will occur. The multiuser detector comprises a set of single-user turbo decoders and an iterative decoding algorithm that exploits a collaboration between the decoders and the demodulator [27]. If the number of signals colliding during an interval is too large, then the symbols during that interval are erased. The detector decodes the user's signal with the smallest number of erasures among the undecoded signals during a particular iteration. Likelihood ratios associated with the symbols of a decoded codeword are fed back to the demodulator to enable an improved demodulation of those unerased symbols occurring during a collision with the decoded signal. Simulation results indicate that an excellent performance may be possible, but the computational complexity is high, and an implementation is impractical for most networks.

6.6   Frequency-Hopping Multiple Access

Two major advantages of frequency hopping are that it can be implemented over a much larger frequency band than it is possible to implement direct-sequence spreading, and that the band can be divided into noncontiguous segments. Another major advantage is that frequency hopping provides resistance to multiple-access interference, while not requiring power control to prevent the near-far problem. Since direct-sequence systems cannot escape the near-far problem by hopping, accurate power control is crucial but becomes much less effective as the fading rate increases. These advantages of frequency hopping will be decisive in many applications. For example, the Bluetooth system and combat net radios use frequency hopping to avoid the near-far problem.

Frequency-hopping systems are usually part of a *frequency-hopping code-division multiple-access* (FH/CDMA) network in which all systems share the same M frequency channels. In a *synchronous* FH/CDMA network, the systems coordinate their frequency transitions and hopping patterns. Consequently, as many as M frequency-hopping signals can be simultaneously accommodated by the network with insignificant multiple-access interference at any of the active receivers. Network coordination is much simpler to implement than for a DS/CDMA network because the timing alignments must be within a small fraction of a hop duration instead of a small fraction of a spreading-sequence chip. Multipath signals and errors in range estimates can be accommodated at some cost in the energy per information bit by increasing the switching time between frequency-hopping pulses. However, some type of centralized or cellular architecture is required, and such an architecture is often unavailable.

Asynchronous FH/CDMA Networks

An *asynchronous* FH/CDMA network has systems that transmit and receive autonomously and asynchronously. When two or more frequency-hopping signals using the same frequency channel are received simultaneously, they are said to *collide*. Since the probability of a collision in an asynchronous network is decreased by increasing the number of frequency channels in the hopset, it is highly desirable to choose a data modulation that has a compact spectrum. Good candidates are FH/CPFSK systems that use a frequency discriminator for demodulation. As explained in Chapter 3, binary CPFSK with $h = 0.7$ and bandwidth such that $BT_s = 1$ provides excellent potential performance if the spectral splatter and intersymbol interference generated by this modulation are negligible. However, for approximately the same degree of spectral splatter and intersymbol interference as MSK with $BT_s = 1$, the bandwidth must be increased so that $BT_s = 1.4$, which reduces the number of frequency channels M in a fixed hopping band. This much reduction in M is enough to completely offset the intrinsic performance advantage of binary CPFSK with $h = 0.7$. Thus, the choice between the latter and MSK or GMSK will depend on the details of the impact of the spectral splatter and intersymbol interference.

Let d represent the *duty factor*, which is defined as the probability that an interferer using the same frequency channel will degrade the reception of a symbol. Thus, $d = q_1 q_2$ is the product of the probability q_1 that an interferer is transmitting and the probability q_2 that a significant portion of the interferer's transmitted waveform occurs during the symbol interval. The probability q_2 is upper bounded and well approximated by the probability that there is any overlap in time of the interference and the symbol interval. For synchronous frequency hopping, $q_2 = 1$. Since $T_{sw} > T_s$, it follows from elementary probability that for asynchronous frequency hopping, $q_2 \approx (T_d + T_s)/T_h$. For voice communications with voice-activity detection, $q_1 = 0.4$ is a typical value.

For asynchronous frequency hopping, the fact that $T_{sw} > T_s$ ensures that each potentially interfering frequency-hopping signal transmits power in at most one frequency channel during the reception of one symbol of a desired signal. Therefore, assuming that an interferer may transmit in any frequency channel with equal probability, the probability that a potentially interfering signal collides with the desired signal during a symbol interval is

$$c = \frac{d}{M} \tag{6-209}$$

When a collision occurs, the symbol is said to be *hit* by the interfering signal. For MFSK, M is given by (3-71).

Consider an FH/CDMA network of K asynchronous systems with negligible spectral splatter and intersymbol interference. The code symbols are interleaved so that each code symbol of a codeword is transmitted in a separate dwell interval. Test symbols are used to determine erasures of all the symbols in a dwell interval (Chapter 3). The $2N_t$ test symbols are split into separate sets of $N_t \geq 1$ test symbols at each end of a dwell interval [28]. Thus, if a code symbol is hit by one or more of the $K - 1$ interfering signals, then at least one set of the test symbols in that same interval is also hit. For analytical simplicity, we make the following assumptions:

1. If at least one of the two test symbols at the opposite ends of a dwell interval is hit, then an erasure is always made. Thus, if a code symbol is hit, an erasure is always made.

2. If a code symbol is not hit, then this condition has a negligible influence on the probability that one of the two end test symbols is hit.

3. The probability that both end test symbols are hit is negligible.

These assumptions are approximately valid if $N_h >> N_t \geq 1$ and the K-1 interfering signals have approximately the same or more power than the desired signal. The first assumption implies that the probability of the erasure of a code symbol is

$$P_\epsilon = [1 - (1 - c)^{K-1}] + (1 - c)^{K-1} P_{\epsilon 0} \tag{6-210}$$

where $P_{\epsilon 0}$ is the erasure probability given that no hit of the code symbol occurred. Observe that if neither of the end test symbols is hit, then no test symbol is hit. Therefore, the assumptions imply that

$$P_{\epsilon 0} = 2[1 - (1 - c)^{K-1}] + [2(1 - c)^{K-1} - 1]\left\{1 - \left[1 - F\left(\frac{\mathcal{E}_s}{N_0}\right)\right]^{2N_t}\right\} \quad (6\text{-}211)$$

where the first term is the probability that one of the two end test symbols is hit, and the term in braces is the probability that although no test symbols are hit, an erasure occurs because at least one of the detected test symbols is incorrect. For MFSK modulation, each channel symbol is a code symbol and the energy per symbol is $\mathcal{E}_s = mr\mathcal{E}_b$, where $m = \log_2 q$ is the number of bits in a q-ary symbol, r is the code rate, and $\mathcal{E}_b$ is the energy per bit. Under the first assumption, the code-symbol error probability is

$$P_s = (1 - c)^{K-1}(1 - P_{\epsilon 0})F\left(\frac{mr\mathcal{E}_b}{N_0}\right) \quad (6\text{-}212)$$

where $F(x)$ is given by (3-64) in the absence of fading and by (3-66) in the presence of Ricean fading.

Suppose that each q-ary code symbol is mapped into q_1-ary channel symbols with $q_1 = 2^{m_1}$ and m/m_1 chosen to be an integer. The channel symbols are interleaved over m/m_1 dwell intervals to ensure independence of symbol errors when the fading in each dwell interval, if present, is independent. Since all m/m_1 channel symbols must be received correctly for there to be no code-symbol error and the channel-symbol errors are independent, (1-32) implies that

$$P_s = 1 - \left[1 - (1 - c)^{K-1}(1 - P_{\epsilon 0})F\left(\frac{m_1 r\mathcal{E}_b}{N_0}\right)\right]^{m/m_1} \quad (6\text{-}213)$$

where $F(x)$ is given by (3-67) for binary modulations with no fading and by (3-68) when the channel symbols experience independent Ricean fading. Equation (3-78) gives P_b for errors-and-erasures decoding.

Let W denote the bandwidth of the hopping band and B_u denote the bandwidth of binary FSK in the absence of coding. For MFSK channel symbols, (3-71) indicates that the number of disjoint frequency channels available for frequency hopping is

$$M = \left\lfloor \frac{2(\log_2 q_1)rW}{q_1 B_u} \right\rfloor \quad (6\text{-}214)$$

which decreases with the channel-symbol alphabet size. The fundamental advantage of MSK is the reduced bandwidth per frequency channel. The number of available frequency channels is

$$M = \left\lfloor \frac{rW}{B_{MSK}} \right\rfloor \geq \left\lfloor \frac{2rW}{B_u} \right\rfloor \quad (6\text{-}215)$$

since $B_{MSK} \leq B_u/2$.

Figure 6.27 illustrates P_b versus K-1 for FH/MFSK and FH/MSK systems that use a Reed-Solomon (64, 24) code with errors-and-erasures decoding

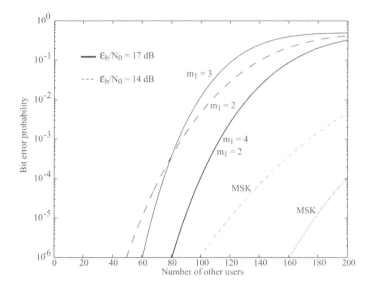

Figure 6.27: Performance of FH/MFSK and FH/MSK systems with Reed-Solomon (64, 24) code, various alphabet sizes, erasures, $W/B_u = 1000$, $d = 1$, and no fading. $N_t = 6$ for binary modulations; $N_t = 3$ for 4-ary FSK; and $N_t = 2$ for 8-ary FSK.

against asynchronous multiple-access interference in the absence of fading. The graphs are computed from (6-210) through (6-215) with M given by the lower bound in (6-215) for MSK. In all cases, $W/B_u = 1000$ and $d = 1$. It is assumed that N_h is sufficiently large that there is no expansion of required bandwidths, as illustrated in Table 3.1. The 8-ary MFSK channel symbols have $m_1 = 3$ and $N_t = 2$, the 4-ary MFSK channel symbols have $m_1 = 2$ and $N_t = 3$, and the binary channel symbols have $m_1 = 1$ and $N_t = 10$. If $\mathcal{E}_b/N_0$ is sufficiently large, the substantial benefits obtained from using binary or quaternary MFSK channel symbols and the further benefit from using MSK are apparent in the figure. These increases in the number of other users that can be accommodated must be weighed against the disadvantage of binary channel symbols in the presence of partial-band interference, as shown in Section 3.3. The figure illustrates that as $\mathcal{E}_b/N_0$ drops from 17 dB to 14 dB, the FH/FSK and FH/MSK systems degrade substantially while the nonbinary FH/MFSK systems degrade imperceptibly. This result is due to the larger symbol energy of nonbinary MFSK.

The results in the figure do not depend on N_h primarily because $T_{sw} > T_s$. If one makes the unrealistic assumption that $T_{sw} = 0$, then $N_h - 1$ symbols are hit by an interfering signal with probability d/M, and the frequency transition of the interfering signal causes one symbol to be hit with probability $2d/M$. Thus, $c = (N_h + 1)d/N_h M$, which does exhibit a dependence on N_h.

To obtain good performance against both partial-band interference and multiple-access interference, a turbo code and binary channel symbols are needed. However, even if $\mathcal{E}_s$ is known, perhaps through power control, the turbo decoder computation must be modified to account for the fluctuations from symbol-to-symbol in the interference-plus-noise variance caused by multiple-access interference [29]. When DPSK is the modulation, a suitable modification uses (1-145).

If a turbo code is not feasible, then a Reed-Solomon code with errors-and-erasures decoding is a good choice. However, for low to moderate thermal-noise levels, a trade-off is necessary in the choice of the modulation. If one is primarily interested in avoiding multiple-access interference, then binary channel symbols are desirable. If stronger protection against partial-band interference but weaker protection against multiple-access interference is needed, then non-binary channel symbols are preferable.

The results in Figure 6.27 are based on the practical assumption of a fixed bandwidth W. If this bandwidth constraint is dropped and W is optimized to produce the maximum network throughput for each channel-symbol alphabet size, then it is found that 4-ary or 8-ary channel symbols produce higher throughputs than FSK in a frequency-hopping network [30].

Mobile Peer-to-Peer and Cellular Networks

Mobile FH/CDMA systems [31] are suitable for both peer-to-peer and cellular communication networks. Mobile peer-to-peer communications are used in mobile communication networks that possess no supporting infrastructure, fixed or mobile; each user has identical signal processing capability. Peer-to-peer communications have both commercial applications and important military applications, the latter primarily because of their robustness in the presence of node losses. Power control and, hence, current DS/CDMA are not viable for peer-to-peer communications because of the lack of a centralized architecture. Current plans to use multiuser detection in direct-sequence CDMA systems still require power control, which is highly desirable for the synchronization.

A unified evaluation of the potential performance of both mobile peer-to-peer and sectorized FH/CDMA systems is provided by analysis and simulation. The propagation path losses are modeled as the result of power-law losses, shadowing, and fading. In Chapter 5, it is shown that the probability distribution function of the normalized local-mean power p_l/p_0 is

$$F(x) = 1 - Q\left\{ \frac{a}{\sigma_s} \ln\left[x\left(\frac{r}{R_0} \right)^\beta \right] \right\} \qquad (6\text{-}216)$$

where $a = (10 \log_{10} e)$, p_0 is the average received power when the distance is $r = R_0$, β is the *attenuation power law*, and σ_s is the standard deviation in decibels. The fading causes a power fluctuation about the local-mean power.

One method of combining antenna outputs is predetection combining, which requires the estimation of the signal and interference-plus-noise power levels at each antenna for maximal-ratio combining or selection diversity and requires the cophasing of the L antenna outputs for maximal-ratio or coherent equal-gain combining. Since the relative phases and power levels of the signals at the L antennas change after every hop, it is almost always impractical to implement predetection combining. As a much more practical alternative, a receiver can combine the demodulated outputs rather than the signals from the L antennas. This postdetection combining eliminates the cophasing and does not require the time alignment of L signals in practical applications because any misalignment is much smaller than a symbol duration. The estimation of power levels can be eliminated by the use of a fixed combining rule, such as equal-gain or square-law combining.

In the receiver of a frequency-hopping system, each antenna output is dehopped and filtered. The interference plus noise in each dehopped signal is approximated by independent bandlimited white Gaussian noise, with equivalent power given by

$$\sigma_1^2 = \sigma_n^2 + \sum_{i=1}^{K-1} p_{ui} \qquad (6\text{-}217)$$

where σ_n^2 is the thermal noise power, $K-1$ is the number of active frequency-hopping interference signals , and p_{ui} is the local-mean interference power received from source i. The Gaussian model is reasonable for large numbers of interference signals that generally fade independently and experience different Doppler shifts. The total interference power is approximately uniform (white) over the receiver passband following dehopping if $BT_s = \zeta \leq 1$. The L diversity antennas are assumed to be close enough to each other that the power-law losses and shadowing are nearly the same, and thus the local-mean power from a source is the same at each antenna. Each active interfering mobile may actually represent a cluster of mobiles. In this cluster, some discipline such as carrier-sense multiple access is used to ensure that there is at most one transmitted signal at any time.

The desired signal is assumed to experience frequency-nonselective Rayleigh fading. The Rayleigh fading model is appropriate under the pessimistic assumption that the propagation paths are often obstructed, and thus, the power of the direct line-of-sight signal is small compared with the reflected signal power. Frequency-nonselective fading occurs if $B < B_{\text{coh}}$. Rayleigh fading may be negligible if mobile speeds are very low, which would occur if each mobile consisted of a person walking. Shadowing would still occur but would be slowly varying over time.

Spectrally compact CPFSK or GMSK signals do not have enough frequency shift to be demodulated by classical noncoherent demodulators with parallel matched filters and envelope detectors, but can be demodulated by a frequency discriminator. We consider binary MSK with discriminator demodulation. For postdetection diversity, the outputs of L discriminators are weighted and combined. The weighting is by the square of the envelope at the input to each discriminator. When the desired signal undergoes independent Rayleigh fading at each antenna and the channel parameters remain constant for at least one symbol duration, a calculation using the results of [32] yields the symbol error probability:

$$P_s = \binom{2L-1}{L} \left(\frac{1}{4} + \frac{1}{3}\zeta^2 \right)^L (\bar{\rho})^{-L} \qquad (6\text{-}218)$$

where $\zeta = BT_s$, $\bar{\rho} = p_s/\sigma_1^2 \gg 1$, and p_s is the local-mean power of the desired signal. A comparison of this equation with (5-135) and (5-169) when $\zeta = 1$ so that $\bar{\rho} = \bar{\gamma}$ verifies that MSK with discriminator demodulation and square-law postdetection combining provides nearly the same P_s as ideal DPSK. The slowly varying shadowing in practical networks ensures that P_s is almost always nearly constant over an interleaved codeword or constraint length. The information-bit error rate following hard-decision decoding can be calculated from P_s with the equations of Chapter 1. The theoretical loss due to using postdetection rather than predetection combining is less than a decibel [32].

Peer-to-Peer Networks

Consider a peer-to-peer network of independent, identical, frequency-hopping systems that have L omnidirectional antennas, generate the same output power, share the same carriers and frequency channels, and are nearly stationary in location over a single symbol duration. The antennas are separated from each other by several wavelengths, so that the fading of both the desired signal and the interfering signals at one antenna is independent of the fading at the other antennas. A few wavelengths are adequate because mobiles, in contrast to base stations, tend to receive superpositions of reflected waves arriving from many random angles. Because of practical physical constraints, spatial diversity will ordinarily be effective only if the carrier frequencies exceed roughly 1 GHz. Polarization diversity and other forms of adaptive array processing are alternatives.

Since for peer-to-peer communications it is assumed that an interfering mobile may transmit in any frequency channel with equal probability, the probability that power from an interferer enters the transmission channel of the desired signal is

$$P_t = \frac{d}{M} \qquad (6\text{-}219)$$

It is assumed that M is sufficiently large that we may neglect the fact that a channel at one of the ends of the hopping band has only one adjacent channel instead of two. Consequently, the probability that the power from an interferer enters one of the two adjacent channels of the desired signal is

$$P_a = \frac{2d}{M} \qquad (6\text{-}220)$$

The probability that the power enters neither the transmission channel nor the adjacent channels is $(1 - 3d/M)$. These equations make it apparent that the performance of a frequency-hopping system depends primarily on the ratio $M_1 = M/d$. This ratio is called the *equivalent number of channels* because any decrease in the duty factor has the same impact as an increase in the number of frequency channels; what matters most for performance is this ratio.

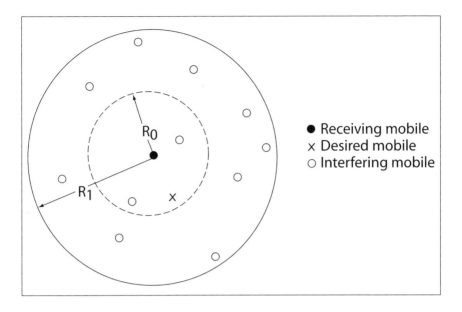

Figure 6.28: Geometry of a peer-to-peer communication network.

In the simulation, the locations of the mobiles are assumed to be uniformly distributed in a circular region surrounding a specific mobile receiver, as illustrated in Figure 6.28. Therefore, the radial distance of a mobile from the receiver has the probability distribution function

$$G(r) = \frac{r^2}{R^2} , \quad 0 < r \leq R \tag{6-221}$$

where R is the radius of the circle. The distance of the desired mobile is randomly selected according to this distribution with $R = R_0$, where R_0 is the maximum communication range and corresponds to a received area-mean signal power equal to p_0. The distance of each interfering mobile is randomly selected according to this distribution with $R = R_1$. The selected distance of the desired mobile is substituted into (6-216) as the value of r, and then (6-216) is used to randomly select the local-mean power of the desired signal at the receiver. The probabilities given by (6-219) and (6-220) are used to determine if an interfering mobile produces power in the transmission channel or in one of the adjacent channels of the desired signal. If the power enters the transmission channel, then the power level is randomly selected according to (6-216) with the distance of the mobile substituted. If the power enters one of the adjacent channels, then the potential local-mean power level is first randomly selected via (6-216) and then multiplied by the adjacent-splatter ratio K_s (Chapter 3) to determine the net interference power p_{ui} that appears in (6-217). The effects of p_0 and σ_n^2 are determined solely by the *minimum area-mean SNR*, which occurs at the maximum range $r = R_0$ of the desired signal and is equal to p_0/σ_n^2.

Once the local-mean power levels and the noise power are calculated, the symbol error probability P_s is calculated with (6-217) and (6-218) subject to the constraint that $P_s \leq 1/2$. Each simulation experiment was repeated for 10,000 trials, with different randomly selected mobile locations in each trial. The performance measure is the *spatial reliability*, which is defined as the fraction of trials for which P_s is less than a specified performance threshold E. The appropriate value of the threshold depends on the desired information-bit error probability and the error-control code. The spatial reliability is essentially the probability that an outage does not occur.

Figures 6.29 to 6.31 depict the results of three simulation experiments for peer-to-peer networks. The figures plot the spatial reliability as a function of K -1 for various values of L, assuming Rayleigh fading, MSK, and (6-218) with the constraint that $P_s \leq 1/2$, . The parameter values are $\beta = 4$, $\sigma_s = 8$ dB, $E = 0.01$, $\zeta = 1$, $K_s = 0.015$, $R_0 = 1$, and $R_1 = 2$. The value of K_s results from assuming contiguous frequency channels with center frequencies separated by B. The units of R_0 and R_1 are immaterial to the calculation of the spatial diversity.

Figure 6.29 provides a baseline with which the other figures may be compared. For this figure, the assumptions are that $M_1 = 250$, and the minimum area-mean SNR $= 20$ dB. The number of equivalent frequency channels M_1 could model voice communications with $M = 90$ channels and $d = 0.36$; alternatively, it could model continuous data communications with $M = 225$ and $d = 0.9$. The figure illustrates the dramatic performance improvement provided by dual spatial diversity when Rayleigh fading occurs. Further increases in diversity yield diminishing returns. One can assess the impact of the spectral splatter in this example by setting $K_s = 0$ and observing the change in the spatial reliability. The change is small, and nearly imperceptible if $K < 25$.

Figure 6.30 illustrates the effect of increasing the number of equivalent channels to $M_1 = 500$. Let the *capacity* of the network be defined as the maximum number of interfering mobiles for which the spatial reliability exceeds 0.95. Figures 6.28 and 6.29 and other simulation results indicate that for the parameter values selected, the capacity C for dual spatial diversity is approximately proportional to M_1; specifically, $C \approx 0.07 M_1$ for $100 \leq M_1 \leq 1000$. If E is increased to 0.02, the capacity for dual spatial diversity increases by approximately 20 percent.

Figure 6.31 illustrates the sensitivity of the network to an increase in the minimum area-mean SNR, which may be due to a change in p_0 or σ_n^2. For no spatial diversity or dual diversity, a substantial performance improvement occurs when the minimum area-mean SNR $= 25$ dB. Other simulation results indicate that a decrease in the minimum area-mean SNR below 20 dB severely degrades performance.

Since (6-218) relates P_s to $\bar{p}$, the local-mean SINR, the spatial reliability has an alternative and equivalent definition as the fraction of trials for which the SINR exceeds a specified threshold Z_l. Thus, the graphs labeled $L = 1, 2, 3$, and 4 in Figures 6.29 to 6.31 (and later in Figures 6.33 to 6.36) correspond to $Z_l = 17.7$ dB, 10.0 dB, 7.7 dB, and 6.5 dB, respectively.

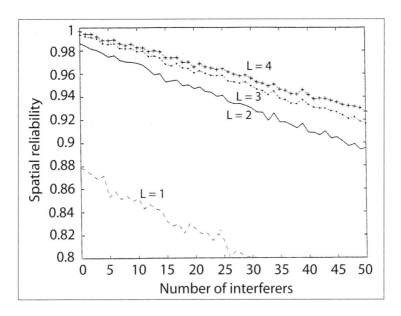

Figure 6.29: Spatial reliability for $M_1 = 250$ and minimum area-mean SNR = 20 dB.

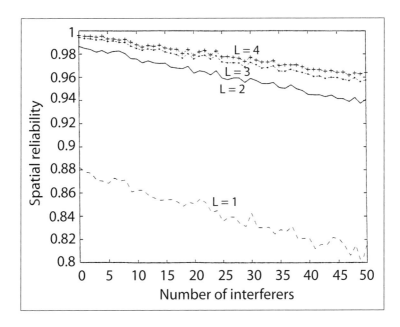

Figure 6.30: Spatial reliability for $M_1 = 500$ and minimum area-mean SNR = 20 dB.

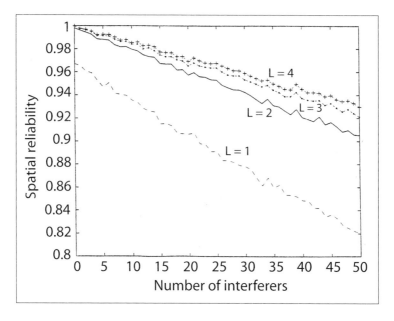

Figure 6.31: Spatial reliability for $M_1 = 250$ and minimum area-mean SNR = 25 dB.

The performance of FH/CDMA communications in a mobile peer-to-peer network is greatly improved by the use of spatial diversity, which usually requires carrier frequencies in excess of 1 GHz. A crucial parameter is the number of equivalent frequency channels, which can be increased not only by an increase in the number of frequency channels, but also by a decrease in the duty factor of the network mobiles. The data modulation method that is most suitable appears to be MSK or some other form of CPM such as GMSK (Chapter 3). For these modulations, $BT_s \approx 1$, and the scenario modeled, the spectral splatter from adjacent channels, is not an important factor if the number of active interferers is much smaller than the number of equivalent channels.

Cellular Networks

In a cellular network, each base station assigns separate directional sector antennas or separate outputs of a phased array to cover disjoint angular sectors in both the transmitting and receiving modes. Typically, there are three sectors, and $2\pi/3$ radians are in each angular sector. The mobile antennas are assumed to be omnidirectional. Ideal sector antennas have uniform gain over the covered sector and negligible sidelobes. With these antennas, only mobiles in the covered sector can cause multiple-access interference on an uplink from a mobile to a base station, and the number of interfering signals on the link is reduced by a factor s approximately equal to the number of sectors. Only the antenna serving a cell sector oriented toward a mobile can cause multiple-access interference on a downlink from the controlling base station to a mobile. Therefore, the number of interfering signals is reduced approximately by a factor s on both the uplinks and downlinks. Practical sector antennas have patterns with sidelobes that extend into adjacent sectors, but the performance degradation due to overlapping sectors is significant only for a small percentage of mobile locations. Ideal sector antennas are assumed in the subsequent simulation.

Spatial diversity may be obtained through the deployment of L antennas in each mobile and L antenna elements for each sector antenna of each base station. The antennas are separated from each other enough that the fading of both the desired signal and the interfering signals at one antenna is independent of the fading at the other antennas. A few wavelengths are adequate for a mobile because it tends to receive superpositions of reflected waves arriving from many random angles. Many wavelengths separation may be necessary for a base station located at a high position, and polarization diversity may sometimes be a more practical means of obtaining diversity.

In a cellular network, the frequency-hopping patterns can be chosen so that at any given instant in time, the frequencies of the mobiles within a cell sector are all different and, hence, the received signals are all orthogonal if the mobile transmissions are properly synchronized. Exact synchronization on a downlink is possible because a common timing is available. The advancing or retarding of the transmit times of the mobiles enables the arrival times at the base station of the uplink signals to be synchronized. The switching time or guard time between frequency-hopping pulses must be large enough to ensure that neither a small synchronization error nor multipath signals can subvert the orthogonality. The appropriate transmit times of a mobile can be determined from position information provided by the Global Positioning System and the known location of the base station. Alternatively, the transmit times can be determined from arrival-time measurements at the base station that are sent to the mobile. These measurements may be based on the adaptive thresholding [33] of the leading and/or trailing edges of a sequence of frequency-hopping pulses.

Let N_s denote the number of mobiles assigned to a cell sector. To ensure orthogonality of N_s received signals within a cell sector, a simple procedure is to generate a periodic frequency-hopping pattern that does not repeat until all the carrier frequencies in a hopset of size $M \geq N_s$ have been used. Mobile n is assigned this pattern with a delay of $n - 1$ hop durations, where $n = 1, 2, \ldots, N_s$. If the patterns associated with different sectors are all drawn from a set of *one-coincidence sequences* [34], then any two signals from different cells or sectors will collide in frequency at a base station at most once during the period of the hopping patterns. However, the use of one-coincidence sequences throughout a network requires frequency planning, which may be too costly in some applications.

It is possible to ensure not only the orthogonality of N_s signals in a sector but also that the received carrier frequencies in any two patterns are separated by at least νB, where ν is a positive integer, so that the spectral splatter is greatly reduced or negligible. Let $k = 0, 1, 2, \ldots, M - 1$ label the hopset frequencies in ascending order. Suppose that a frequency-hopping pattern is generated that does not repeat until all the carrier frequencies in a hopset of size $M \geq \nu N_s$ have been used. When mobile 1 hops to frequency k, mobile n hops to frequency $[k + \nu (n - 1)]$ modulo M. Frequency-hopping signals that use frequencies determined by this procedure are called *separated orthogonal signals*. Choosing $\nu = 2$ will generally be adequate because spectral splatter from channels that are not adjacent will be nearly always insignificant if a spectrally compact data modulation is used.

FH/CDMA networks largely avoid the near-far problem by continually changing the carrier frequencies so that frequency collisions become brief, unusual events. Thus, power control in a FH/CDMA network is unnecessary, and all mobiles may transmit at the same power level. When power control is used, it tends to benefit signals from mobiles far from an associated sector antenna, while degrading signals from mobiles close to it. Simulation results [35] indicate that even perfect power control typically increases system capacity by only a small amount. There are good reasons to forego this slight potential advantage and not use power control. The required overhead may be excessive. If geolocation of mobiles is done by using measurements at two or more base stations, then the power control may result in significantly less signal power arriving at one or more base stations and the consequent loss of geolocation accuracy.

Consider communications between a base station and a mobile assigned to sector A of a particular cell, as illustrated in Figure 6.32 for a hexagonal grid of cells. Because of orthogonality, no other signal in sector A will use the same carrier frequency at the same time and thereby cause interference in the transmission channel (current frequency channel) of either the uplink or downlink. Consider another sector covered by the sector antenna of sector A; an example is sector B. Assuming that an interfering signal may independently use any frequency in the network hopset with equal probability, the probability that a mobile in the covered sector produces interference in the transmission channel of the uplink and degrades a particular symbol is

$$P_m = \frac{dN_s}{M} \tag{6-222}$$

This equation also gives the probability that a sector antenna serving another sector that is oriented toward the desired mobile degrades a symbol by producing interference in the transmission channel of the downlink. Because of orthogonality within each sector, no more than one signal from a sector will produce interference in the transmission channel of either link. A sector with mobiles that may interfere with communications over an uplink or a sector with an antenna that may produce interference over a downlink is called an *interfering sector*.

It is assumed that M is sufficiently large that we may neglect the fact that a channel at one of the ends of the hopping band has only one adjacent channel within the band instead of two. Let $N_1 = 1$ if a signal from an interfering sector uses the transmission channel of communicators in sector A; let $N_1 = 0$ if it does not. The probability that $N_1 = 1$ is N_s/M. The $N_s - N_1$ interference signals from a sector that do not enter the transmission channel are assumed

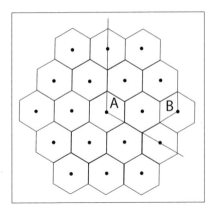

Figure 6.32: Hexagonal grid of cells. Communicators are in sector A. Sector B is an interfering sector.

to be randomly distributed among the $M - 1$ frequency channels excluding the transmission channel. There are $\binom{M-1}{N_s-N_1}$ ways to choose the channels with interference signals. There are $\binom{2}{1}$ ways to choose one of the two adjacent channels to have an interference signal and $\binom{M-3}{N_s-N_1-1}$ ways to choose $N_s - N_1 - 1$ channels with interference signals out of the $M - 3$ channels excluding both the transmission channel and the adjacent channels. The probability that an adjacent channel with an interference signal actually receives interference power is q_1. Similarly, there is one way to choose both adjacent channels with interference signals and $\binom{M-3}{N_s-N_1-2}$ ways to choose $N_s - N_1 - 2$ channels with interference signals out of $M - 3$ channels. The probability that exactly one of the two adjacent channels with interference signals actually receives interference power is $2q_1(1 - q_1)$. Because of the sector synchronization, either all of the signals from a sector overlap a desired symbol with probability q_2 or none of them do. Therefore, the probability that a symbol is degraded by interference in exactly one of the adjacent channels of the communicators is

$$
\begin{aligned}
P_{a1} &= \frac{\binom{2}{1}\binom{M-3}{N_s-N_1-1}}{M-1_s-N_1} q_1 q_2 + \frac{\binom{M-3}{N_s-N_1-2}}{\binom{M-1}{N_s-N_1}} 2\, q_1(1-q_1)q_2 \\
&= \frac{2d(N_s - N_1)}{(M-1)(M-2)}[M - 2 - q_1(N_s - N_1 - 1)], \quad M \geq N_s \qquad (6\text{-}223)
\end{aligned}
$$

Similarly, the probability that a symbol is degraded by interference in both adjacent channels is

$$P_{a2} = \frac{\binom{M-3}{N_s-N_1-2}}{\binom{M-1}{N_s-N_1}} q_1^2 q_2$$
$$= \frac{dq_1 (N_s - N_1)(N_s - N_1 - 1)}{(M-1)(M-2)} , \qquad M \geq N_s \qquad (6\text{-}224)$$

For adjacent-channel interference from within sector A, P_{a1} and P_{a2} are given by the same equations with $N_1 = 1$ to reflect the fact that one of the mobiles is the communicating mobile.

Suppose that separated orthogonal frequency-hopping patterns with $\nu = 2$ are used. There is no adjacent-channel interference from sector A. If a signal from an interfering sector B uses the transmission channel so that $N_1 = 1$, an event with probability N_s/M, then the carrier separation of the signals generated in sector B ensures that there is no adjacent-channel interference from sector B. Suppose that no signal from sector B uses the transmission channel so that $N_1 = 0$. Interference in exactly one adjacent channel results if the transmission channel of the desired signal in sector A, which may be any of $M-N_s$ channels, is located next to one of the two end channels of a set of $N_s \geq 1$ separated channels being used in sector B, neglecting hopset end effects. It also results if the transmission channel is located between two separated channels, of which only one is currently being used in sector B, again neglecting hopset end effects. Therefore, the probability that a symbol is degraded by interference in exactly one of the adjacent channels of the communicators is

$$P_{a1} = \left[\frac{2q_1}{M-N_s} + \frac{N_s - 1}{M-N_s} q_1(1-q_1) \right] q_2$$
$$= \frac{d}{M-N_s} [(N_s - 1)(1-q_1) + 2] , \quad M \geq 2N_s , \; N_1 = 0, \; N_s \geq 1 \quad (6\text{-}225)$$

Interference in both adjacent channels results if the transmission channel is located between two separated channels of sector B and both are being used, neglecting hopset end effects. Therefore, the probability that a symbol is degraded by interference in both adjacent channels is

$$P_{a2} = \frac{dq_1 (N_s - 1)}{M-N_s} , \quad M \geq 2N_s , \; N_1 = 0, \; N_s \geq 1 \qquad (6\text{-}226)$$

If $N_s = 0$, then $P_{a1} = P_{a2} = 0$.

In the simulation, the spatial configuration consists of a hexagonal grid of cells with base stations at their centers. Each cell has a radius R_0 from its center to a corner. A central cell is surrounded by an inner concentric tier of 6 cells and an outer concentric tier of 12 cells, as depicted in Figure 6.32. Other tiers are assumed to generate insignificant interference in the central cell. An equal number of mobiles, each transmitting at the same power level, is located in each sector and served by that sector's antenna. This assumption is pessimistic since slightly improved performance may be possible if a mobile is served by the sector antenna providing a signal with the least attenuation, and if hysteresis effects during handoffs are not too severe. Each signal transmitted by a sector antenna is allocated the same power. The set of frequency-hopping patterns used in each sector is assumed to be selected independently of the other sectors. Since the parameter R_0 in (6-216) is equal to the maximum communication range, p_0 is the minimum received area-mean power of a desired signal. The location of each mobile within a sector is assumed to be uniformly distributed.

In each simulation trial for communications in sector A of the central cell, the location of the desired mobile is randomly selected according to the uniform distribution. The selected distance of the desired mobile is substituted into (6-216) as the value of r, and then (6-216) is used to randomly select the local-mean power of the desired signal at the receiver. Each transmitting and receiving beam produced by a sector antenna is assumed to have a constant gain over its sector and zero gain elsewhere.

For an uplink of sector A, interference is assumed to arrive from mobiles within sector A, mobiles in the 6 sectors of the two cells in the inner tier that were covered by the beam of sector A, and mobiles in the 11 complete sectors and 2 half-sectors of the five cells in the outer tier completely or partially covered by the beam. The 2 half-sectors are approximated by an additional complete sector in the outer tier. Equations (6-222) to (6-226) are used to determine if a sector contains mobiles that produce power in the transmission channel or in one or both of the adjacent channels. If the sector does, then the locations of the three or fewer interfering mobiles are randomly selected according to the uniform distribution, and their distances from the central cell's base station are computed.

For a downlink of sector A, interference is assumed to arrive from the facing sector antenna of each cell in the two surrounding tiers. Equations (6-222) to (6-226) are used to determine if a signal generated by an interfering sector antenna produces power in the transmission channel or the adjacent channels of the desired signal. If so, then the distance between the sector antenna and the desired mobile is computed. The angular location of the desired mobile is randomly selected from a uniform distribution over the $2\pi/3$ radians spanning sector A.

If the power from an interferer enters the transmission channel, then the power level is randomly selected according to (6-216), with the appropriate distance substituted. If the power enters an adjacent channel, then the potential local-mean power level is first randomly selected via (6-216) and then multiplied by K_s to determine the net interference power p_{ui} that appears in (6-217). The shadowing parameter σ_s is assumed to be the same for all signals originating from all cells. The effects of p_0 and σ_n^2 are determined solely by p_0/σ_n^2, the minimum area-mean SNR. Since only ratios affect the performance, the numerical value of R_0 in the simulation is immaterial and is set equal to unity.

Once the local-mean power levels and the noise power are calculated, the symbol error probability is calculated with (6-217) and (6-218) subject to the constraint that $P_s \leq 1/2$. Each simulation experiment was repeated for 20,000 trials, with different randomly selected mobile locations in each trial. The

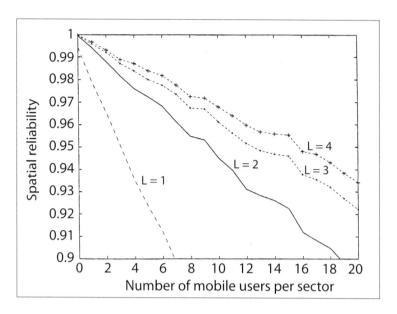

Figure 6.33: Spatial reliability for uplinks, separated orthogonal hopping, $M =$ 100, and minimum area-mean SNR = 30 dB.

performance measure is the spatial reliability, which is a function of $\bar{p}$, the SINR. The appropriate value of the threshold E depends on the desired information-bit error probability and the error-control code.

Figures 6.33 to 6.36 depict the results of four simulation experiments for the uplinks of a cellular network. The figures plot spatial reliability as a function of N_s for various values of L, assuming MSK, three sectors, and that $\beta = 4$, q_1 = 0.4, $q_2 = 1.0$, $\sigma_s = 8$ dB, $E = 0.01$, $\zeta = 1$, and $K_s = 0.015$. The value of K_s results from assuming contiguous frequency channels with the center frequencies separated by the bandwidth of a frequency channel.

Figure 6.33 provides a baseline with which other figures may be compared. For this figure, separated orthogonal frequency hopping with $\nu = 2$, $M = 100$, and minimum area-mean SNR = 30 dB are assumed. The figure illustrates the dramatic performance improvement provided by dual spatial diversity when Rayleigh fading occurs. Further increases in diversity yield diminishing returns. One can assess the impact of the spectral splatter in this example by setting $K_s = 0$ and observing the change in spatial reliability. The change is insignificant because by far the most potentially damaging splatter arises from mobiles in the same sector as the desired mobile, and the separated orthogonality has eliminated it.

Figure 6.34 shows the effect of using orthogonal rather than separated orthogonal frequency hopping. The performance loss is significant in this example and becomes more pronounced as M decreases. When separated orthogonal frequency hopping is used and the spectral splatter is negligible, then the spatial

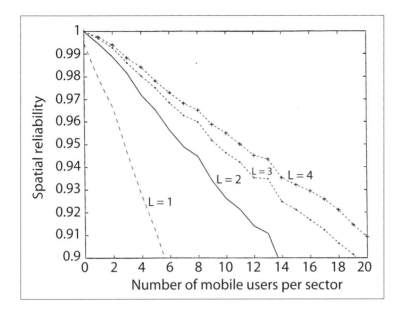

Figure 6.34: Spatial reliability for uplinks, orthogonal hopping, $M = 100$, and minimum area-mean SNR = 30 dB.

reliability depends primarily on $M_1 = M/d$, the equivalent number of channels. In Figure 6.33, $M_1 = 250$.

Figure 6.35 illustrates the effect of increasing M to 200, and hence increasing M_1 to 500. The *uplink capacity* C_u of a cellular network is defined as the maximum number of interfering mobiles per cell for which the spatial reliability exceeds 0.95. Figures 6.33 and 6.35 and other simulation results indicate that for three sectors per cell, dual diversity, and the other parameter values selected, the uplink capacity is $C_u \approx 0.108\,M_1$ for $50 \le M_1 \le 1000$. This equation is sensitive to parameter variations. If the shadowing standard deviation σ_s is lowered to 6 dB, it is found that C_u increases by roughly 57 percent. Alternatively, if the threshold E is raised to 0.04, corresponding to SINR = 7 dB, it is found that C_u increases by roughly 59 percent.

Figure 6.36 illustrates the sensitivity of the network to a decrease in the minimum area-mean SNR, which may be due to a change in either p_0 or σ_n^2. A substantial performance loss occurs when the minimum area-mean SNR is reduced to 20 dB, particularly for no spatial diversity or dual diversity. Other simulation results indicate that an increase in the minimum area-mean SNR beyond 30 dB barely improves performance.

The downlinks of a cellular network are considered in Figure 6.37, where the models and parameter values are otherwise the same as in Figure 6.33. The performance is worse for the downlinks of Figure 6.37 than for the uplinks of Figure 6.33 because of the relative proximity of some of the interfering sector antennas to the desired mobile. The *downlink capacity* C_d, which is defined

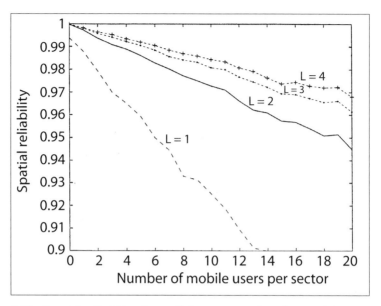

Figure 6.35: Spatial reliability for uplinks, separated orthogonal hopping, $M =$ 100, and minimum area-mean SNR = 20 dB.

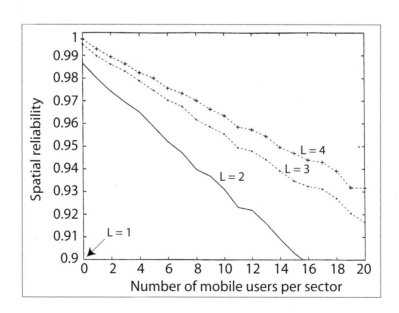

Figure 6.36: Spatial reliability for uplinks, separated orthogonal hopping, $M =$ 200, and minimum area-mean SNR = 30 dB.

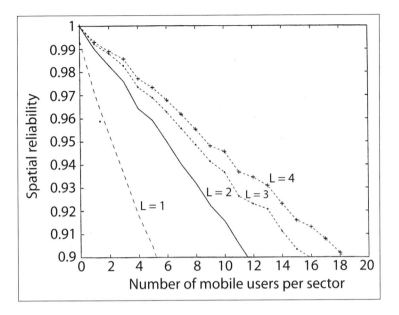

Figure 6.37: Spatial reliability for downlinks, separated orthogonal hopping, M = 100, and minimum area-mean SNR = 30 dB.

analogously to the uplink capacity, is $C_d \approx 0.072\, M_1$ for $50 \leq M_1 \leq 1000$. A more realistic comparison of the downlinks and uplinks must take into account the differences between the high-power amplifiers and low-noise amplifiers in the base station and those in the mobiles. Assuming a net 10 dB advantage in the minimum area-mean SNR for the downlinks, Figures 6.35 and 6.36 provide a performance comparison of the two links. The performance of the downlinks is still slightly worse if $L \geq 2$ and $N_s \geq 4$. The difference in performance is further increased if physical constraints in the mobiles limit the downlinks to L = 1 or 2 while $L = 4$ for the uplinks.

Compared with direct-sequence systems, frequency-hopping systems have a bandwidth advantage in that frequency hopping over a large, possibly non-contiguous, spectral band is as practical as direct-sequence spreading over a much smaller, necessarily contiguous, spectral band. Even deprived of its bandwidth advantage, as well as power control and the use of one-coincidence codes, FH/CDMA can provide nearly the same multiple-access capacity over the uplinks as DS/CDMA subject to realistic power-control imperfections. For a numerical example, consider a cellular network with three sectors, shadowing standard deviation $\sigma_s = 6$ dB, and d = $q_1 = 3/8$ due to the voice activity. A contiguous spectral band of bandwidth W = 1.25 MHz is occupied by the CDMA signals. The symbol rate is $1/T_s = 8$ kb/s so that the processing gain is 156.5 for DS/CDMA, and the number of frequency channels for FH/CDMA with $\zeta = BT_s = 1$ is M = 156. For DS/CDMA, it is assumed that p_0 is the received power at the base station from all associated mobiles when the power control is perfect and that the SNR before the despreading is -1 dB. Therefore, the SNR is 20.94 dB after the despreading. For FH/CDMA without power control, the minimum area-mean SNR is assumed to be 20.94 dB. The uplink capacity C_u is calculated as the number of mobiles per cell that can be accommodated while maintaining an SINR above a specified threshold Z with 95 percent probability. For FH/CDMA with dual diversity and Z = 10 dB, it is found that $C_u \approx 60$. For DS/CDMA with dual diversity and coherent phase-shift keying, a comparison of (6-218) with (5-135) and (5-130) indicates that a comparable performance can be obtained when the SINR is roughly 3 dB less. Thus, the threshold for DS/CDMA is set at Z = 7 dB. Using (6-131) with Z = 7 dB, it is found that $C_u \approx 60$ when the power-control error has $\sigma_e = 2$ dB

For coherent demodulation of a signal that hops over a wide band to be a practical possibility in a fading environment, either a pilot signal must be available, turbo or other iterative decoding must be feasible, or the dwell time must be large enough that a small portion of it can be dedicated to carrier synchronization. In the latter case, the degradation due to the dedicated portion and the occasional failure to achieve carrier synchronization for a frequency-hopping pulse must be less than the potential gain due to the coherent demodulation, which is large. If ideal coherent demodulation is assumed in the preceding example so that Z = 7 dB, then it is found that $C_u \approx 108$, an increase of 80 percent. This uplink capacity is approximately obtained by DS/CDMA with Z = 7 dB when $\sigma_e = 0.4$ dB, an impractically low value.

For a specified sectorization, diversity, and waveform, the capacity of a cellular FH/CDMA network is approximately proportional to the equivalent number of frequency channels. Thus, a desired capacity can be attained by choosing a sufficiently large number of frequency channels. A major advantage of frequency hopping is that these channels do not have to be spectrally contiguous but can be scattered throughout a large spectral band. Another advantage is that power control is not required. Its absence allows a substantial reduction of system complexity and overhead cost and facilitates geolocation. Sectorization, orthogonality, and dual diversity are invaluable, but higher levels of diversity offer sharply decreasing gains. If spectral splatter is a problem, separated orthogonal signaling can be used to eliminate it. The overall limit on the capacity of a FH/CDMA network appears to be set more by the downlinks than the uplinks.

6.7 Problems

1. A Gold sequence is constructed from a maximal sequence with characteristic polynomial $1 + x^2 + x^3$. The second sequence is obtained by decimation of the maximal sequence by q = 3. (a) Find one period of each of the two sequences. Is the second sequence maximal? (b) List the 7 cross-correlation values of this pair of sequences. Show that they are a preferred pair.

2. The characteristic polynomials for generating Gold sequences of length 7 are: $f_1(x) = 1 + x + x^3$ and $f_2(x) = 1 + x^2 + x^3$. (a) What is the generating function for the maximal Gold sequence generated by $f_1(x)$ and initial contents 1 00 ? (b) What is the generating function for the maximal Gold sequence generated by $f_2(x)$ and initial contents 100 ? (c) What is the general expression for the generating function of an arbitrary nonmaximal Gold sequence? (d) What is the generating function for the Gold sequence generated by adding the sequences in (a) and (b)? (e) What is the value at $k = 0$ of the periodic cross correlation between the sequences in (a) and (b)?

3. A small set of Kasami sequences is formed by starting with the maximal sequence generated by the characteristic polynomial $1 + x^2 + x^3 + x^4 + x^8$. After decimation by q, a second sequence with characteristic polynomial $1 + x + x^4$ is found. (a) What is the value of q? How many sequences are in the set? What is the period of each sequence? What is the peak magnitude of the periodic cross-correlation? Draw a block diagram of the generator of the small Kasami set. (b) Prove whether or not the second sequence is maximal.

4. The small set of the preceding problem is extended to a large set of Kasami sequences by a decimation of the original maximal sequence by q_1. A third sequence with characteristic polynomial $1 + x^2 + x^3 + x^4 + x^5 + x^7 + x^8$ is found. (a) What is the value of q_1? How many sequences are in the large set? What is the period of each sequence? What is the peak magnitude of the periodic cross-correlation? Draw a block diagram of the generator of the large Kasami set. (b) Prove whether or not the third sequence is maximal.

5. For the feedback shift register of Figure 6.2 with initial contents of 002, list the successive contents during one period of the output sequence.

6. Derive (6-63) for both rectangular and sinusoidal chip waveforms.

7. Use bounding and approximation methods to establish (6-84).

8. Using the methods outlined in the text, derive (6-142) and then (6-143).

9. Using the methods outlined in the text, derive (6-145) and then (6-146).

10. Consider the decorrelating detector for two synchronous users. (a) Evaluate the two sampled correlator outputs when the received signal is (6-174). (b) Use the linear transformation matrix to construct a detailed block diagram of the receiver. Write equations for the symbol estimates. (c) Evaluate the noise covariance matrix at the input and at the output of the linear transformer.

11. Consider the conventional detector for two synchronous users. (a) Evaluate $P_s(1)$ as $N_0 \to 0$ for the three cases: $\rho^2 \mathcal{E}_2 \leq \mathcal{E}_1$, $\rho^2 \mathcal{E}_2 \geq \mathcal{E}_1$, and $\rho^2 \mathcal{E}_2 = \mathcal{E}_1$. (b) For $\rho^2 \mathcal{E}_2 \leq \mathcal{E}_1$, find the noise level that minimizes P_s.

12. Consider the MMSE receiver for two synchronous users. (a) Evaluate the linear transformation matrix. (b) Construct a detailed block diagram of the receiver. Write equations for the symbol estimates.

13. Consider an FH/CDMA network with two mobiles that communicate with a base station. The modulation is ideal DPSK, the receiver uses EGC, and both the environmental noise and the spectral splatter are negligible. The propagation conditions are such that one signal arriving at the base station is 10 dB stronger than the other one during a time interval without power control. Use (5-133) and (5-169) with $L = 1$ and $L = 2$ to assess the relative merits of introducing power control in the presence of Rayleigh fading. Assume that $M_1 = 10$, that there are many symbols per dwell interval, and that an acceptable average channel-symbol error probability is 0.02.

6.8 References

1. D. V. Sarwate and M. B. Pursley, "Crosscorrelation Properties of Pseudo-random and Related Sequences," *Proc. IEEE*, vol. 68, pp. 593–619, May 1980.

2. M. B. Pursley, "Spread-Spectrum Multiple-Access Communications," in *Multi-User Communications Systems*, G. Longo, ed., New York: Springer-Verlag, 1981.

3. M. B. Pursley, D. V. Sarwate, and W. E. Stark, "Error Probability for Direct-Sequence Spread-Spectrum Multiple-Access Communications—Part 1: Upper and Lower Bounds," *IEEE Trans. Commun.*, vol. 30, pp. 975–984, May 1982.

4. A. R. Hammons and P. V. Kumar, "On a Recent 4-Phase Sequence Design for CDMA," *IEICE Trans. Commun.*, vol. E76-B, pp. 804–813, August 1993.

5. T. G. Macdonald and M. B. Pursley, "The Performance of Direct-Sequence Spread Spectrum with Complex Processing and Quaternary Data Modulation," *IEEE J. Select. Areas Commun.*, vol. 18, pp. 1408–1417, August 2000.

6. D. Torrieri, "Performance of Direct-Sequence Systems with Long Pseudonoise Sequences," *IEEE J. Select. Areas Commun.*, vol. 10, pp. 770–781, May 1992.

7. G. Zang and C. Ling, "Performance Evaluation for Band-limited DS-CDMA Systems Based on Simplified Improved Gaussian Approximation," *IEEE Trans. Commun.*, vol. 51, pp. 1204–1213, July 2003.

8. L. B. Milstein, "Wideband Code Division Multiple Access," *IEEE J. Select. Areas Commun.*, vol. 48, pp. 1318–1327, August 2000.

9. S. Kondo and L. Milstein, "Performance of Multicarrier DS CDMA Systems," *IEEE Trans. Commun.*, vol. 44, pp. 238-246, February 1996.

10. K. Higuchi et al., "Experimental Evaluation of Combined Effect of Coherent Rake Combining and SIR-Based Fast Transmit Power Control for Reverse Link of DS-CDMA Mobile Radio," *IEEE J. Select. Areas Commun.*, vol. 18, pp. 1526–1535, August 2000.

11. L. Hanzo, L.-L. Yang, E.-L. Kuan, and K. Yen, *Single- and Multi-Carrier CDMA Multi-User Detection, Space-Time Spreading, Synchronization, Networking and Standards.* New York: Wiley, 2003.

12. D. R. Barry, E. A. Lee, and D. G. Messerschmitt, *Digital Communication, 3rd ed.* Boston: Kluwer Academic, 2004.

13. C. C. Lee and R. Steele, "Closed-loop Power Control in CDMA Systems," *IEE Proc.-Commun.*, vol. 143, pp. 231–239, August 1996.

14. S. DeFina and P. Lombardo, "Error Probability Analysis for CDMA Systems with Closed-Loop Power Control," *IEEE Trans. Commun.*, vol. 49, pp. 1801–1811, October 2001.

15. D. Torrieri, "Instantaneous and Local-Mean Power Control for Direct-Sequence CDMA Cellular Networks," *IEEE Trans. Commun.*, vol. 50, pp. 1310–1315, August 2002.

16. G. Stuber, *Principles of Mobile Communication, 2nd ed.* Boston: Kluwer Academic, 2001.

17. I. S. Gradsteyn and I. M. Ryzhik, *Tables of Integrals, Series and Products, 6th ed.* San Diego: Academic Press, 2000.

18. M. Zorzi, "On the Analytical Computation of the Interference Statistics with Applications to the Performance Evaluation of Mobile Radio Systems," *IEEE Trans. Commun.*, vol. 45, pp. 103–109, January 1997.

19. A. J. Viterbi, *CDMA Principles of Spread Spectrum Communication.* Reading, MA: Addison-Wesley, 1995.

20. G. E. Corazza, G. DeMaio, and F. Vatalaro, "CDMA Cellular Systems Performance with Fading, Shadowing, and Imperfect Power Control," *IEEE Trans. Veh. Technol.*, vol. 47, pp. 450–459, May 1998.

21. R. N. McDonough and A. D. Whalen, *Detection of Signals in Noise, 2nd ed.* San Diego: Academic Press, 1995.

22. J. G. Proakis, *Digital Communications, 4th ed.* New York: McGraw-Hill, 2001.

23. S. Verdu, *Multiuser Detection.* New York: Cambridge University Press, 1998.

24. U. Madhow, "Blind Adaptive Interference Suppression for Direct-Sequence CDMA," *Proc. IEEE*, vol. 86, pp. 2049–2069, October 1998.

25. S. Buzzi and H. V. Poor, "Channel Estimation and. Multiuser Detection in Long-Code DS/CDMA Systems," *IEEE J. Select. Areas Commun.*, vol. 19, pp. 1476–1487, August 2001.

26. M. Sawahashi et al., "Experiments on Pilot Symbol-Assisted Coherent Multistage Interference Canceller for DS-CDMA Mobile Radio," *IEEE J. Select. Areas Commun.*, vol. 20, pp. 433–449, February 2002.

27. N. Sharma and E. Geraniotis, "Soft Multiuser Demodulation and Iterative Decoding for FH/SSMA with a Block Turbo Code," *IEEE Trans. Commun.*, vol. 51, pp. 1561-1570, September 2003.

28. M. B. Pursley, "The Derivation and Use of Side Information in Frequency-Hop Spread Spectrum Communications," *IEICE Trans. Commun.*, vol. E76-B, pp. 814–824, August 1993.

29. H. El Gamal and E. Geraniotis, "Iterative Channel Estimation and Decoding for Convolutionally Coded Anti-Jam FH Signals," *IEEE Trans. Commun.*, vol. 50, pp. 321–331, February 2002.

30. K. Choi and K.Cheun, "Maximum Throughput of FHSS Multiple-Access Networks Using MFSK Modulation," *IEEE Trans. Commun.*, vol. 52, pp. 426-434, March 2004.

31. D. Torrieri, "Mobile Frequency-Hopping CDMA Systems," *IEEE Trans. Commun.*, vol. 48, pp. 1318–1327, August 2000.

32. F. Adachi and J. D. Parsons, "Unified Analysis of Postdetection Diversity for Binary Digital FM Radio," *IEEE Trans. Veh. Technol.*, vol. 37, pp. 189–198, November 1988.

33. D. Torrieri, *Principles of Secure Communication Systems, 2nd ed.* Boston: Artech House, 1992.

34. A. A. Shaar and P. A. Davies, "Survey of One-Coincidence Sequences for Frequency Hopped Spread Spectrum Systems," *IEE Proc., vol. 131, pt. F,* pp. 719–724, December 1984.

35. S. Chennakeshu et al., "Capacity Analysis of a TDMA-Based Slow-Frequency-Hopped Cellular System," *IEEE Trans. Veh. Technol.*, vol. 45, pp. 531–542, August 1996.

Chapter 7

Detection of Spread-Spectrum Signals

This chapter presents a statistical analysis of the unauthorized detection of spread-spectrum signals. The basic assumption is that the spreading sequence or the frequency-hopping pattern is unknown and cannot be accurately estimated by the detector. Thus, the detector cannot mimic the intended receiver.

7.1   Detection of Direct-Sequence Signals

The results of Section 2.3 indicate that the maximum magnitude of the power spectral density of a direct-sequence signal with a random spreading sequence is $A^2 T_c/4 = \mathcal{E}_s/2G$, where $\mathcal{E}_s$ is the symbol energy and G is the processing gain. A spectrum analyzer usually cannot detect a signal with a power spectral density below that of the background noise, which has spectral density $N_0/2$. Thus, a received $\mathcal{E}_s/N_0 > G$ is an approximate necessary, but not sufficient, condition for a spectrum analyzer to detect a direct-sequence signal. If $\mathcal{E}_s/N_0 < G$, detection may still be probable by other means. If not, the direct-sequence signal is said to have a *low probability of interception*.

Ideal Detection

Detection theory leads to various detection receivers depending on precisely what is assumed to be known about the signal to be detected. We make the idealized assumptions that the chip timing of the spreading waveform is known and that whenever the signal is present, it is present during the entire observation interval. The spreading sequence is modeled as a random binary sequence, which implies that a time shift of the sequence by a chip duration corresponds to the same stochastic process. Thus, to account for uncertainty in the chip timing, one might partition a chip interval of known duration among several parallel detectors each of which implements a different chip timing.

Consider the detection of a direct-sequence signal with PSK modulation:

$$s(t) = \sqrt{2S}p(t)\cos\left(2\pi f_c t + \phi\right) \tag{7-1}$$

where S is the average signal power, f_c is the known carrier frequency, and ϕ is the carrier phase assumed to be constant over the *observation interval* $0 \leq t \leq T$. The spreading waveform $p(t)$, which subsumes the random data modulation, is given by (2-76) with the $\{p_i\}$ modeled as a random binary sequence. To determine whether a signal $s(t)$ is present based on the observation of the received signal, classical detection theory requires that one choose between the hypothesis H_1 that the signal is present and the hypothesis H_0 that the signal is absent. Over the observation interval, the received signal under the two hypotheses is

$$r(t) = \begin{cases} s(t) + n(t), & H_1 \\ n(t), & H_0 \end{cases} \tag{7-2}$$

where $n(t)$ is zero-mean, white Gaussian noise with two-sided power spectral density $N_0/2$.

The coefficients in the expansion of the observed waveform in terms of ν orthonormal basis functions constitute the received vector $\mathbf{r} = [r_1\, r_2\, \ldots\, r_\nu]$. Let $\boldsymbol{\theta}$ denote the vector of parameter values that characterize the signal to be detected. The *average likelihood ratio* [1], which is compared with a threshold for a detection decision, is

$$\Lambda(\mathbf{r}) = \frac{E_{\boldsymbol{\theta}}[f(\mathbf{r}|H_1, \boldsymbol{\theta})]}{f(\mathbf{r}|H_0)} \tag{7-3}$$

where $f(\mathbf{r}|H_1, \boldsymbol{\theta})$ is the conditional density function of $\mathbf{r}$ given hypothesis H_1 and the value of $\boldsymbol{\theta}$, $f(\mathbf{r}|H_0)$ is the conditional density function of $\mathbf{r}$ given hypothesis H_0, and $E_{\boldsymbol{\theta}}$ is the expectation over the random vector $\boldsymbol{\theta}$. The coefficients in the expansion of the Gaussian process $n(t)$ in terms of the orthonormal basis functions are uncorrelated and, hence, statistically independent. Since each coefficient is Gaussian with variance $N_0/2$,

$$f(\mathbf{r}|H_1, \boldsymbol{\theta}) = \prod_{i=1}^{\nu} \frac{1}{\sqrt{\pi N_0}} \exp\left[-\frac{(r_i - s_i)^2}{N_0}\right] \tag{7-4}$$

$$f(\mathbf{r}|H_0) = \prod_{i=1}^{\nu} \frac{1}{\sqrt{\pi N_0}} \exp\left(-\frac{r_i^2}{N_0}\right) \tag{7-5}$$

where the $\{s_i\}$ are the coefficients of the signal. Substituting these equations into (7-3) yields

$$\Lambda(\mathbf{r}) = E_{\boldsymbol{\theta}}\left\{\exp\left[\frac{2}{N_0}\sum_{i=1}^{\nu} r_i s_i - \frac{1}{N_0}\sum_{i=1}^{\nu} s_i^2\right]\right\} \tag{7-6}$$

Expansions in the orthonormal basis functions indicate that if $\nu \to \infty$, the average likelihood ratio may be expressed in terms of the signal waveforms as

$$\Lambda[r(t)] = E_{\boldsymbol{\theta}}\left\{\exp\left[\frac{2}{N_0}\int_0^T r(t)s(t)dt - \frac{\mathcal{E}}{N_0}\right]\right\} \tag{7-7}$$

where $\mathcal{E}$ is the energy in the signal waveform over the observation interval of duration T.

If N is the number of chips, each of duration T_c, received in the observation interval, then there are 2^N equally likely patterns of the spreading sequence. For *coherent detection*, we set $\phi = 0$ in (7-1), substitute it into (7-7), and then evaluate the expectation to obtain

$$\Lambda(r(t)) = \exp\left(-\frac{\mathcal{E}}{N_0}\right) \sum_{j=1}^{2^N} \exp\left[\frac{2\sqrt{2S}}{N_0} \sum_{i=0}^{N-1} p_i^{(j)} r_i'\right] \quad \text{(coherent)} \qquad (7\text{-}8)$$

where $p_i^{(j)}$ is chip i of pattern j and

$$r_i' = \int_{iT_c}^{(i+1)T_c} r(t)\psi(t - iT_c) \cos\left(2\pi f_c t\right) dt \qquad (7\text{-}9)$$

These equations indicate how $\Lambda(r(t))$ is to be calculated by the ideal coherent detector. The factor $\exp\left(-\mathcal{E}/N_0\right)$ is irrelevant in the sense that it can be merged with the threshold level with which the average likelihood ratio is compared.

For the more realistic *noncoherent detection* of a direct-sequence signal, the received carrier phase is assumed to be uniformly distributed over $[0, 2\pi)$. Substituting (7-1) into (7-7), using a trigonometric expansion, dropping the irrelevant factor that can be merged with the threshold level, and then evaluating the expectation over the random spreading sequence, we obtain

$$\Lambda(r(t)) = E_\phi\left\{\sum_{j=1}^{2^N} \exp\left[\frac{2\sqrt{2S}}{N_0} \sum_{i=0}^{N-1} p_i^{(j)} \left(r_{ic} \cos\phi - r_{is} \sin\phi\right)\right]\right\} \qquad (7\text{-}10)$$

where

$$r_{ic} = \int_{iT_c}^{(i+1)T_c} r(t)\psi\left(t - iT_c\right) \cos\left(2\pi f_c t\right) dt \qquad (7\text{-}11)$$

$$r_{is} = \int_{iT_c}^{(i+1)T_c} r(t)\psi\left(t - iT_c\right) \sin\left(2\pi f_c t\right) dt \qquad (7\text{-}12)$$

and $E_\phi\{\ \}$ denotes the expectation with respect to ϕ.

The *modified Bessel function of the first kind and order zero* is given by

$$I_0(x) = \frac{1}{2\pi} \int_0^{2\pi} \exp(x \cos u) du \qquad (7\text{-}13)$$

Since the cosine is a periodic function and the integration is over the same period, we may replace $\cos u$ with $\cos(u+\phi)$ for any ϕ in (7-13). A trigonometric expansion with $x_1 = x \cos\phi$ and $x_2 = x \sin\phi$ then yields

$$I_0(x) = \frac{1}{2\pi} \int_0^{2\pi} \exp\left(x_1 \cos u - x_2 \sin u\right) du, \quad x = \sqrt{x_1^2 + x_2^2} \qquad (7\text{-}14)$$

Using this relation and the uniform distribution of ϕ, the average likelihood ratio of (7-10) becomes

$$\Lambda(r(t)) = \sum_{j=1}^{2^N} I_0 \left(\frac{2\sqrt{2SR_j}}{N_0} \right) \quad \text{(noncoherent)} \tag{7-15}$$

where

$$R_j = \left[\sum_{i=0}^{N-1} p_i^{(j)} r_{ic} \right]^2 + \left[\sum_{i=0}^{N-1} p_i^{(j)} r_{is} \right]^2 \tag{7-16}$$

These equations define the optimum noncoherent detector for a direct-sequence signal. The presence of the desired signal is declared if (7-15) exceeds a threshold level.

The implementation of either the coherent or noncoherent optimum detector would be very complicated, and the complexity would grow exponentially with N, the number of chips in the observation interval. Calculations [2] indicate that the ideal coherent and noncoherent detectors typically provide 3 dB and 1.5 dB advantages, respectively, over the far more practical wideband radiometer, which is analyzed subsequently. The use of four or two wideband radiometers, respectively, can compensate for these advantages with less complexity than the optimum detectors. Furthermore, implementation losses and imperfections in the optimum detectors are likely to be significant.

Radiometer

Among the many alternatives [3] to the optimum detector, the *radiometer* is notable in that it requires virtually no detailed information about the signals to be detected other than their rough spectral location. Not even whether the modulation is binary or quaternary is required. Suppose that the signal to be detected is approximated by a zero-mean, white Gaussian process. Consider two hypotheses that both assume the presence of a zero-mean, bandlimited white Gaussian process over an observation interval $0 \leq t \leq T$. Under H_0, only noise is present, and the one-sided power spectral density over the signal band is N_0, while under H_1, both signal and noise are present, and the power spectral density is N_1 over this band. Using ν orthonormal basis functions as in the derivation of (7-4) and (7-5) and ignoring the effects of the bandlimiting, we find that the conditional densities are approximated by

$$f(\mathbf{r}|H_i) = \prod_{l=1}^{\nu} \frac{1}{\sqrt{\pi N_i}} \exp\left(-\frac{r_l^2}{N_i} \right), \quad i = 0, 1 \tag{7-17}$$

Calculating the likelihood ratio, taking the logarithm, and merging constants with the threshold, we find that the decision rule is to compare

$$V = \sum_{l=1}^{\nu} r_l^2 \tag{7-18}$$

to a threshold. If we let $\nu \to \infty$ and use the properties of orthonormal basis functions, then we find that the test statistic is

$$V = \int_0^T r^2(t)dt \tag{7-19}$$

where the assumption of bandlimited processes is necessary to ensure the finiteness of the statistic. A device that implements this test statistic is called an *energy detector* or *radiometer*. Although it was derived for a bandlimited white Gaussian signal, the radiometer is a reasonable configuration for determining the presence of unknown deterministic signals.

A radiometer may have one of the three equivalent forms shown in Figure 7.1. Consider the system of Figure 7.1(a), which gives a direct realization of (7-19). The bandpass filter is assumed to be an ideal rectangular filter that passes the deterministic desired signal $s(t)$ with negligible distortion while limiting the noise. The filter has center frequency f_c, bandwidth W, and produces the output

$$y(t) = s(t) + n(t) \tag{7-20}$$

where $n(t)$ is bandlimited white Gaussian noise with a two-sided power spectral density equal to $N_0/2$. Squaring and integrating $y(t)$, taking the expected value, and observing that $n(t)$ is a zero-mean process, we obtain

$$E[V] = \int_0^T s^2(t)dt + \int_0^T E[n^2(t)]dt$$
$$= \mathcal{E} + N_0 T W \tag{7-21}$$

which indicates that the radiometer output is an unbiased estimate of the total energy after the filtering.

A bandlimited deterministic signal can be represented as (Appendix C.1)

$$s(t) = s_c(t) \cos 2\pi f_c t - s_s(t) \sin 2\pi f_c t \tag{7-22}$$

Since the spectrum of $s(t)$ is confined within the filter passband, $s_c(t)$ and $s_s(t)$ have frequency components confined to the band $|f| \leq W/2$. The Gaussian noise emerging from the bandpass filter can be represented in terms of quadrature components as (Appendix C.2)

$$n(t) = n_c(t) \cos 2\pi f_c t - n_s(t) \sin 2\pi f_c t \tag{7-23}$$

where $n_c(t)$ and $n_s(t)$ have flat power spectral densities, each equal to N_0 over $|f| \leq W/2$. Substituting (7-23) and (7-22) into (7-20), squaring and integrating $y(t)$, and assuming that $f_c \gg W$ and $f_c \gg 1/T$, we obtain

$$V = \frac{1}{2} \int_0^T [s_c(t) + n_c(t)]^2 dt + \frac{1}{2} \int_0^T [s_s(t) + n_s(t)]^2 dt \tag{7-24}$$

A straightforward calculation verifies that the baseband radiometer of Figure 7.1(b) also produces this test statistic.

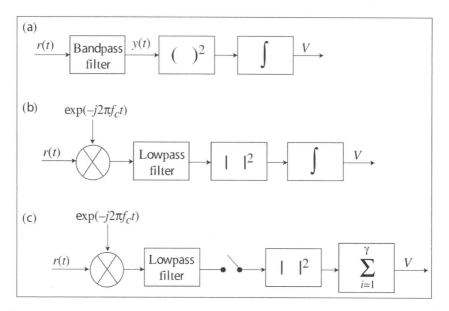

Figure 7.1: Radiometers: (a) passband, (b) baseband with integration, and (c) baseband with sampling at rate $1/W$ and summation.

The *sampling theorems* for deterministic and stochastic processes (Appendix C.3) provide expansions of $s_c(t)$, $s_s(t)$, $n_c(t)$ and $n_s(t)$ that facilitate a statistical performance analysis. For example,

$$s_c(t) = \sum_{i=-\infty}^{\infty} s_c\left(\frac{i}{W}\right) \operatorname{sinc}(Wt - i) \qquad (7\text{-}25)$$

where $\operatorname{sinc} x = (\sin \pi x)/\pi x$. Since the Fourier transform of the sinc function is a rectangular function, using Parseval's theorem from Fourier analysis and evaluating the resulting integral yields the approximations:

$$\int_0^T \operatorname{sinc}^2(Wt - i)dt \approx \int_{-\infty}^{\infty} \operatorname{sinc}^2(Wt - i)dt = \frac{1}{W}, \quad 0 < i \le TW \qquad (7\text{-}26)$$

$$\int_0^T \operatorname{sinc}(Wt-i)\operatorname{sinc}(Wt-j)dt \approx \int_{-\infty}^{\infty} \operatorname{sinc}(Wt-i)\operatorname{sinc}(Wt-j)dt = 0, \quad i \ne j$$
$$(7\text{-}27)$$

The rapid decline of $\operatorname{sinc} x$ for $|x| > 1$ implies that

$$\int_0^T \operatorname{sinc}^2(Wt - i)dt \approx 0, \quad i \le 0 \text{ or } i > TW \qquad (7\text{-}28)$$

We define $\gamma = \lfloor TW \rfloor$, where $\lfloor x \rfloor$ denotes the integer part of x. Substituting expansions similar to (7-25) into (7-24) and then using the preceding approximations, we obtain

$$V \approx \frac{1}{2W} \sum_{i=1}^{\gamma} \left[s_c\left(\frac{i}{W}\right) + n_c\left(\frac{i}{W}\right) \right]^2 + \frac{1}{2W} \sum_{i=1}^{\gamma} \left[s_s\left(\frac{i}{W}\right) + n_s\left(\frac{i}{W}\right) \right]^2$$

(7-29)

where it is always assumed that $TW \geq 1$. The error introduced by (7-28) at $i = 0$ and the error introduced by (7-26) at $i = TW$ are both nearly $1/2W$. For other values of i, the errors caused by the approximations are much less than $1/2W$ and decrease as TW increases. Equation (7-29) becomes an increasingly accurate approximation of (7-24) as γ increases. A test statistic proportional to (7-29) can be derived for the baseband radiometer of Figure 7.1(c) and the sampling rate $1/W$ without invoking the sampling theorems and the accompanying approximations.

Since $n(t)$ is a zero-mean Gaussian process and has a power spectral density that is symmetrical about f_c, $n_c(t)$ and $n_s(t)$ are zero-mean, independent Gaussian processes (Appendix C.2). Thus, $n_c(i/W)$ and $n_s(j/W)$ are zero-mean, independent Gaussian random variables. Equation (C-40) implies that the power spectral densities of $n_c(t)$ and $n_s(t)$ are

$$S_c(f) = S_s(f) = \begin{cases} N_0 , & |f| \leq W/2 \\ 0 , & |f| > W/2 \end{cases}$$

(7-30)

The associated autocorrelation functions are

$$R_c(\tau) = R_s(\tau) = N_0 W \operatorname{sinc}(W\tau)$$

(7-31)

which indicates that $n_c(i/W)$ is statistically independent of $n_c(j/W)$, $i \neq j$, and similarly for $n_s(i/W)$ and $n_s(j/W)$. Therefore, (7-29) becomes

$$V = \frac{N_0}{2} \left\{ \sum_{i=1}^{\gamma} A_i^2 + \sum_{i=1}^{\gamma} B_i^2 \right\}$$

(7-32)

where the $\{A_i\}$ and the $\{B_i\}$ are statistically independent Gaussian random variables with unit variances and means

$$m_{1i} = E[A_i] = \frac{1}{\sqrt{N_0 W}} s_c\left(\frac{i}{W}\right)$$

(7-33)

$$m_{2i} = E[B_i] = \frac{1}{\sqrt{N_0 W}} s_s\left(\frac{i}{W}\right)$$

(7-34)

Thus, $2V/N_0$ has a *noncentral chi-squared* (χ^2) *distribution* (Appendix D.1) with 2γ degrees of freedom and a noncentral parameter

$$\lambda = \sum_{i=1}^{\gamma} m_{1i}^2 + \sum_{i=1}^{\gamma} m_{2i}^2 = \frac{1}{N_0 W} \sum_{i=1}^{\gamma} s_c^2\left(\frac{i}{W}\right) + \frac{1}{N_0 W} \sum_{i=1}^{\gamma} s_s^2\left(\frac{i}{W}\right)$$

$$\approx \frac{1}{N_0} \int_0^T \left[s_c^2(t) + s_s^2(t) \right] dt \approx \frac{2}{N_0} \int_0^T s^2(t) dt = \frac{2\mathcal{E}}{N_0}$$

(7-35)

The probability density function of $Z = 2V/N_0$ is

$$f_Z(x) = \frac{1}{2} \left(\frac{x}{\lambda}\right)^{(\gamma-1)/2} \exp\left(-\frac{x+\lambda}{2}\right) I_{\gamma-1}\left(\sqrt{x\lambda}\right) u(x) \qquad (7\text{-}36)$$

where $I_n(\ )$ is the modified Bessel function of the first kind and order n defined by (D-11), and $u(x) = 1$, $x \geq 0$, and $u(x) = 0$, $x < 0$. Using the series expansion in λ of the Bessel function and then setting $\lambda = 0$ in (7-36), we obtain the probability density function for Z in the absence of the signal:

$$f_Z(x) = \frac{1}{2^\gamma \Gamma(\gamma)} x^{\gamma-1} \exp\left(-\frac{x}{2}\right) u(x), \qquad \lambda = 0 \qquad (7\text{-}37)$$

where $\Gamma(\ )$ is the gamma function defined by (D-12). The direct application of the statistics of Gaussian variables to (7-32) yields

$$E[V] = \mathcal{E} + N_0\gamma \qquad (7\text{-}38)$$

$$\mathrm{var}(V) = 2N_0\mathcal{E} + N_0^2\gamma \qquad (7\text{-}39)$$

Equation (7-38) approaches the exact result of (7-21) as TW increases.

Let V_t denote the threshold level to which V is compared. A false alarm occurs if $V > V_t$ when the signal is absent. Application of (7-37) yields the probability of a false alarm:

$$\begin{aligned}
P_F &= \int_{2V_t/N_0}^{\infty} \frac{1}{2^\gamma \Gamma(\gamma)} v^{\gamma-1} e^{-v/2} dv \\
&= 1 - \Gamma\left(\frac{2V_t}{N_0}, \gamma\right) \qquad (7\text{-}40)
\end{aligned}$$

where the *incomplete gamma function* is defined as

$$\Gamma(x, a) = \frac{1}{\Gamma(a)} \int_0^x e^{-t} t^{a-1} dt \qquad (7\text{-}41)$$

and $\Gamma(\infty, a) = 1$. Integrating (7-40) by parts $\gamma - 1$ times yields the series

$$P_F = \exp\left(-\frac{V_t}{N_0}\right) \sum_{i=0}^{\gamma-1} \frac{1}{i!} \left(\frac{V_t}{N_0}\right)^i \qquad (7\text{-}42)$$

Since correct detection occurs if $V > V_t$ when the signal is present, (7-36) indicates that the probability of detection is

$$P_D = \int_{2V_t/N_0}^{\infty} \frac{1}{2} \left(\frac{v}{\lambda}\right)^{(\gamma-1)/2} \exp\left(-\frac{v+\lambda}{2}\right) I_{\gamma-1}\left(\sqrt{v\lambda}\right) dv \qquad (7\text{-}43)$$

The *generalized Marcum Q-function* is defined as

$$Q_m(\alpha, \beta) = \int_\beta^\infty x \left(\frac{x}{\alpha}\right)^{m-1} \exp\left(-\frac{x^2+\alpha^2}{2}\right) I_{m-1}(\alpha x) dx \qquad (7\text{-}44)$$

where m is a nonnegative integer, and α and β are nonnegative real numbers. A change of variables in (7-43) and the substitution of (7-35) yield

$$P_D = Q_\gamma \left(\sqrt{2\mathcal{E}/N_0}, \sqrt{2V_t/N_0} \right) \qquad (7\text{-}45)$$

The threshold V_t is usually set to a value that ensures a specified P_F. To derive an easily computed closed-form expression for V_t in terms of P_F, we first approximate (7-40). When $TW \gg 1$, $\gamma \approx TW$, and the central limit theorem for the sum of independent, identically distributed random variables with finite means and variances indicates that the distribution of V given by (7-32) is approximately Gaussian. Using (7-38) and (7-39) with $\mathcal{E} = 0$ and the Gaussian distribution, we obtain

$$P_F \approx \frac{1}{(2\pi N_0^2 TW)^{1/2}} \int_{V_t}^{\infty} \exp\left[-\frac{(v - N_0 TW)^2}{2N_0^2 TW} \right] dv$$

$$= Q\left[\frac{V_t - N_0 TW}{(N_0^2 TW)^{1/2}} \right], \quad TW \gg 1 \qquad (7\text{-}46)$$

Inverting this equation, we obtain V_t in terms of P_F and N_0. Accordingly, if the estimate of N_0 is $\hat{N}_0$ and P_F is specified, then the threshold should be

$$V_t \approx \hat{N}_0 \sqrt{TW} Q^{-1}(P_F) + \hat{N}_0 TW , \quad TW \gg 1 \qquad (7\text{-}47)$$

where $Q^{-1}(\ )$ denotes the inverse of the function $Q(\ )$. In the absence of a signal, (7-21) indicates that $N_0 = E[V]/TW$. Thus, N_0 can be estimated by averaging sampled radiometer outputs when it is known that no signal is present.

In some applications, one might wish to specify the *false alarm rate*, which is the expected number of false alarms per unit time, rather than P_F. If successive observation intervals do not overlap each other except possibly at end points, then the false alarm rate is

$$F = \frac{P_F}{T} \qquad (7\text{-}48)$$

For $TW > 100$, the generalized Marcum Q-function in (7-45) is difficult to compute and to invert. If V is approximated by a Gaussian random variable, then (7-38) and (7-39) imply that

$$P_D \approx Q\left[\frac{V_t - N_0 TW - \mathcal{E}}{(N_0^2 TW + 2N_0 \mathcal{E})^{1/2}} \right], \quad TW \gg 1 \qquad (7\text{-}49)$$

Figure 7.2 depicts P_D versus $\mathcal{E}/N_0$ for radiometers with $\hat{N}_0 = N_0$ and $P_F = 10^{-3}$. Equations (7-47) and (7-49) are used to calculate V_t and P_D, respectively. The figure illustrates the increased energy required to maintain a specified P_D as TW increases. The figure also illustrates the impact of the imperfect estimation of N_0 when $P_F = 10^{-3}$ and $TW = 10^7$. When the estimation uncertainty is enough that $\hat{N}_0 = 1.001\, N_0$, the required value of $\mathcal{E}/N_0$ for a specified P_D is increased considerably.

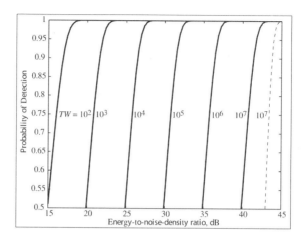

Figure 7.2: Probability of detection versus $\mathcal{E}/N_0$ for wideband radiometer with $P_F = 10^{-3}$ and various values of TW. Solid curves are the $\hat{N}_0 = N_0$; dashed curve is for $\hat{N}_0 = 1.001 N_0$.

The sensitivity of the radiometer to errors in $\hat{N}_0$ when TW is large, which has been observed experimentally [3], is due to the fact that $E[V]$ contains a bias term equal to $N_0 TW$ and $\text{var}(V)$ contains a term equal to $N_0^2 \, TW$, as indicated by (7-38) and (7-39). Setting $\hat{N}_0$ high enough that $\hat{N}_0 \geq N_0$ is certain ensures that V_t will be large enough that the required P_F is achieved regardless of the exact value of N_0. It is important that $\hat{N}_0/N_0$ is as close to unity as possible to avoid degrading P_D when TW is large. Consequently, the radiometer output due to noise alone, which provides $\hat{N}_0$, should be observed often enough that $\hat{N}_0$ closely tracks the changes in N_0 that might result from small changes in the circuitry or the environmental noise.

When V_t is specified, the value of $\mathcal{E}/N_0$ necessary to achieve a specified value of P_D may be obtained by inverting (7-45), which is computationally difficult but can be closely approximated by inverting (7-49). Assuming that $2V_t > N_0 TW$ and $TW >> 1$, we obtain the necessary value:

$$\frac{\mathcal{E}}{N_0} \approx \left[Q^{-1}(P_D)\right]^2 + \frac{V_t}{N_0} - TW - \left[Q^{-1}(P_D)\right] \sqrt{\left[Q^{-1}(P_D)\right]^2 + \frac{2V_t}{N_0} - TW}$$

$$(7\text{-}50)$$

According to (7-47), the condition $2V_t > N_0 TW$ is satisfied if $P_F \leq 1/2$ and $\hat{N}_0 \approx N_0$. The substitution of (7-47) into (7-50) and a rearrangement of terms

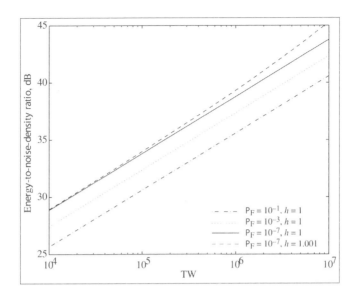

Figure 7.3: Energy-to-noise-density ratio versus TW for wideband radiometer with $P_D = 0.99$ and various values of P_F and h .

yields

$$\frac{\mathcal{E}}{N_0} \approx h\sqrt{TW}\beta + (h-1)TW + \psi(\beta, \xi, TW, h), \quad TW \gg 1 \qquad (7\text{-}51)$$

where

$$\beta = Q^{-1}(P_F), \quad \xi = Q^{-1}(P_D), \quad h = \hat{N}_0/N_0 \qquad (7\text{-}52)$$

$$\psi(\beta, \xi, TW, h) = \xi^2 - \sqrt{TW}\xi \left[2h - 1 + \frac{2\beta h}{\sqrt{TW}} + \frac{\xi^2}{TW}\right]^{1/2} \qquad (7\text{-}53)$$

As TW increases, the significance of the third term in (7-51) decreases, while that of the second term increases if $h > 1$. Figure 7.3 shows $\mathcal{E}/N_0$ versus TW for $P_D = 0.99$ and various values of P_F and h.

If the signal duration is T_1, then the detected signal power is $S = \mathcal{E}/T_1$. Equation (7-51) indicates that the detected power necessary to achieve specified values of P_D and either P_F or F is

$$\frac{S}{N_0} \approx \begin{cases} h\dfrac{\sqrt{TW}}{T_1}\beta + (h-1)W\dfrac{T}{T_1} + \dfrac{\psi}{T_1} &, \quad T_1 < T \\[2ex] h\sqrt{\dfrac{W}{T}}\beta + (h-1)W + \dfrac{\psi}{T} &, \quad T_1 \geq T \end{cases} \qquad (7\text{-}54)$$

This equation indicates that increasing the observation interval T decreases the required power only if $T \leq T_1$. Although a single radiometer is incapable of determining whether one or more than one signal has been detected, narrowband interference can be rejected by the methods of Section 2.7.

7.2   Detection of Frequency-Hopping Signals

An interception receiver intended for the detection of frequency-hopping signals may be designed according to the principles of classical detection theory or according to more intuitive ideas. The former approach is useful in setting limits on what is possible, but the latter approach is more practical and flexible and less dependent on knowledge of the characteristics of the frequency-hopping signals.

Ideal Detection

To enable a tractable analysis, the idealized assumptions are made that the hopset is known and that the hop epoch timing, which includes the hop-transition times is known. Consider slow frequency-hopping signals with CPM (FH/CPM), which includes continuous-phase MFSK. The signal over the ith hop interval is

$$s(t) = \sqrt{2S} \cos\left[2\pi f_j t + \phi(\mathbf{d}_n, t) + \phi_i\right] , \ (i-1)T_h \leq t < iT_h \qquad (7\text{-}55)$$

where S is the average signal power, f_j is the carrier frequency associated with the ith hop, $\phi(\mathbf{d}_n, t)$ is the CPM component that depends on the data sequence $\mathbf{d}_n$, and ϕ_i is the phase associated with the ith hop. The parameters f_j, ϕ_i, and the components of $\mathbf{d}_n$ are modeled as random variables. The derivation of the average likelihood ratio (7-7) is still valid, but the vector $\boldsymbol{\theta}$ has different parameters as components.

The M carrier frequencies in the hopset are assumed to be equally likely over a given hop and statistically independent from hop to hop for N_h hops. Dividing the integration interval in (7-7) into N_h parts, averaging over the M frequencies, and dropping the irrelevant factor $1/M$, we obtain

$$\Lambda[r(t)] = \prod_{i=1}^{N_h} \sum_{j=1}^{M} \Lambda_{ij}[r(t)|f_j] \qquad (7\text{-}56)$$

$$\Lambda_{ij}[r(t)|f_j] = E_{\mathbf{d}_n, \phi_i} \left\{ \exp\left[\frac{2}{N_0} \int_{(i-1)T_h}^{iT_h} r(t)s(t)dt - \frac{\mathcal{E}_h}{N_0} \right] \right\} \qquad (7\text{-}57)$$

where the condition in the argument of $\Lambda_{ij}[\ ]$ indicates that the carrier frequency over the ith hop is f_j, the expectation is over the remaining random parameters $\mathbf{d}_n$ and ϕ_i, and $\mathcal{E}_h$ is the energy per hop. The decomposition in (7-56) indicates that the general structure of the detector has the form illustrated in Figure 7.4. The average likelihood ratio of (7-56) is compared with a threshold to determine whether a signal is present. The threshold may be set to ensure the tolerable false-alarm probability when the signal is absent. Assuming that $\mathcal{E}_h = ST_h$ is

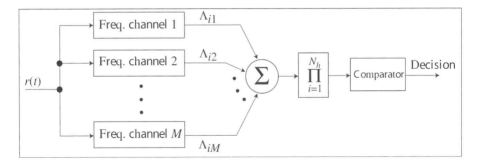

Figure 7.4: General structure of optimum detector for frequency-hopping signal with N_h hops and M frequency channels.

the same for every hop and carrier frequency, we may drop the irrelevant factor $\exp(-\mathcal{E}_h/N_0)$ in (7-57), which only affects the threshold level.

Each of the N_d data sequences that can occur during a hop is assumed to be equally likely. For *coherent detection* of FH/CPM [5], we set $\phi_i = 0$ in (7-55), substitute it into (7-57), and then evaluate the expectation to obtain

$$\Lambda_{ij}[r(t)|f_j] = \sum_{n=1}^{N_d} \exp\left\{ \frac{2\sqrt{2S}}{N_0} \int_{(i-1)T_h}^{iT_h} r(t) \cos\left[2\pi f_j t + \phi(\mathbf{d}_n, t)\right] \right\} \quad \text{(coherent)}$$

(7-58)

where irrelevant factors have been dropped. This equation indicates how Λ_{ij} in Figure 7.4 is to be calculated for each hop i and each frequency channel j corresponding to carrier frequency f_j. Equations (7-56) and (7-58) define the optimum coherent detector for any slow frequency-hopping signal with CPM.

For noncoherent detection of FH/CPM [4], the received carrier phase ϕ_i is assumed to be uniformly distributed over $[0, 2\pi)$ during a given hop and statistically independent from hop to hop. Substituting (7-55) into (7-57), averaging over the random phase in addition to the sequence statistics, and dropping irrelevant factors yields

$$\Lambda_{ij}[r(t)|f_j] = \sum_{n=1}^{N_d} I_0\left(\frac{2\sqrt{2SR_{ijn}}}{N_0} \right) \quad \text{(noncoherent)} \quad (7-59)$$

where

$$R_{ijn} = \left\{ \int_{(i-1)T_h}^{iT_h} r(t) \cos\left[\chi_{jn}(t)\right] dt \right\}^2 + \left\{ \int_{(i-1)T_h}^{iT_h} r(t) \sin\left[\chi_{jn}(t)\right] dt \right\}^2$$

(7-60)

and

$$\chi_{jn}(t) = 2\pi f_j t + \phi(\mathbf{d}_n, t) \quad (7-61)$$

Equations (7-56), (7-59), (7-60), and (7-61) define the optimum noncoherent

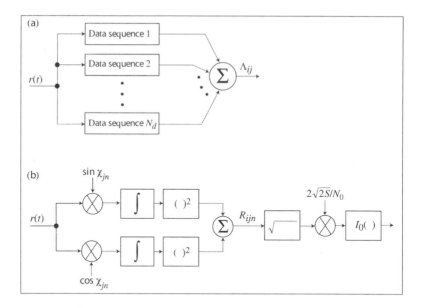

Figure 7.5: Optimum noncoherent detector for slow frequency hopping with CPM: (a) basic structure of frequency channel j for hop i with parallel cells for N_d candidate data sequences, and (b) cell for data sequence n.

detector for any slow frequency-hopping signal with CPM. The means of producing (7-59) is diagrammed in Figure 7.5.

A major contributor to the huge computational complexity of the optimum detectors is the fact that with N_s data symbols per hop and an alphabet size q, there may be $N_d = q^{N_s}$ data sequences per hop. Consequently, the computational burden grows exponentially with N_s. However, if it is known that the data modulation is CPFSK with a modulation index $h = 1/n$, where n is a positive integer, the computational burden has a linear dependence on N_s [4]. Even then, the optimum detectors are extremely complex when the number of frequency channels is large.

The preceding theory may be adapted to the detection of fast frequency-hopping signals with MFSK as the data modulation. Since there is one hop per MFSK channel symbol, the information is embedded in the sequence of carrier frequencies. Thus, we may set $N_d = 1$ and $\phi(\mathbf{d}_n, t) = 0$ in (7-58) and (7-59). For coherent detection, (7-58) reduces to

$$\Lambda_{ij}[r(t)|f_j] = \exp\left[\frac{2\sqrt{2S}}{N_0}\int_{(i-1)T_h}^{iT_h} r(t)\cos\left(2\pi f_j t\right)dt\right] \qquad \text{(coherent)} \quad (7\text{-}62)$$

Equations (7-56) and (7-62) define the optimum coherent detector for a fast frequency-hopping signal with MFSK. For noncoherent detection, (7-59), (7-60), and (7-61) reduce to

$$\Lambda_{ij}[r(t)|f_j] = I_0 \left(\frac{2\sqrt{2SR_{ij}}}{N_0} \right) \qquad \text{(noncoherent)} \qquad (7\text{-}63)$$

$$R_{ij} = \left[\int_{(i-1)T_h}^{iT_h} r(t) \cos(2\pi f_j t) dt \right]^2 + \left[\int_{(i-1)T_h}^{iT_h} r(t) \sin(2\pi f_j t) dt \right]^2 \qquad (7\text{-}64)$$

Equations (7-56), (7-63), and (7-64) define the optimum noncoherent detector for a fast frequency-hopping signal with MFSK. Performance analyses for the detectors of fast frequency-hopping signals are given in [5].

Instead of basing detector design on the average likelihood ratio, one might apply a composite hypothesis test in which the presence of the signal is detected while simultaneously one or more of the unknown parameters under hypothesis H_1 are estimated. To simultaneously detect the signal and determine the frequency-hopping pattern, (7-56) is replaced by the *generalized likelihood ratio*:

$$\Lambda\left[r(t)\right] = \prod_{i=1}^{N_h} \max_{i \le j \le M} \{\Lambda_{ij}\left[r(t)|f_j\right]\} \qquad (7\text{-}65)$$

where the equations and subsystems for $\Lambda_{ij}[r(t)|f_j]$ remain the same. Equation (7-65) indicates that a maximum-likelihood estimate of f_j is made for each hop. Thus, an optimum test to determine the frequency channel occupied by the frequency-hopping signal is conducted during each hop. Although the detection performance is suboptimal when the generalized likelihood ratio is used to design a detector, this detector provides an important signal feature and is slightly easier to implement and analyze [4], [5].

Wideband Radiometer

Among the many alternatives to the optimum detector, two of the most useful are the *wideband radiometer* and the *channelized radiometer*. The wideband radiometer is notable in that it requires virtually no detailed information about the parameters of the frequency-hopping signals to be detected other than their rough spectral location. The price paid for this robustness is much worse performance than more sophisticated detectors that exploit additional information about the signal [4]. The channelized radiometer is designed to explicitly exploit the spectral characteristics of frequency-hopping signals. In its optimal form, the channelized radiometer gives a performance nearly as good as that of the ideal detector. In its suboptimal form, the channelized radiometer trades performance for practicality and the easing of the required *a priori* information about the signal to be detected.

Channelized Radiometer

A *channelized radiometer* comprises K parallel radiometers, each of which has the form of Figure 7.1 and monitors a disjoint portion of the hopping band of a

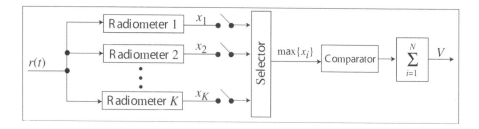

Figure 7.6: Channelized radiometer.

frequency-hopping signal, as depicted in Figure 7.6. The largest of the sampled radiometer outputs is compared to a threshold V_t stored in a comparator. If the threshold is exceeded, the comparator sends a 1 to the summer; otherwise it sends a 0. If the hop dwell epochs are at least approximately known, the channelized radiometer may improve its detection reliability by adding the 1's produced by N consecutive comparator outputs corresponding to multiple frequency hops of the signal to be detected. A signal is declared to be present if the sum V equals or exceeds the integer r, which serves as a second threshold. The two thresholds V_t are r are jointly optimized for the best system performance.

Ideally, $K = M$, the number of frequency channels in a hopset, but many fewer radiometers may be a practical or economic necessity; if so, each radiometer may monitor M_r frequency channels, where $1 \leq M_r \leq M$. Because of insertion losses and the degradation caused by a power divider, it is unlikely that many more than 30 parallel radiometers are practical. An advantage of each radiometer covering many frequency channels is the reduced sensitivity to imprecise knowledge of the spectral boundaries of frequency channels. Since it is highly desirable to implement the parallel radiometers with similar circuitry, their bandwidths are assumed to be identical henceforth.

To prevent steady interference in a single radiometer from causing false alarms, the channelized radiometer must be able to recognize when one of its constituent radiometers produces an output above the threshold for too many consecutive samples. The channelized system may then delete that constituent radiometer's output from the detection algorithm or it may reassign the radiometer to another spectral location.

In the subsequent analysis of the channelized radiometer of Figure 7.6, the observation interval of the parallel radiometers, which is equal to the sampling interval, is assumed to equal the hop duration T_h. The effective observation time of the channelized radiometer, $T = NT_h$, should be less than the minimum expected message duration to avoid processing extraneous noise. Let P_{F1} denote the probability that a particular radiometer output at the sampling time exceeds the comparator threshold V_t when no signal is present. This probability is given by the right-hand side of (7-42). Therefore, a derivation similar to that of (7-47) indicates that if the sampling times are aligned with the frequency transitions, then the threshold necessary to achieve a specified P_{F1} is

$$V_t \approx \hat{N}_0 \sqrt{M_r T_h B} Q^{-1}(P_{F1}) + \hat{N}_0 M_r T_h B , \quad M_r T_h B \gg 1 \quad (7\text{-}66)$$

where B is the bandwidth of each of the M_r frequency channels encompassed by a radiometer passband. The probability that at least one of the K parallel radiometer outputs exceeds V_t is

$$P_{F2} = 1 - (1 - P_{F1})^K \quad (7\text{-}67)$$

assuming that the channel noises are statistically independent because the radiometer passbands are disjoint. The probability of a false alarm of the channelized radiometer is the probability that the output V equals or exceeds a threshold r:

$$P_F = \sum_{i=r}^{N} \binom{N}{i} P_{F2}^i (1 - P_{F2})^{N-i} \quad (7\text{-}68)$$

To solve this equation for P_{F2} in terms of P_F, we observe that the *incomplete beta function* is defined as

$$F(x|a, b) = \frac{1}{B(a, b)} \int_0^x t^{a-1}(1 - t)^{b-1} dt , \quad 0 \le x \le 1 \quad (7\text{-}69)$$

where $B(a, b)$ is the *beta function* and $F(1|a, b) = 1$. In terms of this function, (7-68) becomes

$$P_F = F(P_{F2}|r, N - r + 1) \quad (7\text{-}70)$$

The inverse of the incomplete beta function, which we denote by $F^{-1}(\ )$, may be easily computed by Newton's method or approximations [6]. Therefore, if $M_r T_h B \gg 1$, (7-66), (7-67), and (7-70) may be combined to determine the approximate threshold necessary to achieve a specified P_F:

$$V_t \approx \hat{N}_0 \sqrt{M_r T_h B} Q^{-1} \left\{ 1 - \left[1 - F^{-1}\left(P_F | r, N - r + 1 \right) \right]^{1/K} \right\} + \hat{N}_0 M_r T_h B$$
$$(7\text{-}71)$$

where it is assumed that N_0 does not vary across the hopping band and, hence, there is one $\hat{N}_0$ and one V_t for all the parallel radiometers.

The number of sampling intervals during which the signal is present is $N_1 = T_1/T_h$, where T_1 is the intercepted signal duration. For simplicity, it is assumed that N_1 is an integer. Furthermore, we assume that at most a single radiometer receives significant signal energy during each sampling interval. Let P_{D1} denote the probability that a particular radiometer output exceeds the threshold when a signal is present in that radiometer. Derivations similar to those of (7-45) and (7-49) imply that

$$P_{D1} = Q_L \left(\sqrt{2\mathcal{E}_h/N_0} , \sqrt{2V_t/N_0} \right)$$
$$\approx Q \left[\frac{V_t - N_0 M_r T_h B - \mathcal{E}_h}{(N_0^2 M_r T_h B + 2N_0 \mathcal{E}_h)^{1/2}} \right] , \quad M_r T_h B \gg 1 \quad (7\text{-}72)$$

where $L = \lfloor M_r T_h B \rfloor$ and $\mathcal{E}_h$ is the energy per hop dwell time. Let P_{D2} denote the probability that the threshold is exceeded by the sampled maximum of the parallel radiometer outputs. It is assumed that when a signal is present it occupies any one of M frequency channels with equal probability and that all radiometer passbands are within the hopping band. Consequently, the signal has probability M_r/M of being in the passband of a particular radiometer and probability KM_r/M of being in the passband of some radiometer. Since a detection may be declared in response to a radiometer that does not receive the signal,

$$P_{D2} = \frac{KM_r}{M}\left[1 - (1 - P_{D1})(1 - P_{F1})^{K-1}\right] + \left(1 - \frac{KM_r}{M}\right)P_{F2} \qquad (7\text{-}73)$$

where the second term vanishes if the radiometer passbands cover the hopping band so that $KM_r = M$. The probability of detection associated with the observation interval when the signal is actually present during $N_1 \leq N$ of the hop intervals is

$$P_D = \sum_{i=r}^{N}\sum_{j=0}^{i}\binom{N_1}{j}\binom{N-N_1}{i-j}P_{D2}^{j}(1-P_{D2})^{N_1-j}\,P_{F2}^{i-j}(1-P_{F2})^{N-N_1-i+j}$$

$$(7\text{-}74)$$

If at least the minimum duration of a frequency-hopping signal is known, the overestimation of N might be avoided so that $N_1 = N$. The detection probability then becomes

$$P_D = \sum_{i=r}^{N}\binom{N}{i}P_{D2}^{i}(1-P_{D2})^{N-i}$$

$$= F(P_{D2}|r, N-r+1) \qquad (7\text{-}75)$$

A suitable, but not optimal, choice for the second threshold is $r = \lfloor N/2 \rfloor$ when the full hopping band is monitored by the channelized radiometer. In general, numerical results indicate that

$$r = \left\lfloor \frac{KM_r N}{2M} \right\rfloor \qquad (7\text{-}76)$$

is a good choice for partial-band monitoring.

If detection decisions are made in terms of fixed observation intervals of duration $T = NT_h$, and successive intervals do not overlap except possibly at end points, then the false alarm rate defined in (7-48) is an appropriate design parameter. This type of detection is called *block detection*, and the false-alarm rate is

$$F = \frac{P_F}{NT_h} \qquad (7\text{-}77)$$

To prevent the risk of major misalignment of the observation interval with the time the signal is being transmitted, either block detection must be supplemented with hardware for arrival-time estimation or the duration of successive observation intervals should be less than roughly half the anticipated signal duration.

A different approach to mitigating the effect of a misalignment, called *binary moving-window detection*, is for the observation interval to be constructed by dropping the first sampling interval of the preceding observation interval and adding a new sampling interval. A false alarm is considered to be a detection declaration at the end of the new interval when no signal is actually present. Thus, a false alarm occurs only if the comparator input for an added sampling interval exceeds the threshold, the comparator input for the discarded sampling interval did not, and the count for the preceding observation interval was $r - 1$. Therefore, the probability of a false alarm is

$$P_{F0} = C(0,1)C(r-1, N-1)C(1,1) \tag{7-78}$$

where

$$C(i, N) = \binom{N}{i} P_{F2}^i (1 - P_{F2})^{N-i} , \quad i \leq N \tag{7-79}$$

It follows that the false-alarm rate is

$$F = \frac{P_{F0}}{T_h} = \frac{r}{NT_h} \binom{N}{r} P_{F2}^r (1 - P_{F2})^{N+1-r} \tag{7-80}$$

Since the right-hand side of this equation is proportional to the first term of the series in (7-68),

$$F \leq \frac{rP_F}{NT_h} \tag{7-81}$$

This inequality indicates that the false alarm rate is nearly r times as large for moving-window detection as it is for block detection. Thus, moving-window detection usually requires a higher comparator threshold for the same false-alarm rate and, hence, more signal power to detect a frequency-hopping signal. However, moving-window detection with $N \approx N_1 >> 1$ inherently limits the misalignment between the occurrence of the intercepted signal and some observation interval. If the signal occurs during two successive obsevation intervals, then for one of the observation intervals, the misalignment is not more than $T_h/2$.

As an example, it is assumed that there are $M = 2400$ frequency channels, the signal duration is known, and there is no misalignment so that $N_1 = N$. Block detection is used so that (7-77) is applicable, $F = 10^{-7}/T_h$, $B = 250/T_h$, and $\hat{N}_0 = N_0$. Figure 7.7 plots P_D versus $\mathcal{E}_h/N_0$ for the channelized radiometer with full hopping-band coverage so that $M_r = M/K$, and several values of K and N. The figure also shows the results for a wideband radiometer with $TW = BT_h M$ $N = 6 \cdot 10^5 \cdot N$, and $N = 150$ or 750. It is observed that the channelized radiometer with $K = 30$ is much better than the wideband radiometer when $N = 150$, but loses its advantage for $P_D \leq 0.995$ when $N = 750$. The substantial advantage of the channelized radiometer with $K = M$ and $M_r = 1$ is apparent. As N increases, the channelized radiometer can retain its advantage over the wideband radiometer by increasing K accordingly.

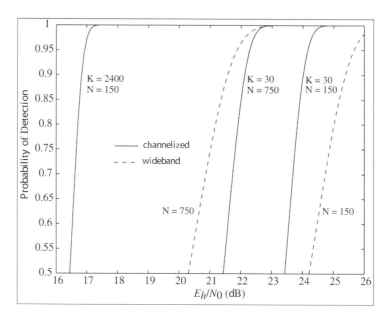

Figure 7.7: Probability of detection versus $\mathcal{E}_h/N_0$ for channelized and wideband radiometers with full coverage, $N_1 = N$, $\hat{N}_0 = N_0$, $M = 2400$, $F = 10^{-7}/T_h$, and $B = 250/T_h$.

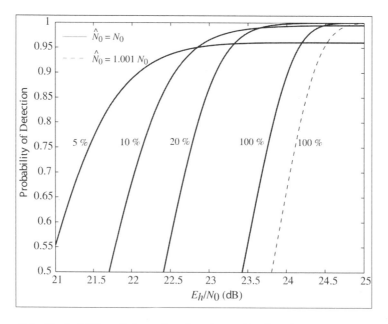

Figure 7.8: Probability of detection for channelized radiometer with different percentages of coverage, $N_1 = N = 150$, $K = 30$, $M = 2400$, $F = 10^{-7}/T_h$, $B = 250/T_h$, and $\hat{N}_0 = N_0$ or $\hat{N}_0 = 1.001\, N_0$.

In Figure 7.8, $N = N_1 = 150$ and $K = 30$, but M_r and $\hat{N}_0/N_0$ are variable. The fraction of the hopping band monitored by the channelized radiometer is denoted by the *monitored fraction* $\beta = KM_r/M$. It is observed that when $\hat{N}_0 = 1.001 \ N_0$, there is only a small performance loss for the channelized radiometer despite the fact that $TW = 9 \cdot 10^7$. The relative insensitivity of the channelized radiometer to small errors in $\hat{N}_0$ is a major advantage over the wideband radiometer. The figure illustrates the following trade-off when K and M are fixed: as M_r decreases, fewer frequency channels are monitored, but less noise enters a radiometer. The net result is beneficial when β is reduced to 20 percent. However, the figure indicates that for $\beta = 10$ percent or 5 percent, the hopping-band coverage becomes inadequate to enable a P_D greater than 0.995 and 0.96, respectively, regardless of $\mathcal{E}_h/N_0$. Thus, there is a minimum fraction of the hopping band that must be monitored to ensure a specified P_D.

As $\mathcal{E}_h/N_0 \to \infty$, (7-72) indicates that $P_{D1} \to 1$. Therefore, (7-73) implies that $P_{D2} \to \beta + (1 - \beta)P_{F2}$. Suppose that V_t is raised to a sufficiently high level that $P_{F2} \ll \beta$ and, hence, $P_{D2} \approx \beta$. If detection is to be accomplished for the minimum monitored fraction, then $r = 1$ is the best choice for the second threshold. For $r = 1$ and $N_1 = N$, (7-75) yields

$$P_D = 1 - (1 - P_{D2})^N \qquad (7\text{-}82)$$

Since $P_{D2} \approx \beta$, (7-82) implies that even if $\mathcal{E}_h/N_0 \to \infty$, the realization of a specified P_D requires the minimum monitored fraction

$$\beta_{\min} = 1 - (1 - P_D)^{1/N} \qquad (7\text{-}83)$$

Thus, if $P_D = 0.99$ and $N = N_1 = 150$, then $\beta_{\min} = 0.03$. Many other aspects of the channelized radiometer, including the effects of timing misalignments, are examined in [6].

7.3 Problems

1. If the form and the parameters of a signal $s(t)$ to be detected are known, optimum detection in white Gaussian noise $n(t)$ can be accomplished by an ideal matched filter or correlator. Let $r(t)$ denote the received signal and T denote the observation interval. (a) Write an expression for the sampled output of an ideal matched filter. (b) This output r_1 is compared to a threshold V_t to determine whether the target signal is present. Suppose that it is present and coincides with an observation interval. Assuming that $n(t)$ is zero-mean and has two-sided power spectral density $N_0/2$ and that the signal energy is $\mathcal{E}$, what are the mean and the variance of r_1? (c) What is the probability of detection P_D? What is the probability of false alarm P_F? Express P_D in terms of P_F. (d) What is the value of $\mathcal{E}/N_0$ necessary to ensure specified values of P_F and P_D?

2. An unbiased estimate of N_0 may be obtained from the wideband radiometer output when the target signal is absent. How long must the obser-

vation interval T_n be if it is required that 3 standard deviations of the estimate do not exceed $0.001N_0$?

3. The *receiver operating characteristic* (ROC) is a traditional plot depicting P_D versus P_F for various values of TW or $\mathcal{E}/N_0$. The ROC may be calculated from (7-49) and (7-46). Plot the ROC for the wideband radiometer with $\mathcal{E}/N_0 = 20 \ dB$ and no noise-measurement error. Let $TW = 104$ and 105.

4. Derive (7-50) using the method described in the text.

5. Find conditions under which (7-51) indicates that a negative energy is required. What is the physical implication of this result?

6. Show that the first two terms in (7-54) give the intercepted power necessary to achieve $P_D = 1/2$ and the specified value of either P_F or F.

7. Consider a channelized radiometer that is to detect a frequency-hopping signal with $N = 1$ hop. (a) Find V_t in terms of P_F. (b) If $KM_r = M$, $\hat{N}_0 = N_0$, $N_1 = N$, and $\mathcal{E}_h/N_0$ is small, derive the S/N_0 required for specified values of P_D and P_F.

7.4   References

1. M. D. Srinath, P. K. Rajasekaran, and R. Viswanathan, *Introduction to Statistical Signal Processing with Applications*. Englewood Cliffs, NJ: Prentice Hall, 1996.

2. A. Polydoros and C. L. Weber, "Detection Performance Considerations for Direct-Sequence and Time-Hopping LPI Waveforms," *IEEE J. Select. Areas Commun.*, vol. 3, pp. 727–744, September 1985.

3. D. A. Hill and E. B. Felstead, "Laboratory Performance of Spread Spectrum Detectors," *IEE Proc.-Commun.*, vol. 142, pp. 243–249, August 1995.

4. B. K. Levitt et al., "Optimum Detection of Slow Frequency-Hopped Signals," *IEEE Trans. Commun.*, vol. 42, pp. 1990–2000, Feb./March/April 1994.

5. N. C. Beaulieu, W. L. Hopkins, and P. J. McLane, "Interception of Frequency-Hopped Spread-Spectrum Signals," *IEEE. J. Select. Areas Commun.*, vol. 8, pp. 853–870, June 1990.

6. L. E. Miller, J. S. Lee, and D. J. Torrieri, "Frequency-Hopping Signal Detection Using Partial Band Coverage," *IEEE Trans. Aerosp. Electron. Syst.*, vol. 29, pp. 540–553, April 1993.

7. D. Torrieri, *Principles of Secure Communication Systems, 2nd ed.* Boston: Artech House, 1992.

Appendix A

Inequalities

A.1 Jensen's Inequality

A function $g(x)$ defined on an open interval I is *convex* if

$$g(px + (1-p)y) \leq pg(x) + (1-p)g(y) \tag{A-1}$$

for x, y in I and $0 \leq p \leq 1$. Suppose that $g(x)$ has a continuous, nondecreasing derivative $g'(x)$ on I. The inequality is valid if $p = 0$ or 1. If $x \geq y$ and $0 \leq p < 1$,

$$g(px + (1-p)y) - g(y) = \int_{y}^{px+(1-p)y} g'(z)dz \leq p(x-y)g'(px + (1-p)y)$$

$$\leq \frac{p}{1-p} \int_{px+(1-p)y}^{x} g'(z)dz$$

$$= \frac{p}{1-p}[g(x) - g(px + (1-p)y)]$$

Simplifying this result, we obtain (A-1). If $y \geq x$, a similar analysis again yields (A-1). Thus, *if $g(x)$ has a continuous, nondecreasing derivative on I, it is convex.*

Lemma. If $g(x)$ is a convex function on the open interval I, then

$$g(y) \geq g(x) + g^-(x)(y - x) \tag{A-2}$$

for all y, x in I, where $g^-(x)$ is the left derivative of $g(x)$.

Proof: If $y - x \geq z > 0$, then substituting $p = 1 - z/(y-x)$ into (A-1) gives

$$g(x + z) \leq \left(1 - \frac{z}{y-x}\right)g(x) + \frac{z}{y-x}g(y)$$

which yields

$$\frac{g(x+z) - g(x)}{z} \leq \frac{g(y) - g(x)}{y - x}, \quad y - x \geq z > 0 \tag{A-3}$$

If $v > 0$ and $z > 0$, then (A-1) implies that

$$g(x) \leq \frac{z}{v+z}g(x-v) + \frac{v}{v+z}g(x+z)$$

which yields

$$\frac{g(x)-g(x-v)}{v} \leq \frac{g(x+z)-g(x)}{z} \,, \quad v,z > 0 \qquad (A\text{-}4)$$

Inequality (A-3) indicates that the ratio $[g(y)-g(x)]/(y-x)$ decreases monotonically as $y \to x$ from above and (A-4) implies that this ratio has a lower bound. Therefore, the right derivative $g^+(x)$ exists on I. If $x - y \geq v > 0$, then (A-1) with $p = 1 - v/(x-y)$ implies that

$$g(x-v) \leq \left(1 - \frac{v}{x-y}\right)g(x) + \frac{v}{x-y}g(y)$$

which yields

$$\frac{g(x)-g(y)}{x-y} \leq \frac{g(x)-g(x-v)}{v} \,, \quad x - y \geq v > 0 \qquad (A\text{-}5)$$

This inequality indicates that the ratio $[g(x)-g(y)]/(x-y)$ increases monotonically as $y \to x$ from below and (A-4) implies that this ratio has an upper bound. Therefore, the left derivative $g^-(x)$ exists on I, and (A-4) yields

$$g^-(x) \leq g^+(x) \qquad (A\text{-}6)$$

Taking the limits as $z \to 0$ and $v \to 0$ in (A-3) and (A-5), respectively, and then using (A-6), we find that (A-2) is valid for all y, x in I.$\Box$

Jensen's inequality.   If X is a random variable with a finite expected value $E[X]$, and $g(\ )$ is a convex function on an open interval containing the range of X, then

$$E[g(X)] \geq g(E[X]) \qquad (A\text{-}7)$$

Proof: Set $y = X$ and $x = E[X]$ in (A-2), which gives $g(X) \geq g(E[X]) + g^-(E[X])(X - E[X])$. Taking the expected values of the random variables on both sides of this inequality gives Jensen's inequality.$\Box$

A.2   Chebyshev's Inequality

Consider a random variable X with distribution $F(x)$. Let $E[X] = m$ denote the expected value of X and $P[A]$ denote the probability of event A. From elementary probability it follows that

$$E[|X-m|^k] = \int_{-\infty}^{\infty} |x-m|^k dF(x) \geq \int_{|x-m|\geq\alpha}^{\infty} |x-m|^k dF(x)$$

$$\geq \alpha^k \int_{|x-m|\geq\alpha}^{\infty} dF(x) = \alpha^k P[|X-m| \geq \alpha]$$

Therefore,

$$P[|X - m| \geq \alpha] \leq \frac{1}{\alpha^k} E[|X - m|^k] \tag{A-8}$$

Let $\sigma^2 = E[(X - m)^2]$ denote the variance of X. If $k = 2$, then (A-8) becomes *Chebyshev's inequality*:

$$P[|X - m| \geq \alpha] \leq \frac{\sigma^2}{\alpha^2} \tag{A-9}$$

As an application, let $\alpha = 3\sigma$. Then Chebyshev's inequality indicates that $P[|X - m| < 3\sigma] \geq 8/9$. Therefore, the probability that a random variable is within three standard deviations of its mean value is at least $8/9$.

Appendix B

Adaptive Filters

The input and weight vectors of an adaptive filter are

$$\mathbf{x} = [x_1 \ x_2 \ldots x_N]^T, \quad \mathbf{W} = [W_1 \ W_2 \ldots W_N]^T \tag{B-1}$$

where T denotes the transpose and the components of the vectors may be real or complex. The filter output is the scalar

$$y = \mathbf{W}^T \mathbf{x} \tag{B-2}$$

The derivation of the optimal filter weights depends on the specification of a performance criterion or estimation procedure. A number of different estimators of the desired signal can be implemented by linear filters that produce (B-2). Unconstrained estimators that depend only on the second-order moments of $\mathbf{x}$ can be derived by using performance criteria based on the mean square error or the signal-to-noise ratio of the filter output. Similar estimators result from using the maximum-*a-posteriori* or the maximum-likelihood criteria, but the standard application of these criteria includes the restrictive assumption that any interference in $\mathbf{x}$ has a Gaussian distribution.

The difference between the desired response d and the filter output is the *error signal*:

$$\epsilon = B - \mathbf{W}^T \mathbf{x} \tag{B-3}$$

The most widely used method of estimating the desired signal is based on the minimization of the expected value of the squared error magnitude, which is proportional to the mean power in the error signal. Let H denote the conjugate transpose and an asterisk denote the conjugate. We obtain

$$\begin{aligned}
E[|\epsilon|^2] = E[\epsilon^* \epsilon] &= E\left[\left(d^* \mathbf{W}^H \mathbf{x}^*\right)\left(B - \mathbf{x}^T \mathbf{W}\right)\right] \\
&= E[|d|^2] - \mathbf{W}^H \mathbf{R}_{xd} - \mathbf{R}_{xd}^H \mathbf{W} + \mathbf{W}^H \mathbf{R}_{xx} \mathbf{W}
\end{aligned} \tag{B-4}$$

where

$$\mathbf{R}_{xx} = E\left[\mathbf{x}^* \mathbf{x}^T\right] \tag{B-5}$$

is the $N \times N$ Hermitian *correlation matrix* of $\mathbf{x}$ and

$$\mathbf{R}_{xd} = E\left[\mathbf{x}^* d\right] \tag{B-6}$$

is the $N \times 1$ *cross-correlation* vector. If we assume that $y \neq \mathbf{0}$ when $\mathbf{W} \neq \mathbf{0}$, then $\mathbf{R}_{xx}$ must be positive definite.

In terms of its real part $\mathbf{W}_R$, and its imaginary part $\mathbf{W}_I$, a complex weight vector is defined as

$$\mathbf{W} = \mathbf{W}_R - j\mathbf{W}_I \tag{B-7}$$

The *gradient* of f with respect to the n-dimensional, real-valued vector $\mathbf{x}$ is defined as the column vector $\nabla_{\mathbf{x}} f$ with components $\partial f/\partial x_i, i = 1, 2, \ldots, n$. Let ∇_{wr} and ∇_{wi} denote the $N \times 1$ gradient vectors with respect to $\mathbf{W}_R$ and $\mathbf{W}_I$ respectively. The *complex gradient* with respect to $\mathbf{W}$ is defined as

$$\bar{\nabla}_w = \nabla_{wr} - j\nabla_{wi} \tag{B-8}$$

Let W_i, W_{Ri} and $W_{Ii}, i = 1, 2, \ldots, N$, denote the components of $\mathbf{W}, \mathbf{W}_R$, and $\mathbf{W}_I$, respectively. Let $g(\mathbf{W}, \mathbf{W}^*)$ denote a real-valued function of $\mathbf{W}$ and $\mathbf{W}^*$. Regarding $\mathbf{W}$ and $\mathbf{W}^*$ as independent variables, we assume that g is an analytic function of each W_i when $\mathbf{W}^*$ is held constant and an analytic function of each W_i^* when $\mathbf{W}$ is held constant. We define ∇_{w*} as the gradient with respect to $\mathbf{W}^*$. Since $W_i = W_{Ri} - jW_{Ii}$,

$$\frac{\partial W_i}{\partial W_{Ri}} = 1, \quad \frac{\partial W_i}{\partial W_{Ii}} = -j, \quad \frac{\partial W_i^*}{\partial W_{Ri}} = 1, \quad \frac{\partial W_i^*}{\partial W_{Ii}} = j \tag{B-9}$$

The chain rule of calculus then implies that

$$\begin{aligned}\bar{\nabla}_w g\left(\mathbf{W}, \mathbf{W}^*\right) &= \nabla_{wr} g - j\nabla_{wi} g \\ &= \nabla_w g + \nabla_{w*} g - j\left(-j\nabla_w g + j\nabla_{w*} g\right)\end{aligned} \tag{B-10}$$

Thus,

$$\bar{\nabla}_w g\left(\mathbf{W}, \mathbf{W}^*\right) = 2\nabla_{w*} g\left(\mathbf{W}, \mathbf{W}^*\right) \tag{B-11}$$

This result allows a major simplification in calculations.

Since $\nabla_{\mathbf{x}}\left(\mathbf{x}^{\mathbf{T}}\mathbf{y}\right) = \nabla_{\mathbf{x}}\left(\mathbf{y}^{\mathbf{T}}\mathbf{x}\right) = \mathbf{y}$, (B-11) and (B-4) yield

$$\bar{\nabla}_w E\left[|\epsilon|^2\right] = 2\mathbf{R}_{xx}\mathbf{W} - 2\mathbf{R}_{xd} \tag{B-12}$$

Since $\nabla_{wr} g = 0$ and $\nabla_{wi} g = 0$ imply that $\bar{\nabla}_w g = 0$, a necessary condition for the optimal weight is obtained by setting $\bar{\nabla}_w E[|\epsilon|^2] = \mathbf{0}$. Thus, if $\mathbf{R}_{xx}$ is positive definite and hence nonsingular, the necessary condition provides the *Wiener-Hopf equation* for the optimal weight vector:

$$\mathbf{W}_0 = \mathbf{R}_{xx}^{-1}\mathbf{R}_{xd} \tag{B-13}$$

To prove the optimality, we substitute $\mathbf{W} = \mathbf{W}_0$ into (B-4) to obtain the mean square error

$$\epsilon_m^2 = E[|d|^2] - \mathbf{R}_{xd}^H \mathbf{R}_{xx}^{-1}\mathbf{R}_{xd} \tag{B-14}$$

Equations (B-4), (B-13), and (B-14) imply that

$$E[|\epsilon|^2] = \epsilon_m^2 + (\mathbf{W} - \mathbf{W}_0)^H \mathbf{R}_{xx} (\mathbf{W} - \mathbf{W}_0) \qquad (\text{B-15})$$

Since $\mathbf{R}_{xx}$ is positive definite, this equation shows that the Wiener-Hopf equation provides a unique optimal weight vector and that (B-14) gives the minimum mean square error.

Since the computational difficulty of inverting the correlation matrix is considerable when the number of weights is large, and insofar as time-varying signal statistics may require frequent computations, adaptive algorithms not entailing matrix inversion have been developed. Suppose that a performance measure, $P(\mathbf{W})$, is defined so that it has a minimum value when the weight vector has its optimal value. In the *method of steepest descent*, the weight vector is changed along the direction of the negative gradient of the performance measure. This direction gives the largest decrease in $P(\mathbf{W})$. If the signals and weights are complex, separate steepest-descent equations can be written for the real and imaginary parts of the weight vector. Combining these equations, we obtain

$$\mathbf{W}(k+1) = \mathbf{W}(k) - \mu \bar{\nabla}_w P(\mathbf{W}(k)) \qquad (\text{B-16})$$

where the *adaptation constant* μ controls the rate of convergence and the stability. For complex signals and weights, a suitable performance measure is $P(\mathbf{W}) = E[|\epsilon|^2]$. The application of (B-12) and (B-16) leads to the *steepest-descent algorithm*:

$$\mathbf{W}(k+1) = \mathbf{W}(k) - 2\mu \left[\mathbf{R}_{xx} \mathbf{W}(k) - \mathbf{R}_{xd} \right] \qquad (\text{B-17})$$

This ideal algorithm produces a deterministic sequence of weights and does not require a matrix inversion, but it requires the knowledge of $\mathbf{R}_{xx}$ and $\mathbf{R}_{xd}$. However, the possible presence of interference means that $\mathbf{R}_{xx}$ is unknown. In the absence of information about the direction of the desired signal, $\mathbf{R}_{xd}$ is also unknown.

The *least-mean-square (LMS) algorithm* is obtained when $\mathbf{R}_{xx}$ is estimated by $\mathbf{x}^*(k)\mathbf{x}^T(k)$, $\mathbf{R}_{xd}$ is estimated by $\mathbf{x}^*(k)d(k)$, and (B-3) is applied in (B-17). The LMS algorithm is

$$\mathbf{W}(k+1) = \mathbf{W}(k) + 2\mu\epsilon(k)\mathbf{x}^*(k) \qquad (\text{B-18})$$

For a fixed value of $\mathbf{W}(k)$, the product $\epsilon(k)\mathbf{x}^*(k)$ is an unbiased estimate of $\bar{\nabla}_w E[\epsilon^2]$. According to this algorithm, the next weight vector is obtained by adding to the present weight vector the input vector scaled by the amount of error. It can be shown that, for an appropriate value of μ, the mean of the weight vector converges to the optimal value given by the Wiener-Hopf equation.

Appendix C

Signal Characteristics

C.1 Bandpass Signals

The *Hilbert transform* provides the basis for signal representations that facilitate the analysis of bandpass signals and systems. The Hilbert transform of a real-valued function $g(t)$ is

$$H[g(t)] = \hat{g}(t) = \frac{1}{\pi} \int_{-\infty}^{\infty} \frac{g(u)}{t - u} \, du \tag{C-1}$$

Since its integrand has a singularity, the integral is defined as its Cauchy principal value:

$$\int_{-\infty}^{\infty} \frac{g(u)}{t - u} \, du = \lim_{\epsilon \to 0} \left[\int_{-\infty}^{t-\epsilon} \frac{g(u)}{t - u} \, du + \int_{t+\epsilon}^{\infty} \frac{g(u)}{t - u} \, du \right] \tag{C-2}$$

provided that the limit exists. Since (C-1) has the form of the convolution of $g(t)$ with $1/\pi t$, $\hat{g}(t)$ results from passing $g(t)$ through a linear filter with an impulse response equal to $1/\pi t$. The transfer function of the filter is given by the Fourier transform

$$\mathcal{F}\left\{\frac{1}{\pi t}\right\} = \int_{-\infty}^{\infty} \frac{\exp(-j2\pi ft)}{\pi t} \, dt \tag{C-3}$$

where $j = \sqrt{-1}$. This integral can be rigorously evaluated by using contour integration. Alternatively, we observe that since $1/t$ is an odd function,

$$\mathcal{F}\left\{\frac{1}{\pi t}\right\} = -2j \int_{0}^{\infty} \frac{\sin 2\pi ft}{\pi t} \, dt$$

$$= -j \, \text{sgn}(f) \tag{C-4}$$

where $\text{sgn}(f)$ is the *signum function* defined by

$$\text{sgn}(f) = \begin{cases} 1, & f > 0 \\ 0, & f = 0 \\ -1, & f < 0 \end{cases} \tag{C-5}$$

Let $G(f) = \mathcal{F}\{g(t)\}$, and let $\hat{G}(f) = \mathcal{F}\{\hat{g}(t)\}$. Equations (C-1) and (C-4) and the convolution theorem imply that

$$\hat{G}(f) = -j\,\mathrm{sgn}(f)G(f) \tag{C-6}$$

Because $H[\hat{g}(t)]$ results from passing $g(t)$ through two successive filters, each with transfer function $-j\,\mathrm{sgn}(f)$,

$$H[\hat{g}(t)] = -g(t) \tag{C-7}$$

provided that $G(0) = 0$.

Equation (C-6) indicates that taking the Hilbert transform corresponds to introducing a phase sift of $-\pi$ radians for all positive frequencies and $+\pi$ radians for all negative frequencies. Consequently,

$$H[\cos 2\pi f_0 t] = \sin 2\pi f_0 t \tag{C-8}$$
$$H[\sin 2\pi f_0 t] = -\cos 2\pi f_0 t \tag{C-9}$$

These relations can be formally verified by taking the Fourier transform of the left-hand side of (C-8) or (C-9), applying (C-6), and then taking the inverse Fourier transform of the result. If $G(f) = 0$ for $|f| > W$ and $f_c > W$, the same method yields

$$H[g(t)\cos 2\pi f_0 t] = g(t)\sin 2\pi f_c t \tag{C-10}$$
$$H[g(t)\sin 2\pi f_0 t] = -g(t)\cos 2\pi f_c t \tag{C-11}$$

A *bandpass signal* is one with a Fourier transform that is negligible except for $f_c - W/2 \le |f| \le f_c + W/2$, where $0 \le W < 2f_c$ and f_c is the center frequency. If $W << f_c$, the bandpass signal is often called a *narrowband signal*. A complex-valued signal with a Fourier transform that is nonzero only for $f > 0$ is called an *analytic signal*.

Consider a bandpass signal $g(t)$ with Fourier transform $G(f)$. The analytic signal $g_a(t)$ associated with $g(t)$ is defined to be the signal with Fourier transform

$$G_a(f) = [1 + \mathrm{sgn}(f)]G(f) \tag{C-12}$$

which is zero for $f \le 0$ and is confined to the band $|f - f_c| \le W/2$ when $f > 0$. The inverse Fourier transform of (C-12) and (C-6) imply that

$$g_a(t) = g(t) + j\hat{g}(t) \tag{C-13}$$

The *complex envelope* of $g(t)$ is defined by

$$g_l(t) = g_a(t)\exp[-j2\pi f_c t] \tag{C-14}$$

where f_c is the center frequency if $g(t)$ is a bandpass signal. Since the Fourier transform of $g_l(t)$ is $G_a(f+f_c)$, which occupies the band $|f| \le W/2$, the complex envelope is a baseband signal that may be regarded as an *equivalent lowpass representation* of $g(t)$. Equations (C-13) and (C-14) imply that $g(t)$ may be expressed in terms of its complex envelope as

$$g(t) = \text{Re}[g_l(t)\exp(j2\pi f_c t)] \tag{C-15}$$

The complex envelope can be decomposed as

$$g_l(t) = g_c(t) + jg_s(t) \tag{C-16}$$

where $g_c(t)$ and $g_s(t)$ are real-valued functions. Therefore, (C-15) yields

$$g(t) = g_c(t)\cos(2\pi f_c t) - g_s(t)\sin(2\pi f_c t) \tag{C-17}$$

Since the two sinusoidal carriers are in phase quadrature, $g_c(t)$ and $g_s(t)$ are called the *in-phase* and *quadrature* components of $g(t)$, respectively. These components are lowpass signals confined to $|f| \le W/2$.

Applying Parseval's identity from Fourier analysis and then (C-6), we obtain

$$\int_{-\infty}^{\infty} \hat{g}^2(t)\,dt = \int_{-\infty}^{\infty} |\hat{G}(f)|^2\,df = \int_{-\infty}^{\infty} |G(f)|^2\,df = \int_{-\infty}^{\infty} g^2(t)\,dt \tag{C-18}$$

Therefore,

$$\int_{-\infty}^{\infty} |g_l(t)|^2\,dt = \int_{-\infty}^{\infty} |g_a(t)|^2\,dt = \int_{-\infty}^{\infty} g^2(t)\,dt + \int_{-\infty}^{\infty} \hat{g}^2(t)\,dt$$

$$= 2\int_{-\infty}^{\infty} g^2(t)\,dt = 2\mathcal{E} \tag{C-19}$$

where $\mathcal{E}$ denotes the energy of the bandpass signal $g(t)$.

C.2 Stationary Stochastic Processes

Consider a stochastic process $n(t)$ that is a zero-mean, wide-sense stationary process with autocorrelation

$$R_n(\tau) = E[n(t)n(t + \tau)] \tag{C-20}$$

where $E[x]$ denotes the expected value of x. The Hilbert transform of this process is the stochastic process defined by

$$\hat{n}(t) = \frac{1}{\pi}\int_{-\infty}^{\infty} \frac{n(u)}{t - u}\,du \tag{C-21}$$

where it is assumed that the Cauchy principal value of the integral exists for almost every sample function of $n(t)$. This equation indicates that $\hat{n}(t)$ is a zero-mean stochastic process. The zero-mean processes $n(t)$ and $\hat{n}(t)$ are *jointly wide-sense stationary* if their correlation and cross-correlation functions are not functions of t. A straightforward calculation using (C-21) and (C-20) gives the cross correlation

$$R_{n\hat{n}}(\tau) = E[n(t)\hat{n}(t+\tau)] = \frac{1}{\pi} \int_{-\infty}^{\infty} \frac{R_n(u)}{\tau - u} du = \hat{R}_n(\tau) \qquad \text{(C-22)}$$

A similar derivation using (C-7) yields the autocorrelation

$$R_{\hat{n}}(\tau) = E[\hat{n}(t)\hat{n}(t+\tau)] = R_n(\tau) \qquad \text{(C-23)}$$

Equations (C-20), (C-22), and (C-23) indicate that $n(t)$ and $\hat{n}(t)$ are jointly wide-sense stationary.

The *analytic signal* associated with $n(t)$ is the zero-mean process defined by

$$n_a(t) = n(t) + j\hat{n}(t) \qquad \text{(C-24)}$$

The autocorrelation of the analytic signal is defined as

$$R_a(\tau) = E[n_a^*(t)n_a(t+\tau)] \qquad \text{(C-25)}$$

where thee asterisk denotes the complex conjugate. Using (C-20) and (C-22) to (C-25), we obtain

$$R_a(\tau) = 2R_n(\tau) + 2j\hat{R}_n(\tau) \qquad \text{(C-26)}$$

which establishes the wide-sense stationarity of the analytic signal.

Since (C-20) indicates that $R_n(\tau)$ is an even function, (C-22) yields

$$R_{n\hat{n}}(0) = \hat{R}_n(0) = 0 \qquad \text{(C-27)}$$

which indicates that $n(t)$ and $\hat{n}(t)$ are uncorrelated. Equations (C-23), (C-26), and (C-27) yield

$$R_{\hat{n}}(0) = R_n(0) = 1/2 R_a(0) \qquad \text{(C-28)}$$

The *complex envelope* of $n(t)$ or the *equivalent lowpass representation* of $n(t)$ is the zero-mean stochastic process defined by

$$n_l(t) = n_a(t) \exp(-j2\pi f_c t) \qquad \text{(C-29)}$$

where f_c is an arbitrary frequency usually chosen as the center or carrier frequency of $n(t)$. The complex envelope can be decomposed as

$$n_l(t) = n_c(t) + j n_s(t) \qquad \text{(C-30)}$$

where $n_c(t)$ and $n_s(t)$ are real-valued, zero-mean stochastic processes.

Equations (C-29) and (C-30) imply that

$$\begin{aligned} n(t) &= \text{Re}[n_l(t) \exp(j2\pi f_c t)] \\ &= n_c(t) \cos(2\pi f_c t) - n_s(t) \sin(2\pi f_c t) \end{aligned} \qquad \text{(C-31)}$$

Substituting (C-24) and (C-30) into (C-29) we find that

$$n_c(t) = n(t) \cos(2\pi f_c t) + \hat{n}(t) \sin(2\pi f_c t) \qquad \text{(C-32)}$$
$$n_s(t) = \hat{n}(t) \cos(2\pi f_c t) - n(t) \sin(2\pi f_c t) \qquad \text{(C-33)}$$

The *autocorrelations* of $n_c(t)$ and $n_s(t)$ are defined by

$$R_c(\tau) = E[n_c(t)n_c(t+\tau)] \tag{C-34}$$

and

$$R_s(\tau) = E[n_s(t)n_s(t+\tau)] \tag{C-35}$$

Using (C-32) and (C-33) and then (C-20), (C-23), and (C-24) and trigonometric identities, we obtain

$$R_c(\tau) = R_s(\tau) = R_n(\tau)\cos(2\pi f_c \tau) + \hat{R}_n(\tau)\sin(2\pi f_c \tau) \tag{C-36}$$

which shows explicitly that if $n(t)$ is wide-sense stationary, then $n_c(t)$ and $n_s(t)$ are wide-sense stationary with the same autocorrelation function. The variances of $n(t)$, $n_c(t)$, and $n_s(t)$ are all equal because

$$R_c(0) = R_s(0) = R_n(0) \tag{C-37}$$

A derivation similar to that of (C-36) gives the cross correlation

$$R_{cs}(\tau) = E[n_c(t)n_s(t+\tau)] = \hat{R}_n(\tau)\cos(2\pi f_c \tau) - R_n(\tau)\sin(2\pi f_c \tau) \tag{C-38}$$

Equations (C-36) and (C-38) indicate that $n_c(t)$ and $n_s(t)$ are jointly wide-sense stationary. Equations (C-28) and (C-38) give

$$R_{cs}(0) = 0 \tag{C-39}$$

which implies that $n_c(t)$ and $n_s(t)$ are uncorrelated.

Equation (C-21) indicates that $\hat{n}(t)$ is generated by a linear operation on $n(t)$. Therefore, if $n(t)$ is a zero-mean Gaussian process, $\hat{n}(t)$ and $n(t)$ are zero-mean jointly Gaussian processes. Equations (C-32) and (C-33) then imply that $n_c(t)$ and $n_s(t)$ are zero-mean jointly Gaussian processes. Since they are uncorrelated, $n_c(t)$ and $n_s(t)$ are statistically independent, zero-mean Gaussian processes.

The *power spectral density* of a signal is the Fourier transform of its autocorrelation. Let $S(f)$, $S_c(f)$, and $S_s(f)$ denote the power spectral densities of $n(t)$, $n_c(t)$, and $n_s(t)$, respectively. We assume that $S_n(f)$ occupies the band $f_c - W/2 \le |f| \le f_c + W/2$ and that $f_c > W/2 \ge 0$. Taking the Fourier transform of (C-36), using (C-6), and simplifying, we obtain

$$S_c(f) = S_s(f) = \begin{cases} S_n(f - f_c) + S_n(f + f_c), & |f| \le W/2 \\ 0, & |f| > W/2 \end{cases} \tag{C-40}$$

Thus, if $n(t)$ is a passband process with one-sided bandwidth W, then $n_c(t)$ and $n_s(t)$ are baseband processes with one-sided bandwidths $W/2$. This property and the statistical independence of $n_c(t)$ and $n_s(t)$ when $n(t)$ is Gaussian make (C-31) a very useful representation of $n(t)$.

Similarly, the cross-spectral density of $n_c(t)$ and $n_s(t)$ can be derived by taking the Fourier transform of (C-38) and using (C-6). After simplification, the result is

$$S_{cs}(f) = \begin{cases} j[S_n(f - f_c) - S_n(f + f_c)], & |f| \le W/2 \\ 0, & |f| > W/2 \end{cases} \tag{C-41}$$

If $S_n(f)$ is locally symmetric about f_c, then

$$S_n(f_c + f) = S_n(f_c - f), \quad |f| \le W/2 \qquad \text{(C-42)}$$

Since a power spectral density is a real-valued, even function, $S_n(f_c - f) = S_n(f - f_c)$. Equation (C-42) then yields $S_n(f + f_c) = S_n(f - f_c)$ for $|f| \le W/2$. Therefore, (C-41) gives $S_{cs}(f) = 0$, which implies that

$$R_{cs}(\tau) = 0 \qquad \text{(C-43)}$$

for all τ. Thus, $n_c(t)$ and $n_s(t+\tau)$ are uncorrelated for all τ, and if $n(t)$ is a zero-mean Gaussian process, then $n_c(t)$ and $n_s(t + \tau)$ are statistically independent for all τ.

The autocorrelation of the complex envelope is defined by

$$R_l(\tau) = \frac{1}{2} E[n_l^*(t) n_l(t + \tau)] \qquad \text{(C-44)}$$

where the $1/2$ is inserted so that

$$R_l(0) = R_n(0) \qquad \text{(C-45)}$$

which follows from (C-28) and (C-29). Substituting (C-30) into (C-44) and using (C-36) and (C-38), we obtain

$$R_l(\tau) = R_c(\tau) + j R_{cs}(\tau) \qquad \text{(C-46)}$$

The power spectral density of $n_l(t)$, which we denote by $S_l(f)$, can be derived from (C-46), (C-41), and (C-40). If $S_n(f)$ occupies the band $f_c - W/2 \le |f| \le f_c + W/2$ and $f_c > W/2 \ge 0$, then

$$S_l(f) = \begin{cases} 2S_n (f + f_c), & |f| \le W/2 \\ 0, & |f| > W/2 \end{cases} \qquad \text{(C-47)}$$

Equations (C-36) and (C-38) yield

$$R_n(\tau) = R_c(\tau) \cos(2\pi f_c \tau) - R_{cs}(\tau) \sin(2\pi f_c \tau) \qquad \text{(C-48)}$$

Equations (C-48) and (C-46) imply that

$$R_n(\tau) = \text{Re} \left[R_l(\tau) \exp \left(j2\pi f_c \tau \right) \right] \qquad \text{(C-49)}$$

We expand the right-hand side of this equation by using the fact that $\text{Re}[z] = (z + z^*)/2$. Taking the Fourier transform and observing that $S_l(f)$ is a real-valued function, we obtain

$$S_n(f) = \frac{1}{2} S_l(f - f_c) + \frac{1}{2} S_l(-f - f_c) \qquad \text{(C-50)}$$

If $S_n(f)$ is locally symmetric about f_c, then (C-47) and (C-42) imply that $S_l(-f) = S_l(f)$, and (C-50) becomes

$$S_n(f) = \frac{1}{2} S_l(f - f_c) + \frac{1}{2} S_l(f + f_c) \qquad \text{(C-51)}$$

Power Spectral Densities of Communication Signals

Many useful communication signals are modeled as having the form

$$s(t) = Ad_1(t)\cos(2\pi f_c t + \theta) + Ad_2(t)\sin(2\pi f_c t + \theta) \qquad \text{(C-52)}$$

where θ is an independent random variable that is uniformly distributed over $0 \le \theta < 2\pi$. The modulations have the form

$$d_i(t) = \sum_{k=-\infty}^{\infty} a_{ik}\psi(t - kT - T_0 - t_i), \quad i = 1, 2 \qquad \text{(C-53)}$$

where $\{a_{ik}\}$ is a sequence of independent, identically distributed random variables, $a_{ik} = +1$ with probability $1/2$ and $a_{ik} = -1$ with probability $1/2$, $\psi(t)$ is a pulse waveform, T is the pulse duration, t_i is the relative pulse offset, and T_0 is an independent random variable that is uniformly distributed over the interval $(0, T)$ and reflects the arbitrariness of the origin of the coordinate system. Since a_{ik} is independent of a_{in} when $n \ne k$, it follows that $E[a_{ik}a_{in}] = 0$, $n \ne k$. Therefore, the autocorrelation of $d_i(t)$ is

$$R_{di}(\tau) = E[d_i(t)d_i(t + \tau)]$$

$$= \sum_{k=-\infty}^{\infty} E[\psi(t - kT - T_0 - t_i)\psi(t - kT - T_0 - t_i + \tau)] \qquad \text{(C-54)}$$

Expressing the expected value as an integral over the range of T_0 and changing variables, we obtain

$$R_{di}(\tau) = \sum_{k=-\infty}^{\infty} \frac{1}{T} \int_{t-kT-T-t_i}^{t-kT-t_i} \psi(x)\psi(x + \tau)dx$$

$$= \frac{1}{T} \int_{-\infty}^{\infty} \psi(x)\psi(x + \tau)dx, \quad i = 1, 2 \qquad \text{(C-55)}$$

This equation indicates that $d_1(t)$ and $d_2(t)$ are wide-sense stationary processes with the same autocorrelation.

If the sequences $\{a_{1k}\}$ and $\{a_{2k}\}$ are statistically independent, then the autocorrelation of $s(t)$ is

$$R_s(\tau) = \frac{A^2}{2} R_{d1}(\tau)\cos(2\pi f_c \tau) + \frac{A^2}{2} R_{d2}(\tau)\sin(2\pi f_c \tau) \qquad \text{(C-56)}$$

where $R_{d1}(\tau)$ and $R_{d2}(\tau)$ are the autocorrelations of $d_1(t)$ and $d_2(t)$, respectively. This equation indicates that $s(t)$ is wide-sense stationary. If the sample functions of $d_1(t)$ and $d_2(t)$ have Fourier transforms that vanish for $|f| \ge f_c$, then (C-10), (C-11), (C-24), and (C-29) indicate that the complex envelope of $s(t)$ is

$$s_l(t) = Ad_1(t) - jAd_2(t) \qquad \text{(C-57)}$$

Equation (C-44) and the independence of $d_1(t)$ and $d_2(t)$ imply that the auto-correlation of $s_l(t)$ is

$$R_l(\tau) = \frac{A^2}{2} R_{d1}(\tau) + \frac{A^2}{2} R_{d2}(\tau) \tag{C-58}$$

The power spectral density of $s_l(t)$ is the Fourier transform of $R_l(\tau)$. From (C-58) and (C-55), we obtain the density

$$S_l(f) = A^2 \frac{|G(f)|^2}{T} \tag{C-59}$$

where $G(f)$ is the Fourier transform of $\psi(t)$.

In a quadriphase-shift-keying (QPSK) signal, $d_1(t)$ and $d_2(t)$ are usually modeled as independent random binary sequences with pulse duration $T = 2T_b$, where T_b is a bit duration. The component amplitude is $A = \sqrt{\mathcal{E}_b/T_b}$, where $\mathcal{E}_b$ is the energy per bit. If $\psi(t)$ is rectangular with unit amplitude over $[0, 2T_b]$, then (C-59) yields the power spectral density for QPSK:

$$S_l(f) = 2\mathcal{E}_b \, \text{sinc}^2 2T_b f \tag{C-60}$$

which is the same as the density for PSK. For a binary minimum-shift-keying (MSK) signal with the same component amplitude,

$$\psi(t) = \sqrt{2} \sin\left(\frac{\pi t}{2T_b}\right) , \quad 0 \le t < 2T_b \tag{C-61}$$

Therefore, the power spectral density for MSK is

$$S_l(f) = \frac{16\mathcal{E}_b}{\pi^2} \left[\frac{\cos(2\pi T_b f)}{16 T_b^2 f^2 - 1}\right]^2 \tag{C-62}$$

C.3   Sampling Theorems

Consider the Fourier transform $G(f)$ of an absolutely integrable function $g(t)$. The periodic extension of $G(f)$ is defined as

$$\bar{G}(f) = \sum_{i=-\infty}^{\infty} G(f + iW) \tag{C-63}$$

where W is the period of $\bar{G}(f)$ and it is assumed that the series converges uniformly. Suppose that $\bar{G}(f)$ has a piecewise continuous derivative so that it can be represented as a uniformly convergent complex Fourier series:

$$\bar{G}(f) = \sum_{k=-\infty}^{\infty} c_k \exp\left(-j2\pi k \frac{f}{W}\right) \tag{C-64}$$

where the Fourier coefficient c_k is given by

$$c_k = \frac{1}{W} \int_{-W/2}^{W/2} \bar{G}(f) \exp\left(j2\pi k \frac{f}{W}\right) df \tag{C-65}$$

Substituting (C-63) into (C-65) and interchanging the order of the summation and the integration, which is justified because of the uniform convergence, we obtain

$$c_k = \frac{1}{W} \sum_{i=-\infty}^{\infty} \int_{-W/2}^{W/2} G(f + iW) \exp\left(j2\pi k \frac{f}{W}\right) df \qquad \text{(C-66)}$$

We change variables and observe the $\exp(j2\pi ki) = 1$ to obtain

$$c_k = \frac{1}{W} \sum_{i=-\infty}^{\infty} \int_{-W/2+iW}^{W/2+iW} G(f) \exp\left(j2\pi k \frac{f}{W} - j2\pi ki\right) df$$

$$= \frac{1}{W} \int_{-\infty}^{\infty} G(f) \exp\left(j2\pi k \frac{f}{W}\right) df \qquad \text{(C-67)}$$

Since $g(t)$ is absolutely integrable, the last integral is the inverse Fourier transform of $G(f)$ evaluated at $t = k/W$, and

$$c_k = \frac{1}{W} g\left(\frac{k}{W}\right) \qquad \text{(C-68)}$$

Substituting (C-68) into (C-64) yields one version of the *Poisson sum formula*:

$$\bar{G}(f) = \frac{1}{W} \sum_{k=-\infty}^{\infty} g\left(\frac{k}{W}\right) \exp\left(-\frac{j2\pi kf}{W}\right) \qquad \text{(C-69)}$$

where the series converges uniformly. If we define $T = 1/W$, then the right-hand side of (C-69) is proportional to the discrete-time Fourier transform of the sequence $g(kT)$.

Suppose that the Fourier transform vanishes outside a frequency band:

$$G(f) = 0, \quad |f| > W/2 \qquad \text{(C-70)}$$

It follows that

$$g(t) = \int_{-W/2}^{W/2} G(f) \exp(j2\pi ft) df \qquad \text{(C-71)}$$

Since $G(f) = \bar{G}(f)$ for $|f| < W/2$, (C-71) and (C-69) and the interchange of a summation and integration yield

$$g(t) = \sum_{k=-\infty}^{\infty} g\left(\frac{k}{W}\right) \frac{1}{W} \int_{-W/2}^{W/2} \exp\left[j2\pi f\left(t - \frac{k}{W}\right)\right] df \qquad \text{(C-72)}$$

Evaluating this integral and defining sinc $x = \sin(\pi x)/\pi x$, we obtain the *sampling theorem* for deterministic signals:

$$g(t) = \sum_{k=-\infty}^{\infty} g\left(\frac{k}{W}\right) \operatorname{sinc}(Wt - k) \qquad \text{(C-73)}$$

where the samples are separated by $T = 1/W$.

Consider a wide-sense stationary stochastic process $n(t)$ with autocorrelation $R_n(\tau)$ and power spectral density $S_n(f)$, which is the Fourier transform of $R_n(\tau)$. If

$$S_n(f) = 0, \quad |f| > W/2 \tag{C-74}$$

then it follows from the sampling theorem that

$$R_n(\tau) = \sum_{k=-\infty}^{\infty} R_n\left(\frac{k}{W}\right) \text{sinc}(W\tau - k) \tag{C-75}$$

For an arbitrary constant α, the Fourier transform of $R(\tau-\alpha)$ is $S_n(f)\exp(-j2\pi f\alpha)$, which is zero for $|f| > W/2$. Therefore, (C-75) can be applied to $R'_n(\tau) = R_n(\tau - \alpha)$, which gives

$$R_n(\tau - \alpha) = \sum_{k=-\infty}^{\infty} R_n\left(\frac{k}{W} - \alpha\right) \text{sinc}(W\tau - k) \tag{C-76}$$

We define the stochastic process

$$n_\nu(t) = \sum_{k=-\nu}^{\nu} n\left(\frac{k}{W}\right) \text{sinc}(Wt - k) \tag{C-77}$$

The mean square difference between $n(t)$ and $n_\nu(t)$ is

$$E\{[n(t) - n_\nu(t)]^2\} = R_n(0) - 2\sum_{k=-\nu}^{\nu} R_n\left(t - \frac{k}{W}\right) \text{sinc}(Wt - k)$$
$$+ \sum_{i=-\nu}^{\nu} \text{sinc}(Wt - i) \sum_{k=-\nu}^{\nu} R_n\left(\frac{i - k}{W}\right) \text{sinc}(Wt - k) \tag{C-78}$$

Since $R_n(\tau) = R_n(-\tau)$, the repeated use of (C-76) yields

$$\lim_{\nu \to \infty} E\{[n(t) - n_\nu(t)]^2\} = 0 \tag{C-79}$$

which states that the mean square difference between $n(t)$ and $n_\nu(t)$ approaches zero. With equality interpreted in the sense of this limit, the *sampling theorem* for stationary stochastic process is

$$n(t) = \sum_{k=-\infty}^{\infty} n\left(\frac{k}{W}\right) \text{sinc}(Wt - k) \tag{C-80}$$

C.4   Direct-Conversion Receiver

Receivers often extract the complex envelope of the desired signal before applying it to a matched filter. The main components in a *direct-conversion* receiver are shown in Figure C.1(a). The spectra of the received signal $g(t)$, the input to the baseband filter $g'(t) = g(t)\exp(-j2\pi f_c t)$, and the complex envelope $g_l(t)$ are depicted in Figure C.1(b). Let $2h(t)$ denote the impulse response of the filter. The output of the filter is

$$y(t) = \int_{-\infty}^{\infty} 2\,g(\tau)\exp(-j2\pi f_c\tau)h(t-\tau)\,d\tau \qquad \text{(C-81)}$$

Using (C-15) and the fact that $\mathrm{Re}(x) = (x+x^*)/2$, where x^* denotes the complex conjugate of x, we obtain

$$y(t) = \int_{-\infty}^{\infty} g_l(\tau)h(t-\tau)\,d\tau + \int_{-\infty}^{\infty} g_l(\tau)h(t-\tau)\exp(-j4\pi f_c\tau)\,d\tau \qquad \text{(C-82)}$$

The second term is the Fourier transform of $g_l(\tau)h(t-\tau)$ evaluated at frequency $-2f_c$. Assuming that $g_l(\tau)$ and $h(t-\tau)$ have transforms confined to $|f| < f_c$, their product has a transform confined to $|f| < 2f_c$, and the second term in (C-82) vanishes. If the Fourier transform of $h(t)$ is a constant over the passband of $g_l(t)$, then (C-82) implies that $y(t)$ is proportional to $g_l(t)$, as desired. Figure C.1(c) shows the direct-conversion receiver for real-valued signals.

The direct-conversion receiver alters the character of the noise $n(t)$ entering it. Suppose that $n(t)$ is a zero-mean, white Gaussian noise process with autocorrelation

$$R_n(\tau) = E[n(t)n(t+\tau)] = \frac{N_0}{2}\delta(\tau) \qquad \text{(C-83)}$$

where $\delta(\tau)$ denotes the Dirac delta function, and $N_0/2$ is the two-sided noise-power spectral density. The complex-valued noise at the output of Figure C.1(a) is

$$z(t) = \int_{-\infty}^{\infty} 2n(u)e^{-j2\pi f_c u}h(t-u)\,du \qquad \text{(C-84)}$$

Since it is a linear function of $n(t)$, $z(t)$ is zero-mean and its real and imaginary parts are jointly Gaussian. The autocorrelation of a wide-sense stationary, complex-valued process $z(t)$ is defined as

$$R_z(\tau) = \frac{1}{2}E[z^*(t)z(t+\tau)] \qquad \text{(C-85)}$$

Substituting (C-84), interchanging the expectation and integration operations, using (C-83) to evaluate one of the integrals, and then changing variables, we obtain

$$R_z(\tau) = N_0 \int_{-\infty}^{\infty} h(u)h^*(u+\tau)\,du \qquad \text{(C-86)}$$

If the filter is an ideal bandpass filter with Fourier transform

$$H(f) = \begin{cases} 1, & |f| \le W \\ 0, & otherwise \end{cases} \qquad \text{(C-87)}$$

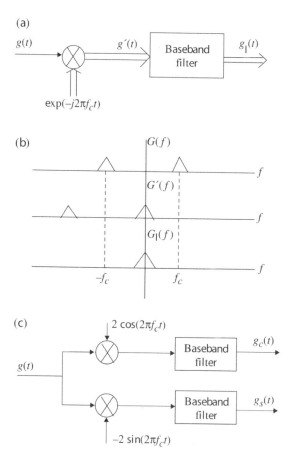

Figure C.1: Envelope extraction: (a) direct-conversion receiver, (b) associated spectra, and (c) implementation with real-valued signals.

then evaluating the Fourier transform of both sides of (C-86) gives

$$S_n(f) = \begin{cases} N_0, & |f| \leq W \\ 0, & otherwise \end{cases} \tag{C-88}$$

Thus, if the subsequent filters have narrower bandwidths than W or if W$\to \infty$, then the autocorrelation of z(t) may be approximated by

$$R_z(\tau) = N_0 \delta(\tau) \tag{C-89}$$

This approximation permits major analytical simplifications. Equations (C-84) and (C-83) imply that

$$E[z(t)z(t+\tau)] = 2N_0 e^{-j4\pi f_c t} \int_{-\infty}^{\infty} e^{j4\pi f_c u} f(u+\tau)f(u)\, du \tag{C-90}$$

Reasoning similar to that following (C-82) leads to

$$E[z(t)z(t+\tau)] = 0 \tag{C-91}$$

A complex-valued stochastic process $z(t)$ that satisfies (C-91) is called a *circularly symmetric* process. Let $z^R(t)$ and $z^I(t)$ denote the real and imaginary parts of $z(t)$, respectively. Setting $\tau = 0$ in (C-91) and (C-85), and then using (C-86), Parseval's identity, and (C-87), we obtain

$$E[(z^R(t))^2] = E[(z^I(t))^2] = 2N_0 W \tag{C-92}$$

$$E[(z^R(t)z^I(t))] = 0 \tag{C-93}$$

Thus, $z^R(t)$ and $z^I(t)$ are zero-mean, independent Gaussian processes with the same variance.

$$Z = \sum_{i=1}^{N} A_i^2 \tag{D-1}$$

Appendix D

Probability Distributions

D.1 Chi-Square Distribution

Consider the random variable

$$Z = \sum_{i=1}^{N} A_i^2 \qquad \text{(D-1)}$$

where the $\{A_i\}$ are independent Gaussian random variables with means $\{m_i\}$ and common variance σ^2. The random variable Z is said to have a *noncentral chi-square* (χ^2) *distribution* with N degrees of freedom and a *noncentral parameter*

$$\lambda = \sum_{i=1}^{N} m_i^2 \qquad \text{(D-2)}$$

To derive the probability density function of Z, we first note that each A_i has the density function

$$f_{A_i}(x) = \frac{1}{\sqrt{2\pi}\sigma} \exp\left[-\frac{(x - m_i)^2}{2\sigma^2}\right] \qquad \text{(D-3)}$$

From elementary probability, the density of $Y_i = A_i^2$ is

$$f_{Y_i}(x) = \frac{1}{2\sqrt{x}}[f_{A_i}(\sqrt{x}) + f_{A_i}(-\sqrt{x})]\, u(x) \qquad \text{(D-4)}$$

where $u(x) = 1, x \geq 0$, and $u(x) = 0, x < 0$. Substituting (D-3) into (D-4), expanding the exponentials, and simplifying, we obtain the density

$$f_{Y_i}(x) = \frac{1}{\sqrt{2\pi x}\sigma} \exp\left(-\frac{x + m_i^2}{2\sigma^2}\right) \cosh\left(\frac{m_i\sqrt{x}}{\sigma^2}\right) u(x) \qquad \text{(D-5)}$$

The characteristic function of a random variable X is defined as

$$C_X(j\nu) = E[e^{j\nu X}] = \int_{-\infty}^{\infty} f_X(x) \exp(j\nu x)\, dx \qquad \text{(D-6)}$$

where $j = \sqrt{-1}$, and $f_X(x)$ is the density of X. Since $C_X(j\nu)$ is the conjugate Fourier transform of $f_X(x)$,

$$f_X(x) = \frac{1}{2\pi} \int_{-\infty}^{\infty} C_X(j\nu) \exp(-j\nu x)\, d\nu \tag{D-7}$$

From Laplace or Fourier transform tables, it is found that the characteristic function of $f_{Y_i}(x)$ is

$$C_{Y_i}(j\nu) = \frac{\exp[jm_i^2\nu/(1 - j2\sigma^2\nu)]}{(1 - j^2\sigma^2\nu)^{1/2}} \tag{D-8}$$

The characteristic function of a sum of independent random variables is equal to the product of the individual characteristic functions. Because Z is the sum of the Y_i, the characteristic function of Z is

$$C_Z(j\nu) = \frac{\exp[j\lambda\nu/(1 - j2\sigma^2\nu)]}{(1 - j^2\sigma^2\nu)^{N/2}} \tag{D-9}$$

where we have used (D-2). From (D-9), (D-7), and Laplace or Fourier transform tables, we obtain the probability density function of *noncentral χ^2 random variable* with N degrees of freedom and a noncentral parameter λ:

$$f_Z(x) = \frac{1}{2\sigma^2} \left(\frac{x}{\lambda}\right)^{(N-2)/4} \exp\left[-\frac{x+\lambda}{2\sigma^2}\right] I_{N/2-1}\left(\frac{\sqrt{x\lambda}}{\sigma^2}\right) u(x) \tag{D-10}$$

where $I_n(\ )$ is the modified Bessel function of the first kind and order n. This function may be represented by

$$I_n(x) = \sum_{i=0}^{\infty} \frac{(x/2)^{n+2i}}{i!\,\Gamma(n+i+1)} \tag{D-11}$$

where the gamma function is defined as

$$\Gamma(x) = \int_0^{\infty} y^{x-1} \exp(-y)dy\ ,\quad x > 0 \tag{D-12}$$

The probability distribution function of a noncentral χ^2 random variable is

$$F_Z(x) = \int_0^x \frac{1}{2\sigma^2} \left(\frac{y}{\lambda}\right)^{(N-2)/4} \exp\left(-\frac{y+\lambda}{2\sigma^2}\right) I_{N/2-1}\left(\frac{\sqrt{y\lambda}}{\sigma^2}\right) dy\ ,\quad x \geq 0 \tag{D-13}$$

If N is even so that $N/2$ is an integer, then $F_Z(\infty) = 1$ and a change of variables in (D-13) yield

$$F_Z(x) = 1 - Q_{N/2}\left(\frac{\sqrt{\lambda}}{\sigma}\ ,\ \frac{\sqrt{x}}{\sigma}\right)\ ,\quad x \geq 0 \tag{D-14}$$

where the *generalized Marcum Q-function* is defined as

$$Q_m(\alpha, \beta) = \int_\beta^\infty x \left(\frac{x}{\alpha}\right)^{m-1} \exp\left(-\frac{x^2 + \alpha^2}{2}\right) I_{m-1}(\alpha x)\, dx \qquad \text{(D-15)}$$

and m is an integer. Since $Q_m(\alpha, 0) = 1$, it follows that $1 - Q_m(\alpha, \beta)$ is an integral with finite limits that can be numerically integrated. However, the numerical computation of the generalized Q-function is simplified if it is expressed in alternative forms [2]. The mean, variance, and moments of Z can be easily obtained by using (D-1) and the properties of independent Gaussian random variables. The mean and variance of Z are

$$E[Z] = N\sigma^2 + \lambda \qquad \text{(D-16)}$$
$$\sigma_z^2 = 2N\sigma^4 + 4\lambda\sigma^2 \qquad \text{(D-17)}$$

where σ^2 is the common variance of the $\{A_i\}$.

From (D-9), it follows that the sum of two independent noncentral χ^2 random variables with N_1 and N_2 degrees of freedom, noncentral parameters λ_1 and λ_2, respectively, and the same parameter σ^2 is a noncentral χ^2 random variable with $N_1 + N_2$ degrees of freedom and noncentral parameter $\lambda_1 + \lambda_2$.

D.2 Central Chi-Square Distribution

To determine the probability density function of Z when the $\{A_i\}$ have zero means, we substitute (D-11) into (D-10) and then take the limit as $\lambda \to 0$. We obtain

$$f_Z(x) = \frac{1}{(2\sigma^2)^{N/2}\Gamma(N/2)} x^{N/2-1} \exp\left(-\frac{x}{2\sigma^2}\right) u(x) \qquad \text{(D-18)}$$

Alternatively, this equation results if we substitute $\lambda = 0$ into the characteristic function (D-9) and then use (D-7). Equation (D-18) is the probability density function of a *central χ^2 random variable* with N degrees of freedom. The probability distribution function is

$$F_Z(x) = \int_0^x \frac{1}{(2\sigma^2)^{N/2}\Gamma(N/2)} y^{N/2-1} \exp\left(-\frac{y}{2\sigma^2}\right) dy, \quad x \geq 0 \qquad \text{(D-19)}$$

If N is even so that $N/2$ is an integer, then integrating this equation by parts $N/2 - 1$ times yields

$$F_Z(x) = 1 - \exp\left(-\frac{x}{2\sigma^2}\right) \sum_{i=0}^{N/2-1} \frac{1}{i!} \left(\frac{x}{2\sigma^2}\right)^i, \quad x \geq 0 \qquad \text{(D-20)}$$

By direct integration using (D-18) and (D-12) or from (D-16) and (D-17), it is found that the mean and variance of Z are

$$E[Z] = N\sigma^2 \qquad \text{(D-21)}$$
$$\sigma_z^2 = 2N\sigma^4 \qquad \text{(D-22)}$$

D.3 Rice Distribution

Consider the random variable

$$R = \sqrt{A_1^2 + A_2^2} \qquad (D\text{-}23)$$

where A_1 and A_2 are independent Gaussian random variables with means m_1 and m_2, respectively, and a common variance σ^2. The probability distribution function of R must satisfy $F_R(r) = F_Z(r^2)$, where $Z = A_1^2 + A_2^2$ is a χ^2 random variable with two degrees of freedom. Therefore, (D-14) with $N = 2$ implies that

$$F_R(r) = 1 - Q_1\left(\frac{\sqrt{\lambda}}{\sigma}, \frac{r}{\sigma}\right), \quad r \geq 0 \qquad (D\text{-}24)$$

where $\lambda = m_1^2 + m_2^2$. This function is called the *Rice probability distribution function*. The *Rice probability density function*, which may be obtained by differentiation of (D-24), is

$$f_R(r) = \frac{r}{\sigma^2} \exp\left(-\frac{r^2 + \lambda}{2\sigma^2}\right) I_0\left(\frac{r\sqrt{\lambda}}{\sigma^2}\right) u(r) \qquad (D\text{-}25)$$

The moments of even order can be derived from (D-23) and the moments of the independent Gaussian random variables. The second moment is

$$E[R^2] = 2\sigma^2 + \lambda \qquad (D\text{-}26)$$

In general, moments of the Rice distribution are given by an integration over the density in (D-25). Substituting (D-11) into the integrand, interchanging the summation and integration, changing the integration variable, and using (D-12), we obtain a series that is recognized as a special case of the confluent hypergeometric function. Thus,

$$E[R^n] = (2\sigma^2)^{n/2} \exp\left(-\frac{\lambda}{2\sigma^2}\right) \Gamma\left(1 + \frac{n}{2}\right) {}_1F_1\left(1 + \frac{n}{2}, 1; \frac{\lambda}{2\sigma^2}\right), \quad n \geq 0 \qquad (D\text{-}27)$$

where the *confluent hypergeometric function* is defined as

$${}_1F_1(\alpha, \beta; x) = \sum_{i=0}^{\infty} \frac{\Gamma(\alpha + i)\Gamma(\beta)x^i}{\Gamma(\alpha)\Gamma(\beta + i)i!}, \quad \beta \neq 0, -1, -2, \ldots \qquad (D\text{-}28)$$

The Rice density function often arises in the context of a transformation of variables. Let A_1 and A_2 represent independent Gaussian random variables with common variance σ^2 and means λ and zero, respectively. Let R and Θ be implicitly defined by $A_1 = R\cos\Theta$ and $A_2 = R\sin\Theta$. Then (D-23) and $\Theta = \tan^{-1}(A_2/A_2)$ describes a transformation of variables. A straightforward calculation yields the joint density function of R and Θ:

$$f_{R,\Theta}(r, \theta) = \frac{r}{2\pi\sigma^2} \exp\left(-\frac{r^2 - 2r\lambda\cos\theta + \lambda^2}{2\sigma^2}\right), \quad r \geq 0, \quad |\theta| \leq \pi \qquad (D\text{-}29)$$

The density function of the envelope R is obtained by integration over θ. Since the *modified Bessel function of the first kind and order zero* satisfies

$$I_0(x) = \frac{1}{2\pi} \int_0^{2\pi} \exp(x \cos u) \, du \qquad \text{(D-30)}$$

this density function reduces to the Rice density function (D-25). The density function of the angle Θ is obtained by integrating (D-29) over r. Completing the square of the argument in (D-29), changing variables, and defining

$$Q(x) = \frac{1}{\sqrt{2\pi}} \int_{-x}^{\infty} \exp\left(-\frac{y^2}{2}\right) dy = \frac{1}{2} \operatorname{erfc}\left(\frac{x}{\sqrt{2}}\right) \qquad \text{(D-31)}$$

where erfc( ) is the complementary error function, we obtain

$$f_\Theta(\theta) = \frac{1}{2\pi} \exp\left(-\frac{\lambda^2}{2\sigma^2}\right) + \frac{\lambda \cos \theta}{\sqrt{2\pi}\sigma} \exp\left(-\frac{\lambda^2 \sin^2 \theta}{2\sigma^2}\right) \left[1 - Q\left(\frac{\lambda \cos \theta}{\sigma}\right)\right],$$
$$|\theta| \le \pi \qquad \text{(D-32)}$$

Since (D-29) cannot be written as the product of (D-25) and (D-32), the random variables R and Θ are not independent.

Since the density function of (D-25) must integrate to unity, we find that

$$\int_0^{\infty} r \exp\left(-\frac{r^2}{2b}\right) I_0\left(\frac{r\sqrt{\lambda}}{b}\right) dr = b \exp\left(\frac{\lambda}{2b}\right) \qquad \text{(D-33)}$$

where λ and b are positive constants. This equation is useful in calculations involving the Rice density function.

D.4 Rayleigh Distribution

A Rayleigh-distributed random variable is defined by (D-23) when A_1 and A_2 are independent Gaussian random variables with zero means and a common variance σ^2. Since $F_R(r) = F_Z(r^2)$, where Z is a central χ^2 random variable with two degrees of freedom, (D-20) with $N = 2$ implies that the *Rayleigh probability distribution function* is

$$F_R(r) = 1 - \exp\left(-\frac{r^2}{2\sigma^2}\right), \quad r \ge 0 \qquad \text{(D-34)}$$

The *Rayleigh probability density function*, which may be obtained by differentiation of (D-34), is

$$f_R(r) = \frac{r}{\sigma^2} \exp\left(-\frac{r^2}{2\sigma^2}\right) u(r) \qquad \text{(D-35)}$$

By a change of variables in the defining integral, any moment of R can be expressed in terms of the gamma function defined in (D-12). Therefore,

$$E[R^n] = (2\sigma^2)^{n/2} \Gamma\left(1 + \frac{n}{2}\right) \qquad \text{(D-36)}$$

Certain properties of the gamma function are needed to simplify (D-36). An integration by parts of (D-12) indicates that $\Gamma(1 + x) = x\Gamma(x)$. A direct integration yields $\Gamma(1) = 1$. Therefore, when n is an integer, $\Gamma(n) = (n - 1)!$. Changing the integration variable by substituting $y = z^2$ in (D-12), it is found that $\Gamma(1/2) = \sqrt{\pi}$.

Using these properties of the gamma function, we obtain the mean and the variance of a Rayleigh-distributed random variable:

$$E[R] = \sqrt{\frac{\pi}{2}}\sigma \tag{D-37}$$

$$\sigma_R^2 = \left(2 - \frac{\pi}{2}\right)\sigma^2 \tag{D-38}$$

Since A_1 and A_2 have zero means, the joint probability density function of the random variables $R = \sqrt{A_1^2 + A_2^2}$ and $\Theta = \tan^{-1}(A_2/A_1)$ is given by (D-29) with $\lambda = 0$. Therefore,

$$f_{R,\Theta}(r, \theta) = \frac{r}{2\pi\sigma^2}\exp\left(-\frac{r^2}{2\sigma^2}\right), \quad r \geq 0, \quad |\theta| \leq \pi \tag{D-39}$$

Integration over θ yields (A-35), and integration over r yields the uniform probability density function:

$$f_\Theta(\theta) = \frac{1}{2\pi}, \quad |\theta| \leq \pi \tag{D-40}$$

Since (D-39) equals the product of (D-35) and (D-40), the random variables R and Θ are independent. In terms of these random variables, $A_1 = R\cos\Theta$ and $A_2 = R\sin\Theta$. A straightforward calculation using the independence and densities of R and Θ verifies that A_1 and A_2 are zero-mean, independent, Gaussian random variables with common variance σ^2. Since the square of a Rayleigh-distributed random variable may be expressed as $R^2 = A_1^2 + A_2^2$, where A_1 and A_2 are zero-mean, independent, Gaussian random variables with common variance σ^2, R^2 has the distribution of a central chi-square random variable with 2 degrees of freedom. Therefore, (D-18) with $N = 2$ indicates that the square of a Rayleigh-distributed random variable has an exponential probability density function with mean $2\sigma^2$.

D.5   Exponentially Distributed Random Variables

Consider the random variable

$$Z = \sum_{i=1}^{N} Y_i \tag{D-41}$$

where the $\{Y_i\}$ are independent, exponentially distributed random variables with unequal positive means $\{m_i\}$. The exponential probability density function of Y_i is

$$f_{Y_i}(x) = \frac{1}{m_i}\exp\left(-\frac{x}{m_i}\right)u(x) \tag{D-42}$$

A straightforward calculation yields the characteristic function

$$C_{Y_i}(j\nu) = \frac{1}{1 - j\nu m_i} \tag{D-43}$$

Since Z is the sum of independent random variables, (D-43) implies that its characteristic function is

$$C_Z(j\nu) = \prod_{i=1}^{N} \frac{1}{1 - j\nu m_i} \tag{D-44}$$

To derive the probability density function of Z, (D-7) is applied after first expanding the right-hand side of (D-44) in a partial-fraction expansion. The result is

$$f_Z(x) = \sum_{i=1}^{N} \frac{B_i}{m_i} \exp\left(-\frac{x}{m_i}\right) u(x) \tag{D-45}$$

where

$$B_i = \begin{cases} \displaystyle\prod_{\substack{k=1 \\ k \neq i}}^{N} \frac{m_i}{m_i - m_k} & , \quad N \geq 2 \\[4mm] 1 & , \quad N = 1 \end{cases} \tag{D-46}$$

and $m_i \neq m_k$, $i \neq k$. A direct integration and algebra yields the probability distribution function

$$F_Z(x) = 1 - \sum_{i=1}^{N} B_i \exp\left(-\frac{x}{m_i}\right) , \quad x \geq 0 \tag{D-47}$$

Equations (D-45) and (D-12) give

$$E[Z^n] = \Gamma(1 + n) \sum_{i=1}^{N} B_i m_i^n , \quad n \geq 0 \tag{D-48}$$

When the $\{m_i\}$ are equal so that $m_i = m$, $1 \leq i \leq N$, then $C_Z(j\nu) = (1 - j\nu m)^{-N}$. Therefore, the probability density function of Z is

$$f_Z(x) = \frac{1}{(N-1)!m^N} x^{N-1} \exp\left(-\frac{x}{m}\right) u(x) \tag{D-49}$$

which is a special case of the *gamma density function*. Successive integration by parts yields

$$F_Z(x) = 1 - \exp\left(-\frac{x}{m}\right) \sum_{i=0}^{N-1} \frac{1}{i!}\left(\frac{x}{m}\right)^i \tag{D-50}$$

From (D-49) and (D-12), the mean and variance of Z are found to be

$$E[Z] = Nm \tag{D-51}$$

$$\sigma_Z^2 = Nm^2 \tag{D-52}$$